Lactic Acid Bacteria

FOOD SCIENCE AND TECHNOLOGY

A Series of Monographs, Textbooks, and Reference Books

1. Flavor Research: Principles and Techniques, *R. Teranishi, I. Hornstein, P. Issenberg, and E. L. Wick*
2. Principles of Enzymology for the Food Sciences, *John R. Whitaker*
3. Low-Temperature Preservation of Foods and Living Matter, *Owen R. Fennema, William D. Powrie, and Elmer H. Marth*
4. Principles of Food Science
   Part I: Food Chemistry, *edited by Owen R. Fennema*
   Part II: Physical Methods of Food Preservation, *Marcus Karel, Owen R. Fennema, and Daryl B. Lund*
5. Food Emulsions, *edited by Stig E. Friberg*
6. Nutritional and Safety Aspects of Food Processing, *edited by Steven R. Tannenbaum*
7. Flavor Research: Recent Advances, *edited by R. Teranishi, Robert A. Flath, and Hiroshi Sugisawa*
8. Computer-Aided Techniques in Food Technology, *edited by Israel Saguy*
9. Handbook of Tropical Foods, *edited by Harvey T. Chan*
10. Antimicrobials in Foods, *edited by Alfred Larry Branen and P. Michael Davidson*
11. Food Constituents and Food Residues: Their Chromatographic Determination, *edited by James F. Lawrence*
12. Aspartame: Physiology and Biochemistry, *edited by Lewis D. Stegink and L. J. Filer, Jr.*
13. Handbook of Vitamins: Nutritional, Biochemical, and Clinical Aspects, *edited by Lawrence J. Machlin*
14. Starch Conversion Technology, *edited by G. M. A. van Beynum and J. A. Roels*
15. Food Chemistry: Second Edition, Revised and Expanded, *edited by Owen R. Fennema*

16. Sensory Evaluation of Food: Statistical Methods and Procedures, *Michael O'Mahony*
17. Alternative Sweetners, *edited by Lyn O'Brien Nabors and Robert C. Gelardi*
18. Citrus Fruits and Their Products: Analysis and Technology, *S. V. Ting and Russell L. Rouseff*
19. Engineering Properties of Foods, *edited by M. A. Rao and S. S. H. Rizvi*
20. Umami: A Basic Taste, *edited by Yojiro Kawamura and Morley R. Kare*
21. Food Biotechnology, *edited by Dietrich Knorr*
22. Food Texture: Instrumental and Sensory Measurement, *edited by Howard R. Moskowitz*
23. Seafoods and Fish Oils in Human Health and Disease, *John E. Kinsella*
24. Postharvest Physiology of Vegetables, *edited by J. Weichmann*
25. Handbook of Dietary Fiber: An Applied Approach, *Mark L. Dreher*
26. Food Toxicology, Parts A and B, *Jose M. Concon*
27. Modern Carbohydrate Chemistry, *Roger W. Binkley*
28. Trace Minerals in Foods, *edited by Kenneth T. Smith*
29. Protein Quality and the Effects of Processing, *edited by R. Dixon Phillips and John W. Finley*
30. Adulteration of Fruit Juice Beverages, *edited by Steven Nagy, John A. Attaway, and Martha E. Rhodes*
31. Foodborne Bacterial Pathogens, *edited by Michael P. Doyle*
32. Legumes: Chemistry, Technology, and Human Nutrition, *edited by Ruth H. Matthews*
33. Industrialization of Indigenous Fermented Foods, *edited by Keith H. Steinkraus*
34. International Food Regulation Handbook: Policy · Science · Law, *edited by Roger D. Middlekauff and Philippe Shubik*
35. Food Additives, *edited by A. Larry Branen, P. Michael Davidson, and Seppo Salminen*
36. Safety of Irradiated Foods, *J. F. Diehl*
37. Omega-3 Fatty Acids in Health and Disease, *edited by Robert S. Lees and Marcus Karel*
38. Food Emulsions: Second Edition, Revised and Expanded, *edited by Kåre Larsson and Stig E. Friberg*
39. Seafood: Effects of Technology on Nutrition, *George M. Pigott and Barbee W. Tucker*
40. Handbook of Vitamins: Second Edition, Revised and Expanded, *edited by Lawrence J. Machlin*
41. Handbook of Cereal Science and Technology, *Klaus J. Lorenz and Karel Kulp*
42. Food Processing Operations and Scale-Up, *Kenneth J. Valentas, Leon Levine, and J. Peter Clark*
43. Fish Quality Control by Computer Vision, *edited by L. F. Pau and R. Olafsson*
44. Volatile Compounds in Foods and Beverages, *edited by Henk Maarse*

45. Instrumental Methods for Quality Assurance in Foods, *edited by Daniel Y. C. Fung and Richard F. Matthews*
46. Listeria, Listeriosis, and Food Safety, *Elliot T. Ryser and Elmer H. Marth*
47. Acesulfame-K, *edited by D. G. Mayer and F. H. Kemper*
48. Alternative Sweeteners: Second Edition, Revised and Expanded, *edited by Lyn O'Brien Nabors and Robert C. Gelardi*
49. Food Extrusion Science and Technology, *edited by Jozef L. Kokini, Chi-Tang Ho, and Mukund V. Karwe*
50. Surimi Technology, *edited by Tyre C. Lanier and Chong M. Lee*
51. Handbook of Food Engineering, *edited by Dennis R. Heldman and Daryl B. Lund*
52. Food Analysis by HPLC, *edited by Leo M. L. Nollet*
53. Fatty Acids in Foods and Their Health Implications, *edited by Ching Kuang Chow*
54. *Clostridium botulinum*: Ecology and Control in Foods, *edited by Andreas H. W. Hauschild and Karen L. Dodds*
55. Cereals in Breadmaking: A Molecular Colloidal Approach, *Anne-Charlotte Eliasson and Kåre Larsson*
56. Low-Calorie Foods Handbook, *edited by Aaron M. Altschul*
57. Antimicrobials in Foods: Second Edition, Revised and Expanded, *edited by P. Michael Davidson and Alfred Larry Branen*
58. Lactic Acid Bacteria, *edited by Seppo Salminen and Atte von Wright*

Additional Volumes in Preparation

Rice Science and Technology, *edited by Wayne E. Marshall and James I. Wadsworth*

Principles of Food Enzymology for the Food Sciences: Second Edition, *John R. Whitaker*

Food Biosensor Analysis, *edited by Gabriele Wagner and George G. Guilbault*

Lactic Acid Bacteria

edited by

Seppo Salminen

Valio Ltd. and
University of Helsinki
Helsinki, Finland

Atte von Wright

University of Kuopio
Kuopio, Finland

University of Helsinki
Helsinki, Finland

Royal Veterinary and Agricultural University
Frederiksberg, Denmark

Marcel Dekker, Inc. **New York • Basel • Hong Kong**

Library of Congress Cataloging-in-Publication Data

Lactic acid bacteria / edited by Seppo Salminen, Atte von Wright.
p. cm. -- (Food science and technology ; 58)
Includes bibliographical references and index.
ISBN 0-8247-8907-5
1. Lactic acid bacteria. I. Salminen, Seppo. II. von Wright, Atte.
III. Series: Food science and technology (Marcel Dekker, Inc.) ; 58.
QR121.L333 1993 93-3880
660'.62--dc20 CIP

The publisher offers discounts on this book when ordered in bulk quantities. For more information, write to Special Sales/Professional Marketing at the address below.

This book is printed on acid-free paper.

MARCEL DEKKER, INC.
270 Madison Avenue, New York, New York 10016

Current printing (last digit):
10 9 8 7 6 5 4 3 2 1

PRINTED IN THE UNITED STATES OF AMERICA

Preface

Lactic acid–producing fermentation is an old invention. Many different cultures in various parts of the world have used it to improve the storage qualities, palatability, and nutritive value of perishable foods such as milk, vegetables, meat, fish, legumes, and cereals. The organisms that produce this type of fermentation, lactic acid bacteria, have had an important role in preserving foods, preventing food poisoning, and indirectly feeding the hungry on every continent.

In the developed world, lactic acid bacteria are mainly associated with fermented dairy products such as cheese, buttermilk, and yogurt. The use of dairy starter cultures has become an industry during this century. Because of this, the technological aspects of lactic acid fermentation have been well covered in both research and training in food sciences.

Since the days of the Russian scientist Metchnikoff, lactic acid bacteria have also been associated with beneficial health effects. Today, an increasing number of health foods and so-called functional foods as well as pharmaceutical preparations are promoted with health claims based on the characteristics of certain strains of lactic acid bacteria. Most of these strains, however, have not been thoroughly studied, and consequently the claims are not well substantiated. Moreover, the accepted standards of clinical protocols, including double-blind randomized study designs, have not been applied in most "health-claim" studies—health benefits are judged mainly using subjective criteria. Additionally, the specific bacterial strains used in the studies are often poorly identified. Most information about the health effects of lactic acid bacteria is thus anecdotal. There is a clear need for critical study of the effects on health of strain selection and the quality of fermented foods and their ingredients. Clinical studies should be properly conducted as double-blind, placebo-controlled randomized trials. Both the defined bacterial strains and the proposed products should be studied to verify results. Only such studies produce the solid data that can back up health claims.

This book reviews current developments in the study of lactic acid bacteria using the above-mentioned criteria. An overview of the taxonomy and general physiology of lactic acid bacteria is given. A discussion of the genetics of lactic acid bacteria as a future area of interest is included as well as a chapter on the technological aspects of manufacturing functional lactic acid bacteria starters. Many chapters consider our present knowledge of the effects of lactic acid bacteria in human health and disease and as animal probiotics.

One chapter of particular interest describes the development of individual lactic acid microflora. It was written by an Estonian research group that worked in association with the former Soviet space program. These results have not been previously published in the West.

Thus, this book attempts to shed light on little-known and controversial aspects of lactic acid bacteria and their applications. As new techniques as well as new interest in these organisms develop, the anecdotal evidence on the health benefits of specific strains of lactic acid bacteria is slowly being replaced by a more scientific outlook. This book should serve as an important introduction to any student or scientist interested in these developments.

In particular, those working with lactic acid bacteria and fermented foods or feed products within universities and the food industry should find this book most interesting. It will also be helpful to dairy scientists and technologists, both as a textbook and as a handbook for product development. It will be useful to government organizations developing regulatory policies for products based on lactic acid fermentation and bacteria, especially when health claims are concerned. Finally, consumer groups interested in the effects of lactic acid bacteria may benefit from the comprehensive reviews in this volume.

Readers are referred to most recent literature in the area, covering the subject well from various aspects. Our aim has been to give an overview of a rapidly changing and extremely important area of food and nutrition research.

Seppo Salminen
Atte von Wright

Contents

Contributors

J. T. Ahokas Royal Melbourne Institute of Technology, Melbourne, Victoria, Australia

Lars T. Axelsson* Swedish University of Agricultural Sciences, Uppsala, Sweden

Jean Ballongue Laboratoire de Chimie Biologique, Université de Nancy 1, Vandoeuvre-les-Nancy, and Centre de Recherche International André Gaillard, Ivry-sur-Seine, France

Marc Bigret Cultures and Enzymes Division, Sanofi Bio-Industries, Paris, France

P. Michael Davidson Department of Food Science and Toxicology, University of Idaho, Moscow, Idaho

Margaret Deighton Department of Applied Biology and Biotechnology, Royal Melbourne Institute of Technology, Melbourne, Victoria, Australia

D. C. Donohue Toxicology Centre, Royal Melbourne Institute of Technology, Melbourne, Victoria, Australia

Rangne Fonden Panova Ltd., Stockholm, Sweden

**Present affiliation*: MATFORSK, Norwegian Food Research Institute, Osloveien, Norway

Barry R. Goldin Department of Community Health, Tufts University School of Medicine, Boston, Massachusetts

Sherwood Gorbach Department of Community Medicine and Unit of Infectious Diseases, Tufts University School of Medicine, Boston, Massachusetts

Dallas G. Hoover Department of Food Science, University of Delaware, Newark, Delaware

Yuan-Kun Lee Department of Microbiology, National University of Singapore, Kent Ridge, Singapore

Alice H. Lichtenstein Department of Nutrition, USDA Human Nutrition Research Center on Aging at Tufts University, Boston, Massachusetts

Reet Mändar Institute of General and Molecular Pathology, University of Tartu, and Laboratory of Microbiology, Tartu University Hospital, Tartu, Estonia

Annika Mäyrä-Mäkinen Starter Culture Division, Department of Research and Development, Valio Ltd., Helsinki, Finland

Marika Mikelsaar Institute of General and Molecular Pathology, University of Tartu, and Laboratory of Microbiology, Tartu University Hospital, Tartu, Estonia

Juha Nousiainen Farm Services, Valio Ltd., Helsinki, Finland

Patricia Ramos Department of Microbiology, Sodima Research and Development, Sodima Centre de Recherche International André Gaillard, Ivry-sur-Seine, France

Seppo Salminen Research and Development Centre, Valio Ltd., Helsinki, and Department of Applied Chemistry and Microbiology, University of Helsinki, Helsinki, Finland

Hannu Salovaara Department of Food Technology, University of Helsinki, Helsinki, Finland

Jouko Setälä Farm Services, Valio Ltd., Helsinki, Finland

Mervi Sibakov Starter Cultures Division, Department of Research and Development, Valio Ltd., Helsinki, Finland

Atte von Wright Department of Pharmaceutical Chemistry and Department of Biochemistry and Biotechnology, University of Kuopio, Kuopio, Finland; Department of General Microbiology, University of Helsinki, Helsinki, Finland; and Department of Dairy and Food Science, The Royal Veterinary and Agricultural University, Fredericksberg, Denmark

Siew-Fai Wong Department of Quality Control and Department of Research and Development, Malaysia Dairy Industries, Singapore

1

Lactic Acid Bacteria: Classification and Physiology

Lars T. Axelsson*
Swedish University of Agricultural Sciences, Uppsala, Sweden

I. SUMMARY

Lactic acid bacteria are a group of Gram-positive bacteria united by a constellation of morphological, metabolic, and physiological characteristics. The general description of the bacteria included in the group is Gram-positive, nonsporing, nonrespiring cocci or rods, which produce lactic acid as the major end product during the fermentation of carbohydrates. The boundaries of the group have been subject to some controversy, but there has been general agreement that the genera *Lactobacillus, Leuconostoc, Pediococcus,* and *Streptococcus* form the core of the group. Recent taxonomic revisions of these genera suggest that the lactic acid bacteria comprise the following: *Aerococcus, Carnobacterium, Enterococcus, Lactobacillus, Lactococcus, Leuconostoc, Pediococcus, Streptococcus, Tetragenococcus,* and *Vagococcus*. The classification of lactic acid bacteria into different genera is largely based on morphology, mode of glucose fermentation, growth at different temperatures, configuration of the lactic acid produced, ability to grow at high salt concentrations, and acid or alkaline tolerance. For some of the newly described genera, additional characteristics such as fatty acid composition

**Present affiliation*: MATFORSK, Norwegian Food Research Institute, Osloveien, Norway

and motility are used in classification. The measurements of true phylogenetic relationships with rRNA sequencing have aided the classification of lactic acid bacteria (LAB) and clarified the phylogeny of the group. Most genera in the group form phylogenetically distinct groups, but some, in particular *Lactobacillus* and *Leuconostoc*, are very heterogeneous and the phylogenetic clusters do not correlate with the current classification based on phenotypic characters. New tools for classification and identification of LAB are underway. The most promising for routine use are nucleic acid probing techniques, partial rRNA gene sequencing using the polymerase chain reaction, and soluble protein patterns.

Two main sugar fermentation pathways can be distinguished among lactic acid bacteria. Glycolysis (Embden-Meyerhof pathway) results in almost exclusively lactic acid as end product under standard conditions, and the metabolism is referred to as homolactic fermentation. The 6-phosphogluconate/phosphoketolase pathway results in significant amounts of other end products, such as ethanol, acetate, and CO_2 in addition to lactic acid, and the metabolism is referred to as heterolactic fermentation. Various growth conditions may significantly alter the end-product formation by some lactic acid bacteria. These changes can be attributed to an altered pyruvate metabolism and/or the use of external electron acceptors such as oxygen or organic compounds.

Lactic acid bacteria creates a proton motive force mainly by means of a membrane-located H^+ATPase at the expense of ATP. The proton motive force drives the uphill transport of metabolites and ions into the cell. End-product efflux may contribute to the formation of a proton motive force, thus sparing ATP. Sugar transport is mediated mainly by proton-motive-force-dependent permease systems or phosphoenolpyruvate–sugar phosphotransferase systems. The latter are tightly coupled to and regulated with sugar metabolism. Transport of amino acids is mediated by proton-motive-force-dependent systems, antiport systems, or phosphate-bond-linked systems.

II. GENERAL INTRODUCTION

What is a lactic acid bacterium? Asking the question to scientists in the field would probably result in a fairly uniform answer. This is due more to historic tradition, dating to the turn of the century, than to the existence of an unequivocal definition of the term. The term *lactic acid bacteria* was then used synonymously with "milk-souring organisms." Important progress in the classification of these bacteria was made when the similarity between milk-souring bacteria and other lactic-acid-producing bacteria of other habitats was recognized (Henneberg, 1904; Löhnis, 1907). However, confusion was still prevalent when the monograph of Orla-Jensen (1919) appeared. This work has had a large impact on the systematics of LAB, and, although revised to some extent, it is still valid and the classification basis is remarkably unchanged. Orla-Jensen used a few characters as classification basis: morphology (cocci or rods, tetrad formation), mode of glucose fermentation (homo- or heterofermentation), growth at certain "cardinal" temperatures (e.g., 10°C and 45°C), and form of lactic acid produced (D, L, or both). As will be seen, these characters are still very important in current LAB classification. After the work by Orla-Jensen, the view emerged that the core of LAB comprised four genera: *Lactobacillus, Leuconostoc, Pediococcus*, and *Streptococcus*. There has always been some controversy on what the boundaries of the group are (Ingram, 1975), but this will not be dealt with here. The classification section of this chapter will concentrate on these four genera, or rather what used to be these genera, since major taxonomic revisions have recently resulted in the description of new genera.

Orla-Jensen regarded LAB as a "great natural group," indicating a belief that the bacteria included were phylogenetically related and separated from other groups. At that time, only phenotypic characters could be examined and evaluated as phylogenetic markers. Today, we have the means to examine, in detail, macromolecules of the cell, believed to be more accurate in defining relationships and phylogenetic positions. These are, of course, the nucleic acids. Fortunately, nature has provided us with different kinds of nucleic acids for different types of taxonomic studies. Close relations (at species and subspecies level) can be determined with DNA-DNA homology studies (Johnson, 1984). For determining phylogenetic positions of species and genera, ribosomal RNA (rRNA) is more suitable, since the sequence contains both well-conserved and less-conserved regions. It is now possible to determine the sequence of long stretches of rRNA (~1500 bases of 16S rRNA) from bacteria (Lane et al., 1985). Comparisons of these sequences are currently the most powerful and accurate technique for determining phylogenetic relationships of microorganisms (Woese, 1987). With this technique, a clearer picture of the phylogeny of LAB is emerging, and the ideas of Orla-Jensen can be examined with some accuracy. In addition, rRNA sequencing is becoming an important aid in the classification of LAB, as exemplified by the descriptions of new genera (Collins et al., 1990; Wallbanks et al., 1990). The classification section of this chapter will deal with both the "classical" classification schemes and the current phylogenetic status of LAB.

The physiology of LAB has been of interest ever since it was recognized that these bacteria are involved in the acidification of food and feed products. Increased knowledge of the LAB physiology, such as metabolism, nutrient utilization, etc., has been one way to achieve more controlled processes. Today, modern genetic techniques are considered to be promising in this regard (see Chapter 6). However, efforts in this direction will not be fruitful unless there is a sound understanding of the physiology of these bacteria. The fermentative nature of LAB is also of considerable academic interest, since this makes them excellent model systems for the study of energy transduction, solute transport, and membrane biology (Kashket, 1987; Konings et al., 1989; Maloney, 1990).

The designation *lactic acid bacteria* perhaps implies that these bacteria have a somewhat "simple" metabolism, resulting in one or a few fermentation end products. This may also be the case in the laboratory environment that we often impose on them. However, it is clear that LAB have a very diverse metabolic capacity, which enables them to adapt to a variety of conditions. The physiology section of this chapter will describe the main features of LAB, such as carbohydrate metabolism and bioenergetics. However, some of the emphasis will be on the different variations of the general "theme" of metabolism that may occur under certain conditions.

This volume concerns the technological, nutritional, and health aspects of LAB. This reflects the intimate association of the term with food and feed manufacture. Again, this is perhaps more of a historic tradition than a scientifically reached position, since the group includes bacteria which are highly pathogenic and therefore undesirable in food (e.g., many streptococci). In addition, lactobacilli generally associated with food have been implicated in disease (Kandler and Weiss, 1986); carnobacteria are normal inhabitants in meat, but are also fish pathogens (Collins et al., 1987). There are more examples of the "dual" nature of LAB as a group. The main emphasis in this chapter will, however, be on LAB that are *normally* associated with food manufacture and positive health aspects.

III. CLASSIFICATION OF LACTIC ACID BACTERIA

A. General Description and Included Genera

An unequivocal definition of the term *lactic acid bacteria* does not exist. Inevitably, most characteristics that would be used in such a definition are subject to qualification (Ingram, 1975), meaning that they are accurate only under conditions that might be termed "normal" or "standard" and that exceptions to the definition can be found. Therefore, it is more appropriate to describe the *typical* lactic acid bacterium, which is Gram-positive, nonsporing, catalase-negative, devoid of cytochromes, of nonaerobic habit but aerotolerant, fastidious, acid-tolerant, and strictly fermentative with lactic acid as the major end product during sugar fermentation. LAB are generally associated with habitats rich in nutrients, such as various food products (milk, meat, vegetables), but some are also members of the normal flora of the mouth, intestine, and vagina of mammals. Variations of this general theme are common, excluding the Gram-positive and nonsporing characters, which cannot be disputed (spore-forming bacteria that resemble LAB, e.g., *Sporolactobacillus*, are more related to bacilli). In any case, the "definition" is useful in being a core or center around which the actual descriptions of genera or species are formulated. A key feature of LAB that must be emphasized is the inability to synthesize porphyrin groups (e.g., heme). This is the actual physiological background for some of the characteristics mentioned. This makes LAB devoid of a "true" catalase and cytochromes when grown in laboratory growth media, which lack hematin or related compounds. Under these conditions, which are "normal" in most studies of these bacteria, LAB do not possess the mechanism of an electron transport chain and rely on fermentation, i.e., substrate-level phosphorylation, for generating energy. Since catalase activity, mediated by a non-heme "pseudocatalase," can occur in some LAB (Whittenbury, 1964; Kono and Fridovich, 1983), the lack of cytochromes may be a more reliable characteristic in preliminary diagnosing than the commonly used catalase test (Ingram, 1975). However, it is important to note that the situation may be totally different if hematin (or hemoglobin) is added to the growth medium. A true catalase, and even cytochromes, may be formed by some LAB, in some cases resulting in respiration with a functional electron transport chain (Whittenbury, 1964, 1978; Bryan-Jones and Whittenbury, 1969; Ritchey and Seeley, 1976; Wolf et al., 1991).

The genera that, in most respects, fit the general description of the typical LAB are (as they appear in the latest edition of *Bergey's Manual* from 1986) *Aerococcus* (*A.*), *Lactobacillus* (*Lb.*), *Leuconostoc* (*Ln.*), *Pediococcus* (*P.*), and *Streptococcus* (*S.*). Major revisions of the taxonomy of LAB, in particular of the streptococci, was anticipated in *Bergey's Manual* of 1986 (Hardie, 1986b; Schleifer, 1986) and to some extent already realized by the year of that issue. Thus, the former genus *Streptococcus* was first divided into three: *Enterococcus* (*E.*), *Lactococcus* (*Lc.*), and *Streptococcus sensu stricto* (Schleifer and Kilpper-Bälz, 1984, 1987; Schleifer et al., 1985). Later, some motile LAB, otherwise resembling lactococci, were suggested to form a separate genus, *Vagococcus* (*V.*) (Collins et al., 1989). The genera *Lactobacillus, Leuconostoc*, and *Pediococcus* have largely remained unchanged, but some rod-shaped LAB, previously included in *Lactobacillus*, are now forming the genus *Carnobacterium* (*C.*) (Collins et al., 1987), and the former species *Pediococcus halophilus* has been raised to genus level, forming the genus *Tetragenococcus* (*T.*) (Collins et al., 1990). Revisions after 1986 are supported by extensive chemotaxonomic and genetic data. Further revisions are to be expected with regard to the genera *Lactobacillus, Leuconostoc*, and *Pediococcus*.

B. Classification at the Genus Level

As mentioned, the basis for the classification of LAB in different genera has essentially remained unchanged since the work of Orla-Jensen (1919). Although morphology is regarded as questionable as a key character in bacterial taxonomy

(Woese, 1987), it is still very important in the current descriptions of the LAB genera. Thus, LAB can be divided into rods (*Lactobacillus* and *Carnobacterium*) and cocci (all other genera). Furthermore, cell division in two planes, leading to tetrad formation, is used as a key characteristic in the differentiation of the cocci. The tetrad-forming genera are *Aerococcus, Pediococcus*, and *Tetragenococcus*.

An important characteristic used in the differentiation of the LAB genera is the mode of glucose fermentation under standard conditions, i.e., nonlimiting concentrations of glucose and growth factors (amino acids, vitamins, and nucleic acid precursors) and limited oxygen availability. Under these conditions, LAB can be divided into two groups: homofermentative, which convert glucose almost quantitatively to lactic acid, and heterofermentative, which ferment glucose to lactic acid, ethanol/acetic acid, and CO_2 (Sharpe, 1979). In practice, a test for gas production from glucose will distinguish between the groups (Sharpe, 1979). (For a more detailed discussion concerning the metabolic pathways, see Section IV.A.) Leuconostocs and a subgroup of lactobacilli are heterofermentative; all other LAB are homofermentative.

Growth at certain temperatures is mainly used to distinguish between some of the cocci. Enterococci grow at both 10°C and 45°C, lactococci and vagococci at 10°C, but not at 45°C. Streptococci do not grow at 10°C, while growth at 45°C is dependent on the species. Salt tolerance (6.5% NaCl) may also be used to distinguish among enterococci, lactococci/vagococci, and streptococci, although variable reactions can be found among streptococci (Mundt, 1986). Extreme salt tolerance (18% NaCl) is confined to the genus *Tetragenococcus*. Tolerance to acid and/or alkaline conditions are also useful characteristics. Enterococci are characterized by growth at both high and low pH. The formation of the different isomeric forms of lactic acid during fermentation of glucose can be used to distinguish between leuconostocs and most heterofermentative lactobacilli, as the former produce only D-lactic acid and the latter a racemate (DL-lactic acid).

A summary of the differentiation of the LAB genera with classical phenotypic tests is shown in Table 1. The newly described genus *Carnobacterium* is indistinguishable from *Lactobacillus* with these tests, as is *Vagococcus* from *Lactococcus. Vagococcus* and *Carnobacterium* have a unique fatty acid composition, which clearly separates these genera from other LAB (Collins et al., 1987, 1989). Furthermore, species of *Vagococcus* are generally motile (Collins et al., 1989), which is an unusual property among LAB. A typical character of carnobacteria is the failure to grow on acetate media, commonly used as selective media for lactobacilli (Collins et al., 1987), but this property is actually shared with some species and strains of lactobacilli. Pediococci can be confused with aerococci, since the morphology is similar. However, most pediococci are more acid-tolerant than aerococci and grow well anaerobically, contrary to the

Table 1 Differential Characteristics of Lactic Acid Bacteria[a]

	Rods		Cocci						
Characteristic	*Carnob.*	*Lactob.*	*Aeroc.*	*Enteroc.*	*Lactoc. Vagoc.*	*Leucon.*	*Pedioc.*	*Streptoc.*	*Tetragenoc.*
Tetrad formation	–	–	+	–	–	–	+	–	+
CO_2 from glucose[b]	–	±	–	–	–	+	–	–	–
Growth at 10°C	+	±	+	+	+	+	±	–	+
Growth at 45°C	–	±	–	+	–	–	±	±	–
Growth in 6.5% NaCl	ND[d]	±	+	+	–	±	±	–	+
Growth in 18% NaCl	–	–	–	–	–	–	–	–	+
Growth at pH 4.4	ND	±	–	+	±	±	+	–	–
Growth at pH 9.6	–	–	+	+	–	–	–	–	+
Lactic acid[c]	L	D,L,DL[e]	L	L	L	D	L, DL[e]	L	L

[a]+, positive; –, negative; ±, response varies between species; ND, not determined.
[b]Test for homo- or heterofermentation of glucose; negative and positive denotes homofermentative and heterofermentative, respectively.
[c]Configuration of lactic acid produced from glucose.
[d]No growth in 8% NaCl has been reported.
[e]Production of D-, L-, or DL-lactic acid varies between species.

microaerophilic nature of aerococci (Evans, 1986). A former species of *Pediococcus, P. urinaeequi*, earlier the most likely to be confused with aerococci, is now considered to be *A. viridans*, a suggestion mainly based on rRNA sequence (Collins et al., 1990).

It should be noted that there are phenotypic overlaps between genera, and exceptions to the general rules outlined in Table 1 can be found. Classification of LAB is becoming dependent on more sophisticated methods, of which rRNA sequencing probably is the most accurate at the genus level.

C. Classification at Species Level

Within the scope of this chapter, it is impossible to describe the classification of all species of LAB. For instance, only the genus *Lactobacillus* comprises about 50 recognized species (Collins et al., 1991). Therefore, the following section will only be a summary, concentrating on the means by which classification within a genus can be done and the most interesting species from a food technology point of view.

As indicated previously, proper classification of LAB is beginning to rely on molecular biology methods, although Orla-Jensen's concepts are still viable. This is perhaps truer regarding classification at species level than at genus level. In some cases, only DNA-DNA homology studies can resolve problems in classification (Kandler, 1984). Still, the classical phenotypic/biochemical characterization is important for a preliminary classification as well as learning about the properties of the strains. Some of the characters listed in Table 1 are useful also in the classification at species level, e.g., salt and pH tolerance, growth at certain temperatures, and configuration of the lactic acid produced. Other characters used in the phenotypic/biochemical characterization of strains are range of carbohydrates fermented, arginine hydrolysis, acetoin formation (Vogues-Proskauer test), bile tolerance, type of hemolysis, production of extracellular polysaccharides, growth factor requirements, presence of certain enzymes (e.g., β-galactosidase and β-glucoronidase), growth characteristics in milk, and serological typing. Further characterization includes more molecular/chemotaxonomic approaches, including type of diamino acid in the peptidoglycan, presence and type of teichoic acid, presence and type of menaquinones, guanine + cytosine (G + C) ratio of the DNA, fatty acid composition, and electrophoretic mobility of the lactate dehydrogenase (LDH).

1. *Enterococcus, Lactococcus, Streptococcus,* and *Vagococcus*

As mentioned, the genera *Enterococcus, Lactococcus, Streptococcus,* and *Vagococcus* were earlier included in one genus, *Streptococcus*. For details regarding the major taxonomic revision of the "streptococci" and the current classification of enterococci, lactococci, and streptococci, the reader is referred to an excellent review by Schleifer and Kilpper-Bälz (1987), which summarizes the phenotypical, biochemical, and molecular characteristics of the species included in the genera. The summary presented here is based on that review.

Historically, serological typing with the Lancefield grouping (Lancefield, 1933) has been very important in the classification of streptococci. The method is now considered to be less important in classification, but still very useful in the rapid identification of major pathogens (Sharpe, 1979; Hardie, 1986b; Schleifer and Kilpper-Bälz, 1987). However, there is undoubtedly a strong (but not 100%) correlation between presence of the group D antigen and enterococci (previously designated "group D streptococci" or "fecal streptococci"). Similarly, the group N antigen is correlated with lactococci (previously "group N streptococci" or "lactic streptococci"), but note that the vagococci also possess the group N antigen (Collins et al., 1989).

Despite the formation of new genera, the genus *Streptococcus sensu stricto* is still very large and the classification is difficult. The genus is broadly divided into three groups: pyogenic, oral, and "other" streptococci. Some anaerobic cocci, previously included in the genus as the group "anaerobic streptococci" (Hardie, 1986a), were shown to be unrelated to all other streptococci and have been excluded (Schleifer and Kilpper-Bälz, 1987). The pyogenic group contains some famous pathogens, e.g., *S. pyogenes*. Another pathogen, *S. pneumoniae*, was earlier also included in this group, but has been transferred to the oral group, which otherwise contains species mostly associated with the oral cavity of man and animals. Some of the oral streptococci, e.g., *S. mutans*, can be causative agents of dental caries (Hardie, 1986c). As a general rule, the pyogenic streptococci are β-hemolytic, while the oral streptococci are α- or nonhemolytic.

The only streptococcal species that is associated with food technology is *S. thermophilus*, which is used in the manufacture of yogurt (in coculture with *Lb. delbrückii* subsp. *bulgaricus*). *S. thermophilus* was included in the group other streptococci by Schleifer and Kilpper-Bälz (1987) and Hardie (1986d), but could conceivably be included in the oral group, since a close relationship at the DNA level with *S. salivarius* has been established (Farrow and Collins, 1984; Hardie, 1986d). Farrow and Collins (1984) proposed that *S. thermophilus* should be considered a subspecies of *S. salivarius*, since DNA-DNA homology values of greater than 70% were determined. Thus, the name *S. salivarius* subsp. *thermophilus* became valid, and, to my knowledge, still is. However, the suggestion was later rejected, since an investigation of a large number of strains revealed much lower DNA-DNA homology values for some strains. In addition, the large phenotypic differences would justify two separate species (Schleifer and Kilpper-Bälz, 1987). For this reason, and for convenience, I will use the name *S. thermophilus* throughout this chapter. The heat resistance, the ability to grow at 52°C, and the rather limited number of carbohydrates attacked distinguish *S. thermophilus* from most other streptococci (Hardie, 1986d).

Lactococci are intimately associated with dairy products, but out of the five species currently recognized (Schleifer et al., 1985; Williams et al., 1990) only one, *Lc. lactis*, is actually used in dairy technology. Three subspecies can be distinguished: *Lc. lactis* subsp. *lactis*, *Lc. lactis* subsp. *cremoris*, and *Lc lactis* subsp. *hordniae*. Only the first two are important in dairy manufacture. *Lc lactis* subsp. *lactis* includes species formerly designated *S. lactis* subsp. *lactis*, *S. lactis* subsp. *diacetylactis* and *Lactobacillus xylosus* (Schleifer et al., 1985). The latter illustrates the fact that morphology sometimes can be deceptive in classification of LAB; i.e., the distinction between cocci and rods is not always an easy task. *Lc. lactis* subsp. *cremoris* includes species previously designated *S. cremoris* or *S. lactis* subsp. *cremoris*. *Lc. lactis* subsp. *cremoris* is distinguished from *Lc. lactis* subsp. *lactis* by the inability to (i) grow at 40°C, (ii) grow in 4% NaCl, (iii) hydrolyze arginine, and (iv) ferment ribose (Schleifer et al., 1985).

As noted, species of the newly described genus *Vagococcus* are easily confused with lactococci, differing mainly in fatty acid composition and motility (Collins et al., 1989).

Enterococci are not considered to be of particular importance in food technology. Some species, in particular *E. faecalis* (previously *S. faecalis*), can be opportunistic pathogens (Parker, 1978) and are, therefore, generally undesirable in food. However, preparations of *E. faecium* (previously *S. faecium*) and *E. faecalis* have been used as probiotics (Fuller, 1986; Tournut, 1989) and as silage inoculants (Seale, 1986). The probioic approach is not farfetched, since the natural habitat of most enterococci is the intestine of man and animals (Mundt, 1986). The species of enterococci are differentiated mainly by carbohydrate fermentation patterns, arginine and hippurate hydrolysis, and presence and/or type of menaquinones (Schleifer and Kilpper-Bälz, 1987). *E. faecalis* has been of great value for general LAB research in being a model organism in physiological studies of, for instance, bioenergetics and membrane biology (Kashket, 1987; Maloney, 1990).

2. *Aerococcus, Pediococcus, and Tetragenococcus*

Pediococci are important in food technology, in both a negative and positive sense. *P. damnosus* is a major spoilage organism in beer manufacture, since growth may lead to diacetyl/acetoin formation, resulting in a buttery taste (Garvie, 1986b). *P. acidilactici* and *P. pentosaceus* are used as starter cultures for sausage making and as silage inoculants (Seale, 1986; Hammes et al., 1990). Pediococci may also be important constituents of the complex known as the nonstarter lactic acid bacteria (NSLAB), which is involved in the ripening of cheese (Fox et al., 1990; Olson, 1990). The main characters for distinguishing between the species are the range of sugars fermented, hydrolysis of arginine, growth at different pH levels (7.0 and 4.5), and the configuration of lactic acid produced (Garvie, 1986b). *P. pentosaceus* and *P. acidilactici* cannot be separated using these characteristics, but have been shown to be distinct species with DNA-DNA homology studies (Garvie, 1986b). These species are also similar in that they may produce a non-heme pseudocatalase (Whittenbury, 1964; Garvie, 1986b).

Pediococci may be confused with aerococci (see above). The genus *Aerococcus* is monospecific (*A. viridans*) and will not be dealt with further.

As mentioned, the genus *Tetragenococcus* contains strains previously regarded as *P. halophilus*. Only one species, *T. halophilus*, is currently recognized (Collins et al., 1990). In addition to its extreme salt tolerance (>18% NaCl), which distinguish it from other LAB, *T. halophilus* has a salt requirement for growth, generally 5% NaCl (Garvie, 1986b). *T. halophilus* is an important species in lactic fermentations of foods containing high concentrations of salt, e.g., soy sauce (Garvie, 1986; Abe and Uchida, 1989).

3. *Leuconostoc*

The leuconostocs are separated from other cocci of the LAB by their heterofermentative metabolism. They can, however, easily be confused with "coccoid rods" of the heterofermentative lactobacilli. The configuration of the lactic acid formed separate leuconostocs from most heterofermentative lactobacilli (see above), but some species of the latter may also produce D-lactic acid (Garvie, 1986a; Kandler and Weiss, 1986). Leuconostocs are also unable to hydrolyze arginine, contrary to most heterofermentative lactobacilli, but, again, there are exceptions among the latter (Garvie, 1986a; Kandler and Weiss, 1986). A number of investigations have been made concerning the relationship between leuconostocs and the species of heterofermentative lactobacilli that resemble the leuconostocs phenotypically (Garvie, 1986a; Yang and Woese, 1989; Martinez-Murcia and Collins, 1990). The current view is that *Ln. paramesenteroides*, together with some heterofermentative lactobacilli, e.g., *Lb. confusus* and *Lb. viridescens*, may represent the nucleus of a new genus, separated from both other heterofermentative lactobacilli and other leuconostocs (Martinez-Murcia and Collins, 1990). However, it is at present not possible to distinguish between the *Ln. paramesenteroides* group and the main group of leuconostocs (designated *Leuconostoc sensu stricto*; Martinez-Murcia and Collins, 1990) using physiological/chemotaxonomic characters.

Leuconostocs may form significant amounts of diacetyl from citrate in milk and some species, mainly *Ln. mesenteroides* subsp. *cremoris*, have been used in the dairy industry for this purpose. Leuconostocs are also important in spontaneous vegetable fermentations, e.g., sauerkraut, where they often initiate the lactic fermentation (Daeschel et al., 1987).

Species and subspecies are distinguished by characteristics such as carbohydrate fermentation patterns, dextran formation (from sucrose), growth requirements, salt and pH tolerance, and dissimilation of citrate and/or malate, but classification is difficult (Garvie, 1986a). The dissimilation of malate requires attention, as this metabolism results in L-lactic acid (Radler, 1975). Hence, care must be taken that malate is not present in the growth medium when the configuration of the lactic acid (from glucose fermentation) is to be determined (Garvie, 1986a). The malate-dissimilating, ethanol- and acid-tolerant species, *Ln. oenos*, characteristically found in wine, is phenotypically very different from other leuconostocs. Phylogenetic investigations indicate that *Ln. oenos* may warrant a separate genus (Yang and Woese, 1989; Martinez-Murcia and Collins, 1990).

4. *Lactobacillus and Carnobacterium*

The genus *Lactobacillus* is by far the largest of the genera included in LAB. It is also very heterogeneous, encompassing species with a large variety of phenotypic, biochemical, and physiological properties. The heterogeneity is reflected by the range of mol% G+C of the DNA of species included in the genus. This range is 32–53%, which is twice the span usually accepted for a single genus (Schleifer and Stackebrandt, 1983). The heterogeneity and the large number of species are due to the definition of the genus, which essentially is rod-shaped lactic acid bacteria. Such a definition is comparable to an arrangement where all the coccoid LAB were included in one genus. However, among the cocci, phenotypic traits were early recognized, which made differentiation into several genera possible. Even if the situation was more difficult for the rod-shaped LAB, Orla-Jensen (1919) essentially tried to divide this group in a similar way as with the cocci. Thus, the subgenera of *Lactobacillus* were created: *Thermobacterium*, *Streptobacterium*, and *Betabacterium*. Remarkably, this division is still valid to a considerable degree, although the designations have been dropped and some modifications of the definitions of the subgroups have been made (Kandler, 1984; Kandler and Weiss, 1986). Table 2 summarizes the characters used to distinguish among the three groups and some of the more well-known species included in each group. The physiological basis for the division is (generally) the presence or absence of the key enzymes of homo- and heterofermentative sugar metabolism, fructose-1,6-diphosphate adolase, and phosphoketolase, respectively (Kandler, 1983, 1984; Kandler and Weiss, 1986). (For further details regarding the division of LAB in homo- and heterofermentative, see Section IV.A.)

The classical ways of distinguishing among species of lactobacilli have been carbohydrate fermentation patterns, configuration of lactic acid produced, hydrolysis of arginine, growth requirements, and growth at certain temperatures (Sharpe, 1979, 1981). These characters are still useful, but proper classification may also require analysis of the peptidoglycan, electrophoretic mobility of the

Table 2 Arrangement of the Genus *Lactobacillus*

Characteristic	Group I: Obligately homofermentative	Group II: Facultatively heterofermentative	Group III: Obligately heterofermentative
Pentose fermentation	–	+	+
CO_2 from glucose	–	–	+
CO_2 from gluconate	–	+[a]	+[a]
FDP aldolase present	+	+	–
Phosphoketolase present	–	+[b]	+
	Lb. acidophilus	*Lb. casei*	*Lb. brevis*
	Lb. delbrückii	*Lb. curvatus*	*Lb. buchneri*
	Lb. helveticus	*Lb. plantarum*	*Lb. fermentum*
	Lb. salivarius	*Lb. sake*	*Lb. reuteri*

[a]When fermented.
[b]Inducible by pentoses.
Source: Adapted from Sharpe (1981) and Kandler and Weiss (1986).

LDH, mol% G+C of the DNA, and even DNA-DNA homology studies (Kandler, 1984; Kandler and Weiss, 1986).

Lactobacilli are widespread in nature, and many species have found applications in the food industry. They are generally the most acid-tolerant of the LAB (Kashket, 1987) and will, therefore, terminate many spontaneous lactic fermentations such as silage and vegetable fermentations (Daeschel et al., 1987). Lactobacilli are also associated with the oral cavity, gastrointestinal tract, and vagina of humans and animals (Sharpe, 1981; Kandler and Weiss, 1986). Some species, e.g., *Lb. brevis, Lb. casei*, and *Lb. plantarum*, can be found in many habitats. Others are more specialized and are found only in certain niches, e.g., the sourdough organism *Lb. sanfrancisco* and the yogurt-associated *Lb. delbrückii* subsp. *bulgaricus* (previously *Lb. bulgaricus*).

Within the scope of this chapter, details in the classification of species of lactobacilli are not possible to discuss further. The reader is referred to a review by Kandler (1984), which covers the taxonomy of industrial important species or, for a detailed presentation of the genus, to *Bergey's Manual* (Kandler and Weiss, 1986). Note, however, that new species are described continuously (e.g., see Farrow and Collins, 1988; Embley et al., 1989; Collins et al., 1991). It has been known for some time that the genus *Lactobacillus* and its division into the three groups shown in Table 2 are not in accord with natural relations revealed by phylogenetic analysis (Kandler and Weiss, 1986; Stackebrandt and Teuber, 1988; Collins et al., 1991, and see below). However, it is not easy to envisage how a new classification system would appear, since it would be difficult to find properties, other than rRNA sequences, that would constitute a basis for a phylogenetically correct taxonomy.

Species of the genus *Carnobacterium* were originally classified as group III lactobacilli under the designations *Lb. divergens, Lb. carnis*, and *Lb. piscicola* (Kandler and Weiss, 1986; Collins et al., 1987). Later studies showed that these bacteria were separate from lactobacilli and warranted a separate genus (Collins et al., 1987) and that the metabolism of glucose was predominantly homofermentative (De Bruyn et al., 1988). The fatty acid composition of carnobacteria differ from lactobacilli (Collins et al., 1987). In addition, the combination of *meso*-diaminopimelic acid-direct-type peptidoglycan and production of predominantly L-lactic acid, is shared by only a few species of lactobacilli (Kandler and Weiss, 1986). Carnobacteria are characteristically found in meat and meat products, where they are able to proliferate even at low temperatures (Collins et al., 1987). A simple identification key, confirmed by DNA-DNA hybridization, for distinguishing among *C. divergens, C. piscicola*, and typical meat-associated lactobacilli (e.g., *Lb. curvatus, Lb. sake*, and *Lb. viridescens*) has been published (Montel et al., 1991).

D. Phylogeny of the Lactic Acid Bacteria

Comparisons of the sequence of rRNAs is now regarded to be the optimal measure for determining true phylogenetic relations among bacteria (Woese, 1987). Initially, these comparisons were made by DNA-rRNA hybridizations or oligonucleotide cataloging (i.e., sequencing of cleavage products of rRNA), both of which are labor-intensive methods and not as accurate as one would wish. With the recent advances in molecular genetics techniques, it is now possible to determine the sequence of rather long stretches of rRNA, essentially the whole 16S molecule, by the use of reverse transcriptase (Lane et al., 1985, 1988). The computerized methods now available for handling large amounts of sequence data have made it possible to construct meaningful phylogenetic trees of the entire bacterial kingdom as well as details of certain parts of it (Woese, 1987).

From the data obtained, both from oligonucleotide cataloging and rRNA sequencing, it has been shown that the Gram-positive-type cell wall has a strong phylogenetic relevance. All Gram-positive bacteria cluster in 1 out of the 11 major eubacterial phyla (however, not all bacteria in this phylum have a Gram-positive cell wall; Woese, 1987). The Gram-positive bacteria can be further divided into two main groups or clusters. One cluster consists of bacteria with a mol% G+C of the DNA of above 55%, often designated the *Actinomycetes* subdivision or high–G+C subdivision. This cluster encompasses genera such as *Bifidobacterium, Arthrobacter, Micrococcus, Propionibacterium, Microbacterium, Corynebacterium,*

Actinomyces, and *Streptomyces* (Woese, 1987; Stackebrandt and Teuber, 1988). The other cluster, designated the *Clostridium* subdivision or low-G+C subdivision, contains species with a G+C content in the DNA of 55% or less. All LAB are included in this subdivision, together with aerobes and facultative anaerobes like *Bacillus, Staphylococcus, Listeria*, and anaerobes such as *Clostridium, Peptococcus*, and *Ruminococcus* (Woese, 1987; Stackebrandt and Teuber, 1988).

The details of the phylogenetic relationships between the LAB genera and also between LAB and other genera of the low–G+C subdivision, have been revealed by the extensive work by Collins and co-workers, using the reverse transcriptase rRNA sequencing technique (Collins et al., 1989, 1999, 1991; Martinez-Murcia and Collins, 1990; Wallbanks et al., 1990; Williams et al., 1991). The LAB have been considered to form a "supercluster," which phylogenetically lies in between the strictly anaerobic species (e.g., clostridia) and facultatively or strictly aerobic species (e.g., staphylococci and bacilli), in accord with their life-style, i.e., "on the threshold of anaerobic and aerobic life" (Kandler, 1984; Kandler and Weiss, 1986). This is only partly true. The schematic phylogenetic tree of the LAB as a group, shown in Fig. 1, and the following discussion are based on the work by Collins and co-workers.

As indicated, some genera of the LAB are now partly defined and based on phylogenetic measurements. This is the case for *Aerococcus, Carnobacterium, Enterococcus, Lactococcus, Streptococcus, Tetragenococcus*, and *Vagococcus*, which each form coherent phylogenetic units. Among these, a certain clustering is evident. Thus, *Aerococcus, Carnobacterium, Enterococcus, Tetragenococcus*, and *Vagococcus* are more related to each other than to any other LAB. Furthermore, this group, designated the *Enterococcus* group in Fig. 1, is more related to *Listeria*, and possibly also to *Bacillus*, than to the remaining LAB. This shows that the concept of a LAB supercluster is not entirely correct. The genera *Lactococcus* and *Streptococcus* are also more related to each other than to other LAB, although this tendency is not as strong as for the *Enterococcus* group.

For the remaining LAB genera, *Lactobacillus, Leuconostoc*, and *Pediococcus*, the phylogenetic relationships are not consistent with the current classification schemes. Species of *Leuconostoc* group into three phylogenetic units, each worthy of genus status. The *Leuconostoc sensu stricto* group contains the majority of leuconostocs, including the type species *Ln. mesenteroides*. The second group, designated the paramesenteroides group, contains *Ln. paramesenteroides* and some heterofermentative lactobacilli, e.g., *Lb. confusus, Lb. kandleri*, and *Lb. viridescens*. The species *Ln. oenos* is grouped by itself and represents a case of a rapidly evolving organism, shown by its large evolutionary distance to other Gram-positives examined and unusual phenotype (Yang and Woese, 1989). However, its closest relatives are the two other leuconostoc groups (Martinez-Murcia and Collins, 1990), which may indicate that the adaption to an unusual habitat

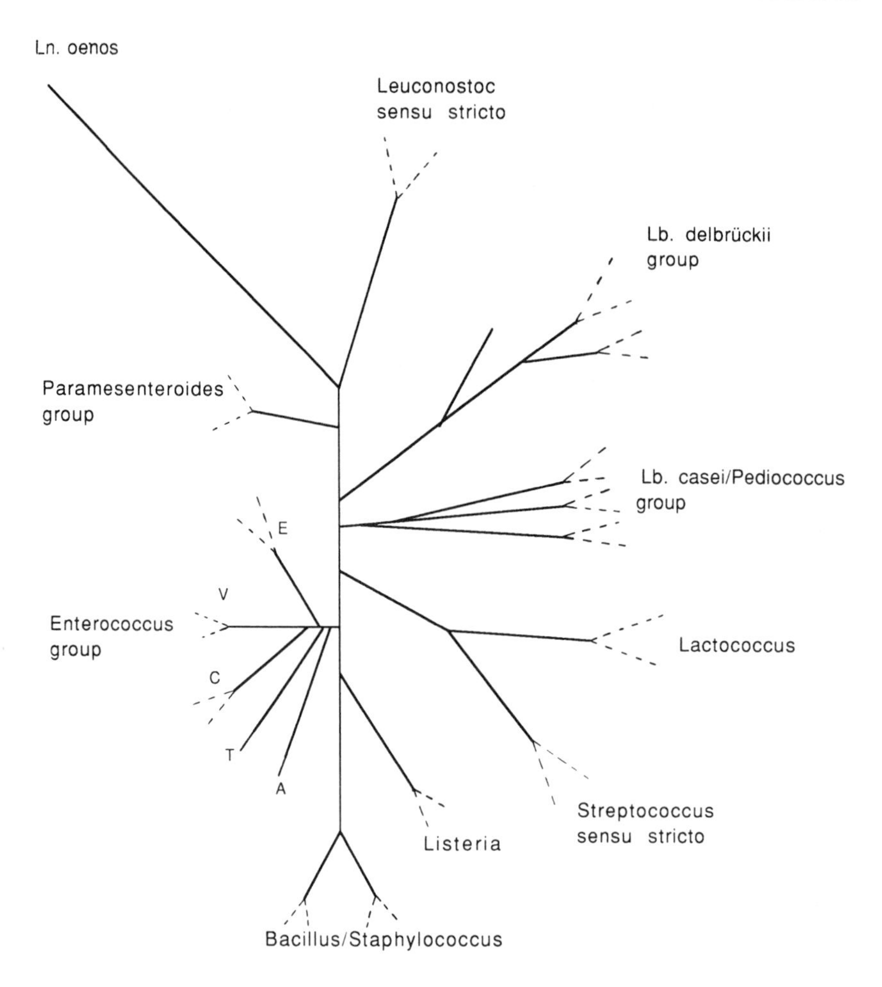

Figure 1 Schematic, unrooted phylogenetic tree of the lactic acid bacteria. Note: evolutionary distances are approximate. Compiled from data by M. D. Collins and co-workers (Collins et al., 1989, 1990, 1991; Martinez-Murcia and Collins, 1990; Wallbanks et al., 1990; Williams et al., 1991). A, *Aerococcus*; C, *Carnobacterium*; E, *Enterococcus*; T, *Tetragenococcus*; V, *Vagococcus*.

(wine) has required extensive changes in the genome, resulting in a rapid divergence from related species.

The species currently recognized as lactobacilli group into three clusters. Unfortunately, these clusters do not correlate with the subdivision of the genus described above (see Table 2). One of the clusters has already been mentioned, the paramesenteroides group, in which some heterofermentative lactobacilli are grouped (see above). The other two are (i) the *Lb. delbrückii* group, which contains many, but notably not all, of the obligately homofermentative (group I) lactobacilli, e.g., *Lb. delbrückii* (with subspecies), *Lb. acidophilus, Lb. helveticus*, and *Lb. jensenii*, and (ii) the *Lb. casei–Pediococcus* group, comprising the remaining obligately homofermentative, the remaining heterofermentative, and all facultatively heterofermentative lactobacilli. In addition, as the designation indicates, included in the latter group are all species of *Pediococcus*. All pediococci except *P. dextrinicus* form a tight cluster within the *Lb. casei–Pediococcus* group.

What are the conclusions that can be drawn from the phylogenetic investigations of LAB? One obvious conclusion, is that morphology is a poor indicator of relatedness, as shown by the phylogenetic structure of *Lactobacillus-Pediococcus-Leuconostoc*. Another conclusion is that future taxonomic revisions are needed, but the question remains how to proceed. As pointed out by Woese (1987), there has to be some correlation between the phylogenetic position of a certain strain or species and its phenotype. The problem lies in how to find the right characteristic(s) that will correlate with true relationships. As a group, the LAB is not clearly separated from all other groups of the low G+C subdivision of the Gram-positives. In particular, it seems that some LAB "overlap" with some of the more aerobic genera such as *Bacillus* and *Staphylococcus*. No comprehensive rRNA sequence comparisons between LAB and the strict anaerobes have been made. However, earlier oligonucleotide cataloging suggests that the evolutionary distance between these groups is very large (Stackebrandt and Teuber, 1988). Therefore, it may be relevant to state that the phylogenetic position of LAB do indeed reflect their life-style of being on the threshhold of anaerobic and aerobic life, but with a slight tendency to group together with aerobes. This statement can be illustrated as shown in Fig. 2.

E. New Tools for Classification and Identification

The classification of LAB, described above, is largely based on phenotypical and biochemical characters. In practice, meaning the routine identification of isolates, these characters may not be enough to definitely assign a strain to a particular species. In fact, DNA-DNA homology studies have, in some cases, been the only way to resolve identification problems (Kandler and Weiss, 1986).

A much more attractive way of identifying strains than hybridizations with total DNA is the use of specific DNA probes, directed at nucleic acid targets of the cells. The main advantage with this is, once the probe has been designed, that time-consuming and laborous DNA preparations can be avoided. Instead, colonies on an agar medium can be tested directly with common colony hybridization

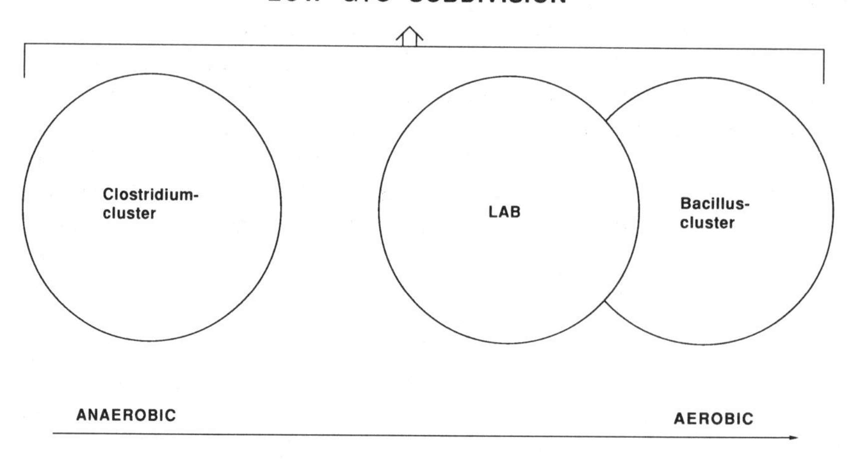

Figure 2 Schematic representation of the phylogenetic position of lactic acid bacteria in relation to anaerobic and aerobic genera of the low G+C subdivision of G^+ bacteria.

methods (Sambrook et al., 1989). However, a specific probe will only detect one species. This means that the probing technique must be used in combination with other identification criteria to narrow the number of possible species that a strain can belong to. The probing technique is perhaps more useful in being the optimal tool for answering questions like, how many bacteria of species xx contains this sample? and so on. Strain specific probes may be of importance for the study of LAB starter cultures in unsterilized food or feed.

One fundamental problem with DNA probes is to find a DNA (or RNA) stretch that is specific for one species (or maybe one particular strain). Two approaches to this problem can be distinguished. The first is an empirical method of trial-and-error character. A DNA library from a particular species is screened for DNA fragments that, when tested as probes, show specificity to that species. The method has been used for developing DNA probes for *Lb. curvatus* (Petrick et al., 1988) and *Lb. helveticus* (Pilloud and Mollet, 1990). The second method starts from known nucleic acid sequencies, and oligonucleotide probes are designed and synthesized after examinations of the sequencies. The nucleic acid of choice is rRNA. The rRNA molecules, in particular 16S and 23S rRNA, contain alternating sequencies of more or less conserved regions. Thus, probes can be designed for different levels of phylogenetic groups, from kingdom to species (Giovannoni et al., 1988; Betzl et al., 1990). Another advantage of rRNA is that these molecules are present in several copies (up to 10^4) in each cell. Therefore, a method employing a probe with an rRNA target will be more sensitive than if a DNA (plasmid or chromosome) directed probe is used.

Identification of LAB with the use of 16S or 23S rRNA-targeted probes has been used for lactococci and enterococci (Betzl et al., 1990), for meat lactobacilli (*Lb. curvatus, Lb. pentosus/plantarum*, and *Lb. sake*; Hertel et al., 1991) and even for distinguishing between the subspecies *lactis* and *cremoris* of *Lc. lactis*. (Salama et al., 1991). 16S rRNA sequence data from LAB are now rapidly accumulating (Collins et al., 1989, 1990, 1991; Martinez-Murcia and Collins, 1990; Wallbanks et al., 1990; Williams et al., 1991). With the aid of computers, these sequence data can be carefully examined, and probes can be designed to fit various applications.

The novel and powerful technique, polymerase chain reaction (PCR; Saiki et al., 1988), may also become useful for identification purposes. With this technique, it is possible to amplify a gene or a part of a gene from a very limited number of cells for subsequent DNA sequencing. The obvious target genes for such an amplification are rRNA genes. The areas of rRNA genes to be amplified and sequenced can be chosen to give maximum resolution with regard to unequivocal identification. Once a complete database of LAB rRNA sequences is constructed, the PCR approach could become very useful. PCR can also be used in combination with probing techniques with the obvious advantage of being much more sensitive than direct hybridization.

The use of restriction endonuclease analysis (REA) is another interesting approach for classifying LAB (Ståhl et al., 1990). Chromosomal DNA is cleaved with a set of restriction endonucleases, and the fragments are electrophoresed in agarose gels and visualized. The banding patterns obtained from different strains are then subject to multivariate data analysis. This method apparently has a very high resolution capacity and can be used as a classification tool, both at species and genus level (Ståhl et al., 1990).

Another new technique, which holds some promise of being simple and rapid, is soluble protein patterns (Dicks and van Vuuren, 1987). Total proteins are liberated from the cells and electrophoresed in gels. The banding patterns are subject to various data analyses. Recent results with this technique suggest high correlations between similarity in banding patterns and DNA-DNA homology values. The technique should thus be useful in classification at species level. A data base with a considerable number of LAB species (currently 400) is under construction (Pot et al., 1991).

Gas chromatography (GC) analysis of cellular fatty acids, combined with data base libraries with type species profiles, has been considered to be a powerful tool in classification and identification of bacteria (e.g., Lechevalier, 1977; Moss, 1981; Miller, 1984). The technique has proven useful for LAB in connection with the descriptions of the new genera *Carnobacterium* and *Vagococcus* (see above).

With regard to species identification, the picture is not that clear. Rizzo et al. (1987) used GC analysis of fatty acids and cell-wall monosaccharides to distinguish between *Lactobacillus* species and suggested the method to be useful. However, strains of *Lb. fermentum* and *Lb. cellobiosus* (now considered to be one species, *Lb. fermentum*; Kandler and Weiss, 1986) fell in two separate clusters with little similarity. Still, with more standardized systems and fatty acid profiling of a greater number of type species, the method will become useful in certain applications as one of several screening methods for classification (R. Dainty, personal communication).

IV. METABOLISM OF LACTIC ACID BACTERIA

The essential feature of LAB metabolism is efficient carbohydrate fermentation coupled to substrate-level phosphorylation. The generated ATP is subsequently used for biosynthetic purposes. LAB as a group exhibit an enormous capacity to degrade different carbohydrates and related compounds. Generally, the predominant end product is, of course, lactic acid (>50% of sugar carbon). It is clear, however, that LAB adapt to various conditions and change their metabolism accordingly. This may lead to significantly different end-product patterns. This section will describe the well-known fermentation pathways and how various sugars are fermented. It is also attempting to show some of the more unusual features of LAB metabolism, which may be of importance in their natural habitat.

A. Major Fermentation Pathways

1. Hexose Fermentation

As mentioned in Section III, there are two major pathways for hexose (e.g., glucose) fermentation occurring within LAB. The pathways are shown in Fig. 3. The transport and phosphorylation of glucose may occur as outlined in the figure, i.e., transport of free sugar and phosphorylation by an ATP-dependent glucokinase. Some species use the phosphoenolpyruvate–sugar phosphotransferase system (PTS), in which phosphoenolpyruvate is the phosphoryl donor (see Section V.B.4). In either case, a high-energy phosphate bond is required for activation of the sugar.

Glycolysis (Embden-Meyerhof pathway), used by all LAB except leuconostocs and group III lactobacilli, is characterized by the formation of fructose-1,6-diphosphate (FDP), which is split by an FDP aldolase into dihydroxyacetonephosphate (DHAP) and glyceraldehyde-3-phosphate (GAP). GAP (and DHAP via GAP) is then converted to pyruvate in a metabolic sequence including substrate-level phosphorylation at two sites. Under normal conditions, i.e., excess sugar and limited access to oxygen, pyruvate is reduced to lactic acid by a NAD^+-dependent lactate dehydrogenase (nLDH), thereby reoxidizing the NADH formed during the earlier glycolytic steps. A redox balance is thus obtained, lactic acid is virtually the only end product, and the metabolism is referred to as a homolactic fermentation.

The other main fermentation pathway has had several designations, such as the pentose phosphate pathway, the pentose phosphoketolase pathway, the hexose monophosphate shunt, and, as used by Kandler and Weiss (1986) in *Bergey's Manual*, the 6-phosphogluconate pathway. I will refer to it as the 6-phosphogluconate/phosphoketolase (6-PG/PK) pathway, thereby recognizing a key step in the metabolic sequence (the phosphoketolase split) and at the same time distinguish it from the bifidum pathway, which also involves phosphoketolase, but does not have 6-phosphogluconate as intermediate (Gottschalk, 1986). It is characterized by initial dehydrogenation steps with the formation of 6-phosphogluconate, followed by decarboxylation. The remaining pentose-5-phosphate is split by phosphoketolase into GAP and acetyl phosphate. GAP is metabolized in the same way as for the glycolytic pathway, resulting in lactic acid formation. When no additional electron acceptor is available (see Section IV.D), acetyl phosphate is reduced to ethanol via acetyl CoA and acetaldehyde. Since this metabolism leads to significant amounts of other end products (CO_2, ethanol) in addition to lactic acid, it is referred to as a heterolactic fermentation.

The terminology regarding these pathways and the bacteria that use them is rather confusing, and it is perhaps appropriate to add a note of caution. In general, the term *homofermentative LAB* refers to those in the group that use the glycolytic pathway for glucose fermentation, whereas *heterofermentative LAB* are those that use the 6-PG/PK pathway. However, it should be noted that glycolysis may lead to a heterolactic fermentation (meaning significant amounts of end products other than lactic acid) under certain conditions and that some LAB, regarded as homofermentative, use the 6-PG/PK pathway when metabolizing certain substrates. This will be discussed in more detail later.

In theory, homolactic fermentation of glucose results in 2 moles of lactic acid and a net gain of 2 ATP per mole glucose consumed. Heterolactic fermentation of glucose through the 6-PG/PK pathway gives 1 mole each of lactic acid, ethanol, and CO_2 and 1 ATP/glucose. In practice, these theoretical values are seldom obtained. A conversion factor of 0.9 from sugar carbon to end product is common and probably reflects an incorporation of sugar carbon into the biomass, even though most growth factors (e.g., amino acids, nucleotides, and vitamins) are supplied in excess in the rich media that is frequently used. These complex media may also contribute to other fermentation balances and to the formation of other end products, in particular acetic acid, since compounds like organic acids, amino acids, and sugar residues can alter the fermentation (Kandler, 1983). The presence

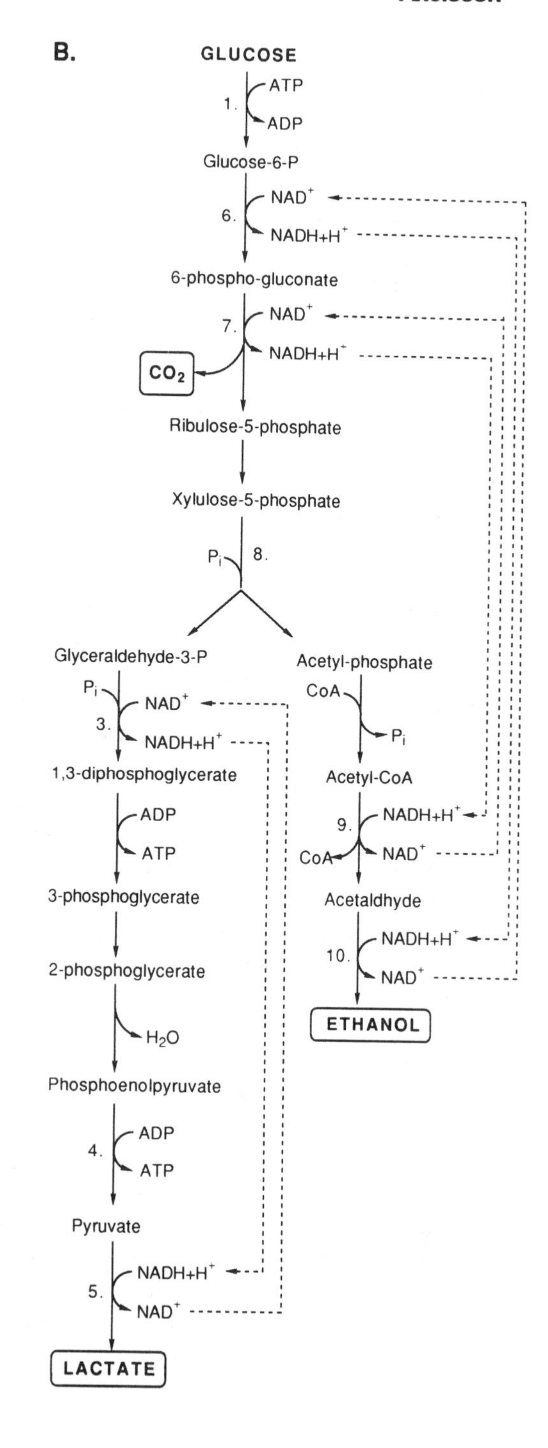
A.
GLUCOSE
1.
ATP
ADP
Glucose-6-P
Fructose-6-P
ATP
ADP
Fructose-1,6-DP
2.
Glyceraldehyde-3-P
Dihydroxy-acetone-P
2 P_i
3.
2 NAD^+
2 $NADH+2H^+$
2 1,3-diphosphoglycerate
2 ADP
2 ATP
2 3-phosphoglycerate
2 2-phosphoglycerate
H_2O
2 phosphoenolpyruvate
4.
2 ADP
2 ATP
2 pyruvate
5.
2 $NADH+2H^+$
2 NAD^+
2 LACTATE
B.
GLUCOSE
1.
ATP
ADP
Glucose-6-P
6.
NAD^+
$NADH+H^+$
6-phospho-gluconate
7.
NAD^+
$NADH+H^+$
CO_2
Ribulose-5-phosphate
Xylulose-5-phosphate
P_i
8.
Glyceraldehyde-3-P
Acetyl-phosphate
P_i
3.
NAD^+
$NADH+H^+$
1,3-diphosphoglycerate
ADP
ATP
3-phosphoglycerate
2-phosphoglycerate
H_2O
Phosphoenolpyruvate
4.
ADP
ATP
Pyruvate
5.
$NADH+H^+$
NAD^+
LACTATE
CoA
P_i
Acetyl-CoA
9.
$NADH+H^+$
CoA
NAD^+
Acetaldhyde
10.
$NADH+H^+$
NAD^+
ETHANOL

of oxygen may also have a significant effect on the metabolism (Condon, 1987, and see Section IV.C).

Hexoses other than glucose, such as mannose, galactose, and fructose are fermented by many LAB. The sugars enter the major pathways at the level of glucose-6-phosphate or fructose-6-phosphate after isomerization and/or phosphorylation. One important exception is galactose metabolism in LAB which uses a PTS for uptake of this sugar, e.g., *Lc. lactis, E. faecalis*, and *Lb. casei*. In these species, the galactose-6-phosphate formed by the PTS is metabolized through the tagatose-6-phosphate pathway (Bissett and Anderson, 1974), which is shown in Fig. 4A. Tagatose is a stereoisomer of fructose, but separate enzymes are required for the metabolism of the tagatose derivatives. The tagatose pathway coincides with glycolysis at the level of GAP. Many strains in this category do also have the capacity to transport galactose with a permease and convert it to glucose-6-phosphate via the Leloir pathway (Fig. 4B; Thomas et al., 1980). This pathway is also used by galactose-fermenting LAB that transport galactose with a permease and lack a galactose PTS (Kandler, 1983; Konings et al., 1989; Fox et al., 1990).

2. *Disaccharide Fermentation*

Depending on the mode of transport, disaccharides enter the cell either as free sugars or as sugar phosphates. In the former case, the free disaccharides are split by specific hydrolases to monosaccharides, which then enter the major pathways described above. In the latter case, i.e., when sugar PTSs are involved, specific phosphohydrolases split the disaccharide phosphates into one part free monosaccharides and one part monosaccharide phosphates.

By far the most studied disaccharide metabolism in LAB is lactose fermentation. Most strains of *Lc. lactis*, at least those used as dairy starters, contain a lactose PTS (Lawrence and Thomas, 1979; Thompson, 1979). A lactose PTS from *Lb. casei* is also well characterized (Chassy and Alpert, 1989). In these strains, lactose enters the cytoplasm as lactose phosphate, which is cleaved by phospho-β-D-galactosidase (P-β-gal) to yield glucose and galactose-6-phosphate. Glucose is phosphorylated by glucokinase and metabolized through the glycolytic pathway, whereas galactose-6-phosphate is metabolized through the tagatose-6-phosphate

---

Figure 3 Major fermentation pathways of glucose (A) homolactic fermentation (glycolysis, Embden-Meyerhof pathway); (B) heterolactic fermentation (6-phosphogluconate/phosphoketolase pathway). Selected enzymes are numbered: 1. glucokinase; 2. fructose-1,6-diphosphate aldolase; 3. glyceraldehyde-3-phosphate dehydrogenase; 4. pyruvate kinase; 5. lactate dehydrogenase; 6. glucose-6-phosphate dehydrogenase; 7. 6-phosphogluconate dehydrogenase; 8. phosphoketolase; 9. acetaldehyde dehydrogenase; 10. alcohol dehydrogenase.

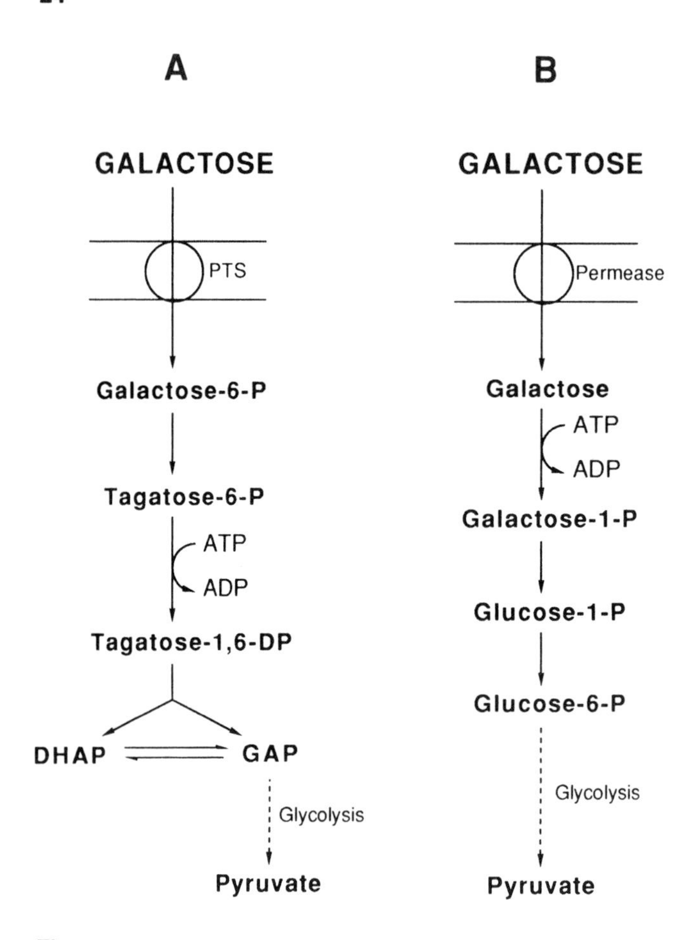

Figure 4 Galactose metabolism in lactic acid bacteria: (A) tagatose-6-phosphate pathway; (B) Leloir pathway.

pathway. The enzyme system of the lactose PTS and P-β-gal are generally inducible and repressed by glucose (Kandler, 1983).

The lactose metabolism of *Lc. lactis* and *Lb. casei* is the best-studied system of sugar fermentation occurring in LAB. The knowledge is relevant for understanding sugar metabolism in general (Thompson, 1988) and milk fermentation in particular. However, an equally common way to metabolize lactose among LAB is by means of a lactose carrier (permease) and subsequent cleavage by β-galactosidase (β-gal) to yield glucose and galactose (Premi et al., 1972; Bhowmik and Marth, 1990; Fox et al., 1990), which may then enter the major pathways. Some reports on this subject would suggest that many LAB contain both a lactose PTS and a lactose permease system for lactose metabolism, since both P-β-gal and β-gal activity were found in the same strains (Premi et al., 1972; Hickey et al., 1986). However, low-P-β-gal activity in strains with high-β-gal activity may represent an artefact, since the artificial substrate used for P-β-gal (*ortho*-nitrophenylgalactose phosphate) may be hydrolyzed by β-gal or by a phosphatase yielding the substrate for β-gal (Hickey et al., 1986; Fox et al., 1990).

Some of the "thermophilic" LAB, e.g., *S. thermophilus, Lb. delbrückii* subsp. *bulgaricus, Lb. delbrückii* subsp. *lactis*, and *Lb. acidophilus* only metabolize the glucose moiety after transport of lactose and cleavage by β-gal, while galactose is excreted into the medium (Hickey et al., 1986; Hutkins and Morris, 1987). This has been attributed to a low galactokinase activity (Thomas and Crow, 1984; Hickey et al., 1986), but may also be an energetically favorable reaction of the lactose transport system (Hutkins and Ponne, 1991, and see Section V.B.2.b).

Maltose fermentation among LAB has been studied most extensively in lactococci. A permease system for transport seems to be operational (Konings et al., 1989; Sjöberg and Hahn-Hägerdahl, 1989). An interesting feature of maltose metabolism in *Lc. lactis* 65.1 is that maltose is cleaved by a maltose phosphorylase into glucose and β-glucose-1-phosphate. Only the glucose moiety is used in glycolysis, whereas β-glucose-1-phosphate probably is a precursor for cell-wall synthesis (Sjöberg and Hahn-Hägerdahl, 1989).

Sucrose fermentation mediated by a permease system is initiated by the cleavage of the sugar by sucrose hydrolase to yield glucose and fructose, which enter the major pathways. In some lactococci, sucrose is transported by sucrose PTS and a specific sucrose-6-phosphate hydrolase cleaves the sucrose-6-phosophate to glucose-6-phosphate and fructose (Thompson and Chassy, 1981). The sucrose PTS and sucrose-6-phosphate hydrolase are induced by the presence of sucrose in the medium (Thompson and Chassy, 1981). Sucrose may also act as a donor of monosaccharides for exopolysaccharide formation in certain LAB. In dextran production by *Ln. mesenteroides*, sucrose is cleaved by a cell wall associated enzyme, dextransucrase. The glucose moiety is used for dextran synthesis and fructose is fermented in the usual manner (Cerning, 1990).

Fermentation of other disaccharides, such as cellobiose, melibiose, and trehalose, has not been studied to any large extent. The ability to ferment these sugars differ between different species of LAB. Presumably, the metabolism is mediated by specific transport systems and hydrolases, resulting in the respective monosaccharides (or monosaccharide phosphates), which enter the common pathways.

3. Fermentation of Pentoses and Related Compounds

Pentoses are readily fermented by many LAB. In general, specific permeases is used to transport the sugars into the cells. Inside, the pentoses are phosphorylated and converted to ribulose-5-phosphate or xylulose-5-phosphate by epimerases or isomerases (Kandler, 1983). These compounds can then be metabolized by the lower half of the 6-PG/PK pathway (Fig. 3B). This would imply that only heterofermentative LAB can utilize pentoses, but this is not the case. In fact, disregarding some strain and species differences, all genera of LAB are pentose positive with one exception, the group I lactobacilli. Homofermentative LAB that utilize pentoses do so in the same was as heterofermentative LAB. The phosphoketolase of these species is generally induced by pentoses (Kandler, 1983). The heterolactic fermentation of pentoses results in a different end-product pattern compared to glucose fermentation. No CO_2 is formed, and since no dehydrogenation steps are necessary to reach the intermediate xylulose-5-phosphate, the reduction of acetyl phosphate to ethanol becomes redundant. Instead, acetyl phosphate is used by the enzyme acetate kinase in a substrate-level phosphorylation step yielding acetate and ATP. Fermentation of pentoses thus leads to production of equimolar amounts of lactic and acetic acid.

Although not a pentose, this may be an appropriate place to mention the fermentation of gluconate, which can be performed by some LAB. Similar to pentoses, gluconate is fermented by the 6-PG/PK pathway, yielding a heterolactic fermentation by species regarded as homofermentative. The metabolism has been studied in *E. faecalis*, but a similar pathway may also exist in *Lb. casei* (London, 1990). In *E. faecalis*, gluconate is transported by an inducible gluconate PTS and the resulting 6-phosphogluconate enters the 6-PG/PK pathway (London, 1990). Many heterofermentative lactobacilli also ferment gluconate (Kandler and Weiss, 1986), presumably in a similar metabolic sequence as for *E. faecalis*, although transport may be mediated by a permease, since heterofermentative LAB apparently lack PTS systems (Romano et al., 1979; Reizer et al., 1988b). Since a dehydrogenation step is necessary before the phosphoketolase reaction, some acetyl phosphate has to be reduced to ethanol in order to maintain the redox balance.

A few species of LAB, e.g., *Lb. casei*, can grow at the expense of pentitols. These are translocated through the membrane by specific pentitol PTS. The resulting pentitol phosphates are oxidized to pentose phosphates by dehydrogenases and subsequently metabolized through the 6-PG/PK pathway (London, 1990). Similar to gluconate fermentation, some ethanol is produced from acetyl phosphate because of the need to reoxidize the NADH formed during the initial dehydrogenation.

4. Sugar Fermentation and Metabolic Categories of LAB

From the descriptions of sugar fermentation patterns of LAB presented above it is possible to divide the group broadly into three metabolic categories. The genus *Lactobacillus* contain species which can be placed in all three categories, and this is in fact the basis for the division of the genus in three groups (Kandler and Weiss, 1986). The first category includes the group I lactobacilli, which are obligately homofermentative, meaning that sugars only can be fermented by glycolysis. The second category includes leuconostocs and group III lactobacilli, which are obligately heterofermentative, meaning that only the 6-PG/PK pathway is available for sugar fermentation. The apparent difference on the enzyme level between these two categories is the presence or absence of the key enzymes of glycolysis and 6-PG/PK pathway, FDP aldolase, and phosphoketolase, respectively. Obligately homofermentative species possess a constitutive FDP aldolase and lack phosphoketolase, whereas the opposite holds for obligately heterofermentative species (Kandler, 1983; Kandler and Weiss, 1986). This leads to the obvious difference in end-product formation from glucose, but also to the inability of the group I lactobacilli to attack pentoses (and gluconate). The third category, including most of the remaining LAB (i.e., group II lactobacilli, enterococci, lactococci, pediococci, streptococci, tetragenococci, and vagococci) holds an intermediate position. They resemble the obligately homofermentative LAB in that they possess a constitutive FDP aldolase, resulting in the use of glycolysis for hexose fermentation. As mentioned previously, pentoses (and presumably gluconate and pentitols when fermented) induce the synthesis of phosphoketolase, resulting in a heterolactic fermentation. These LAB are thus homofermentative with regard to hexoses and heterofermentative with regard to pentoses and some other substrates and should, therefore, be termed facultatively heterofermentative (Kandler, 1983; Kandler and Weiss, 1986).

The position of species in the genus *Carnobacterium* is somewhat unclear. These bacteria were first classified as heterofermentative (then under the designation *Lb. divergens*), since gas and acetic acid were produced in significant amounts (Kandler and Weiss, 1986). Later, more detailed studies have shown that glucose is fermented to almost entirely lactic acid, probably by the glycolytic pathway, the other products arising from metabolism of other components in the medium than glucose (De Bruyn et al., 1988). Since carnobacteria generally ferment ribose and gluconate (Kandler and Weiss, 1986; Collins et al., 1987), they should probably be regarded as faculatively heterofermentative.

5. Configuration of Lactic Acid

During the fermentation of sugars, different species of LAB produce either exclusively L-lactic acid, exclusively D-lactic acid, approximately equal amounts of both or predominantly one form, but measurable amounts of the other (Garvie, 1980; Kandler and Weiss, 1986; Schleifer, 1986). This depends on the presence of specific NAD^+-dependent lactate dehydrogenases (nLDH) and their respective activities. Thus, if both D- and L-lactic acids are formed, there are generally one D-nLDH and one L-nLDH present. Only a few species, e.g., *Lb. curvatus* and *Lb. sake*, produce an enzyme, termed a racemase, which converts L-lactic acid to D-lactic acid (Garvie, 1980). In this case, the L-lactic acid initially produced induce the racemase, which results in a mixture of D- and L-lactic acid. *P. pentosaceus* and many lactobacilli change the ratio of the isomers during batch growth. Generally, L-lactic acid is the major form produced in the early growth phase and D-lactic acid in the late to stationary phase (Garvie, 1980). The pH and internal pyruvate concentration have been thought to influence the activities of the LDHs and thus the ratio of the isomers at different growth phases (Garvie, 1980), but not much is known in this regard. *Lb. casei* and lactococci possess an allosteric L-nLDH, which is activated by FDP (see Section IV.B.2). Enterococci and streptococci also possess this type of LDH (Garvie, 1980).

B. Fates of Pyruvate

It is well known that LAB may change their metabolism in response to various conditions, resulting in a different end-product pattern than seen with glucose fermentation under normal conditions. In most of these cases, the change can be attributed to an altered pyruvate metabolism, the use of external electron acceptors, or both, as these may be connected to each other. This will be discussed below.

The essential feature of most bacterial fermentations is the oxidation of a substrate to generate energy-rich intermediates, which subsequently can be used for ATP production by substrate-level phosphorylation. The oxidation results in the formation of NADH from NAD^+, which has to be regenerated in order for the cells to continue the fermentation. Pyruvate holds a key position in many fermentations in serving as an electron (or hydrogen) acceptor for this regeneration step. Indeed, this is true in both major fermentation pathways used by LAB (Fig. 3). Under certain circumstances, LAB use alternative ways of utilizing pyruvate than the reduction to lactic acid. The alternative fates of pyruvate are depicted in Fig. 5. All these reactions are not used by a single strain, but represent a summary of the LAB group as a whole (Kandler, 1983). Different species may use different pathways, depending on conditions and enzymatic capacity. Some of these reactions may be operational even under normal glucose fermentation, but then serving an anabolic role. For instance, the formation of acetyl CoA can be required for lipid biosynthesis (Cogan et al., 1989).

1. *The Diacetyl/Acetoin Pathway*

The pathway(s) leading to diacetyl (butter aroma) and acetoin (Fig. 5) is common among LAB (Kandler, 1983) and is very significant technologically in the fermentation of milk (Lawrence and Thomas, 1979; Cogan, 1985). However, this metabolism proceeds to a significant degree only if there is a pyruvate surplus in the cell relative to the need for NAD^+ regeneration. A pyruvate surplus can be created in two ways: (i) an alternative source of pyruvate than the fermented carbohydrate exists in the growth medium and (ii) another compound act as electron acceptor, thus sparing the pyruvate formed by carbohydrate fermentation.

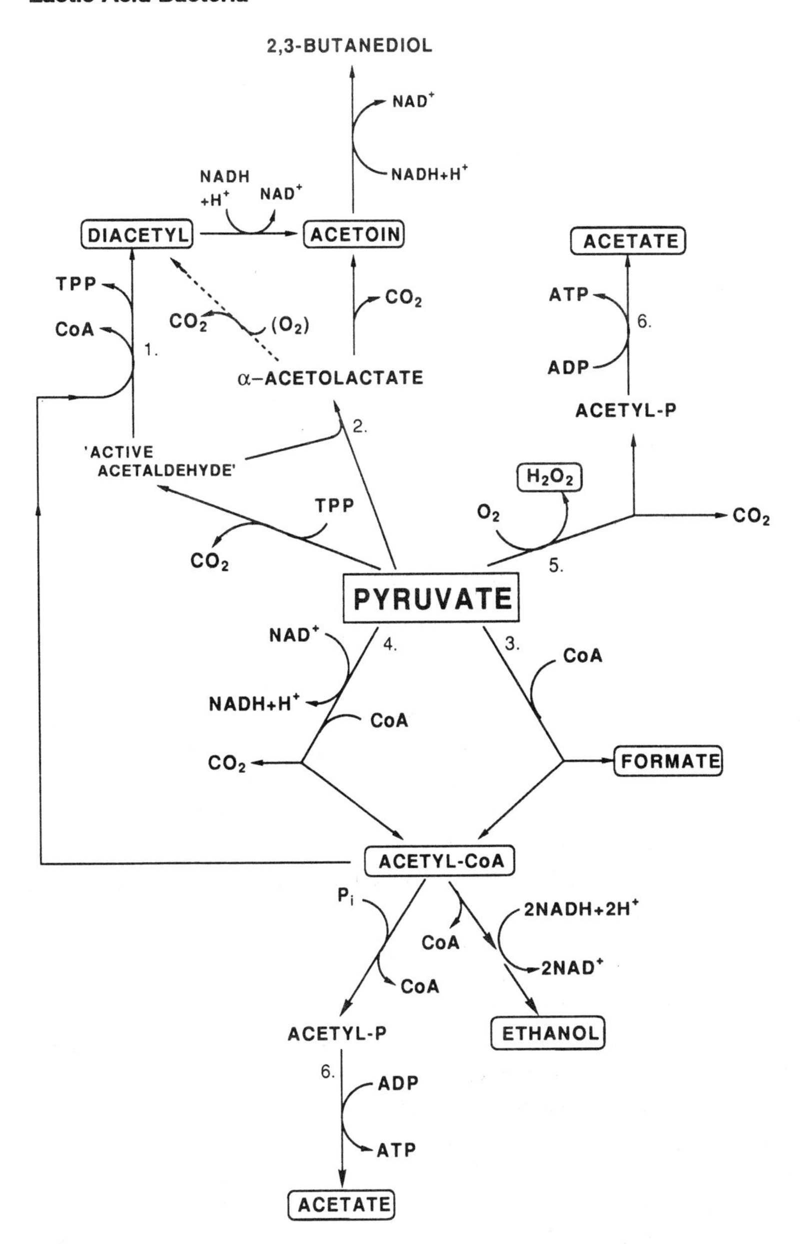

Figure 5 Pathways for the alternative fates of pyruvate. Dashed arrows denote a nonenzymatic reaction. Important metabolites and end products are framed. Selected enzymatic reactions are numbered: 1. diacetyl synthase; 2. acetolactate synthase; 3. pyruvate-formate lyase; 4. pyruvate dehydrogenase; 5. pyruvate oxidase; 6. acetate kinase.

The former way is what occurs in milk, where additional pyruvate originates from the breakdown of citrate, which is present in fairly large amounts (~ 1.5 mg/mL). Citrate is cleaved by citrate lyase to oxaloacetate and acetate. Oxaloacetate is further decarboxylated to pyruvate and CO_2 by oxaloacetate decarboxylase (Harvey and Collins, 1961). Species used in the dairy industry for the purpose of diacetyl production are *Lc. lactis* subsp. *lactis* (biovar *diacetylactis*) and *Ln. mesenteroides* subsp. *cremoris* (former name *Ln. cremoris* or *Ln. citrovorum*). These species may vary somewhat in induction pattern and pH optimum of the enzymes of citrate dissimilation and diacetyl/acetoin production (Harvey and Collins, 1961; Cogan et al., 1981; Cogan, 1985; Marshall, 1987). In general, low sugar concentrations and low pH favor diacetyl/acetoin formation. There has been some debate as to which of the two possible pathways (Fig. 5) is the most important. It has been argued that the pathway involving acetyl CoA would be unlikely in *Lc. lactis*, since this species has limited ability to produce acetyl CoA (Cogan, 1985). α-Acetolactate can be detected in cultures producing diacetyl and the compound decomposes (nonenzymatically) to CO_2 and diacetyl, in particular at low pH (Cogan, 1985). Acetoin is produced in much larger amounts than diacetyl, but does not contribute to the aroma (Marshall, 1987).

2. *The Pyruvate-Formate Lyase System*

Another branch of pyruvate metabolism shown in Fig. 5 consists of the pyruvate-formate lyase system. The enzyme pyruvate-formate lyase catalyzes the reaction of pyruvate and CoA to formate and acetyl CoA. Acetyl CoA may be used either as electron acceptor, resulting ultimately in ethanol formation, as a precursor for substrate-level phosphorylation via acetyl phosphate, or both. The system has been shown to be operational in several species of LAB (Kandler, 1983). Most notably, the pathway is used by some strains of *Lb. casei* and *Lc. lactis*, grown in anaerobic continuous culture under substrate limitation, resulting in a change from a homolactic to a heterolactic fermentation (De Vries et al., 1970; Thomas et al., 1979). End products formed are lactate, acetate, formate, and ethanol. Larger amounts of the products of the pyruvate-formate lyase system (acetate, formate, and ethanol) are formed with decreasing growth rate, i.e., with lowering the dilution rate in the continuous culture system (Thomas et al., 1979). A similar end-product pattern is found when some strains of *Lc. lactis* ferment galactose (Thomas et al., 1980) or maltose (Sjöberg and Hahn-Hägerdahl, 1989).

It has been shown that the change to a heterolactic fermentation by these strains is due to a reduction of the glycolytic rate. This affects the levels of glycolytic intermediates and subsequently the activities of the enzymes that compete for pyruvate, nLDH, and pyruvate-formate lyase. The nLDHs of *Lc. lactis* and *Lb. casei* are allosteric enzymes with a specific requirements for the key glycolytic intermediate FDP (see Fig. 3) for activity (Garvie, 1980). In general, lower intracellular levels of FDP are found in cells of *Lc. lactis* performing heterolactic fermentation, compared to those found in cells during homolactic fermentation (Thomas et al., 1979, 1980; Sjöberg and Hahn-Hägerdahl, 1989). Similarly, the levels of triose phosphates, known inhibitors of pyruvate-formate lyase (Takahashi et al., 1982), are also lower in "heterolactic" cells (Thomas et al., 1980). Thus, during a "semistarved" state, caused by either substrate limitation or the nature of the substrate, the cells respond by regulating certain enzyme activities so as to partly prevent pyruvate from being reduced to lactic acid. Instead, energy can be gained by using the pyruvate-formate lyase pathway, since there is a substrate-level phosphorylation site involving acetyl phosphate and acetate kinase (Fig. 5). Indeed, increased molar growth yield for glucose (Y_{glc}) was obtained in the continuous culture system at low compared to high growth rates (Thomas et al., 1979), indicating that more ATP/glucose was formed during heterolactic fermentation.

The pyruvate-formate lyase system is only active anaerobically, which is consistent with the enzyme being extremely oxygen-sensitive (Thomas and Turner, 1981; Takahashi et al., 1982) and presumably inactivated when cells are exposed to air.

3. The Pyruvate Oxidase Pathway

Oxygen has a profound effect on the fate of pyruvate in LAB. This effect may be direct, mediated by the enzyme pyruvate oxidase (Fig. 5), or indirect through reactions of oxygen with the flavin containing enzymes NADH–H_2O_2 oxidase and NADH–H_2O oxidase (Condon, 1987). Pyruvate oxidase converts pyruvate to CO_2 and acetyl phosphate with the formation of H_2O_2 (Fig. 5). It has been suggested that the enzyme is involved in aerobic metabolism of *Lb. plantarum*, which forms significant amounts of acetic acid aerobically (Sedewitz et al., 1984). In a study by Sedewitz et al. (1984), pyruvate oxidase had the highest activity in the early stationary phase of growth, and lactose-grown cells had generally higher levels of the enzyme than did glucose-grown cells. Since this strain of *Lb. plantarum* had much lower growth rate on lactose than on glucose, a possible explanation of the elevated pyruvate oxidase level could be that the levels of glycolytic intermediates in some way regulated enzyme synthesis (Sedewitz et al., 1984). A certain similarity to the regulation of the pyruvate lyase system in *Lc. lactis*, described above, can thus be noted.

4. The Pyruvate Dehydrogenase Pathway

Evidence have been obtained that a pyruvate dehydrogenase enzyme complex is active in lactococci (Broome et al., 1980; Smart and Thomas, 1987; Cogan et al., 1989). The enzyme complex produces acetyl CoA (Fig. 5) and thus resembles the pyruvate-formate lyase system. The study by Cogan et al. (1989) indicated that the pyruvate dehydrogenase has an anabolic role in producing acetyl CoA for lipid synthesis under aerobic conditions. Anaerobically, this role is probably played by pyruvate-formate lyase, which has the advantage of not reducing NAD^+ in the process. Exposed to air, the cells are dependent on a functional pyruvate dehydrogenase for acetyl CoA production, since pyruvate-formate lyase is inactivated by oxygen. The excess NADH formed can be reoxidized by NADH oxidases. Similar to the pyruvate-formate lyase system, the pyruvate dehydrogenase can also play a role in catabolism, but then primarily under aerobic conditions. The effect can be rather dramatic, however. Aerated cultures of non-growing cells of *Lc. lactis* can perform a homoacetic fermentation under substrate limitation (Smart and Thomas, 1987). In this case, all the pyruvate generated from glycolysis is channeled through the pyruvate dehydrogenase complex with acetic acid (and presumably CO_2) as final product. Under these conditions, the nLDH has probably very low activity and cannot effectively compete for pyruvate. A prerequisite for this metabolism to occur is that the NADH formed during both glycolysis and in the pyruvate dehydrogenase reaction can be reoxidized by NADH oxidases (Smart and Thomas, 1987).

C. Oxygen as Electron Acceptor

In reactions with NADH oxidases, oxygen acts as an external electron acceptor. In many cases, this can be advantageous to LAB. A thorough review on this topic has been provided by Condon (1987).

NADH oxidases seem to be widespread among LAB, and the systems are often induced by oxygen. The products of the reactions are either NAD^+ and H_2O_2 or NAD^+ and H_2O, depending on whether the enzyme mediates a two- or four-electron transfer. Most LAB also possess a NADH peroxidase, which uses H_2O_2 as electron acceptor with the formation of H_2O. This reaction may mask initial NADH–H_2O_2 oxidase activity (Smart and Thomas, 1987). NADH oxidases may compete efficiently with nLDH in homofermentative LAB. This may create a situation similar to the breakdown of citrate described earlier, i.e., a pyruvate surplus available for metabolism through the diacetyl/acetoin pathway. Increased production of acetoin in aerated cultures, compared to unaerated, has been shown for *Lc. lactis* (Cogan et al., 1989), homofermentative lactobacilli, and carnobacteria (Borch and Molin, 1989). Worthy of note is that heterofermentative LAB (leuconostocs and group III lactobacilli) did not respond to aeration with acetoin production, whereas most of the homofermentative did (Borch and Molin, 1989). The probable explanation is that in heterofermentative LAB, the NADH oxidases do not compete with nLDH, but rather with acetaldehyde dehydrogenase and alcohol dehydrogenase, the enzymes of the ethanol "branch" of the 6-PG/PK pathway (Fig. 3 and see below). Accordingly, a pyruvate surplus is not created, and lactic acid remains the main product of pyruvate metabolism.

LAB that ferment glucose by the 6-PG/PK pathway use acetyl phosphate (or more accurately, acetyl CoA) as electron acceptor in addition to pyruvate. Reduction of acetyl phosphate (via acetyl CoA) constitutes the ethanol branch of the pathway. This route is in a way a "waste" of acetyl phosphate, since this compound can be used in substrate-level phosphorylation with the production of ATP. Heterofermentative LAB may circumvent this "waste" by using external electron acceptors. As hinted above, oxygen can play an active role. Lucey and Condon (1986) showed that strains of *Leuconostoc* sp. doubled the Y_{glc} in aerated cultures compared to unaerated. In addition, the growth rate was higher with aeration. The effect was attributed to an active NADH oxidase, which efficiently prevented ethanol formation. Instead, acetate was formed from acetyl phosphate, additional ATP could be produced (by the acetate kinase reaction), and, as a consequence, Y_{glc} increased. A NADH-oxidase-deficient mutant did not shift from ethanol to acetate production and Y_{glc} was the same aerobically and anaerobically, thus supporting the role of NADH oxidase in aerobic metabolism. The shutoff of the ethanol branch of the 6-PG/PK pathway in the presence of oxygen seems to be very common among heterofermentative LAB. However, an increase of Y_{glc} is not always seen (Borch and Molin, 1989). The reason for this is not clear.

The active role of oxygen as electron acceptor in the metabolism of LAB is further illustrated by the fact that certain substrates are fermented only when oxygen is available. This is especially true for the fermentation of reduced compounds, such as polyols. Examples are oxygen-dependent glycerol fermentation by *P. pentosaceus* (Dobrogosz and Stone, 1962) and mannitol fermentation by *Lb. casei* (Brown and VanDemark, 1968). It is also well known that some heterofermentative lactobacilli, most notably strains of *Lb. brevis*, are almost unable to ferment glucose anaerobically, but will ferment it aerobically (Stamer and Stoyla, 1967). This is due to a deficiency in the ethanol branch of the 6-PG/PK pathway, more specifically a lack of acetaldehyde dehydrogenase (Eltz and VanDemark, 1960). The deficiency creates an absolute requirement for an external electron acceptor (e.g., oxygen) in the fermentation of glucose.

Even lactate, the end product of "normal" metabolism, can be fermented to acetate and CO_2 by some LAB (Murphy et al., 1985; Thomas et al., 1985). A pathway for this oxygen-dependent lactate fermentation of *Lb. plantarum* has been proposed, involving NAD^+-dependent and/or NAD^+-independent LDH, pyruvate oxidase, and acetate kinase (Murphy et al., 1985). The ATP yield would be 1 ATP per lactate consumed.

In the presence of hematin or hemoglobin in the growth medium, some LAB, in particular enterococci, may change their otherwise fermentative mode of metabolism to a respiratory one, including the formation of "true" catalase and cytochromes, the use of oxygen as terminal electron acceptor and increased ATP production, presumably by oxidative phosphorylation (Bryan-Jones and Whittenbury, 1969; Ritchey and Seeley, 1976; Whittenbury, 1978). Apparently, these LAB have the capacity to synthesize the apoenzymes of cytochromes and catalase), but, as mentioned previously, they are unable to synthesize porphyrins.

In summary, oxygen interacts with LAB metabolism in a very active way, and it has been argued that it is perhaps inaccurate to designate them as merely "aerotolerant anaerobs" (Whittenbury, 1978; Götz et al., 1980; Condon, 1987), an opinion which has some support, both with regard to the actual relation to oxygen as described above and in the phylogenetic position of these bacteria (see Section III.D).

However, it should be noted that LAB generally do not have the same potential to protect themselves against the toxic effect of oxygen as aerobic organisms. LAB are generally devoid of catalase, although H_2O_2 may be decomposed by a pseudocatalase in some strains (Whittenbury, 1964; Kono and Fridovich, 1983). Since H_2O_2 is produced by NADH oxidases, accumulation may reach autoinhibitory levels, depending on strain (Condon, 1987). Superoxide dismutase is present in some LAB, e.g., lactococci and enterococci, and absent in others, e.g., lactobacilli. Some lactobacilli have developed a unique system for protection against superoxide. This system is based on specific accumulation of Mn^{2+} to high intracellular concentrations (30–35 mM), which have a scavenging effect on superoxide (for a review, see Archibald, 1986). *Lb. plantarum* is the most studied species in this regard, but it appears that other LAB associated with plant material (rich in Mn^{2+}), such as leuconostocs and pediococci, also possess this system (Archibald, 1986).

D. Other Electron Acceptors

LAB are not restricted to oxygen as external electron acceptors. Anaerobically, several organic compounds can serve the same purpose. This is especially true for the heterofermentative LAB. It is perhaps a fair statement to say that this group of LAB has a more obvious use for external electron acceptors, even under "normal" glucose fermentation, since this (in theory) can double the amount of ATP produced per glucose consumed. In this regard, acetyl phosphate holds a key position as intermediate in the 6-PG/PK pathway. The presence or absence of an external electron acceptor will decide whether ethanol (no ATP) or acetate (1 ATP) is formed. Lucey and Condon (1986) have suggested that the ethanol branch of the 6-PG/PK pathway in no more than a salvage route, permitting growth when an external electron acceptor is missing. The fact that some heterofermentative LAB have a defect ethanol branch (see above) and are, therefore, dependent on an external electron acceptor (at least for glucose fermentation) may support this theory.

The changes that occur in the end-product formation of heterofermentative LAB in the presence of an external electron acceptor, compared to in the absence, do not have a counterpart with homofermenters. Under normal growth conditions, only minor deviations of the homolactic fermentation may occur as a result of the use of external electron acceptors. Such a deviation may be production of some acetoin as a result of a buildup of excess pyruvate. However, organic electron acceptors can play an essential role for homofermentative LAB in anaerobic metabolism of certain substrates. Examples of organic external electron acceptors that can be used by LAB are acetaldehyde, a-ketoacids, citrate, fructose, fumarate, and glycerol (Eltz and VanDemark, 1960; Schütz and Radler, 1984a; Chen and McFeeters, 1986a,b; McFeeters and Chen, 1986; Schmitt and Divies, 1990; Talarico et al., 1990). Further details on the metabolism of some of these will be discussed below.

1. Citrate

Citrate is not used directly as an electron acceptor, but act as a precursor to one. The essential step is a cleavage of citrate by citrate lyase, to form acetate and oxaloacetate. Different LAB use different pathways in the further metabolism of these products. Decarboxylation of oxaloacetate, yielding pyruvate, has already been mentioned in connection with the diacetyl/acetoin pathway (see Section IV.B). Growing cells of heterofermentative LAB dissimilating citrate in a cofermentation with carbohydrate do not form any significant amounts of diaceyl or acetoin (Drinan et al., 1976; Cogan, 1987). Rather, the excess pyruvate is reduced to lactic acid. This spares acetyl phosphate from being reduced to ethanol, and more ATP can be formed through the acetate kinase reduction, resulting in a more efficient glucose utilization and increased growth rate (Cogan, 1987).

The products of the citrate lyase reaction are used differently in other LAB. A pathway for succinic acid formation from oxaloacetate was proposed and proven by Chen and McFeeters (1986a). This metabolism was found in the anaerobic fermentation of mannitol by *Lb. plantarum*, in which an external electron acceptor was required for growth (Chen and McFeeters, 1986a; McFeeters and Chen, 1986). This way of utilizing citrate may be more common among LAB, especially heterofermentative, than the special case of *Lb. plantarum* mannitol fermentation would suggest. It may even explain some of the confusing results with regard to carbon recoveries and fermentation balances in studies of LAB and citrate dissimilation (Drinan et al., 1976). Due to the industrial importance, most studies on this subject have been aimed at the understanding of acetoin and diacetyl formation (Drinan et al., 1976), and other fates of citrate may have been overlooked. In studies of intestinal lactobacilli in our laboratory, we initially noticed that some heterofermentative species produced succinic acid in normal, MRS-like media (Axelsson and Lindgren, 1987). Later, it was shown that this was due to a utilization of citrate in a cofermentation with glucose (Axelsson, 1990). Although enzymatic evidence is still missing, the end-product formation for one of these strains, shown in Fig. 6, is consistent with the operation of a citrate lyase and the succinic acid pathway. In addition, the typical stimulation of growth rate and glucose utilization occurred, indicating increased ATP production (Axelsson, unpublished). A hypothetical pathway of this metabolism is depicted in Fig. 7. We have noticed that succinic acid production from citrate is fairly common among heterofermentative lactobacilli isolated from the intestine of several animals (Axelsson and Lindgren, unpublished). The property seems to be common in plant-associated heterofermentative lactobacilli as well (Kaneuchi et al., 1988).

Citrate is also used as an electron acceptor in an anaerobic degradation of lactate, which can be performed by some strains of *Lb. plantarum* (Lindgren et al., 1990). This metabolism is very slow and can only be observed after prolonged incubation. In fact, no growth is evident, but cells performing the metabolism (as evidenced by HPLC) have significantly higher ATP content than control cells (Lindgren and Axelsson, unpublished). This may point to an importance of this

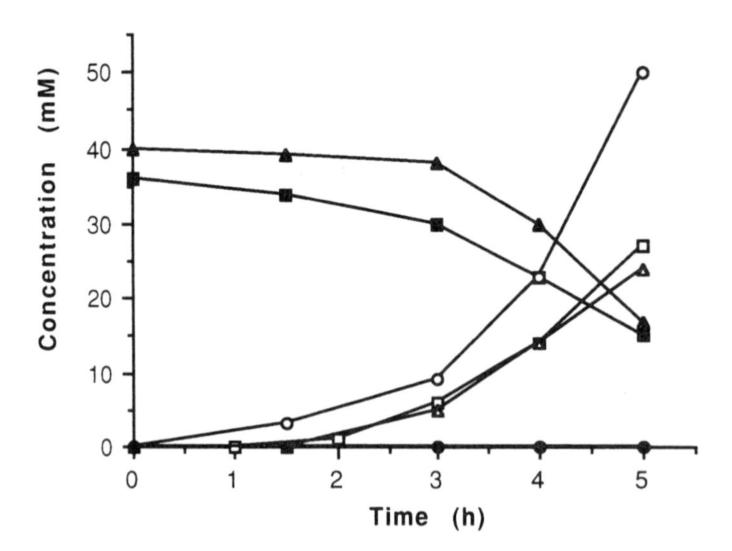

Figure 6 End product formation by *Lactobacillus* Lbp 1031 during growth on glucose and citrate in a modified MRS medium. Samples were withdrawn from the culture and subjected to HPLC analysis. During the sampling time, OD_{600} increased from 0.2 (0 h) to 1.7 (5 h). Symbols: ■, glucose; ▲, citrate; □, lactate; ○, acetate; Δ, succinate; ●, ethanol.

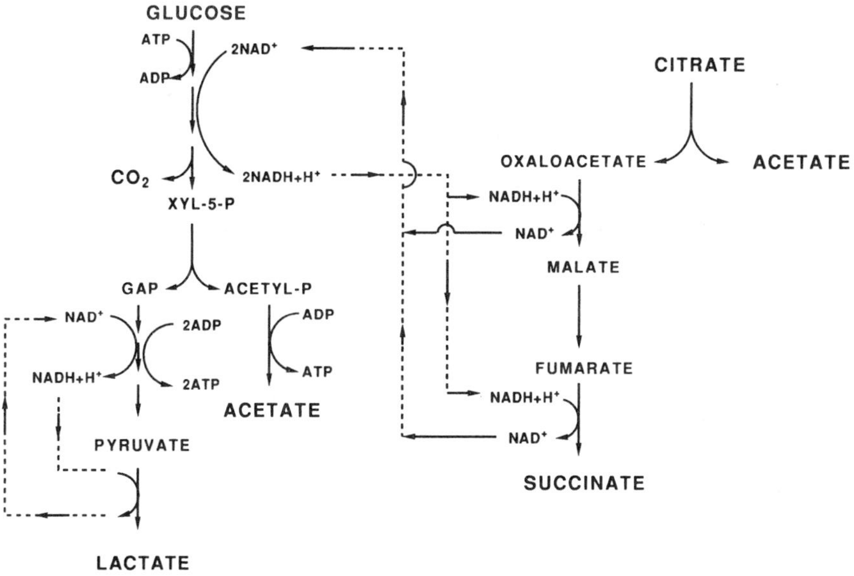

GLUCOSE
ATP
ADP
2NAD$^+$
CITRATE
OXALOACETATE
ACETATE
CO_2
2NADH+H$^+$
XYL-5-P
NADH+H$^+$
NAD$^+$
MALATE
GAP
ACETYL-P
NAD$^+$
2ADP
ADP
NADH+H$^+$
2ATP
ATP
FUMARATE
ACETATE
NADH+H$^+$
PYRUVATE
NAD$^+$
SUCCINATE
LACTATE

Sum: 1 Glucose + 1 Citrate + 2 ADP + 2 P_i ⟶ 1 Lactate + 2 Acetate + 1 CO_2 + 1 Succinate + 2 ATP

Figure 7 Proposed pathway for succinic acid production in heterofermentative lactobacilli growing on glucose and citrate.

metabolism for survival and maintenance. The products of the cometabolism of lactate and citrate are succinic acid, acetate, formate, and CO_2, indicating the operation of both the succinic acid pathway and a pyruvate-formate lyase (Lindgren et al., 1990).

The differences between LAB regarding the use of citrate may depend on the presence or absence of the enzyme oxaloacetate decarboxylase. If present, citrate utilization (the primary step being the lyase reaction) results in an increase in the pyruvate pool, which may lead to an altered end-product pattern, e.g., acetoin production. If oxaloacetate decarboxylase is absent, an alternative route is the succinic acid pathway. A similar comparison was made between *Enterobacter* sp. (citrate-fermenting) and *Escherichia coli* (citrate-nonfermenting), when it was shown that *E. coli* could dissimilate citrate, but only in the presence of a cosubstrate, with the formation of succinic acid (Lütgens and Gottschalk, 1980).

2. Glycerol

Strains of *Lb. brevis, Lb. buchneri,* and *Lb. reuteri,* all heterofermentative, can use glycerol as an electron acceptor in an anaerobic cofermentation with glucose. As mentioned previously, many strains of *Lb. brevis* ferments glucose poorly anaerobically. Some strains do ferment glucose, however, if glycerol is added (Schütz and Radler, 1984a). The products of the cofermentation are lactate, acetate, CO_2, and 1,3-propanediol. Again, the NADH formed during glucose fermentation is not reoxidized by the ethanol pathway, but rather by using glycerol as electron acceptor. Glycerol is first dehydrated to 3-hydroxypropionaldehyde (3-HPA) and further reduced to 1,3-propanediol by a NAD^+–1,3-propanediol dehydrogenase.

Lb. reuteri ferments glucose alone with lactate, ethanol, and CO_2 as end products, but changes to more acetate/less ethanol when glycerol is added (Talarico et al., 1990). The addition of glycerol also stimulates growth rate, and Y_{glc} is increased. The same pathway for glycerol reduction to 1,3-propanediol as for *Lb. brevis* was shown to be functional in *Lb. reuteri* (Talarico et al., 1990; Talarico and Dobrogosz, 1990).

Some differences in response to glycerol between these two species can be noted. Resting cells of *Lb. brevis* metabolizing glycerol, accumulate 1,2-propanediol (Schütz and Radler, 1984a), whereas *Lb. reuteri* under these conditions accumulates and excretes the intermediate 3-HPA. This compound is a potent antimicrobial substance, initially termed *reuterin* (Axelsson et al., 1989; Chung et al., 1989; Talarico and Dobrogosz, 1989). The enzymes of the glycerol pathway seem to be under some kind of regulation in *Lb. brevis,* since cells grown on pentoses do not contain any glycerol dehydratase activity (Schütz and Radler, 1984b). This is logical, considering that pentose fermentation does not generate excess NADH. On the contrary, *Lb. reuteri* has about the same levels of this enzyme whether grown on pentose or glucose, despite the fact that the glycerol pathway is not used when the cells ferment pentoses (Fiuzat, Axelsson, and Dobrogosz, unpublished). The growth response of *Lb. reuteri* with the addition of glycerol (increased growth rate and Y_{glc}) is very similar to that observed when oxygen is used as electron acceptor by *Ln. mesenteroides* (Lucey and Condon, 1986). The increase in Y_{glc} is easy to understand, since more ATP per glucose can be formed through the acetate kinase reaction. But why is the growth rate increased? No clear-cut answer can be given, but Lucey and Condon (1986) suggested that ATP formation is the rate-limiting step in these bacteria.

3. Fructose

It was early noted that fructose fermentation by heterofermentative LAB resulted in mannitol formation (Peterson and Fred, 1920). This metabolism represent an interesting example of a case where the same compound acts as both the growth substrate and as electron acceptor (Eltz and VanDemark, 1960). Fructose is fermented by the normal 6-PG/PK pathway, but some of the sugar is reduced to mannitol by a NAD^+–mannitol dehydrogenase (Eltz and VanDemark, 1960). Similar to the use of other external electron acceptors, this enables the cells to produce ATP through the acetate kinase reaction. Assuming no ethanol formation, the overall equation for fructose fermentation would be

$$3\ \text{Fructose} + 2\ \text{ADP} + 2\text{P}_i \rightarrow 1\ \text{lactate} + 1\ \text{acetate} + 1\ \text{CO}_2 + 2\ \text{mannitol} + 2\ \text{ATP} \quad (1)$$

As can be seen, this is less efficient than glucose fermentation in terms of ATP formed per sugar consumed. However, for some heterofermentative lactobacilli, the growth rate is higher on fructose than on glucose (Axelsson, unpublished). This indicates that under conditions of substrate excess, a priority is given to growth rate rather than efficiency of substrate utilization. The reduction of some of the fructose to mannitol could play a role, since the cells may be able to form more ATP per time unit. The metabolism may be of great importance in natural plant fermentations, where glucose, fructose, and sucrose (glucose + fructose moieties!) are the main sugars (Fleming et al., 1985). Heterofermentative LAB can use glucose as energy source and fructose as electron acceptor, thus obtaining optimal growth rate.

E. The Malo-Lactic Fermentation

The metabolism of L-malic acid (the D-form is not attacked) by LAB has been extensively studied. This has been due to a technological significance, mainly in wine manufacturing, but also because it has presented some interesting physiological problems with regard to metabolism and bioenergetics.

Few LAB, e.g., *E. faecalis* and *Lb. casei*, can use malate as the sole energy source. The NAD^+-dependent malic enzyme catalyzes the decarboxylation of malate to pyruvate and CO_2 (London, 1990), as shown in Eq. (2).

$$\text{Malate} + \text{NAD}^+ \rightarrow \text{Pyruvate} + \text{CO}_2 + \text{NADH} \quad (2)$$

Pyruvate is converted to acetate, ethanol and CO_2 with ATP generation presumably through the acetate kinase reaction.

The more interesting, and much more common, malate metabolism is the conversion of malate to lactate and CO_2, which proceeds in a cofermentation with a fermentable carbohydrate. This fermentation, often referred to as the malo-lactic fermentation (MLF), can be performed by many LAB, in particular those associated with plant material, e.g., lactobacilli, pediococci, and leuconostocs. MLF is significant in the fermentation of vegetables and fruits, where malate is present in fairly high concentrations (Daeschel et al., 1987). The conversion of malate to lactate in the late stages of wine making is well known, where it may be desirable or undesirable, depending on the wine variety (Radler, 1975).

The pathway for malate dissimilation was first believed to be a sequence of reactions, the first being identical to Eq. (2), resulting in pyruvate as free intermediate. Pyruvate would subsequently be reduced to lactic acid by the LDH. An indication of another mechanism was that MLF leuconostocs produced exclusively L-lactic acid from malate (Radler, 1975). Pyruvate could thus not be an intermediate, since leuconostocs only possess a D-LDH (see Section III.B). Subsequently, purification to near homogeneity of the enzyme catalyzing the complete reaction (Eq. (3)) was achieved (Radler, 1975):

$$\text{Malate} \xrightarrow{\text{NAD}^+} \text{Lactate} + CO_2 \tag{3}$$

To distinguish the enzyme from the malic enzyme and malate dehydrogenase, it was given the trivial name malo-lactic enzyme (the proper name being L-malate–NAD^+ carboxylase; Kunkee, 1991). Curiously, although NAD^+ was required for enzymatic activity, no NADH was detected in early studies of the enzyme. This would indicate that hydrogen exchange reactions occurred within the enzyme and that oxaloacetate and/or pyruvate may be intermediates, but bound to the enzyme (Gottschalk, 1986).

LAB performing the MLF in cofermentation with a carbohydrate generally benefit from this in a way which resembles the use of external electron acceptors (see Sections IV.C and D), i.e., increased growth rate and a higher Y_{glc} compared to growth on solely glucose (Pilone and Kunkee, 1976). This is difficult to explain, since apparently no potential electron acceptors such as oxaloacetate or pyruvate is produced by the reaction. The stimulatory effect was first attributed to the deacidification of the external medium (Radler, 1975), which is a consequence of the reaction since malic acid has a lower pK_a than lactic acid (Daeschel, 1988). This is certainly a major factor, but cannot fully account for the stimulatory effect. Later studies have shown that the reaction is not stoichiometrically complete and that there are small but significant amounts of pyruvate and NADH released (Kunkee, 1991). The reaction can thus provide additional electron acceptors (Kunkee, 1991). Measurements of internal ATP concentrations in cells performing the MLF have also indicated that the reaction confers benefits in the form of energy (Cox and Henick-Kling, 1989). These authors suggest that this effect is indirect in that MLF relieves the cells from some of the energy requirements, in particular that required for generating and maintaining a proton gradient. This could be achieved by the energy-recycling model of lactate efflux (for further discussion on energetics; see Section V.A). The MLF could itself be an "indirect" proton pump. There are striking similarities between the malate^{2-}–lactate^{1-} exchange and the oxalate^{2-}–formate^{1-} exchange found in *Oxalobacter formigenes*

(Maloney, 1990). The latter exchange reaction is known to be electrogenic and functions as a proton pump in the formation of a membrane potential (Maloney, 1990). The operation of an electrogenic precursor–product antiporter for the MLF remains, however, to be proven.

V. ENERGY TRANSDUCTION AND SOLUTE TRANSPORT

Members of LAB have been extensively studied with regard to mechanisms of transport and energetics, in particular species of the genera *Enterococcus, Lactococcus*, and *Streptococcus* (i.e., the former streptococci). Model systems, which use these bacteria, are easy to control and manipulate. Aerotolerance, an efficient fermentative metabolism with no oxidative phosphorylation, and a cell wall without outer membrane are valuable properties in this regard. The results obtained with these model systems have been important for the understanding of living cells in general (Maloney, 1990). Sugar transport is of course connected to the carbohydrate metabolism described in previous sections, but the discussion on the subject has been placed here, since transport systems in general are tightly coupled to the bioenergetics of the cells.

There are excellent reviews covering this vast field, e.g., Kashket (1987), Thompson (1987, 1988), Reizer et al. (1988b), Konings et al. (1989), and Maloney (1990), and the presentation here can only be a short summary.

A. Bioenergetics of LAB

1. *ATP, the Proton Motive Force, and Internal pH*

The metabolism of LAB is aimed at the generation of ATP, the universal energy carrier in all living cells. ATP, or high-energy compounds interconvertible with ATP, is needed for the thermodynamically unfavorable "reaction" of building cells. The most important energy-requiring events in cells are the synthesis of macromolecules and the transport of essential solutes against a concentration gradient. The so-called chemiosmotic theory in bioenergetics (Mitchell, 1961, 1972) is now generally accepted. Cellular metabolism leads to an electrochemical proton gradient across the cytoplasmic membrane. The most commonly known system for creating a proton gradient is the membrane-linked electron transport chain, present in respiring organisms. The flow of electrons through the system via different carriers in effect pumps protons out of the cell. The proton gradient across the cytoplasmic membrane is composed of two components: an electrical potential ($\Delta\Psi$), inside negative, and a pH gradient (ΔpH), inside alkaline. $\Delta\Psi$ and ΔpH exert an inwardly directed force termed the proton motive force (PMF) (Konings and Otto, 1983; Konings et al., 1989). In organisms with an electron transport chain, this force is big enough to be converted into chemical energy, i.e.,

ATP. This is accomplished by a membrane-located enzyme, the H^+-translocating ATPase, or in this function also known as the ATP synthase. The energy of the reversal flow of protons "through" the enzyme, into the cell, is used to form ATP from ADP and phosphate.

LAB do not possess an electron transport chain (at least not in the absence of preformed heme) and are hence not able to form ATP in this way. Instead, ATP is generated by substrate-level phosphorylation, which is characteristic for all fermentative organisms. LAB do, however, possess an enzyme very similar to the ATP synthase, but the major role of this enzyme is the reverse reaction, i.e., the hydrolysis of ATP with concomitant pumping of protons out of the cells (Konings et al., 1989; Maloney, 1990). LAB (and fermentative bacteria in general) thus establish a PMF, which can drive energy-consuming reaction such as the uphill transport of metabolites and ions.

The difference in function between the H^+ATPase of LAB and the ATP synthase of respiring organisms is merely a reflection of the different modes of metabolism. The H^+ATPase of *Lc. lactis* is capable of acting as an ATP synthase, provided a large, artificially created PMF is imposed on the cells (Maloney et al., 1974). It is also clearly established that the H^+ATPases of *E. faecalis* and *Lc. lactis* subsp. *cremoris* have the same basic structure as the ATP synthase from mitochondria, chloroplasts, and respiring or photosynthesizing eubacteria (Futai and Kanazawa, 1983; Rimpiläinen et al., 1988). A recent report also suggest that the ATPase of *Lb. casei* is of this type (Muntyan et al., 1990), although a simpler subunit composition of the enzyme was first reported (Biketov et al., 1982).

It is well known that bacteria attempt to maintain the cytoplasmic pH (pH_i) at a certain level or interval (Booth, 1985). LAB are to a certain degree exceptions to this, since they tolerate a lower and wider range of internal pHs. There are, however, differences within the LAB group. In general, lactobacilli are significantly more tolerant to low pH_i than enterococci, leuconostocs (excluding *Ln. oenos*), and streptococci (Kashket, 1987; McDonald et al., 1990). It is logical that the enzymes and the general machinery of the cells have threshold levels, below (and above) which they cannot function. It appears that some lactobacilli have developed a very relaxed system, which works even at a pH of 4.2–4.4. In contrast, enterococci, lactococci, and streptococci rarely tolerate a pH_i lower than 5.0 (Kashket, 1987; Konings et al., 1989). Since the external pH (pH_0) falls well below these values, due to massive acid production, a mechanism must exist to maintain pH_i above the threshold levels. Studies of *E. faecalis* suggest that the H^+ATPase plays a crucial role and that this role is the main function of the enzyme (Kobayashi, 1985). The extrusion of protons by the H^+ATPase and the electrogenic uptake of K^+ maintain the cytoplasm more alkaline than the outside medium. Both the activity and the synthesis of the H^+ATPase are regulated by

pH_i. The mechanism of pH homeostasis in lactococci seems to be very similar to the *E. faecalis* system (Konings et al., 1989).

2. Energy Recycling

The maintenance of a pH_i above the threshold level and the generation of a PMF require the use of a substantial part of the ATP generated by substrate-level phosphorylation. This makes less ATP available for biosynthetic purposes. Any other means of generating or maintaining the PMF than the H^+ATPase would save energy. Some LAB have developed such systems, which to some degree compensate for the drain of ATP.

A general scheme for PMF-driven transport is depicted in Fig. 8A. The inwardly directed gradient of protons is the driving force for the influx of a solute X, which enters the cell together with a proton (proton symport). The question is whether this process, in its principles, is reversible, as shown in Fig. 8B? Here, the outwardly directed gradient of solute Y drives the efflux in symport with a proton, thus creating a PMF. The answer to the question is yes, under certain conditions. Based on theoretical calculations, where solute Y (Fig. 8B) was represented by a fermentation end product (such as lactate), a model was proposed on how energy could be conserved by end-product efflux, the so-called energy recycling model (Michels et al., 1979). If the efflux is electrogenic, i.e., a net charge leaves the cell together with the end product, a PMF is generated. In the case of lactate, more than one proton (on average) per lactate molecule has to be exported to obtain an electrogenic efflux. Experimental results with *Lc. lactis* subsp. *cremoris* have supported the model (Konings and Otto, 1983; ten Brink et al., 1985). A carrier-mediated electrogenic lactate efflux was shown to occur at pH_0 above 6.3 and low external lactate concentrations (<10 mM). At lower pH_0 and higher external lactate concentrations, the lactate efflux was electroneutral and subsequently not contributing to the generation of a PMF. In practice, this energy-saving process is only operational at the initial stage of growth in a batch culture. In an ecological context, the significance of the process may be substantial, since an initial high growth rate (which presumably would be the result of more ATP available for biosynthesis) could be advantageous in competition with other microorganisms. A dramatic effect was observed when *Lc. lactis* subsp. *cremoris* was grown in coculture with a lactate-consuming organism, *Pseudomonas stutzeri*. The conditions for maximum advantage of lactate efflux (low external lactate concentration and high pH_0) were thus maintained throughout growth and a 60–70% increase of the molar growth yield for lactose was obtained (Otto et al., 1980).

The energy-recycling model has recently been shown to be applicable also for acetate efflux in *Lb. plantarum* (Tseng et al., 1991). This species respond to a shift from anaerobic to aerobic conditions by producing some acetate at the expense of lactate and by increasing the molar growth yield for glucose (Tseng and

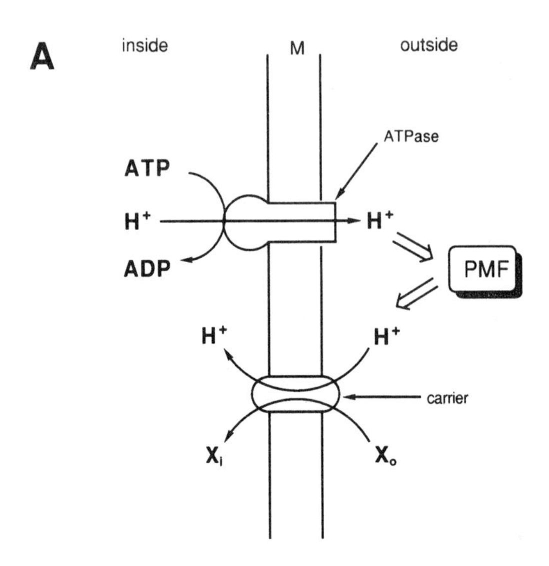

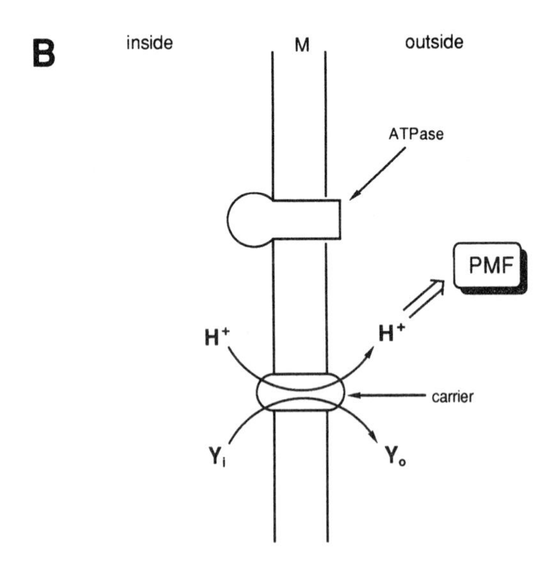

Figure 8 Schematic presentation of proton motive force (PMF) formation by a H^+ATPase and PMF-driven transport (A), and electrogenic end-product efflux (B). M denotes the cytoplasmic membrane.

Montville, 1990). An electrogenic acetate efflux was demonstrated, and this, together with additional ATP formation via the acetate kinase reactions, was suggested to contribute to the energy economy of the cells.

The system of energy recycling by lactate efflux is not general for all LAB. A carrier-mediated, electroneutral export of lactate was shown for *Lb. helveticus* (Gätje et al., 1991). This species is able to produce very large amounts of lactic acid (>200 mM), and the pH_0 drops to levels well below 4.0. The massive lactic acid production and acid tolerance of *Lb. helveticus* is very different from lactococci, and it is perhaps not surprising that the lactate export systems differ. It is not known if the system of *Lb. helveticus* is typical for the more acid-tolerant LAB.

As mentioned previously, energy recycling by lactate efflux has been suggested to partly explain the benefits of the malo-lactic fermentation (see Section IV.E).

B. Transport of Solutes

LAB are generally very fastidious and require amino acids, nucleotides or nucleotide precursors, and vitamins in addition to an energy source, generally a carbohydrate. A prerequisite for rapid growth is efficient transport systems for the uptake of essential nutrients. The transport of solutes is highly connected to the bioenergetics of the cells.

LAB use different types of transport systems, which can be broadly divided into four categories: (i) PMF-driven symport (active or secondary transport), (ii) precursor–product antiport (exchange transport), (iii) phosphate-bond-linked transport, and (iv) group translocation, i.e., phosphoenolpyruvate–sugar phosphotransferase system (sugar PTS).

It should be emphasized that the most extensive research on transport systems among LAB has been done with lactococci (see Konings et al., 1989), in particular *Lc. lactis*. For this species, a nearly complete picture of the transport of sugars, amino acids, peptides, and ions has emerged (Konings et al., 1989). Much of the discussion below is based on these results.

1. PMF-Driven Symport

The PMF-driven transport systems are perhaps the most common and general among LAB (as in most bacteria). The principles of this transport system has already been mentioned and is schematically drawn in Fig. 8A. A specific, membrane-associated protein (carrier, permease) translocates the solute across the membrane in symport with one proton. Presumably, many sugars are transported in this way (see also Section IV.A), although the actual mechanisms, specificities, etc., has not been studied to any large extent, with the exception of lactose transport. PMF-driven lactose transport has been shown in lactococci and in several lactobacilli (Romano et al., 1987; Reizer et al., 1988b; Konings et al., 1989; Fox et al., 1990; Jefferey and Dobrogosz, 1990). In certain lactococci, the permease-mediated transport of lactose coexists with a lactose PTS system (Konings et al., 1989). Similarly, it has been shown that glucose can be transported either by a PMF-driven system or a glucose PTS in a strain of *T. halophilus* (Abe and Uchida, 1990). The reason for having two separate transport systems for the same solute is not fully understood, but difference in affinity may allow the cells to use either system at different substrate concentrations (Konings et al., 1989).

It has been suggested that heterofermentative LAB use exclusively PMF-driven sugar transport, since no sugar PTS have been proven for this group (Romano et al., 1979; Reizer et al., 1988a,b).

In lactococci, most amino acids are also transported by PMF-driven symport systems (Konings et al., 1989). Structurally similar amino acids, e.g., leucine, isoleucine, and valine, may share the same carrier protein. Differences between carriers with regard to affinity and pH regulation has been shown (Konings et al., 1989).

Small peptides are produced by the proteolytic and peptidolytic activity of *Lc. lactis* in milk. The transport of peptides has been shown to be PMF-driven, with broad-specificity carriers (Smid et al., 1989). For the amino acid proline, peptides seem to be the preferred source. This has been attributed to an efficient transport system for proline-containing di- or tripeptides (in the form of Pro-X or Pro-X-X), while an active transport system for the free amino acid is missing (i.e., a dependence on passive diffusion). Moreover, optimal growth of the bacteria in milk is apparently dependent on the balance of peptides produced by proteolysis and peptidolysis. Excess of one particular peptide blocks the transport systems, resulting in impaired growth (Smid and Konings, 1990). These findings are significant, since they imply limitations in the possibilities of manipulating the proteinase/peptidase systems.

2. *Precursor–Product Antiport*

a. *Arginine "Fermentation."* Many LAB can derive energy through substrate-level phosphorylation by the metabolism of arginine (Abdelal, 1979; Crow and Thomas, 1982; Manca De Nadra et al., 1988). The pathway of this metabolism, the arginine deiminase (ADI) pathway, is shown in Fig. 9. For unknown reasons, most arginine-metabolizing LAB cannot use arginine as the sole energy source (Kandler and Weiss, 1986; Konings et al., 1989), but catabolize it simultaneously with a fermentable carbohydrate. Ornithine, one of the end products of the metabolism, is excreted into the medium (Driessen et al., 1987). It has been established, at least for lactococci, that the driving force for arginine uptake and ornithine excretion is the concentration gradients of these compounds (Konings et al., 1989) with no involvement of PMF. The stoichiometry for the arginine–ornithine exchange is strictly 1:1, and the exchange has been shown to be mediated by a single, membrane-associated protein, the arginine/ornithine

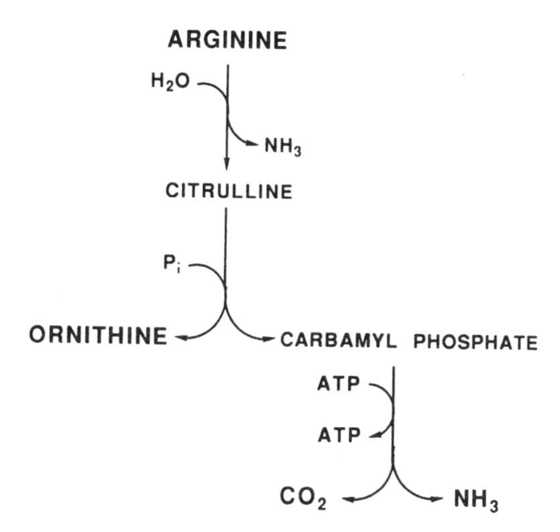

Figure 9 The arginine deiminase pathway.

antiporter (Poolman et al., 1987a). The enzymes of the ADI pathway, except carbamate kinase (which presumably have other, anabolic roles), as well as the antiporter itself is repressed by glucose and induced by arginine (Poolman et al., 1987a; Konings et al., 1989). The genes for these enzymes may be coded by a single operon (Konings et al., 1989). The arginine metabolism leaves the cells without any net accumulation of amino acid, and the question arises as to how the biosynthetic need for arginine is supplied. Some lactococci have the ability to synthesize arginine from ornithine. Ornithine might be recaptured by the PMF-dependent lysine carrier, which has some affinity for ornithine (Konings et al., 1989).

b. *Lactose Transport in* S. Thermophilus. As noted earlier, *S. thermophilus* ferments only the glucose moiety of lactose, while galactose is excreted (Hutkins and Morris, 1987), a property attributed to low galactokinase activity (Thomas and Crow, 1984; Hickey et al., 1986). The transport of lactose in *S. thermophilus* has been believed to be mediated by an ATP-energized, PMF-dependent system (Poolman et al., 1989; Fox et al., 1990). However, recent evidence suggests that galactose efflux is coupled to lactose influx in an antiport-like manner (Hutkins and Ponne, 1991). This could be energetically favorable, i.e., a sparing effect on the ATP needed to establish a PMF. Attempts have been made to obtain galactose-fermenting (Gal^+) mutants of *S. thermophilus* (Thomas and Crow, 1984). Such mutants may suffer from an energetically unfavorable situation, which could explain the difficulties in obtaining stable Gal^+ variants (Thomas and Crow, 1984). It is not known whether similar systems are operational in other LAB excreting galactose during lactose fermentation, e.g., *Lb. delbrückii* and *Lb. acidophilus* (Hickey et al., 1986).

3. Phosphate-Bond-Linked Transport

Glutamate and glutamine transport in lactococci is mediated by a system independent of a PMF, but dependent on the production of metabolic energy, either by glycolysis or the ADI pathway (Poolman et al., 1987b,c). The activity of the transport system decrease if internal ATP levels are lowered. This suggests direct involvement of ATP or another high-energy compound, and the transport has been characterized as being phosphate-bond-linked. Contrary to the PMF-driven systems, the phosphate-bond-linked transport is essentially irreversible. A single transport system for glutamate and glutamine has been identified (Poolman et al., 1987c). The system has an absolute preference for the undissociated form of glutamate (glutamic acid), which leads to growth limitations at alkaline pH values (Poolman and Konings, 1988). Since glutamine transport is independent of pH, the cells can be relieved from this effect if higher concentrations of glutamine are included in the medium as a source of glutamate.

4. The Phosphoenolpyruvate–Sugar Phosphotransferase System

The phosphoenolpyruvate–sugar phosphotransferase system (PTS) is a complex enzyme machinery, whose main function is to translocate a sugar across a membrane with simultaneous phosphorylation (for a comprehensive review, see Reizer et al., 1988b). Since there are two separate molecular species outside (sugar) and inside (sugar phosphate) the membrane, the translocation does not involve any concentration gradients. The energy for the process is provided by the high-energy phosphate bond of phosphoenolpyruvate (PEP). The energy of the phosphoryl group is transferred in a series of reactions, via PTS-specific proteins, to a membrane-located enzyme, which mediates transport and phosphorylation of the sugar.

The components and reactions of the system are depicted in Fig. 10. The first two proteins in the series, Enzyme I (EI) and heat-stable protein (HPr) are sugar-nonspecific (can be shared by several PTSs), whereas Enzyme II (EII) and Enzyme III (EIII) are sugar-specific (denoted by a superscript s in Fig. 10). EII is a membrane-located enzyme, and it acts in concert with the phosphorylated EIII to mediate recognition, translocation, and phosphorylation. EII and EIII are sugar-specific, but the specificity may not be absolute. Thus, the glucose PTS in most LAB also recognizes mannose and is often designated man PTS to distinguish it from the glc PTS, first described in *E. coli*, which has a different substrate specificity (Postma and Lengeler, 1985).

The distribution of PTSs in bacteria has been discussed by several authors (Saier, 1977; Romano et al., 1979; Postma and Lengeler, 1985; Reizer et al., 1988b). There has been general agreement that the presence of PTS is highly

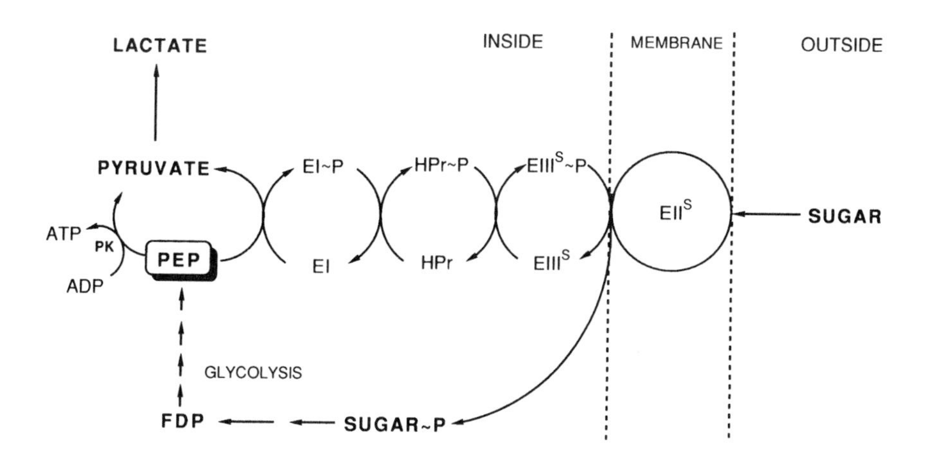

Figure 10 Sugar transport mediated by the phosphoenolpyruvate–sugar phosphotransferase system (PTS) and relation to glycolysis. PK, pyruvate kinase. See text for details.

correlated to the ability to ferment substrates through the Embden-Meyerhof pathway, i.e., glycolysis. LAB have provided some support for this in that heterofermentative LAB seem to be devoid of sugar-PTS activity (Romano et al., 1979; Reizer et al., 1988b), whereas homofermentative LAB possess the system. It has been argued that the production of 2 moles of PEP per hexose consumed (as in glycolysis) is needed for a functional metabolism including a PTS (Romano et al., 1979). However, PTSs for gluconate and pentitols have been described in LAB (London, 1990, and see Section IV.A.3). These compounds are fermented by the 6-PG/PK pathway and hence do not result in two PEP per substrate. As a generalization, however, the sugar PTS is tightly coupled to glycolysis in most bacteria. This is schematically shown in Fig. 10. The system actually constitutes a cycle, where PEP holds a key position (Thompson, 1987). The transport of the sugar is directly coupled to the subsequent metabolism, which in turn provides the PEP needed for a new cycle to begin.

LAB, in particular *Lc. lactis* and *Lb. casei*, have been valuable as model systems for basic research on PTS-mediated transport (Thompson, 1987, 1988; Reizer et al., 1988b). The system is complex and features intriguing regulatory circuits (Reizer et al., 1988b). Within the scope of this chapter, only a few of those are discussed.

As depicted in Fig. 10, PEP has two alternative fates. It can either donate the phosphoryl group to EI and initiate the PTS cycle, or it is used by pyruvate kinase to form ATP. Pyruvate kinase is subject to regulation, where FDP act as activator and P_i as inhibitor (Thompson, 1987). Under optimum glycolyzing conditions, when the FDP level is high and the P_i level low, pyruvate kinase is most active and the concentration of PEP is low (Thompson, 1987). A decrease in the glycolytic rate, as a result of limiting concentration of sugar, will result in a decrease in FDP levels and an increase in P_i (Fordyce et al., 1984). Consequently, pyruvate kinase activity decreases and the concentration of PEP increases. During complete starvation, when both pyruvate kinase and the PTS are inoperative, the cells contain high concentrations of PEP and the preceding intermediates in the glycolytic pathway, 3-phosphoglycerate and 2-phosphoglycerate (Thompson, 1987). This pool of metabolites, designated the *PEP potential*, enables the cells to quickly resume transport and glycolysis once a sugar becomes available. The PEP potential may also be important in providing maintenance energy during starvation by a residual activity of pyruvate kinase (Thompson, 1987).

Sugar-specific components of the PTS are often inducible with the respective sugars (or sugar analogs) acting as inducers (Thompson, 1979; Konings et al., 1989). This "coarse control" of sugar uptake avoids unnecessary enzyme synthesis. LAB also possess mechanisms to avoid uptake of one sugar when a preferential substrate is present. This effect, known as inducer exclusion, has been shown by the use of a galactoside analog, thiomethyl-β-D-galactopyranoside (TMG). This compound is a substrate for the lactose PTS, but is not metabolized beyond TMG phosphate. It has been shown that cells induced for the lactose PTS do not accumulate TMG if glucose or glucose analogs are transported by the mannose PTS (Thompson et al., 1978; Thompson and Saier, 1981; Thompson, 1987). This has been attributed to the preferential use of HPr phosphate by the mannose PTS, which translocates glucose (Thompson et al., 1978; Thompson, 1987).

Another situation arises when the cells are first allowed to accumulate TMG and then presented to glucose. Under these conditions, the accumulated TMG phosphate is rapidly dephosphorylated and TMG exits the cells (Thompson and Saier, 1981). This phenomenon has been designated inducer expulsion. The mechanisms of inducer expulsion is not fully understood, but it is clear that metabolic energy (ATP), HPr, FDP, and a sugar phosphatase participate in the process (Thompson, 1987; Reizer, 1989). A significant finding was that heterofermentative lactobacilli possessed a mechanism, which resembled inducer expulsion (Romano et al., 1987). This was surprising, since inducer expulsion was previously thought to be connected with PTS activity, apparently missing in heterofermentative LAB (see above). Further studies led to the discovery of HPr and ATP-dependent HPr kinase in the heterofermentative species *Lb. brevis* and *Lb. buchneri* (Reizer et al., 1988a). Thus, an important role, beyond its function in PTS, has been assigned to HPr in the overall regulation of sugar uptake and metabolism (Reizer, 1989).

VI. CONCLUDING REMARKS

The aim of this chapter on lactic acid bacteria has been to give a summary of what they are (general description), who they are (classification), what they do (metabolism), and how they do it (metabolism, energetics, and transport). This has not been an easy task. LAB comprise a very diverse group of organisms, but have sufficient characteristics in common that some generalizations can be made. The overall view I would like to pass on is that LAB are more than just lactic acid producers. Lactic acid production is merely a reflection of an underlying metabolism, which is far more complex and, most importantly, more adaptive than one could imagine.

LAB are perfectly adapted to environments rich in nutrients and energy sources. They have, therefore, dispensed with biosynthetic capability. Apparently, some residual genetic material for biosynthesis is still present, since mutagenesis can render some lactobacilli prototrophic for some amino acids (Morishita et al., 1981). However, if these mutants are returned to a rich medium, they readily revert to the auxotrophic state (Morishita et al., 1981). This may have evolutionary implications, but more important it shows that there has been a strong selection for cells that are committed to life in rich environments. These environments are of course excellent for supporting growth of other microorganisms. LAB have therefore developed strategies to efficiently compete with these organisms. One important strategy is acid production and acid tolerance (Lindgren and Dobrogosz, 1990). Perhaps this is the reason why the term LAB arose. A group of bacteria were isolated from similar niches and turned out to be lactic acid producers. It was natural to group them together. Another way to reason is the following: the commitment to life in rich environments demands a simple but effective way of outcompeting other microorganisms. Solution: acid production! Therefore, we stand with, in reality (phylogenetically), a diverse group of organisms, but physiologically similar, since they are specialized for nutrient-rich environments (limited biosynthetic ability) and their metabolism is aimed at acid production.

Although the reasoning above may be somewhat oversimplified, it is clear that the classification problems that always have been evident with regard to LAB may stem from an (historical) overemphasis of a few characters. We now have the means to determine natural relations more objectively and more accurately than ever before. These relations can probably be assessed more easily than to define common phenotypic characters for a particular natural group. For example, the genera *Carnobacterium, Enterococcus, Tetragenococcus*, and *Vagococcus* form a phylogenetic group (see Fig. 1). Do they possess any common characteristics that would suggest this? According to Woese (1987), they probably do, but from current knowledge about these genera we are not able to see them. In future classification of bacteria in general, it will be necessary to do "reverse phylogenetics," i.e., first define the natural relations among bacteria with rRNA sequencing and then search for the unifying phenotypic characters.

The general metabolism and physiology of LAB reflects their adaptation to niches rich in nutrients. They have developed very efficient transport systems, which enable them to quickly take up the necessary solutes. The extreme saccharolytic nature is another example. Very few LAB can use compounds other than carbohydrates (and related substances such as mannitol, etc.) as sole energy sources. However, as I have tried to emphasize, they have developed systems that will allow them to derive more energy from a rich medium than just the carbohydrate. One of these systems are the cofermentations that have been mentioned several times. By using a substrate, otherwise nonfermentable, as an electron acceptor during carbohydrate fermentation, they indirectly derive some energy from that substrate that otherwise would be lost. The malo-lactic fermentation serves the same purpose, whatever the actual mechanism for energy transduction may be. The use of cofermentations may also enable LAB to extend the range of fermentable substrates. For example, *Lb. plantarum* is unable to ferment either citrate or mannitol anaerobically. However, added together, both substrates are used, mannitol as the main energy source and citrate as electron acceptor (see Section IV.D.1).

As indicated above and in Section III, Orla-Jensen's concept of LAB being a "great natural group" may not be entirely correct. However, the term *lactic acid bacteria* will probably be used in the foreseeable future, since it is useful to describe a group of organisms that have many physiological properties in common and, as a generalization, have similarities in their ecological behavior.

REFERENCES

Abdelal, A. T. (1979). Arginine catabolism by microorganisms. *Ann. Rev. Microbiol., 33*:139–168.

Abe, K., and Uchida, K. (1989). Correlation between derepression of catabolite control of xylose metabolism and a defect in the phosphoenolpyruvate:mannose phosphotransferase system in *Pediococcus halophilus, J. Bacteriol., 171*:1793–1800.

Abe, K., and Uchida, K. (1990). Non-PTS uptake and subsequent metabolism of glucose in *Pediococcus halophilus* as demonstrated with a double mutant defective in phosphoenolpyruvate:mannose phosphotransferase system and in phosphofructokinase, *Arch. Microbiol., 153*:537–540.

Archibald, F. (1986). Manganese: its acquisition by and function in the lactic acid bacteria, *Crit. Rev. Microbiol., 13*:63–109.

Axelsson, L. (1990). *Lactobacillus reuteri*, a member of the gut bacterial flora. Studies on antagonism, metabolism and genetics, Ph.D. thesis, Swedish University of Agricultural Sciences.

Axelsson, L. T., Chung, T. C., Dobrogosz, W. J., and Lindgren, S. E. (1989). Production of a broad spectrum antimicrobial substance by *Lactobacillus reuteri, Microb. Ecol. Health Dis., 2*:131–136.

Axelsson, L., and Lindgren, S. (1987). Characterization and DNA homology of *Lactobacillus* strains isolated from pig intestine, *J. Appl. Bacteriol., 62*:433–438.

Betzl, D., Ludwig, W., and Schleifer, K. H. (1990). Identification of lactococci and enterococci by colony hybridization with 23S rRNA-targeted oligonucleotide probes, *Appl. Environ. Microbiol., 56*:2927–2929.

Bhowmik, T., and Marth, E. H. (1990). β-Galactosidase of *Pediococcus* species: induction, purification and partial characterization, *Appl. Microbiol. Biotechnol., 33*: 317–323.

Biketov, S. F., Kasho, V. N., Kozlov, I. A., Mileykovskaya, Y. I., Ostrovsky, D. N., Skulachev, V. P., Tikhonova, G. V., and Tsuprun, V. L. (1982). F_1-like ATPase from anaerobic bacterium *Lactobacillus casei* contains six similar subunits, *Eur. J. Biochem., 129*:241–250.

Bissett, D. L., and Anderson, R. L. (1974). Lactose and D-galactose metabolism in group N streptococci: presence of enzymes for both the D-galactose 1-phosphate and D-tagatose 6-phosphate pathways, *J. Bacteriol., 117*:318–320.

Booth, I. R. (1985). Regulation of cytoplasmic pH in bacteria, *Microbiol. Rev., 49*:359–378.

Borch, E., and Molin, G. (1989). The aerobic growth and product formation of *Lactobacillus, Brochothrix*, and *Carnobacterium* in batch cultures, *Appl. Microbiol. Biotechnol., 30*:81–88.

Broome, M. C., Thomas, M. P., Hillier, A. J., and Jago, G. R. (1980). Pyruvate dehydrogenase activity in group N streptococci, *Aust. J. Biol. Sci., 33*:15–25.

Brown, J. P., and VanDemark, P. J. (1968). Respiration of *Lactobacillus casei, Can. J. Microbiol., 14*:829–835.

Bryan-Jones, D. G., and Whittenbury, R. (1969). Haematin-dependent oxidative phosphorylation in *Streptococcus faecalis, J. Gen. Microbiol., 58*:247–260.

Cerning, J. (1990). Exocellular polysaccharides produced by lactic acid bacteria, *FEMS Microbiol. Rev., 87*:113–130.

Chassy, B. M., and Alpert, C.-A. (1989). Molecular characterization of the plasmid-encoded lactose-PTS of *Lactobacillus casei, FEMS Microbiol. Rev., 63*:157–166.

Chen, K.-H., and McFeeters, R. F. (1986a). Utilization of electron-acceptors for anaerobic mannitol metabolism by *Lactobacillus plantarum*. Enzymes and intermediates in the utilization of citrate, *Food Microbiol., 3*:83–92.

Chen, K.-H., and McFeeters, R. F. (1986b). Utilization of electron acceptors for anaerobic mannitol metabolism by *Lactobacillus plantarum*. Reduction of alpha-keto acids, *Food Microbiol., 3*:93–99.

Chung, T. C., Axelsson, L., Lindgren, S. E., and Dobrogosz, W. J. (1989). In vitro studies on reuterin synthesis by *Lactobacillus reuteri, Microb. Ecol. Health Dis., 2*:137–144.

Cogan, J. F., Walsh, D., and Condon, S. (1989). Impact of aeration on the metabolic end-products formed from glucose and galactose by *Streptococcus lactis, J. Appl. Bacteriol., 66*:77–84.

Cogan, T. M. (1985). The leuconostocs: milk products, in *Bacterial Starter Cultures for Food* (S. E. Gilliland, ed.), CRC Press, Boca Raton, FL, pp. 25–40.

Cogan, T. M. (1987). Co-metabolism of citrate and glucose by *Leuconostoc* spp.: effects on growth, substrates and products, *J. Appl. Bacteriol., 63*:551–558.

Cogan, T. M., O'Dowd, M., and Mellerick, D. (1981). Effects of pH and sugar on acetoin production from citrate by *Leuconostoc lactis, Appl. Environ. Microbiol., 41*:1–8.

Collins, M. D., Ash, C., Farrow, J. A. E., Wallbanks, S., and Williams, A. M. (1989). 16S ribosomal ribonucleic acid sequence analysis of lactococci and related taxa. Description of *Vagococcus fluvialis* gen. nov., sp. nov., *J. Appl. Bacteriol., 67*:453–460.

Collins, M. D., Farrow, J. A. E., Phillips, B. A., Ferusu, S., and Jones, D. (1987). Classification of *Lactobacillus divergens, Lactobacillus piscicola* and some catalase-negative, asporogenous, rod-shaped bacteria from poultry in a new genus, *Carnobacterium, Int. J. Syst. Bacteriol., 37*:310–316.

Collins, M. D., Rodrigues, U., Ash, C., Aguirre, M., Farrow, J. A. E., Martinez-Murcia, A., Phillips, B. A., Williams, A. M., and Wallbanks, S. (1991). Phylogenetic analysis of the genus *Lactobacillus* and related lactic acid bacteria as determined by reverse transcriptase sequencing of 16S rRNA, *FEMS Microbiol. Lett., 77*:5–12.

Collins, M. D., Williams, A. M., and Wallbanks, S. (1990). The phylogeny of *Aerococcus* and *Pediococcus* as determined by 16S rRNA sequence analysis: description of *Tetragenococcus* gen. nov., *FEMS Microbiol. Lett., 70*:255–262.

Condon, S. (1987). Responses of lactic acid bacteria to oxygen, *FEMS Microbiol. Rev., 46*:269–280.

Cox, D. J., and Henick-Kling, T. (1989). Chemiosmotic energy from malolactic fermentation, *J. Bacteriol., 171*:5750–5752.

Crow, V. L., and Thomas, T. D. (1982). Arginine metabolism in lactic streptococci, *J. Bacteriol., 150*:1024–1032.

Daeschel, M. A. (1988). A pH control system based on malate decarboxylation for the cultivation of lactic acid bacteria, *Appl. Environ. Microbiol., 54*:1627–1629.

Daeschel, M. A., Andersson, R. E., and Fleming, H. P. (1987). Microbial ecology of fermenting plant materials, *FEMS Microbiol. Rev., 46*:357–367.

De Bruyn, I. N., Holzapfel, W. H., Visser, L., and Louw, A. I. (1988). Glucose metabolism by *Lactobacillus divergens, J. Gen. Microbiol., 134*:2103–2109.

De Vries, W., Kapteijn, W. M. C., Van der Beek, E. G., and Stouthamer, A. H. (1970). Molar growth yields and fermentation balances of *Lactobacillus casei* 13 in batch cultures and in continuous cultures, *J. Gen. Microbiol., 63*:333–345.

Dicks, L. M. T., and van Vuuren, H. J. J. (1987). Relatedness of heterofermentative *Lactobacillus* species revealed by numerical analysis of total soluble protein patterns, *Int. J. Syst. Bacteriol., 37*:437–440.

Dobrogosz, W. J., and Stone, R. W. (1962). Oxidative metabolism in *Pediococcus pentosaceus*. I. Role of oxygen and catalase, *J. Bacteriol., 84*:716–723.

Driessen, A. J. M., Poolman, B., Kiewiet, R., and Konings, W. N. (1987). Arginine transport in *Streptococcus lactis* is catalyzed by a cationic exchanger, *Proc. Nat. Acad. Sci. USA, 84*:6093–6097.

Drinan, D. F., Tobin, S., and Cogan, T. M. (1976). Citric acid metabolism in hetero- and homofermentative lactic acid bacteria, *Appl. Environ. Microbiol., 31*:481–486.

Eltz, R. W., and VanDemark, P. J. (1960). Fructose dissimilation by *Lactobacillus brevis, J. Bacteriol., 79*:763–776.

Embley, T. M., Faquir, N., Bossart, W., and Collins, M. D. (1989). *Lactobacillus vaginalis* sp. nov. from the human vagina, *Int. J. Syst. Bacteriol., 39*:368–370.

Evans, J. B. (1986). Genus *Aerococcus*, in *Bergey's Manual of Systematic Bacteriology*, Vol. 2 (P. H. A. Sneath, N. S. Mair, M. E. Sharpe, and J. G. Holt, eds.), Williams and Wilkins, Baltimore, p. 1080.

Farrow, J. A. E., and Collins, M. D. (1984). DNA base composition, DNA-DNA homology and long-chain fatty acid studies on *Streptococcus thermophilus* and *Streptococcus salivarius, J. Gen. Microbiol., 130*:357–362.

Farrow, J. A. E., and Collins, M. D. (1988). *Lactobacillus oris* sp. nov. from the human oral cavity, *Int. J. Syst. Bacteriol., 38*:116–118.

Fleming, H. P., McFeeters, R. F., and Daeschel, M. A. (1985). The lactobacilli, pediococci and leuconostocs: vegetable products, in *Bacterial Starter Cultures for Food* (S. E. Gilliland, ed.), CRC Press, Boca Raton, FL, pp. 97–118.

Fordyce, A. M., Crow, V. L., and Thomas, T. D. (1984). Regulation and product formation during glucose or lactose limitation in nongrowing cells of *Streptococcus lactis, Appl. Environ. Microbiol., 48*:332–337.

Fox, P. F., Lucey, J. A., and Cogan, T. M. (1990). Glycolysis and related reactions during cheese manufacture and ripening, *Crit. Rev. Food Sci. Nutr., 29*:237–253.

Fuller, R. (1986). Probiotics, *J. Appl. Bacteriol. Symp. Suppl., 61*:1S–7S.

Futai, M., and Kanazawa, H. (1983). Structure and function of proton-translocating adenosine triphosphatase (F_0F_1): biochemical and molecular biological approaches, *Microbiol. Rev., 47*:285–312.

Garvie, E. I. (1980). Bacterial lactate dehydrogenases, *Microbiol. Rev., 44*:106–139.

Garvie, E. I. (1986a). Genus *Leuconostoc*, in *Bergey's Manual of Systematic Bacteriology, Vol. 2* (P. H. A. Sneath, N. S. Mair, M. E. Sharpe, and J. G. Holt, eds.), Williams and Wilkins, Baltimore, pp. 1071–1075.

Garvie, E. I. (1986b). Genus *Pediococcus*, in *Bergey's Manual of Systematic Bacteriology*, Vol. 2 (P. H. A. Sneath, N. S. Mair, M. E. Sharpe, and J. G. Holt, eds.), Williams and Wilkins, Baltimore, pp. 1075–1079.

Gätje, G., Müller, V., and Gottschalk, G. (1991). Lactic acid excretion via carrier-mediated facilitated diffusion in *Lactobacillus helveticus, Appl. Microbiol. Biotechnol., 34*:778–782.

Giovannoni, S. J., DeLong, E. F., Olsen, G. J., and Pace, N. R. (1988). Phylogenetic group-specific oligodeoxynucleotide probes for identification of single microbial cells, *J. Bacteriol., 170*:720–726.

Gottschalk, G. (1986). *Bacterial Metabolism*, Springer-Verlag, New York.

Götz, F., Elstner, E. F., Sedewitz, B., and Lengfelder, E. (1980). Oxygen utilization by *Lactobacillus plantarum*. II. Superoxide and superoxide dismutation, *Arch. Microbiol., 125*:215–220.

Hammes, W. P., Bantleon, A., and Min, S. (1990). Lactic acid bacteria in meat fermentation, *FEMS Microbiol. Rev., 87*:165–173.

Hardie, J. M. (1986a). Anaerobic streptococci, in *Bergey's Manual of Systematic Bacteriology*, Vol. 2 (P. H. A. Sneath, N. S. Mair, M. E. Sharpe, and J. G. Holt, eds.), Williams and Wilkins, Baltimore, pp. 1066–1068.

Hardie, J. M. (1986b). Genus *Streptococcus* Rosenbach 1884, in *Bergey's Manual of Systematic Bacteriology*, Vol. 2 (P. H. A. Sneath, N. S. Mair, M. E. Sharpe, and J. G. Holt, eds.), Williams and Wilkins, Baltimore, pp. 1043-1047.

Hardie, J. M. (1986c). Oral streptococci, in *Bergey's Manual of Systematic Bacteriology*, Vol. 2 (P. H. A. Sneath, N. S. Mair, M. E. Sharpe, and J. G. Holt, eds.), Williams and Wilkins, Baltimore, pp. 1054–1063.

Hardie, J. M. (1986d). Other streptococci, in *Bergey's Manual of Systematic Bacteriology*, Vol. 2 (P. H. A. Sneath, N. S. Mair, M. E. Sharpe, and J. G. Holt, eds.), Williams and Wilkins, Baltimore, pp. 1068–1070.

Harvey, R. J., and Collins, E. B. (1961). Role of citratase in acetoin formation by *Streptococcus diacetilactis* and *Leuconostoc citrovorum*, *J. Bacteriol.*, *82*:954–959.

Henneberg, W. (1904). Zur Kenntnis der Milchsäurebakterien der Brennerei-Maische, der Milch, des Bieres, der Presshefe, der Melasse, des Sauerkohls, der säuren Gurken und des Sauerteigs; sowie einige Bemerkungen über die Milchsäurebakterien des menschlishen Magens, *Zentbl. Bakt. ParasitKde Abt. II.*, *11*: 154–170.

Hertel, C., Ludwig, W., Obst, M., Vogel, R. F., Hammes, W. P., and Schleifer, K. H. (1991). 23S rRNA-targeted oligonucleotide probes for the rapid identification of meat lactobacilli, *Syst. Appl. Microbiol.*, *14*:173–177.

Hickey, M. W., Hillier, A. J., and Jago, G. R. (1986). Transport and metabolism of lactose, glucose and galactose in homofermentative lactobacilli, *Appl. Environ. Microbiol.*, *51*:825–831.

Hutkins, R. W., and Morris, H. A. (1987). Carbohydrate metabolism in *Streptococcus thermophilus*: a review, *J. Food Prot.*, *50*:876–884.

Hutkins, R. W., and Ponne, C. (1991). Lactose uptake driven by galactose efflux in *Streptococcus thermophilus*: evidence for a galactose-lactose antiporter, *Appl. Environ. Microbiol.*, *57*:941–944.

Ingram, M. (1975). The lactic acid bacteria—a broad view, in *Lactic Acid Bacteria in Beverages and Food* (J. G. Carr, C. V. Cutting, and G. C. Whiting, eds.), Academic Press, London, pp. 1–13.

Jefferey, S. R., and Dobrogosz, W. J. (1990). Transport of β-galactosides in *Lactobacillus plantarum* NC2, *Appl. Environ. Microbiol.*, *56*:2484–2487.

Johnson, S. L. (1984). Bacterial classification III. Nucleic acids in bacterial classification, in *Bergey's Manual of Systematic Bacteriology*, Vol. 1. (N. R. Krieg and J. G. Holt, eds.), Williams and Wilkins, Baltimore, pp. 8–11.

Kandler, O. (1983). Carbohydrate metabolism in lactic acid bacteria, *Antonie van Leeuwenhoek*, *49*:209–224.

Kandler, O. (1984). Current taxonomy of lactobacilli, *Dev. Ind. Microbiol.*, *25*:109–123.

Kandler, O., and Weiss, N. (1986). Regular, non-sporing gram-positive rods, in *Bergey's Manual of Systematic Bacteriology*, Vol. 2 (P. H. A. Sneath, N. S. Mair, M. E. Sharpe, and J. G. Holt, eds.), Williams and Wilkins, Baltimore, pp. 1208–1234.

Kaneuchi, C., Seki, M., and Komagata, K. (1988). Production of succinic acid from citric acid and related acids by *Lactobacillus* strains, *Appl. Environ. Microbiol., 54*:3053–3056.

Kashket, E. R. (1987). Bioenergetics of lactic bacteria: cytoplasmic pH and osmotolerance, *FEMS Microbiol. Rev., 46*:233–244.

Kobayashi, H. (1985). A proton-translocating ATPase regulates pH of the bacterial cytoplasm, *J. Biol. Chem., 260*:72–76.

Konings, W. N., and Otto, R. (1983). Energy transduction and solute transport in streptococci, *Antonie van Leeuwenhoek, 49*:247–257.

Konings, W. N., Poolman, B., and Driessen, A. J. M. (1989). Bioenergetics and solute transport in lactococci, *Crit. Rev. Microbiol., 16*:419–476.

Kono, Y., and Fridovich, I. (1983). Isolation and characterization of the pseudocatalase of *Lactobacillus plantarum, J. Biol. Chem., 258*:6015–6019.

Kunkee, R. E. (1991). Some roles of malic acid and the malolactic fermentation in wine making, *FEMS Microbiol. Rev., 88*:55–72.

Lancefield, R. C. (1933). A serological differentiation of human and other groups of hemolytic streptococci, *J. Exp. Med., 59*:571–591.

Lane, D. J., Pace B., Olsen, G. J., Stahl, D. A., Sogin, M. L., and Pace, N. R. (1985). Rapid determination of 16S ribosomal RNA sequences for phylogenetic analysis, *Proc. Nat. Acad. Sci. USA, 82*:6955–6959.

Lane, D. L., Field, K. G., Olsen, G. J., and Pace, N. R. (1988). Reverse transcriptase sequencing of ribosomal RNA for phylogenetic analysis, *Methods Enzymol., 167*: 138–144.

Lawrence, R. C., and Thomas, T. D. (1979). The fermentation of milk by lactic acid bacteria, in *Microbial Technology: Current State, Future Prospects. Soc. Gen. Microbiol., Symp. 29* (T. Bull, D. C. Ellwood, and C. Ratledge, eds.), University Press, Cambridge, pp. 187–219.

Lechevalier, M. P. (1977). Lipids in bacterial taxonomy—a taxonomist's view, *Crit. Rev. Microbiol., 5*:109–210.

Lindgren, S. E., Axelsson, L. T., and McFeeters, R. F. (1990). Anaerobic L-lactate degradation by *Lactobacillus plantarum, FEMS Microbiol. Lett., 66*:209–214.

Lindgren, S. E., and Dobrogosz, W. J. (1990). Antagonistic activities of lactic acid bacteria in food and feed fermentations, *FEMS Microbiol. Rev., 87*:149–163.

London, J. (1990). Uncommon pathways of metabolism among lactic acid bacteria, *FEMS Microbiol. Rev., 87*:103–111.

Lucey, C. A., and Condon, S. (1986). Active role of oxygen and NADH oxidase in growth and energy metabolism of *Leuconostoc, J. Gen. Microbiol., 132*:1789–1796.

Lütgens, M., and Gottschalk, G. (1980). Why a co-substrate is required for anaerobic growth of *Escherichia coli* on citrate, *J. Gen. Microbiol., 119*:63–70.

Löhnis, F. (1907). Versuch einer Gruppierung der Milchsäurebakterien, *Zentbl. Bakt. ParasitKde, Abt II, 18*:97–149.

Maloney, P. C. (1990). Microbes and membrane biology, *FEMS Microbiol. Rev., 87*:91–102.

Maloney, P. C., Kashket, E. R., and Wilson, T. H. (1974). A protonmotive force drives ATP synthesis in bacteria, *Proc. Nat. Acad. Sci. USA, 71*:3896–3900.

Manca De Nadra, M. C., Pesce de Ruiz Holgado, A. A., and Oliver, G. (1988). Arginine dihydrolase pathway in *Lactobacillus buchneri*: a review, *Biochimie, 70*: 367–374.

Marshall, V. M. (1987). Lactic acid bacteria: starters for flavour, *FEMS Microbiol. Rev., 46*:327–336.

Martinez-Murcia, A. J., and Collins, M. D. (1990). A phylogenetic analysis of the genus Leuconostoc based on reverse transcriptase sequencing of 16S rRNA, *FEMS Microbiol. Lett., 70*:73–84.

McDonald, L. C., Fleming, H. P., and Hassan, H. M. (1990). Acid tolerance of *Leuconostoc mesenteroides* and *Lactobacillus plantarum, Appl. Environ. Microbiol., 56*: 2120–2124.

McFeeters, R. F., and Chen, K.-H. (1986). Utilization of electron acceptors for anaerobic metabolism of *Lactobacillus plantarum*. Compounds which serve as electron acceptors, *Food Microbiol., 3*:73–81.

Michels, P. A. M., Michels, J. P. J., Boonstra, J., and Konings, W. N. (1979). Generation of an electrochemical proton gradient in bacteria by the excretion of metabolic end products, *FEMS Microbiol. Lett., 5*:357–364.

Miller, L. T. (1984). Gas-liquid chromatography of cellular fatty acids as a bacterial identification aid, *Hewlett-Packard Application Note, 228*:37.

Mitchell, P. (1961). Coupling of phosphorylation to electron and hydrogen transfer by a chemiosmotic type of mechanism, *Nature, 191*:144–148.

Mitchell, P. (1972). Chemiosmotic coupling in energy transduction: a logical development in biochemical knowledge, *Bioenergetics, 3*:5–24.

Montel, M.-C., Talon, R., Fournoud, J., and Champomier, M.-C. (1991). A simplified key for identifying homofermentative *Lactobacillus* and *Carnobacterium* spp. from meat, *J. Appl. Bacteriol., 70*:469–472.

Morishita, T., Deguchi, Y., Yajima, M., Sakurai, T., and Yura, T. (1981). Multiple nutritional requirements of lactobacilli: genetic lesions affecting amino acid biosynthetic pathways, *J. Bacteriol., 148*:64–71.

Moss, C. W. (1981). Gas-liquid chromatography as an analytical tool in microbiology, *J. Chromatography, 203*:337–347.

Mundt, J. O. (1986). Enterococci, in *Bergey's Manual of Systematic Bacteriology*, Vol. 2 (P. H. A. Sneath, N. S. Mair, M. E. Sharpe, and J. G. Holt, eds.), Williams and Wilkins, Baltimore, pp. 1063–1065.

Muntyan, M. S., Mesyanzhinova, I. V., Milgrom, Y. M., and Skulachev, V. P. (1990). The F_1-type ATPase in anaerobic *Lactobacillus casei, Biochim. Biophys. Acta, 1016*: 371–377.

Murphy, M. G., O'Connor, L., Walsh, D., and Condon, S. (1985). Oxygen dependent lactate utilization by *Lactobacillus plantarum, Arch. Microbiol., 141*:75–79.

Olson, N. F. (1990). The impact of lactic acid bacteria on cheese flavor, *FEMS Microbiol. Rev., 87*:131–147.

Orla-Jensen, S. (1919), *The Lactic Acid Bacteria*, Host, Copenhagen.

Otto, R., Hugenholtz, J., Konings, W. N., and Veldkamp, H. (1980). Increase of molar growth yield of *Streptococcus cremoris* for lactose as a consequence of lactate consumption by *Pseudomonas stutzeri* in mixed culture, *FEMS Microbiol. Lett., 9*:85–88.

Parker, M. T. (1978). The pattern of streptococcal disease in man, in *Streptococci*. Soc. Appl. Bacteriol. Symp Series, No. 7 (F. A. Skinner, and L. B. Quesnel, eds.), Academic Press, London, pp. 71–106.

Peterson, W. H., and Fred, E. B. (1920). Fermentation of fructose by *Lactobacillus pentoaceticus*, n. sp., *J. Biol. Chem.*, *41*:431–450.

Petrick, H. A. R., Ambrosio, R. E., and Holzapfel, W. H. (1988). Isolation of a DNA probe for *Lactobacillus curvatus*, *Appl. Environ. Microbiol.*, *54*:405–408.

Pilloud, N., and Mollet, B. (1990). DNA probes for the detection of *Lactobacillus helveticus*, *Syst. Appl. Microbiol.*, *13*:345–349.

Pilone, G. J., and Kunkee, R. E. (1976). Stimulatory effect of malolactic fermentation on the growth rate of *Leuconostoc oenos*, *Appl. Environ. Microbiol.*, *32*:405–408.

Poolman, B., Driessen, A. J. M., and Konings, W. N. (1987a). Regulation of arginine-ornithine exchange and the arginine deiminase pathway in *Streptococcus lactis*, *J. Bacteriol.*, *169*:5597–5604.

Poolman, B., Hellingwerf, K. J., and Konings, W. N. (1987b). Regulation of the glutamate-glutamine transport system by intracellular pH in *Streptococcus lactis*, *J. Bacteriol.*, *169*:2272–2276.

Poolman, B., and Konings, W. N. (1988). Relation of growth of *Streptococcus lactis* and *Streptococcus cremoris* to amino acid transport, *J. Bacteriol.*, *170*:700–707.

Poolman, B., Royer, T. J., Mainzer, S. E., and Schmidt, B. F. (1989). Lactose transport system of *Streptococcus thermophilus*: a hybrid protein with homology to the melibiose carrier and enzyme III of phosphoenolpyruvate-dependent phosphotransferase systems, *J. Bacteriol.*, *171*:244–253.

Poolman, B., Smid, E. J., and Konings, W. N. (1987c). Kinetic properties of a phosphate-bond-driven glutamate-glutamine transport system in *Streptococcus lactis* and *Streptococcus cremoris*, *J. Bacteriol.*, *169*:2755–2761.

Postma, P. W., and Lengeler, J. W. (1985). Phosphoenolpyruvate:carbohydrate phosphotransferase system of bacteria, *Microbiol. Rev.*, *49*:232–269.

Pot, B., Descheemaeker, P., Roeges, N., and Kersters, K. (1991). Identification and classification of lactic acid bacteria by sodium dodecyl sulphate polyacrylamide gel-electrophoresis of whole-cell proteins, in *Lactic-91: Lactic Acid Bacteria, Research and Industrial Applications in the Agro-Food Industries*, Symposium abstracts, Caen, France, p. T3.

Premi, L., Sandine, W. E., and Elliker, P. R. (1972). Lactose-hydrolyzing enzymes of *Lactobacillus* species, *Appl. Microbiol.*, *24*:51–57.

Radler, F. (1975). The metabolism of organic acids by lactic acid bacteria, in *Lactic Acid Bacteria in Beverages and Food* (J. G. Carr, C. V. Cutting, and G. C. Whiting, eds.), Academic Press, London, pp. 17–27.

Reizer, J. (1989). Regulation of sugar uptake and efflux in Gram-positive bacteria, *FEMS Microbiol. Rev.*, *63*:149–156.

Reizer, J., Peterkofsky, A., and Romano, A. H. (1988a). Evidence for the presence of heat-stable protein (HPr) and ATP-dependent HPr kinase in heterofermentative lactobacilli lacking phospho*enol*pyruvate:glucose phosphotransferase activity, *Proc. Nat. Acad. Sci. USA*, *85*:2041–2045.

Reizer, J., Saier, M. H., Jr., Deutscher, J., Grenier, F., Thompson, J., and Hengstenberg, W. (1988b). The phosphoenolpyruvate:sugar phosphotransferase system in Gram- positive bacteria: properties, mechanism, and regulation, *Crit. Rev. Microbiol., 15*: 297–338.

Rimpiläinen, M. A., Mettänen, T. T., Niskasaari, K., and Forsén, R. I. (1988). The F_1-ATPase from *Streptococcus cremoris*: isolation, purification and partial characterization, *Int. J. Biochem., 20*:1117–1124.

Ritchey, T. W., and Seeley, H. W. J. (1976). Distribution of cytochrome-like respiration in streptococci, *J. Gen. Microbiol., 93*:195–203.

Rizzo, A. F., Korkeala, H., and Mononen, I. (1987). Gas chromatography analysis of cellular fatty acids and neutral monosaccharides in the identification of lactobacilli, *Appl. Environ. Microbiol., 53*:2883–2888.

Romano, A. H., Brino, G., Peterkofsky, A., and Reizer, J. (1987). Regulation of β-galactosidase transport and accumulation in heterofermentative lactic acid bacteria, *J. Bacteriol., 169*:5589–5596.

Romano, A. H., Trifone, J.D., and Brulston, M. (1979). Distribution of the phosphoenol-pyruvate:glucose phosphotransferase system in fermentative bacteria, *J. Bacteriol., 139*:93–97.

Saier, M. H., Jr. (1977). Bacterial phosphoenolpyruvate:sugar phosphotransferase systems: structural, functional, and evolutionary interrelationships, *Bacteriol. Rev., 41*:856–871.

Saiki, R. K., Gelfand, D. H., Stoffel, S., Scharf, S. J., Higuchi, R., Horn, G. T., Mullis, K. B., and Erlich, H. A. (1988). Primer-directed enzymatic amplification of DNA with thermostable DNA polymerase, *Science, 239*:487–494.

Salama, M., Sandine, W., and Giovannoni, S. (1991). Development and application of oligonucleotide probes for identification of *Lactococcus lactis* subsp. *cremoris, Appl. Environ. Microbiol., 57*:1313–1318.

Sambrook, J., Fritsch, E., and Maniatis, T. (1989). *Molecular Cloning, A Laboratory Manual*, 2nd ed., Cold Spring Harbor Laboratory, Cold Spring Harbor, NY.

Schleifer, K. H. (1986). Gram-positive cocci, in *Bergey's Manual of Systematic Bacteriology*, Vol. 2 (P. H. A. Sneath, N. S. Mair, M. E. Sharpe, and J. G. Holt, eds.), Williams and Wilkins, Baltimore, pp. 999–1103.

Schleifer, K. H., and Kilpper-Bälz, R. (1984). Transfer of *Streptococcus faecalis* and *Streptococcus faecium* to the genus *Enterococcus* nom. rev. as *Enterococcus faecalis* comb. nov. and *Enterococcus faecium* comb. nov., *Int. J. Syst. Bacteriol., 34*:31–34.

Schleifer, K. H., and Kilpper-Bälz, R. (1987). Molecular and chemotaxonomic approaches to the classification of streptococci, enterococci and lactococci: a review, *Syst. Appl. Microbiol., 10*:1–19.

Schleifer, K. H., Kraus, J., Dvorak, C., Kilpper-Bälz, R., Collins, M. D., and Fischer, W. (1985). Transfer of *Streptococcus lactis* and related streptococci to the genus *Lactococcus* gen. nov., *Syst. Appl. Microbiol., 6*:183–195.

Schleifer, K. H., and Stackebrandt, E. (1983). Molecular systematics of procaryotes, *Ann. Rev. Microbiol., 37*:143–187.

Schmitt, P., and Divies, C. (1990). Effect of acetaldehyde on growth, substrates, and products by *Leuconostoc mesenteroides* ssp *cremoris, Biotechnol. Progr., 6*:421–424.

Schütz, H., and Radler, F. (1984a). Anaerobic reduction of glycerol to propanediol-1,3 by *L. brevis* and *L. buchneri, Syst. Appl. Microbiol., 5*:169–178.

Schütz, H., and Radler, F. (1984b). Propanediol-1,2-dehydratase and metabolism of glycerol of *Lactobacillus brevis, Arch. Microbiol., 139*:366–370.

Seale, D. R. (1986). Bacterial inoculants as silage additives, *J. Appl. Bacteriol. Symp. Suppl., 61*:9S–26S.

Sedewitz, B., Schleifer, K. H., and Götz, F. (1984). Physiological role of pyruvate oxidase in the aerobic metabolism of *Lactobacillus plantarum, J. Bacteriol., 160*: 462–465.

Sharpe, M. E. (1979). Identification of the lactic acid bacteria, in *Identification Methods for Microbiologists*, 2nd ed., Soc. Appl. Bacteriol. Technical Series, No. 14 (F. A. Skinner, and D. W. Lovelock, eds.), Academic Press, London, pp. 246–255.

Sharpe, M. E. (1981). The genus *Lactobacillus*, in *The Procaryotes: A Handbook on Habitats, Isolation and Identification of Bacteria* (M. P. Starr, H. Stolp, H. G. Trüper, A. Balows, and H. G. Schlegel, eds.), Springer-Verlag, Berlin, pp. 1653–1674.

Sjöberg, A., and Hahn-Hägerdahl, B. (1989). β-Glucose-1-phosphate, a possible mediator for polysaccharide formation in maltose-assimilating *Lactococcus lactis, Appl. Environ. Microbiol., 55*:1549–1554.

Smart, J. B., and Thomas, T. D. (1987). Effect of oxygen on lactose metabolism in lactic streptococci, *Appl. Environ. Microbiol., 53*:533–541.

Smid, E. J., Driessen, A. J. M., and Konings, W. N. (1989). Mechanism and energetics of dipeptide transport in membrane vesicles of *Lactococcus lactis, J. Bacteriol., 171*: 292–298.

Smid, E. J., and Konings, W. N. (1990). Relationship between utilization of proline and proline-containing peptides and growth of *Lactococcus lactis, J. Bacteriol., 172*: 5286–5292.

Stackebrandt, E., and Teuber, M. (1988). Molecular taxonomy and phylogenetic position of lactic acid bacteria, *Biochimie, 70*:317–324.

Stamer, J. R., and Stoyla, B. O. (1967). Growth response of *Lactobacillus brevis* to aeration and organic catalysts, *Appl. Microbiol., 15*:1025–1030.

Ståhl, M., Molin, G., Persson, A., Ahrné, S., and Ståhl, S. (1990). Restriction endonuclease patterns and multivariate analysis as a classification tool for *Lactobacillus* spp., *Int. J. Syst. Bacteriol., 40*:189–193.

Takahashi, S., Abbe, K., and Yamada, T. (1982). Purification of pyruvate formate-lyase from *Streptococcus mutans* and its regulatory properties, *J. Bacteriol., 149*:672–682.

Talarico, T. L., Axelsson, L.T., Novotny, J., Fiuzat, M., and Dobrogosz, W. J. (1990). Utilization of glycerol as a hydrogen acceptor by *Lactobacillus reuteri*: purification of 1,3-propanediol:NAD^+ oxidoreductase, *Appl. Environ. Microbiol., 56*:943–948.

Talarico, T. L., and Dobrogosz, W. J. (1989). Chemical characterization of an antimicrobial substance produced by *Lactobacillus reuteri, Antimicrob. Agents Chemother., 33*: 674–679.

Talarico, T. L., and Dobrogosz, W. J. (1990). Purification and characterization of glycerol dehydratase from *Lactobacillus reuteri, Appl. Environ. Microbiol., 56*: 1195–1197.

ten Brink, B., Otto, R., Hansen, U.-P., and Konings, W. N. (1985). Energy recycling by lactate efflux in growing and nongrowing cells of *Streptococcus cremoris, J. Bacteriol., 162*:383–390.

Thomas, T. D., and Crow, V. L. (1984). Selection of galactose-fermenting *Streptococcus thermophilus* in lactose-limited chemostat cultures, *Appl. Environ. Microbiol., 48*: 186–191.

Thomas, T. D., Ellwood, D. C., and Longyear, V. M. C. (1979). Change from homo- to heterolactic fermentation by *Streptococcus lactis* resulting from glucose limitation in anaerobic chemostat cultures, *J. Bacteriol., 138*:109–117.

Thomas, T. D., McKay, L. L., and Morris, H. A. (1985). Lactate metabolism by pediococci isolated from cheese, *Appl. Environ. Microbiol., 49*:908–913.

Thomas, T. D., and Turner, K. W. (1981). Carbohydrate fermentation by *Streptococcus cremoris* and *Streptococcus lactis* growing in agar gels, *Appl. Environ. Microbiol., 41*:1289–1294.

Thomas, T. D., Turner, K. W., and Crow, V. L. (1980). Galactose fermentation by *Streptococcus lactis* and *Streptococcus cremoris*: pathways, products and regulation, *J. Bacteriol., 144*:672–682.

Thompson, J. (1979). Lactose metabolism in *Streptococcus lactis*: phosphorylation of galactose and glucose moieties in vivo, *J. Bacteriol., 140*:774–785.

Thompson, J. (1987). Regulation of sugar transport and metabolism in lactic acid bacteria, *FEMS Microbiol. Rev., 46*:221–231.

Thompson, J. (1988). Lactic acid bacteria: model systems for in vivo studies of sugar transport and metabolism in Gram-positive organisms, *Biochimie, 70*: 325–336.

Thompson, J., and Chassy, B. M. (1981). Uptake and metabolism of sucrose by *Streptococcus lactis, J. Bacteriol., 147*:543–551.

Thompson, J., and Saier, M. H., Jr. (1981). Regulation of methyl-β-D-thiogalactopyranoside-6-phosphate accumulation in *Streptococcus lactis* by exclusion and expulsion mechanisms, *J. Bacteriol., 146*:885–894.

Thompson, J., Turner, K. W., and Thomas, T. D. (1978). Catabolite inhibition and sequential metabolism of sugars by *Streptococcus lactis, J. Bacteriol., 133*: 1163–1174.

Tournut, J. (1989). Applications of probiotics to animal husbandry, *Rev. Sci. Tech. Off. Int. Epiz., 8*:551–566.

Tseng, C.-P., and Montville, T. J. (1990). Enzyme activities affecting end product distribution by *Lactobacillus plantarum* in response to changes in pH and O_2, *Appl. Environ. Microbiol., 56*:2761–2763.

Tseng, C.-P., Tsau, J.-L., and Montville, T. J. (1991). Bioenergetic consequences of catabolic shifts by *Lactobacillus plantarum* in response to shifts in environmental oxygen and pH in chemostat cultures, *J. Bacteriol., 173*:4411–4416.

Wallbanks, S., Martinez-Murcia, A. J., Fryer, J. L., Phillips, B. A., and Collins, M. D. (1990). 16S rRNA sequence determination for members of the genus *Carnobacterium* and related lactic acid bacteria and description of *Vagococcus salmonarium* sp. nov., *Int. J. Syst. Bacteriol., 40*:224–230.

Whittenbury, R. (1964). Hydrogen peroxide formation and catalase activity in the lactic acid bacteria, *J. Gen. Microbiol., 35*:13–26.

Whittenbury, R. (1978). Biochemical characteristics of *Streptococcus* species, in *Streptococci*. Soc. Appl. Bacteriol. Symp. Series No. 7 (F. A. Skinner, and L. B. Quesnel, eds.), Academic Press, London, pp. 51–69.

Williams, A. M., Fryer, J. L., and Collins, M. D. (1990). *Lactococcus piscium* sp. nov. a new *Lactococcus* species from salmonid fish, *FEMS Microbiol. Lett., 68*:109–114.

Williams, A. M., Rodrigues, U. M., and Collins, M. D. (1991). Intrageneric relationships of enterococci as determined by reverse transcriptase sequencing of small-subunit rRNA., *Res. Microbiol., 142*:67–74.

Woese, C. R. (1987). Bacterial evolution, *Microbiol. Rev., 51*:221–271.

Wolf, G., Strahl, A., Meisel, J., and Hammes, W. P. (1991). Heme-dependent catalase activity of lactobacilli, *Int. J. Food Microbiol., 12*:133–140.

Yang, D., and Woese, C. R. (1989). Phylogenetic structure of the "leuconostocs": an interesting case of a rapidly evolving organism, *Syst. Appl. Microbiol., 12*:145–149.

2

Industrial Use and Production of Lactic Acid Bacteria

Annika Mäyrä-Mäkinen
Valio Ltd., Helsinki, Finland

Marc Bigret
Sanofi Bio-Industries, Paris, France

I. INTRODUCTION

Since the early 1900s there has been a marked worldwide increase in the industrial production of cheeses and fermented milks. Process technology has progressed toward increasing mechanization, increase in factory sizes, shortening processing times, and larger quantities of milk processed daily in the factory. Milk can be fermented to more than 1000 products with demands of their own special flavor, texture, and final product quality.

All this is reflected in enormous demands at the starter cultures, their activity, stable quality, and bacteriophage resistance. The art of making cultured food products by using the former day's whey or fermented product for today's process has been changes to a science with exact knowledge of the factors influencing the specific starter species and strains.

During the 1980s many outstanding reviews were published about the metabolism, physiology, genetics, production, and use of starter cultures. Thus, in this chapter the general view of the most important factors concerning starter activity, its effect on the final product, and industrial starter production are discussed. The discussion of industrial production of starters is mostly based on the author's experience.

II. MESOPHILIC AND THERMOPHILIC STARTERS USED BY DAIRY INDUSTRY

The starters used in dairy products can be divided into mesophilic and thermophilic starters by their optimum growth temperature. Mesophilic cultures grow in temperatures of 10–40°C, with the optimum around 30°C. Thermophilic starter cultures have their optimum growth temperature between 40°C and 50°C.

Mesophilic starter cultures, composed of acid-forming lactococci and often of flavor producers, are used in production of many cheese varieties, fermented milk products, and ripened cream butter (Petersson, 1988). Thermophilic starters are used for yogurt and for cheese varieties with high cooking temperatures (Emmental, Gruyère, Comte, Grana).

The starter cultures are usually composed of different species or of several strains of a single species. Differently composed starters can be categorized as follows:

Single-strain starter: one strain of a certain species
Multiple-strain starter (defined strain starter): different known strains of one species
Multiple-mixed-strain starter: different defined strains of different species
Raw mixed-strain starter: species and strains partly or all unknown

The traditional starters, raw mixed-strain starters, are widely used, especially in fermented milk products and ripened cream butter. All the categories are known as cheese starters, and the trend is toward multiple-strain starters. Mesophilic cultures are either raw mixed-strain or multiple-strain cultures in Europe and North America. The multiple strains are used singly, in pairs, or in multiples, and their use has been pioneered in New Zealand by Whitehead (Whitehead, 1953) to avoid open-texture defects in cheddar cheese caused by flavor-producing organisms in mixed cultures. The defined strain cultures have led also to the understanding of strain interaction with phages (Lawrence and Pearce, 1972). During the 1970s the big cheese plants started to use defined phage-insensitive strain systems with good success (Heap and Lawrence, 1976; Limsowtin et al., 1977; Lawrence et al., 1978). Since this development the multiple-strain starters have become popular also in Australia (Hull, 1977), United States (Huggins and Sandine, 1979; Richardson et al., 1980; Danielle and Sandine, 1981), and Ireland (Daly, 1983). About 85% of all cheddar cheese was produced with these starters in 1985 (Cogan and Daly, 1987).

The Netherlands a totally different starter system is applied. It is based on the noticed difference in phage sensitivity between the starters propagated in the laboratory ("L cultures") and in practice ("P cultures"). Under aseptic conditions the L cultures became dominated by phage-sensitive strains, while P cultures propagated in nonaseptic dairy conditions contained with the optimum balance between phage-sensitive and -insensitive strains. The cultures used in practice are P cultures kept in The Netherlands Dairy Research Institute and distributed to the cheese factories for large-scale production (Stadhouders and Leenders, 1984).

Thermophilic cheese cultures can also be divided into two categories: raw mixed starters and defined strain starters with multiple or single strains. Raw mixed cultures are widely used in traditional cheese making in Switzerland, France, and Italy. The mixed cultures contain *Str. thermophilus* and different species of *Lactobacillus: L. helveticus, L. lactis, L. bulgaris, L. fermentum*, and *L. acidophilus*. The raw mixed cultures can be natural whey cultures or produced by macerating dried calf vells in the previous day's cheese whey. In spite of the use of these unknown mixed cultures the need of knowing the exact properties of starter strains is well recognized in Switzerland. During the 1970s defects of secondary fermentation occurred in Emmental cheese, caused by starters that were too proteolytic, which stimulated propionic acid fermentation and production of CO_2 (Steffen, 1980a). After research on this defect the variety of starter cultures has decreased, and they are distributed weekly to cheese factories by the Research Institute (Steffen, 1980b).

Single-strain starters containing *Str. thermophilus* and *L. helveticus* have been developed in France (Rousseaux et al., 1968) and used for Gruyère and Emmental cheese as frozen concentrated cultures for bulk starter or direct inoculation of cheese milk. Single-strain starters are also used in Finland for Emmental cheese, starting in the 1930s with *Str. thermophilus* strains, and by the 1950s single strains of *L. helveticus* and propionic acid bacteria were also used throughout the production. By the decrease of natural flora in raw milk during the 1970s and 1980s, it was noticed that more species of lactobacilli and propionic acid bacteria were needed for Emmental cheese to accelerate the ripening time and to improve the taste and eye appeal. Consequently, multiple-strain starters have been developed during the 1980s for Finnish Emmental cheese (Mäyrä-Mäkinen, unpublished data).

A. Mesophilic Species and Types of Starters

The species composition of most mesophilic starters include *Lactococcus lactis* ssp. *lactis, Lactococcus lactis* ssp. *cremoris, Lactococcus lactis* ssp. *lactis* var. *diacetylactis, Leuconostoc lactis*, and *Leuconostoc cremoris*. *Lc. lactis* ssp. *lactis* and *Lc. lactis* ssp *cremoris* are acid-producing microorganisms. When starter cultures contain only these species, they are characterized as O type. *Lc lactis* ssp. *lactis* var. *diacetylactis* and *Leuconostoc* sp. are citric-acid-fermenting bacteria. When the only citrate-fermenting species present is *Lc. lactis* ssp. *lactis* var. *diacetylactis*, the culture is of D type. When the *Leuconostoc* sp. is the only aroma producer, the culture is of B (or L) type. When both aroma-forming species are in the culture, it is called BD (or LD) type.

Several combinations of single or multiple strains of lactic acid bacteria are currently used in cheese making. The various starter systems used in the dairy industry are either mixed cultures or in which the composition of the mixture is not defined or cultures containing well-defined single or multiple strains.

1. *Dutch.* Mixed cultures, coming from dairies or butter plants, are propagated, without isolation, in order to keep a composition as close as possible of the original culture. When those cultures are propagated under asceptic conditions, very few bacteriophages attacks are unnoticed.
2. *New Zealand.* This system is used in many Anglo-Saxon countries and utilizes multiple-strain cultures. These cultures are composed of a small number of defined strains. Either the same culture, containing two to six strains, is used alone for a long time or several cultures are used in rotation in order to prevent bacteriophage attacks. In the latter case, the cultures have to have different bacteriophage sensitivity profiles.
3. *Australian.* This system consists of a single strain, replaced as soon as possible in case of a bacteriophage attack. From the sensitive strain screened in the factory, a secondary resistant strain is derived to replace the original strain. A combination of these two last systems has been successfully used in the United States and Ireland by selecting secondary resistant strains and including them afterwards in multiple-strain cultures.

B. Thermophilic Species and Types of Starters

The thermophilic organisms belong to two genera: *Lactobacillus* and *Streptococcus. Lactobacillus* is a large genus with over 50 both homo- and heterofermentative species. Only a few of these are involved in milk fermentations. *Lb. delbrueckii* spp. *lactis* and *Lb. helveticus* are the starter lactobacilli for cheeses with high cooking temperatures, and *Lb. delbrueckii* ssp. *bulgaricus* (formerly known as *Lb. bulgaricus*) is a component in yogurt together with *Str. salivarius* ssp. *thermophilus. Lb. acidophilus* is of intestinal origin and is widely used in different kind of milk products, because of its believed beneficial effects on human and animal health.

Lactobacilli are used in combination with *Str. thermophilus*. This combination is naturally selected because of the high temperatures used in the fermentation of certain cheeses and yogurt. *Lb. lactis, Lb. bulgaricus*, and *Str. thermophilus* do not metabolize galactose, and thus lactose metabolism by *Str. thermophilus* results in the galactose accumulation in the medium (Thomas and Crow, 1984). For this reason only the galactose-fermenting lactobacilli should be used as starters together with *Str. thermophilus* (Turner and Martley, 1983).

A symbiotic relationship exists between *S. thermophilus* and *Lb. delbrueckii* ssp. *bulgaricus*. *Lb. bulgaricus* stimulates *S. thermophilus* by releasing amino acids, while the latter produces formic-acid-like compounds, which promote the growth of *Lb. bulgaricus*. Many outstanding reviews have been written in this area describing the symbiosis in detail (Driessen et al., 1982; Pette and Lolkema, 1950; Botazzi et al., 1971; Bracquart and Lorient, 1979). Some data has been published also about the symbiotic relationship existing between *S. thermophilus* and *Lb. helveticus*, but the actual compounds involved are not known yet (Accolas et al., 1971).

The lactobacilli with lower optimum growth temperatures, *Lb. casei* and *Lb. plantarum*, grow in cheese as natural contaminants. Some strains of *Lb. casei* produce diacetyl from citrate, but this species is only used as a starter by the Japanese in making fermented milk, yakult.

C. Starter Function

Among the physiological functions of lactococci are several of great importance in cheese manufacturing and maturation, influencing the final organoleptic qualities of the cheese:

Fermentation of sugars, leading to a pH decrease is important for the clotting phenomenon and reduction or prevention of the growth of adventitious microflora.

Protein hydrolysis, which causes the texture and, partially, taste of the cheese.

Synthesis of flavor compounds.

Synthesis of texturizing agents, which may influence the consistency of the product.

Production of inhibitory components.

Since the lipolytic activity of lactococci is very low, it has no major influence on the technology and, consequently, is not further treated here.

1. Acid Production

Lactose is the major fermentable sugar of milk, at a level of 40–50 g/L. The glucose moiety of lactose is used faster than galactose moiety by lactococci. At the end of the growth phase, less than 0.5% of the lactose is used by lactococci (Desmazeaud, 1983). The fermentation product of the lactococci is L(+)-lactic acid.

a. *Sugar Transport Across the Cell Membrane.* The bacterial transport of lactose, glucose, and galactose across the cytoplasmic membranes have been well characterized. Two different mechanisms have been found: the permease and the phosphoenol pyruvate–phosphotransferase (PEP/PTS) systems. The permease system is found in thermophilic species and the leuconostocs, and PEP/PTS in the lactococci. Lactose is transported via PTS. This system, composed of two enzymes, and a soluble factor together with a thermoresistant protein, is PEP-dependent. In the PEP/PTS system lactose is transported into the cell via a complex system by which lactose is phosphorylated to lactose phosphate and thus transported across the cell wall. The lactose phosphate is hydrolyzed to glucose and galactose-6-P by phospho-β-galactosidase (p-β-gal). Lactose can also be transported by a permease system. This system, which requires energy, uses the ATP of the cell. Inside the cell lactose is hydrolyzed to glucose and galactose by β-galactosidase (β-gal). There are relatively few studies on the transport of lactose by thermophilic cultures. Contradictory results have been published on the transport of lactose in *Str. thermophilus*. Both permease and PEP/PTS systems (Hemme et al., 1980) and only permease (Tinson et al., 1982) have been suggested. Lactobacilli possess more often β-gal than p-β-gal (Premi et al., 1972), implying that permease is the most important transport system with them.

b. *Carbohydrates Catabolism.* After transportation, the sugars can be either lactose phosphate, glucose phosphate, galactose phosphate, or corresponding free sugars. These molecules can be metabolized by three different pathways. The lactose phosphate is hydrolyzed by a β-gal to give glucose and galactose-6-phosphate. Then, the glucose moiety is catabolized through the glycolysis pathway (Embden-Meyerhof-Parnas pathway), while galactose-6-phosphate is metabolized along the D-tagatose-6-phosphate pathway. Galactose is used by the Leloir pathway. The carbohydrate metabolism is controlled by both repression and retroinhibition. The repression is a mechanism which controls the enzyme synthesis, and the retroinhibition the enzyme activity.

2. Proteolytic Activity

All the starter culture species are nutritionally very fastidious requiring many amino acids and growth factors for adequate growth. Lactic acid bacteria are only mildly proteolytic compared to, e.g., *Bacillus* and *Pseudomonas*. The lactic acid bacteria utilize the polypeptides generated by milclotting enzymes and by bacterial cell-wall proteinases, and therefore are responsible for the casein degradation. The combined action of proteinases and peptidases provides the cells with peptides and free amino acids. Then peptides and amino acids are transported across the membrane, with specific transport systems. The internalized peptides are hydrolyzed by cytoplasmic peptidases.

a. *Proteinases.* All the milk proteins including whey proteins are available for hydrolysis by starter strains. It could be expected that more accessible proteins in the casein micelle, e.g., K- and β-casein, are hydrolyzed before α_s-casein (Thomas and Mills, 1981). This has been shown in lactococci. *Lb. helveticus* has been shown to attack α_s-casein and partly β-cassein, and *Lb. bulgaricus* degrades all the major caseins, β-casein being most susceptible (Ezzat et al., 1985; Chandan et al., 1982). The proteolytic activity of *Str. thermophilus* is lower than that of lactococci and does not affect the casein hydrolysis, for instance, in cheese (Thomas and Pritchard, 1987).

The proteinases of lactococci, involved in the first step of casein breakdown, are high-molecular-weight proteins are located primarily in the cell wall. Their optimum pH is around 5.5 to 6.5, their isoelectric point is between 4.4 and 4.55, and they are either activated or stabilized by Ca^{2+} ions (Kok, 1990). Several studies describe three cell-wall associated proteinases. One of them is supposed to be responsible for the bitterness of cheese. These observations suggest an active role for the lactococcal proteinases in producing bitter peptides of medium size (tri- to hexapeptide) during cheese maturation (Lemieux and Simard, 1991). The identification and characterization of cell-wall associated proteinases are under investigation. So, it is likely that the known number of different proteinases may change, due to improved knowledge of their specificity and their genetic background.

As spontaneous irreversibly proteinase-negative variants appear with a high frequency, the involvement of plasmid DNA has been studied. Plasmids ranging in size from 13.5 to 100 kilobases (kb) are involved in proteinase production. Recent studies on proteinase localization have shown that they are attached to the cell wall by an "anchor" present at the C terminal end of the protein. The removal of this anchor results in the release of the proteinase (Kok, 1990). In addition, it seems that a membrane-located lipoprotein is involved in the activation of the proteinase. The role of this system, under the control of a gene called *prt M*, is not yet understood.

b. *Peptidases.* The casein degradation, initiated with a milk-clotting enzyme, and proteinases, which produce large peptides, continues with peptidases, which produce smaller peptides and amino acids. A number of peptidases has been described. Aminopeptidase, di- and tripeptidase, an arylpeptidyl-amidase, aminopeptidase P, proline-iminopeptidase, prolinase and prolidase, X-prolyl-dipeptidyl aminopeptidases, endopeptidases, and carboxypeptidases have been found in various lactococci.

Unfortunately, the various studies can hardly be compared, because the methods used by the authors differ. A majority of those peptidases are metal enzymes. It has been suggested that citrate and other carboxylic acids affect peptidase activity (Olson, 1990). This is supported by the fact that in cheese, such as Havarti, where BD cultures are used, the amino acid nitrogen level is higher than in the control cheese prepared only with *Lc. lactis* ssp. *lactis* and *Lc. lactis* ssp. *cremoris*. Lactobacilli exhibit a wide range of peptidase activity. Aminopeptidase activity is especially high in *Lb. helveticus* (Ezzat et al., 1986), but dipeptidase and caseinolytic activities do not vary so much between *Lb. helveticus-*, *Lb. lactis-*, and *Lb. bulgaricus* strains (Atlan et al., 1989). Several peptidases with broad specificities have been isolated from *Lb. casei NCDO* 15 and 2 strains of and *Lb. plantarum* (Ebo-Elnaga and Plapp, 1987). Being natural contaminants in cheese, these species might have a considerable effect on proteolysis, texture, and taste of cheeses (Puchades et al., 1989). The cellular localization of the peptidases has not been unequivocally assigned yet.

3. Aroma Formation

The flavor compounds produced by lactococci can be divided in two categories: the compounds in products of fermented milk, and the compounds present mostly in matured cheeses. The aroma compounds in fermented milk are mainly organic acids, lactic and acetic acid, produced by *Lc. lactis* ssp. *lactis, Lc lactis* ssp. *cremoris*. Secondly, *Lc. lactis* ssp. *lactis* var. *diacetylactis* and *Leuconostoc* ssp. produce acetaldehyde, diacetyl, acetoin, and 2–3 butylene-glycol from the citrate of the milk. It has been suggested that these aroma compounds are produced to avoid pyruvate accumulation in the cell. The pyruvate metabolized from citrate by a citrate lyase (or citritase) is toxic to the cell when its intracellular concentration is too high. The pyruvate is degraded in the cell in the presence of Mg^{2+}, Na^{2+}, and thiamine pyrophosphate. The lactococcal citrate metabolism pathway consists of enzymes the genetic determinant are in plasmids. Indeed, the studies carried out on this topic, have shown that citrate-negative variants of a citrate have always lost a 5.5-Mdal plasmid (Desmazeaud, 1983).

The impact of lactococci on the production of flavor compounds in ripened cheese is much more difficult to determine (Olson, 1990). One of the reasons for this is that the lactococci play an indirect role in cheese flavor production. Their peptidases generate di- or tripeptides and free amino acids, which are further metabolized to volatile compounds. No direct relationship has been established between the amino acid nitrogen and cheese flavor, even though it is known that the former influences the latter. Secondly, some key flavor compounds are present at a concentration of some ngs, and analytical methods are inadequate. Neverthe-less, improved knowledge of proteolysis and peptidolysis in the cheese, analysis on enzymatic systems of lactococci, and evaluation of different strains used in cheese production might allow us to establish a better correlation between the lactococcal activity in cheese and flavor development.

The aroma compounds in Swiss cheese have been reported to be produced by reactions between dicarbonyls and amino acids (Griffith and Hammond, 19898). The dicarbonyls, glyoxal, methylglyoxal, dihydrooxyacetone, and diacetyl have been found in Swiss, mozzarella, and cheddar cheeses and in cultures of *Lb. bulgaricus, Lb. casei, S. thermophilus*, and *Propionibacterium shermanii* (Bednarski et al., 1989). Many varieties of cheese contain these species.

4. Exopolysaccharide Formation (Ropiness)

Many strains of lactic acid bacteria produce exopolysaccharides (EPS). The form of EPS can be as a capsule, closely attached to the bacterial cell, or loosely attached or excreted slime (Sutherland, 1977). This phenomenon is only shortly handled here, a more comprehensive review has been recently published by Cerning (1990).

During the last 10 years utilization of slime-forming lactic acid bacteria has been used more widely in the dairy industry. This property has been utilized in Finland since the last century especially in production of a thick, viscose fermented milk product "villi." The starters of this product contain mesophilic, slime-forming lactococcal strains together with aroma-producing lactococci and leuconostocs. At the end of the 1980s, the production of thermophilic, viscose yogurt starter cultures has become more common and they are widely used to increase the rheological quality of yogurt and to prohibit the syneresis of the coagulum. These viscose starters are used, in some cases, to replace stabilizers in yogurt.

There have also been some attempts to investigate the antitumor activity of slime-forming lactic acid bacteria (Forsén et al., 1987; Oda et al., 1983). The role of EPS in this phenomenon has to be elucidated in further studies.

Production of EPS in mesophilic lactococci has been shown to be plasmid encoded (Neve et al., 1988; Vedamuthu and Neville, 1986; von Wright and Tynkkynen, 1987). This may explain the instability of the slime production especially in higher temperatures. However, viscose character is also unstable in thermophilic starter strains, although they do not contain plasmids (Vescovo et al., 1989; Cerning et al., 1990).

EPS-forming bacteria are often considered to be more resistant to bacteriophages than the nonencapsulated ones. This is not the case among mesophilic lactococci, as these viscose strains are hosts for many phages (Saxelin et al., 1986), and a certain phage can also dissolve the capsular material of even nonhost strains (Saxelin et al., 1979).

The chemical composition of EPS of mesophilic and thermophilic lactic acid bacteria seems to vary very much from strain to strain. All of them have been shown to contain galactose and glucose, sometimes hexose-like components and rhamnose (reviewed by Cerning, 1990). In some isolated capsular materials also protein was found, but the amino acid composition of this protein was similar to that of the whey (Macura and Townsley, 1984). In a recent paper by Toba et al. (1991) isolated glucose, galactose, rhamnose, glycerol, and phosphorus from a capsular polysaccharide of *Lactococcus lactis* ssp. *cremoris*. It was supposed to be a deacylated lipoteichoic acid. Also Nakajima et al. (1990) isolated a phosphorus-containing polysacchride that contained rhamnose, glucose, and galactose but not glycerol.

5. *Production of Inhibitory Components*

The observation that lactic acid bacteria have some preserving effect dates back to the turn of the last century. According to the early research the organic acids from sugar fermentation were responsible of the good keeping quality of fermented foods. Thus reduction of pH and production of organic acids (lactate, acetate) are the primary inhibitory actions by these bacteria. Very few other bacteria are able to grow at pH values achieved by the action of lactic acid bacteria.

Lactic acid bacteria produce also other inhibitory substances although in much smaller amounts. These include hydrogen peroxide, diacetyl, bacteriocins, and secondary reaction products such as hypothiocyanate generated by the action of lactoperoxidase on hydrogen peroxide and thiocyanate. Since many reviews have been written on this topic and there is one chapter in this book about the subject, only general remarks are presented here.

Hydrogen peroxide is generated by different mechanisms by certain lactobacilli during the growth (Daeschel, 1989), and accumulation of hydrogen peroxide in growth media can occur because lactobacilli do not possess the catalase enzyme (Kandler and Weiss, 1986). Antagonistic effect has been demonstrated against *Staphylococcus auerus* (Dahiya and Speck, 1968) and *Pseudomonas* spp. (Price and Lee, 1970).

The second system of inhibition, attributed to lactic acid bacteria and linked to hydrogen peroxide production, is the lactoperoxidase system (LPS) (Piard and Desmazeaud, 1991). To make this system efficient, certain components have to be present in milk. An enzyme, the lactoperoxidase, reacts with two substrates: thiocyanate and hydrogen peroxide. The concentration of lactoperoxidase in milk is 10–30 μg/mL. The thiocyanate (SCN^-), widely distributed in animal secretions, is detected in milk at a concentration varying from 1 to 10 ppm. Hydrogen peroxide (H_2O_2) is produced in the milk by lactic acid bacteria, even at low temperatures. Hydrogen peroxide can react with thiocyanate in a reaction catalyzed by lactoperoxidase to form oxidation products, hypothiocyanate, an oxidation product which inhibits microorganisms (Björck, 1985).

Diactyl is known as butter aroma but it is also well recognized for its antimicrobial action. The inhibitory level by Jay (1982) is 200 μg/mL for yeasts and Gram-negative bacteria and 300 μg/mL for nonlactic Gram-positive bacteria. A relatively large amount is needed for inhibitory action, and thus the use of it in foods may be problematic.

Bacteriocins are a heterogeneous group of antimicrobial substances in respect to producing bacteria, antibacterial spectrum, mode of action, and chemical properties (Daeschel, 1989). Bacteriocins of lactic acid bacteria have been the subject of wide research during past decades. The bacteriocins are defined as follows (Juillard et al., 1987):

They generally have a narrow range of action.
Part of the molecule is a peptide and therefore they are sensitive to proteases.
They are thermostable.

Bacteriocins produced by lactobacilli have been characterized from *Lb. fermentum* (De Klerk and Smit, 1967), *Lb. helveticus* (Upreti and Hindsdill, 1975), *Lb. acidophilus* (Muriana and Klaenhammer, 1987), and *Lb. plantarum* (Daeschel et al., 1986).

The two bacteriocins produced by lactococci, nisin and diplococcin, are well characterized. Nisin was found in 1928 by Roger and Whittier. The first application in Swiss cheese was done 1951 by Hirsch et al., and it was found to be effective in preventing blowing (butyric acid fermentation) caused by clostridia. Nisin is effective against Gram-positive species and also against *Clostridium botulinum* spores which has made it useful in thermally processed foods. Recent investigations have also shown nisin to be inhibitory toward *Listeria monocytogenes*, a pathogen of great concern nowadays (Harris et al., 1989).

Diplococcin, produced by *L. lactis* ssp. *cremoris* is active only against *L. lactis* ssp. *lactis* and *L. lactis* ssp. *cremoris*. As diplococcin is more active against cells in exponential growth phase than in stationary phase, it is suggested that its targets are both RNA and DNA of the cell (Klaenhammer, 1988).

III. FACTORS INFLUENCING ACTIVITY OF STARTERS

A. Milk as Growth Medium

Even though lactic acid bacteria are able to grow in milk, milk is not an optimal growth medium. For instance, an addition of yeast extract can stimulate the production of lactic acid (Desmazeaud, 1990). Another indication is the daily variation of the milk, which leads to modification of the physiological reactions of the microorganisms. It is well known that the origin of the cows, the geographic location, the stage of lactation are parameters able to explain the variation of the milk components. The average composition of cow's milk is

Water 905 g/L
Lactose 49 g/L
Lipids 35 g/L
Protein 34 g/L
Salts 9 g/L
Others (vitamins, enzymes, etc.) traces

The impact of the lactic acid bacteria on the main components of the milk was discussed when uptake of fermentable sugars, proteins and peptides, citrate, and their corresponding enzymatic systems were described. It has also been noted that lactic acid bacteria have a very limited effect on milk fats.

The remaining components, of great importance in the nutrition of lactic acid bacteria, are, on one hand, the vitamins, and, on the other hand, the nonprotein nitrogen (NPN). The vitamin requirements vary from species to species.

Lactococci require niacine (PP), pantothenic acid (B_5), pyridoxin (B_6), and biotin (H). Thermophilic streptococci require pantothenic acid (B_5) nitroflavin (B_2), thiamine (B_1), niacin (PP), biotin, and pyridoxin. Lactobacilli require calcium pantothenate (B_5), niacin (PP) and nitroflavine (B_2). In addition, *Lb. lactis, Lb. bulgaricus*, and *Lb. acidophilus* require cobalamin (B_{12}).

The nonprotein nitrogen represents 5 to 7% of the total nitrogen in milk. The constitutive molecules of this fraction have an important role in the nutrition of lactic acid bacteria, because of their direct uptake by the cell. The concentration of these components (containing less than eight amino acids) is usually too low to provide the required nutrients. Consequently free amino acids (including free methionine, which is an essential amino acid) are not present in milk at a sufficient level to allow satisfactory growth of the cells (Desmazeaud, 1990).

B. Inhibitory Compounds in Milk

The antimicrobial effect of milk has long been known, but not until 1927 were they identified by Jones and Little as lactenins, and afterwards divided as lactenin 1 (L_1) and lactenin 2 (L_2) by Auclair and Hirsch (1953). These antimicrobial substances were later referred to as red protein, i.e., lactoferrin (Ashton and Busta, 1968). A wide variation of antimicrobial factors have been identified since then, some of them derived form the blood of the cow.

Endogenous or exogenous factors, can affect the starter activity in the starter tank or in the dairy process. These include variations in milk composition caused by mastitis or seasonal changes, agglutinins, dissolved oxygen, free fatty acids, inhibitory bacteria, the lactoperoxidase system, lysozyme, lactoferrin, residual sanitizers, and bacteriocins (Korhonen, 1973; Stadhouders, 1974; Björck et al., 1975; Lawrence et al., 1976; Hull, 1983; Carlsson et al., 1983).

The heating of milk and good manufacturing practices combined with intensive quality control procedures have minimized the effect of most of the factors mentioned above in the dairy processes. Therefore they are not reviewed here in detail. The most important sources of constant problems in the dairy processes are antibiotic residues and bacteriophages. These are discussed further below.

1. *Antibiotic Residues*

Antibiotics are the most important group of exogenous inhibitory factors because of their common use in the treatment of mastitis in dairy cows. The antibiotics were introduced in the 1940s for mastitis therapy. Today mastitis is still the most serious problem affecting cows, and causes huge economic losses to dairies. There are also several health aspects associated with the problem, e.g., allergic reactions, intestinal disorders, and development of resistant bacteria (Dewdney, 1977; Dewdney and Edwards, 1984).

To avoid the residues in milk the manufacturers of the veterinary products are generally compelled to state a withdrawal period for the specific product. On the other hand, individual differences between cows are known to exist and thus general withdrawal times are not always valid. Strict penalty schemes with improved testing systems have reduced the residue levels significantly. The number of antibiotics used is huge and still increasing varying from country to country (Ziv, 1980). The most widely used group is b-lactam antibiotics and their derivatives either alone or in combinations. Other common antibiotic groups are aminoglycosides, tetracyclines, macrolides, and sulfa drugs (Archimbault, 1983; Carlsson, 1991).

The levels of antibiotics required to inhibit different starter strains are very strain dependent (Tamime and Robinson, 1985; Cogan and Daly, 1987). This can be seen in Table 1.

Mesophilic cultures are less sensitive to penicillin and spiramycin and more susceptible to streptomycin and cloramphenicol than thermophilic cultures, but also streptomycin-sensitive *S. thermophilus* starter strains have been found (Mäyrä-Mäkinen, 1990). Little information is available on the levels of antibiotics required to inhibit leuconostocs or propionic acid bacteria used in hard cheeses. In an experiment to test the actual effect of low antibiotic levels on cheese produced by mesophilic (Edam cheese) and by thermophilic (Emmenthal cheese) starters different antibiotics were added to the cheese vat. The results are presented in Table 2.

As a summary it can be concluded that low levels of antibiotics cause different kinds of defects in cheese: off-flavors, uneven texture, uneven eye formation, and tendency to butyric acid fermentation. Effects on propionic acid bacteria could be seen in eye formation of Emmenthal cheese and in brown spot defect caused by streptomycin. The brown-spot defect is especially interesting, since it has been

Table 1 Sensitivity of Thermophilic and Mesophilic Starters to Different Antibiotics

Antibiotic	Starter cultures	
	Thermophilic[a]	Mesophilic[b]
Penicillin IU/mL	0.004–0.01	0.005–0.01
Tetracycline μg/mL	0.3–0.5	0.05–0.2
Streptomycin μg/mL	0.5–5.0	0.5–1.0
Cloramphenicol μg/mL	0.5–1.0	0.2–0.3
Spiramycin IU/mL	0.3–0.5	2.0–4.0

[a]Thirty-two *S. thermophilus* strains tested.

[b]Single strains of *Lactococcus lactis* ssp. *lactis/cremoris/diacetilactis* and three mixed cultures (DL) tested.

Source: Mäyrä-Mäkinen, unpublished data.

Table 2 Effect of Different Antibiotics at Low Concentrations on Edam and Emmenthal Cheese Quality

Antibiotic	Cheese quality	
	Edam[a]	Emmenthal[b]
Penicillin		
0.003 IU/mL	No defects	No defects
0.005 IU/mL	Tasteless	Off-flavor, abnormal eye formation
0.008 IU/mL	Bitter, uneven body	Strong off-flavor, butyric acid fermentation
0.01 IU/mL	Strong off-flavor, uneven body	Not tested
Spiramycin		
1.0 IU/mL	Tasteless, uneven body, wet surface	Smell of butyric acid, severe off-flavor, uneven eye distribution
5.0 IU/mL	Strong off-flavor uneven body, slimy surface	Not tested
Streptomycin		
1.0 μg/mL	Abnormal cheese, strong off-flavor	Brown spots in the body, strong off-flavor
Tetracycline		
0.3 μg/mL	Tasteless	Off-flavor
0.7 μg/mL	Tasteless, strong smell on surface	Not tested

[a]Evaluation at 14 weeks.
[b]Evaluation at 3 months.
Source: Mäyrä-Mäkinen, unpublished data.

known for a long time, but the actual cause of the change in the metabolism or of the effect on the growth of propionic acid bacteria has not been defined. In fermented milk products the effect of antibiotics is seen in slow or inhibited acid formation and also in decrease of aroma formation.

2. *Bacteriophages*

Bacteriophages are viruses, specifically infecting bacteria. After having infected the bacterial cells, they use cell's enzymes to grow. After a while, cells are lysed and bacterial growth is stopped. In the dairy industry, phages of lactic and bacteria are of considerable economic importance, because they represent one of the main causes of fermentation failure. Due to the economical importance, a lot of work has been done to improve knowledge on phage infection and to know more about bacterial phage mechanisms.

a. *Taxonomy.* In order to differentiate the phages, various taxonomic criteria have been proposed. The most important of them, proposed for a taxonomic classification, are host range, morphology, serotyping, DNA/DNA hybridization, and structural protein profiles.

Host Range. On a technological point of view, the host range is of great interest. The strain rotation is more easily handled when the host range of the phages is known. But as this criteria cannot be correlated to others, it is rather a practical parameter than a real taxonomic characteristic.

Morphology. Phages are submicroscopic particles consisting of a head containing the DNA, and a tail. Morphological classification was proposed on the basis of electron microscopic observation. In the case of phages of lactic acid bacteria, various morphologies are observed. The head, with either prolate, small isometric, or large isometric shape, has a size of 40 to 70 nm. The tail measures between 100 and 500 nm.

Serotyping. Antibodies have been prepared against pure phages in order to classify them in various groups. These groups have been compared with host range classes (Mata and Ritzenthaler, 1988). The results of these studies have been variable indicating that the serological criteria are not sufficient to classify the majority of the phages.

DNA/DNA Hybridization and Structural Protein Profiles. It has been possible to determine five groups of homology for a large number of phages of *Lc. lactis* by DNA/DNA hybridization technique (Prevots et al., 1990). The details of these five groups are given in Table 3. Groups I and III contain almost 80% of the phages and include only virulent phages. Group II contains both virulent and temperate changes. The classification based on structural protein profiles allows the grouping of the phages corresponding to the five DNA homology groups (Relano et al., 1987).

b. *Phage Development and Bacterial Resistance.* The phage resistance mechanisms developed by lactic acid bacteria, are correlated to the various steps

Table 3 Classification of Lactococcal Bacteriophages by DNA/DNA Hybridization

DNA homology group	Percent of phages of the group	Head morphology	Genome size (kb)
I	29	Prolate	19–22
II	21	Small isometric	30–40
III	48	Small isometric	30–35
IV	1	Large isometric	53
V	1	Large isometric	134

Source: From Prevots et al. (1990).

of the phage infection. The sequences of this phenomenon are the following. First, the phage recognizes a molecular structure on the cell wall. This receptor allows the phage to adsorb on the cell surface. Secondly, the phage injects its chromosome inside the bacteria. The DNA penetration is Ca^{2+}-dependent and energy-requiring. In the next step, the bacterial DNA is hydrolyzed, and the bacterial metabolism is used by the phage genome to develop new phage particles. In the last step, a lytic enzyme or lysin is synthetized by the phage to make the cells burst. As a rule an average of 100 phages per bacteria are released in the milk.

Several lactic acid bacteria have developed mechanisms to resist phage attack (Sanders, 1988). Most of the known mechanisms are coded by plasmids and, therefore, may be transferred from one cell to an other. Four mechanisms are well described:

Adsorption. Phage-resistant strains have mutated their cell-wall structures, recognized as receptor by phages. The phage fails to adsorpt on cell surface and therefore does not infect the strain.

Restriction–Modification. The restriction-modification system combines restriction enzymes, capable of specific endonucleolytic activity, with a modification enzyme, generally exhibiting specific DNA methylation activity. This methylation protects the DNA from the corresponding restriction enzyme. The unmodified phage chromosome is hydrolyzed by the restriction enzymes as soon as it enters the cytoplasm and, consequently, is degraded.

Abortive Infection. In case of abortive infection, all the phases of the infection occur. But, due to an unknown phenomenon, the burst size is very low. So, each attacked cell releases very few phages in the medium. Little or no disturbance is observed in the growth rate and acidification during cheese making.

Lysogenic Immunity. The lysogenic immunity of lactic acid bacteria is observed when the bacterial chromosome harbors the DNA of a lysogenic phage. The prophage probably codes for molecules which inhibit the development of other related phages and render the strain resistant.

IV. USE OF STARTERS IN THE DAIRY PROCESSES

There has been considerable changes in the cultivation of starters during the past 20 years at the dairy level. The starters available for the dairy processes are sold in different forms by several starter producers. Starters are the most important factors determining the final quality and properties of the product. Therefore the selection of starter type and form for use is very important for the individual dairy plant.

General practical steps in the preparation of starters are discussed here, including also the traditional systems (liquid starters) since they are still in practice beside the increasing use of concentrated starters. The four alternative commercially available forms of starter cultures are described in Fig. 1.

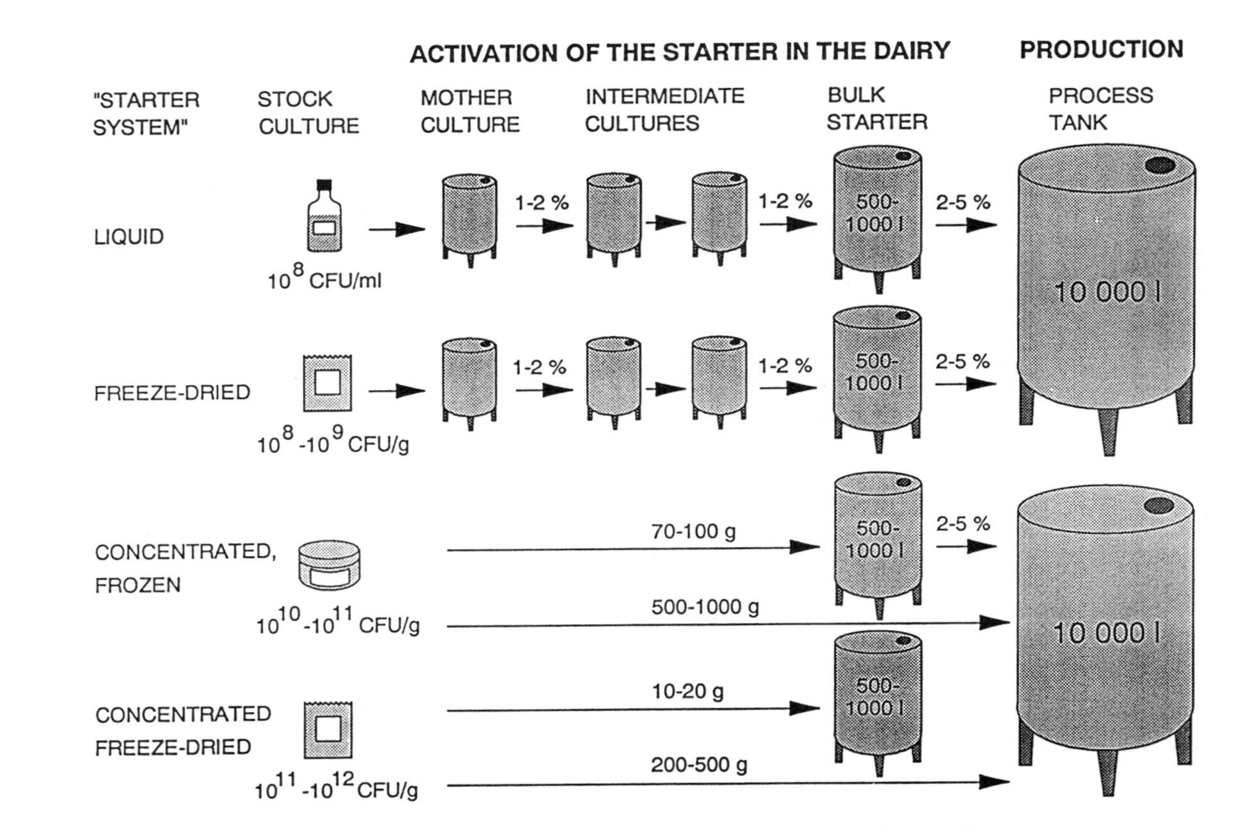

Figure 1 Alternative starter types for the dairy plant and their cultivation steps for the dairy process.

In the traditional liquid starter system the starter is cultivated at first as a liquid stock culture and the sufficient volume is reached by successive subcultures. The stock culture may come weekly from central laboratory or from the plant's own culture. The procedure is thus expensive, laborious, vulnerable, and needs skilled personnel to manage it. The starters can get easily contaminated because of numerous inoculations or get infected by bacteriophages. Still, liquid starters are widely used especially where local special products are made or where the transportation of starters from a central laboratory is easy and regular.

Beside the liquid starters the dairy plant can choose freeze-dried culture for producing mother culture or bulk starter in cases where a small amount of starter is needed. They can be stored for several months at –25°C and thus the plant can use the same production lot for months. In the case of mixed strain cultures the strains are preferred to be freeze-dryed separately in order to avoid the changes in their balance.

The modern systems of concentrated, frozen and concentrated or freeze-dried cultures have made it possible to inoculate directly the bulk starter or the production process itself. This causes significant savings in labor and material costs in the dairy. The production technology of concentrated frozen starters was developed during the 1960s. The use of these starters require low temperatures during shipment and storage. Therefore during 1970 the freeze-drying technique was developed to make their use and transportation still easier.

There are several advantages in the use of concentrated and concentrated freeze-dried starters (Stanley, 1977; Gilliland, 1985): easy to use, the quality is even, the activity is good and can be tested prior to use, its use requires less labor and is easily adapted to five-day production week, and bacteriophage control is easier to manage.

On the other hand, there are also disadvantages (Gilliland, 1985). The shipment of frozen starters is precarious since the temperature changes affect the starter activity, the storage temperature is critical, and thus the activity has to be controlled in the dairy plant. The selection of starters with respect to quality of the final product is not carried out by the dairies but by the starter producers. Not all the good traditional milk starters are suitable for production as concentrated freeze-dried starters. Despite these disadvantages starter research is strongly concentrated on the production techniques and strain selection in order to produce active, directly-to-vat starters.

V. STARTER CULTURES FOR FERMENTED MEAT AND VEGETABLE PRODUCTS

The growth of lactic acid bacteria in milk to produce fermented dairy products is based on a few simple principles: lactic acid bacteria are present, among other bacteria, in several niches. When the physical conditions (such as temperature, AW, pH) allow growth, there is a competition between the various species. The faster development of lactic acid bacteria and the pH decrease by acid production lead to microbiologically stable fermented products. These basic phenomenona of bacterial ecology have also been used to produce meat products such as sausages and fermented vegetables.

For sausage preparation, beside the meat slurry, fermentable sugars, salt and spices, pediococci, such as *P. acidilactici, Lb. plantarum*, and/or *Staphylococcus carnosus* are generally inoculated. The first incubation period (time and temperature are adapted to the various technologies) allows the bacteria to grow, and a pH decrease is measured. After that sausages are cooled and ripened to obtain the final product.

The widely available fermented vegetables are sauerkraut, pickled cucumbers, and green olives. Most of the time, cabbage, cucumbers, or olives are fermented with *Lb. plantarum*. Also *P. cerevisiol, P. pentosaceus, Leuconostoc mesenteroides*, or *Lb. brevis* can be used (Boisen, 1981). During the preparation of these products, NaCl brine is added, the combination of NaCl and lactic acid makes the fermented vegetables stable for a long period of time, at room temperature. The salinity also controls the growth of lactic acid bacteria, thereby influencing the quality of the end product.

VI. PRODUCTION OF STARTERS IN INDUSTRIAL SCALE

The interest on producing concentrated frozen starters arose during 1960s with research on mesophilic single strains and especially cheddar cheese starters. Kosikowski (1966) and Bergere and Hermier (1968) were the first to use neutralization to increase the bacteria count in fermentations. This can be considered as the basis to increase the bacterial counts about 100-fold in the concentrated cultures.

Methods to produce concentrated cultures differ in several ways from the traditional ones. Starter strains are grown under strictly controlled conditions in medium from which the cells are easily harvested into a smaller volume (Gilliland, 1985). The process is very stressful for the starter strains and new kind of selection criteria has to be used. Only about 25–50% of traditionally used strains are suitable for production of concentrated, freeze-dried cultures (Priidak and Bannikova, 1982).

Starter concentrate production can be divided into the following general steps which have been described in detail by Porubcan and Sellars (1979), Gilliland (1985), and Tamine and Robinson (1985):

Preparation of the inoculum
Preparation of the media

Fermentation at constant pH
Harvesting the culture
Adding the cryoprotectant
Freezing
Freeze-drying
Packing and storing

A. Fermentation of Starter Cultures

To produce active and storage stable concentrated cultures several strain-dependent details have to be checked concerning growth medium and growth conditions. The production in the industrial scale of cultures are batch fermentations which are simpler and more convenient comparing to continuous fermentation processes. There are several problems with using continuous fermentations: undesirable contaminations are possible, complex equipment is associated with difficulties in production schedules, and development of bacteriophage problems have also been reported (Lloyd and Pont, 1973).

1. Growth Medium

To choose the growth medium the following aspects have to be considered: costs, ability to produce high number of cells (about 10–15 times higher cell densities that in liquid milk culture) with high activity, and effect on the harvesting methods (Stanley, 1977). It is generally accepted that some milk solids should be included in the growth medium to ensure the synthesis of necessary enzymes for starters to perform well in milk and also to maintain balance between strains in multiple-strain starters when fermented as mixed culture (Stadhouders et al., 1969; Gilliland, 1971).

The use of cheese whey or whey permeate as a growth medium has been investigated primarily because they are inexpensive and contain nutrients which are used by starter strains for growth. Whey alone is not rich enough for maximum growth, and much of the research has involved supplementing whey with extra nutrients. In addition, improper heating causes precipitation in the medium, and clarification might be necessary to avoid this material in concentrated starter culture (Lloyd and Pont, 1973; Mitchell and Gilliland, 1983).

Skim milk has been the most common medium for lactococci, and the difficulties in harvesting have been solved by adding sodium citrate to solubilize milk proteins (Stadhouders et al., 1969).

A variety of different nutrients are needed for starter strains in the culture medium. Porubcan and Sellars (1979) divided them into four groups, and the first group of complex nutrients—skim milk, whey, yeast extract, and peptones—are used to satisfy the complex demands of starters as long as there are no ways to detect the exact growth factors during the fermentation process.

With certain nutrients or additives the resistance of cultures against subsequent concentration/freezing/drying process can be improved. The methods are strain-dependent. The activity and the bacterial count of *L. bulgaricus* after freezing at –196°C can increase considerably if the cells have been grown in an appropriate medium supplemented with Tween 80 (Smittle et al., 1972). The same results have been reported for cultures frozen at –17°C. Oleic acid in Tween 80 was identified as the effective component to increase process stability by raising the levels of a C_{19} cyclopropane fatty acid in the lipid fraction of cells (Goldberg and Eschar, 1977). Most fatty acids of bacterial cells are located in the cell membrane, and it can be assumed that the membrane composition is very important for the cells to survive freezing, maybe by increasing the flexibility of the membrane (Gilliland, 1985). It has also been noticed that lactococci survive freezing at –196°C better than lactobacilli, regardless of the growth medium. This can be caused by the fact that lactococci naturally contain higher levels of C_{19} cyclopropane fatty acid (Gilliland and Speck, 1974). The ratio of unsaturated to saturated fatty acids in cell membranes also seems in some cases to be related to the ability of lactococci and lactobacilli to survive freezing. By increasing the ratio the survival at –17°C has improved (Goldberg and Eschar, 1977).

The addition of 100 μmM–1 mM calcium has been shown to influence the freezing resistance of lactobacilli and change the cell morphology from long chains to short individual cells. Manganese and magnesium did not have this effect (Wright and Klaenhammer, 1983a,b).

2. *Growth Conditions*

The conditions during the fermentation affecting the growth and activity of cultures are temperature, pH, mixing (oxygen content), and type of neutralizer used. Also the optimum cooling/harvesting time relative to the growth curve has to be considered.

Usually the optimum growth temperature of the species is used in the fermentation (Pont and Halloway, 1968; Speckman et al., 1974), but few research results have been published on the effect of temperature on process stability. When growing mixed cultures Bauman and Reinbold (1966) reported better freezing stability at –20°C when a temperature of 32°C had been used during fermentation. Especially with thermophilic cultures it has been noticed that, compared to the optimum growth temperature, a decrease or increase of the growth temperature strain dependence affects the dechaining of cultures and thus also process stability in freezing and freeze-drying (Mäyrä-Mäkinen, unpublished data).

Maintaining the pH of the growth medium at the optimum level of the culture increases the number of cells (Peebles et al., 1969; Petterson, 1975a; Gilliland, 1976). Both the optimum pH area of the culture and type of neutralizer used are of importance. Most of the work has been done with lactococci, and with them the optimum pH is in the range of 6 to 6.5 (Lloyd and Pont, 1973; Cogan et al., 1971; Efstathiou et al., 1975). Ammonium hydroxide seems in general to be the best neutralizer in order to achieve higher cell yields in mixed cultures (Lloyd and Pont, 1973) and also more freezing-resistant cultures at –30°C (Jakubowska et al., 1980). The pH optimum of thermophilic lactobacilli, *L. helveticus, L. lactis*, and *L. bulgaricus*, is 5.4 to 5.8, depending on the strain. By pH and temperature optimization the freezing and freeze-drying stability can be influenced. Also with these species higher yields are obtained with ammonium hydroxide as the neutralizer (Mäyrä-Mäkinen, unpublished data).

The cooling and thus the harvesting time at a certain stage of the growth curve are critical for some cultures. With lactococci harvesting is recommended at the end of the logarithmic growth phase. *Str. thermophilus* cultures lose activity fast if cooling and harvesting is postponed to the stationary phase. On the other hand, many lactobacilli can be harvested irrespective of the growth phases without losing activity (Mäyrä-Mäkinen, unpublished data). Increasing the growth seems to be a combination of several factors of which the formation of lactate salts is considered to be the most important (Porubcan and Sellars, 1979). To reach the maximum bacterial level in the fermentation, the factors limiting the growth have to be considered besides the type and amount of nutrients adequate for optimal growth.

In order to maintain constant pH a continuous agitation is needed. As a result oxygen toxicity has been observed in culturing some lactococci (Keen, 1972). Oxygen can also cause the production of hydrogen peroxide by some starter strains, which can be autoinhibitory. By adding catalase or other reducing agents on the accumulation of H_2O_2 can be prevented (Gilliland and Speck, 1969). Other toxic metabolites can be formed in the growth medium; D-leucine formation has been reported in lactococcal cultures by Gilliland and Speck (1968). Sparging carbon dioxide can be an effective way to avoid the toxicity of oxygen and is actually needed for optimal growth of some starters (Pont and Holloway, 1968; Cogan et al., 1971).

It can be concluded that the basis for active, process-stable culture is built during the fermentation by modifying the growth media and conditions so that they are strain-dependent. Especially important factors to be checked for industrial production are

Dechaining effect of certain components in medium
Temperature optimum to produce process-resistant strains (not always the growth optimum)
pH optimum for growth and further processing
Harvesting time at certain point of growth curve
Process resistance as a selection criteria of starter strains

B. Concentration of Fermented Cultures

Centrifugal separation or membrane processes can be used for harvesting cells from the medium. Centrifugation is mostly used in industrial scale because the low viscosity of the medium, the properties of the cells, big cell size, and higher temperatures favor this technique. The temperature is usually kept between 5 and 15°C, depending on the strain (Porubcan and Sellars, 1979).

Very little has been published about concentration of thermophilic lactobacilli. By Porubcan and Sellars (1979) it is difficult to concentrate *L. bulgaricus, L. lactis*, and *L. acidophilus* from milk-based cultures by centrifugation even with the addition of citrate.

Porubcan and Sellars (1979) also reported industrial scale ultrafiltration of lactococci. About 12-fold concentration can be reached by ultrafiltration without any cell damage caused by heat developing during the process.

New microfiltration processes with ceramic filters have been developed during the 1980s, and these might become an important new method, especially in concentrating process-sensitive cultures.

C. Handling of the Concentrate

Cryoprotectants have been used for years to improve the ability of culture concentrates to survive freezing, frozen storage, and freeze drying. Most of the research of freeze-dried cultures has been done with lactococci. The resistance of cultures against the deleterious effect of freeze-drying can be improved by cryoprotectants (Morichi, 1972). A variety of different cryoprotectants are used (Fennema et al., 1975), but the most common ones in the industrial scale are lactose or sucrose (7%), monosodium glutamate (5%), and ascorbate in milk or water base. Glycerol is widely used in frozen cultures, but there seems to be variation among cultures, and to some cultures glycerol is not effective (Stadhouders et al., 1979). Also lactose (7.5%) has been used with good results, but for some cultures again no effect has been seen (Efstathiou et al., 1975). According to several reports it seems that cryoprotectants are not needed if the concentrate is active, freezing is fast (>1°C/s), and storage is at –196°C (Petterson, 1972; Keogh, 1970). Freezing can affect the activity of cultures also strain-dependently. The most efficient and widely used method is fast-freezing in liquid nitrogen in the form of pellets which are easy to use a frozen concentrate or to freeze-dry (Kilara et al., 1976; Accolas et al., 1972). The use of liquid nitrogen is expensive. Therefore in many cases –40°C freezing and storing temperature is also used.

The extensive report of Morichi (1972) concerning the mechanisms and cryoprotectants involved in freeze-drying gives a view of this complex area. Although possible mechanisms of cryoprotection have been proposed by Fennema et al.

(1975), further research is needed to improve the "freeze-drying resistance" of cultures. As mentioned in the previous chapter, the growth medium and culture conditions affect the strain-dependent freeze-drying resistance.

The pH of concentrate affects the activity during storage. The activity of mesophilic mixed culture concentrate with pH 5.2 was lower than that of a concentrate with pH 6.6 after frozen storage (Stadhouders et al., 1971). The optimum concentrate pH for lactobacilli is 5.4–5.8. Even a lower pH does not seem to have any effect on the activity after freeze-drying. The opposite case is *Str. thermophilus* cultures, which lose activity easily if the pH of the concentrate is below the optimum 6.2–6.6 (Mäyrä-Mäkinen, unpublished data).

The activity of a culture can be well maintained if the storage is carried out under recommended conditions. The shelf-life depends on the form of starter culture (frozen or freeze-dried) and the storage temperature: the lower the temperature, the longer the shelf-life. Frozen concentrates are stored at –40°C, and activity stays very good for at least 6 months. Freeze-dried products are stored at -20 to -40° and a short time in a refrigerator does not usually decrease activity. Again there are great differences between the cultures in storage stability.

VII. FUTURE TRENDS

Current research programs indicate future trends of lactic acid bacteria use in fermented products. As explained in this chapter, as far as technology is concerned, the important characteristics of the strains are more and more understood and measured. The studies carried out in this area are mostly done with pure strains. The first direction where knowledge has to be improved is the understanding of the global behavior of a culture composed of several strains. So, studies must focus on the relationship between the constitutive strains of a mixture. The understanding of the symbiotic and inhibitory phenomena between strains is of prime importance for the control of the cultures used in fermented products.

The second main direction followed by researchers is the development of knowledge in the genetic area. These programs aim at implementing new characteristics in technologically interesting strains.

For example, besides acidification, it would be useful to have strains producing bacteriocins in order to manufacture fermented products where raw materials should not be heat treated, such as sausage, silage or vegetable products and where the contaminant level could be reduced by the action of modified lactic acid bacteria. The problems which have to be solved to make these techniques feasible are the following:

To identify the corresponding genetic determinant of certain physiological characteristics

To be able to use methods which allow a stable transfer of the genes
To dispose of vectors, donor and recipient strains of the same species in order to obtain a "GRAS" microorganism
To solve regulatory problems and constraints, as far as genetically modified microorganisms are concerned

REFERENCES

Abo-Elnaga, I. G., and Plapp, R. (1987). Peptidases of *Lactobacillus casei* and *L. plantarum*, *J. Basic Microbiol.*, *27*:123.

Accolas, J-P., Veaux, M., and Auclair, J. (1971). Etude des interactions entre diverses bactéries lactiques thermophiles et mésophiles, en relation avec la fabrication des fromages á pate cuite, *Le Lait*, *51*:249.

Archimbaut, P. (1983). Persistence in milk of active antimicrobial intramammary substances, in *Proc. 2nd Symp. European Assoc. Veterinary Pharmacology and Toxicology*, Toulouse (V. Ruckebusch, P.-L. Toutain, and G. D. Koritz, eds.), MTP Press, Lancaster.

Ashton, D. H., and Busta, F. F. (1968). Milk components inhibitory to *Bacillus stearothermophilus*, *J. Dairy Sci.*, *51*:842.

Atlan, E., Laloi, P., and Portalier, R. (1989). Isolation and characterization of aminopeptidase-deficient *Lactobacillus bulgaricus* mutants, *Appl. Environ. Microbiol.*, *55*:1717.

Auclair, J. E., and Hirsch, A. (1953). The inhibition of micro-organisms by raw milk. The occurrence of inhibitory and stimulatory phenomena. Method of estimation, *J. Dairy Res.*, *20*:45.

Bauman, D. P., and Reinbold, G. W. (1964). Freezing of lactic cultures, *J. Dairy Res.*, *49*:259.

Bednarski, W., Gedrychowski, L., Hammond, E. G., and Nikolov, Z. L. (1989). A method for determination of α-dicarbonyl compounds, *J. Dairy Sci.*, *72*:2474.

Bergere, J.-L., and Hermier, J. (1968). La production massive de cellules de Streptocoques lactiques. II. Croissance de "Streptococcus lactis" dans un milieu A pH constant, *Le Lait*, *13*:471–472.

Björck, L. (1985). The lactoperoxidase system, in *Natural Antimicrobial Systems*, IDF, Brussels, pp. 18–30.

Björck, L., Rosen, G.-G., Marshall, V., and Reiter, B. (1975). Antibacterial activity of the lactoperoxidase system in milk against *Pseudomonas* and other Gram-negative bacteria, *Appl. Microbiol.*, *30*:199.

Boisen, H. (1981). Selection of bacterial strains for the food industry and their applications in barious food products, *IFST Annual Symp.*

Botazzi, V., Battistotti, B., and Vescovo, M. (1971). Continuous production of yogurt cultures and stimulation of *Lactobacillus bulgaricus* for formic acid, *Milchwissenschaft*, *26*:214.

Bracquart, P., and Lorient, D. (1979). Effet des acides amines et peptides sur la croissance de *Streptococcus thermophilus*. III. Peptides comportant Glu, His et Met, *Milchwissenschaft*, *34*:676.

Carlsson, Å. (1991). Detection of inhibitory substances in milk, Thesis report 6, Swedish University of Agricultural Sciences.

Carlsson, J., Iwami, Y., and Yamada, T. (1983). Hydrogen peroxide excretion by oral streptococci and effect of lactoperoxidase-thiocyanate-hydrogen peroxide, *Infect. Immun., 40*:70.

Cerning, J. (1990). Exocelular polysaccharides produced by lactic acid bacteria, *FEMS Microbiol. Rev., 87*:113–130.

Cerning, J., Bouillanne, C., Landon, M., and Desmazeaud, M. J. (1990). Comparison of exocellular polysaccharide production by thermophilic lactic acid bacteria, *Sci. Aliments, 10*:443–451.

Chandan, R. C., Argyle, P. J., and Mathison, G. E. (1982). Action of *Lactobacillus bulgaricus* proteinase preparations on milk proteins, *J. Dairy Sci., 65*:1408.

Cogan, T. M., Buckley, D. J., and Condon, S. (1971). Optimum growth parameters of lactic streptococci used for the production of concentrated cheese starter culture, *J. Appl. Bacteriol., 34*:403.

Cogan, T. M., and Daly, C. (1987). *Cheese Starter Culture. Cheese: chemistry, physics and microbiology*. (P. F. Fox, ed.), vol. 1, p. 179.

Daeschel, M. A. (1989). Antimicrobial substances from lactic acid bacteria for use as food preservatives, *Food Technol., 43*:164.

Daeschel, M. A., McKenney, M.C., McDonald, L. C. (1986). Characterization of a bacteriocin from *Lactobacillus plantarum*, Abstracts 86th Ann. Meeting, Am. Soc. Microbiol, p. 277.

Dahiya, R. S., and Speck, L. L. (1968). Hydrogen peroxide formation by lactobacilli and its effect on *Staphylococcus aureus, J. Dairy Sci., 51*:1568.

Daly, C. (1983). The use of mesophilic cultures in the dairy industry, *Antonie van Leenwenhoek, 49*:297.

Danielle, S. D. and Sandine, W. E. (1979). Commercial use of a multiple strain starter of known composition in cheddar cheese manufacture, *J. Diary Sci., 62*:70.

De Klerk, H. C. and Smit, J. A. (1967). Properties of a *Lactobacillus fermenti* bacteriocin, *J. Gen. Microbiol., 48*:309.

Desmazeaud, M. (1983). L'etat des connaissances en matiere de nutrition des bacterins lactiques, *Le Lait, 63*:267–316.

Desmazeaud, M. (1990). Le lait: milieu de culture, *Microbiol., Aliment., Nutr., 8*: 313–325.

Dewdney, J. M. (1977). Immunology of the antibiotics, in *The Antigens* IV (M. Sela, ed.), Academic Press, New York.

Dewdney, J. M., and Edwards, R. G. (1984). Penicillin hypersensitivity—is milk a significant hazard? A review, *J.R. Soc. Med., 77*:866.

Driessen, F. M., Kingma, F., and Stadhouders, J. (1982). Evidence that *Lactobacillus bulgaricus* in yogurt is stimulated by carbon dioxide produced by *Streptococcus thermophilus, Neth. Milk Dairy J., 36*:135.

Efstathiou, J. D., McKay, L. L., Morris, H. A., and Zottola, E. A. (1975). Growth and preservation parameters for preparation of a mixed species culture concentrate for cheese manufacture, *J. Milk Food Technol., 38*:444.

Ezzat, N., El-Doda, M., Boullane, D., Zevaco, C., and Blanchard, P. (1985). Cell wall associated proteinases in *Lactobacillus helveticus, Lactobacillus bulgaricus* and *Lactobacillus lactis, Milchwissenschaft, 40*:140.

Ezzat, N., El-Doda, M., Desmazeaud, M. J. and Ismail, A. (1986). Peptide hydrolases from thermobacterium group of lactobacilli. III. Characterization of the intracellular exopeptidases, *Le Lait, 66*:445.

Fennema, D. R., Powrie, W. D., and Marth, E. H. (1975). *Low temperature preservation of foods and living material*, Marcel Dekker, New York.

Forsén, R., Heiska, E., Herva, E., and Arvilommi, H. (1987). Immunobiological effects of *Streptococcus cremoris* from cultured milk "viili," application of human lymphocyte culture techniques, *Int. J. Food Microbiol., 5*:41.

Gilliland, S. E. (1985). Concentrated starter cultures, in *Bacterial Starter Cultures for Foods* (Stanley E. Gilliland, ed.), CRC Press, Boca Raton, FL, p. 145.

Gilliland, S. E., and Speck, M. L. (1968). D-Leucine as an auto-inhibitor of lactic streptococci, *J. Dairy Sci., 51*:1573.

Gilliland, S. E., and Speck, M. L. (1969). Biological response of lactic streptococci and lactobacilli to catalase, *Appl. Microbiol., 17*:797.

Gilliland, S. E., and Speck, M. L. (1974). Relationship of cellular components to the stability of concentrated lactic streptococcus cultures at –17°C, *Appl. Microbiol., 27*:793.

Goldberg, I., and Eschar, L. (1977). Stability of lactic acid bacteria to freezing as related to their fatty acid composition, *Appl. Environ. Microbiol., 33*:489.

Griffith, R., and Hammond, E. G. (1989). Generation of Swiss cheese flavor components by reactions of amino acids with carbonyl compounds, *J. Dairy Sci., 72*:604.

Harris, L. J., Daeschel, M. A., Stiles, M. E., and Klaenhamer, T. R. (1989). Antimicrobial activity of lactic acid bacteria against *Listeria monocytogenes, J. Food Protect., 52*:384.

Heap, H. A., and Lawrence, R. C. (1976). The selection of starter strains for cheesemaking, *N.Z. J. Dairy Sci. Technol., 11*:16.

Hemme, D., Nardi, M., and Jette, D. (1980). Beta-galactosidases et phospho-beta-galactosidases de *Streptococcus thermophilus, Le Lait, 60*:595.

Hirsch, A., Grinsted, E., Chapman, H. R., and Mattick, A. T. R. (1951). A note on the inhibition of an anaerobic sporeformer in Swiss-type cheese by a nisinproducing streptococcus, *J. Dairy Res., 18*:205.

Huggins, A. R., and Sandine, W. E. (1979). Selection and characterization of phage-insensitive lactic streptococci, *J. Dairy Sci., 62*:70.

Hull, R. R. (1977). Control of bacteriophages in cheese factories, *Aust. J. Dairy Technol., 32*:65.

Jakubowska, J., Libudzisz, Z., and Piatriewicz, A. (1980). Evaluation of lactic acid streptococci for the preparation of frozen concentrated starter cultures, *Acta Microbiol. Pol., 29*:135.

Jay, J. M. (1982). Antimicrobial properties of diacetyl, *Appl. Environ. Microbiol., 44*:525.

Juillard, V., Spinner, H. E., Desmazeaud, M. J., and Boquien, C. Y. (1987). Phenomenes de co-operation et d'inhibition entre les bacteries lactiques utilisees en industrie laitiére, *Le Lait, 67(2)*:149.

Kandler, O., and Weiss, N. (1986). Genus *Lactobacillus*, in *Bergey's Manual of Systematic Bacteriology*, vol. 2 (P. H. A., ed.), pp. 1208, Williams and Wilkins, Baltimore, MD.

Keen, A. R. (1972). Growth studies on the lactic streptococci. III. Observation on continuous growth behavior in reconstituted skim-milk, *J. Dairy Res., 39*:151.

Keogh, B. P. (1970). Survival and activity of frozen starter cultures for cheese manufacture, *Appl. Microbiol., 19*:928.

Kilara, A., Shahan, N. K., Das, and Grace, W. R. (1976). Effect of cryoprotective agents on freeze-drying and storage of lactic cultures, *Cultured Dairy Prod. J., 11(2)*:8.

Klaenhammer, T. R. (1988). Bacteriocins of lactic acid bacteria, *Biochemie, 70*:337.

Kok, J. (1990). Genetics of the proteolytic system of lactic-acid bacteria, *FEMS Microbiol. Rev., 87*:15.

Korhonen, H. (1973). Untersuchungen zur Bakterizidie der Milch und Immunisierung der boonier Milch driise, Thesis, University of Helsinki. Meijeritiet, Aikak, 32.

Kosikowski, F. (1966). *Cheese and Fermented Foods*, Edwards Brothers, Ann Arbor, MI, p. 20.

Lawrence, R. C., Heap, H. A., Limsowtin, G. K. Y., and Jarvis, A. W. (1978). Cheddar cheese starters: current knowledge and practices of phage characteristics and strain selection, *J. Dairy Sci., 61*:1181.

Lawrence, R. C., and Pearce, L. E. (1972). Cheese starters under control, *Dairy Ind. Int., 37*:73.

Lemieux, L., and Simard, R. E. (1991). Biter flavour in dairy products. I. A review of the factors likely to influence its development, mainly in cheese manufacture, *Le Lait, 71*:599.

Limsowtin, G. K. Y., Heap, H. A., and Lawrence, R. C. (1977). A multiple starter concept for cheesemaking, *N.Z. J. Dairy Sci. Technol., 12*:101.

Lloyd, G. T., and Pont, E. G. (1973). An experimental continuous culture unit for the production of frozen concentrated cheese starters, *J. Dairy Res., 40*:149.

Macura, D., and Townsley, P. M. (1984). Scandinavian ropy milk-identification and characterization of endogenous ropy lactic streptococci and their extracellular extracellular excretion, *J. Dairy Sci., 67*:735.

Mata, M., and Ritzenthaler, P. (1988). Present state of lactic acid bacteria phage taxonomy, *Biochemie, 70*:395.

Mäyrä-Mäkinen, A. (1990). T101-test for antibiotic residues in milk, *Scandinavian Dairy Inf.*, Vol. 4. *2*:38.

Mitchell, S. L., and Gilliland, S. E. (1983). Pepsinized sweat whey medium for growing *Lactobacillus acidophilus* for frozen concentrated starters, *J. Dairy Res., 40*:149.

Morichi, J. (1972). Mechanism and presentation of cellular injury in lactic acid bacteria subjected to freezing and drying, *56th Ann. Session of IDF*, Japan.

Muriana, P. M., and Klaenhammer, T. R. (1987). Conjugal transfer of plasmid-encoded determinants for bacteriocin production and immunity in *Lactobacillus acidophilus* 88, *Appl. Environ. Microbiol., 53*:553.

Nakajima, H., Toyoda, S., Toba, T., Itoh, T., Mukai, T., Kitazawa, H., and Adachi, S. (1990). A novel phosphopolysaccharide from slime-forming *Lactococcus lactis* subspecies *cremoris* SBT 0495, *J. Dairy Sci., 73*:1472.

Neve, H., Geis, A., and Teuber, M. (1988). Plasmid-encoded functions of ropy lactic acid streptococcal strains from Scandinavian fermented milk, *Biochemie, 70*:437.

Oda, M., Hasegava, H., Komatsu, S., Kambe, M., and Tsuchiya, F. (1983). Antitumor polysaccharide from Lactobacillus sp., *Agric. Biol. Chem., 47*:1623.

Olson, N. F. (1990). The impact of lactic acid bacteria on cheese flavor, *FEMS Microbiol. Rev., 87*:131.

Peebles, M. M., Gilliland, S. E., and Speck, M. L. (1969). Preparation of concentrated lactic streptococcus starters, *Appl. Microbiol., 17*:805.

Petersson, H. E. (1988). Mesophilic starters. *Bull. Int. Dairy Fed., 227*:19.

Pette, J. W., and Lolkema, H. (1950). Yogurt I. Symbiosis and antibiosis in mixed cultures of *Lb. bulgaricus* and *Str. thermophilus, Neth. Milk Dairy J., 4*:197.

Petterson, H.-E. (1972). Behaving av bakteriemassor för meijeri-industriellt bruk, Arbetsutskottet för livsmedels, Jorskning, Kemicentrum, Lund, Sverige, Halvårskrift no. 2.

Piard, J. C., and Desmazeaud, M. J. (1991). Inhibiting factors produced by lactic acid bacteria. I. Oxygen metabolites and catabolism end products, *Le Lait, 71*:525.

Pont, E. G., and Holloway, G. L. (1968). A new approach to the production of cheese starter. Some preliminary investigations, *Aust. J. Dairy Technol., 23*:22.

Porubcan, R. S., and Sellars, R. L. (1979). Lactic starter culture concentrates, in *Microbiology*, (H. T. Peppler, ed.), Van Nostrand Reinhold, Princeton, NJ, p. 59.

Premi, L., Sandine, W. E., and Elliker, P. R. (1972). Lactose-hydrolyzing enzymes of *Lactobacillus* species, *Appl. Microbiol., 24*:51.

Prevots, F., Mata, M., and Ritzentmaler, P. (1990). Taxonomic differentiation of 101 Lactococcal bacteriophages and characterization of bacteriophages with unusually large genomes, *Appl. Environ. Microbiol., 56(7)*:2180.

Price, R. J., and Lee, J. S. (1970). Inhibition of *Pseudomonas* species by hydrogen peroxide producing lactobacilli, *J. Milk Food. Tech., 33*:13.

Priidak, T. A., and Bannikova, L. A. (1982). Selection of lactic acid bacteria for production of cultured concentrate, *Int. Dairy Congr.*, p. 357.

Puchades, R., Lemieux, L., and Simard, R. E. (1989). Evolution of free amino-acids during the ripening of cheddar cheese containing added lactobacilli strains, *J. Food Sci., 54*:885.

Relano, P., Mata, M., Bonneau, M., and Ritzenthaler, P. (1987). Molecular characterization and comparison of 38 virulent and temperate bacteriophages of *Streptococcus lactis, J. Gen. Microbiol., 133*:3053.

Richardson, G. H., Hong, G. L., and Ernstrom, C. A. (1980). Defined single strains of lactic streptococci in bulk culture for cheddar and Monterey cheese manufacture, *J. Dairy Sci., 63*:1981.

Rousseaux, P., Vassal, L., Valles, E., Auclair, J., and Mocquota, G. (1968). The use of concentrated frozen suspensions of thermophilic lactic acid bacteria in making Gruyere cheese, *Le Lait, 48*:241.

Sanders, M. E. (1988). Phage resistance in lactic acid bacteria, *Biochemie, 70*:411.

Saxelin, M. L., Nurmiaho, E. L., Korhola, M. P., and Sundman, V. (1979). Partial characterization of a new C3-type capsule-dissolving phage of *Streptococcus cremoris, Can. J. Microbiol., 25*:1182.

Saxelin, M. L., Nurmiaho, E. L., Meriläinen, V., and Forsén, R. (1986). Ultrastructure and host specificity of bacteriophages of *Streptococcus cremoris, Streptococcus lactis* subsp. *diacetylactis*, and *Leuconostoc cremoris* from Finnish fermented milk "viili," *Appl. Environ. Microbiol., 52*:771.

Smittle, R. B., Gilliland, S. E., and Speck, M. L. (1972). Death of *Lactobacillus bulgaricus* resulting from liquid nitrogen freezing, *Appl. Microbiol., 24*:551.

Speckman, C. A., Sandine, W. E., and Elliker, P. R. (1974). Lyophilized lactic acid starter culture concentrates preparation and use in inoculation of vat milk for cheddar and cottage cheese, *J. Dairy Sci., 57*:165.

Stadhouders, J., (1974). Dairy starter cultures, *Milchwissenschaft, 29(6) :329.*

Stadhouders, J., Hup, G., and Jansen, L.A. (1971). A study of the optimum conditions of freezing and storing concentrated mesophilic starters, *Neth. Milk Dairy J.,*

Stadhouders, J., Hup, G., and Jansen, L. A. (1979). A study of the optimum conditions of freezing and storing concentrated mesophilic starters, *Neth. Milk Dairy J., 25*:229.

Stadhouders, J., Jansen, L. A., and Hup, G. (1969). Preservation of starters and mass production of starter bacteria, *Neth. Milk Dairy J., 23*:182.

Stadhouders, J., and Leenders, G. J. M. (1984). Spontaneously developed mixed-strain cheese starters. Their behaviour towards phages and their use in the Dutch cheese industry, *Neth. Milk Dairy J., 38*:157.

Stanley, G. (1977). The manufacture of starters by batch fermentation and centrifugation to produce concentrates, *J. Soc. Dairy Technol., 30*:36.

Steffen, C. (1980a). Cheesemaking and cheese research in Switzerland, *IDF Doc., 126*:16.

Steffen, C. (1980b). Das neue Kulturenkonzept für die schwizerischen Käsereien, *Deutsche Molkerei-Zeitung, 101*:1186.

Sutherland, I. W. (1977). Bacterial exopolysaccharides, their nature and production, in *Surface Carbohydrates of the Procaryotic Cell* (I. W. Sutherland, ed.), p. 27.

Tamine, A. Y., and Robinson, R. K. (1985). *Yoghurt-Science and Technology*, Pergamon Press, Oxford.

Thomas, T. D., and Mills, D. E. (1981). Nitrogen sources for growth of lactic streptococci in milk, *N.Z. J. Dairy Sci. Technol., 16*:43.

Thomas, T. D., and Pritchard, G. G. (1987). Proteolytic cuzymes of dairy starter cultures, *FEMS Microbiol. Rev., 46*:245.

Tinson, W., Hillier, A. J., and Jago, G. R. (1982). Metabolism of *Streptococcus thermophilus*. Utilization of lactose, glucose and galactose, *Aust. J. Dairy Sci. Technol., 37*:8.

Toba, T., Kotani, T., and S. Adachi. (1991). Capsular polysaccharide of a slime-forming *Lactococcus lactis* ssp. *cremoris* LAPT 3001 isolated from Swedish fermented milk "långfil," *Int. J. Food Microbiol., 12*:167–172.

Turner, K. W., and Martley, F. G. (1983). Galactose fermentation and classification of thermophilic lactobacilli, *Appl. Environ. Microbiol., 45*:1932.

Upreti, G. C., and Hinsdill, R. D. (1975). Production and mode of action of Lactocin 27. Bacteriocin from a homofermentative *Lactobacillus, Antimicrobiol. Agents Chemother., 7*:139.

Vedamuthu, E. R., and Neville, J. M. (1986). Involvement of a plasmid in production or ropiness (mucoidness) in milk cultures by *Streptococcus cremoris* MS, *Appl. Environ. Microbiol., 51*:677.

Vescovo, M., Scolari, G. L., and Bottazzi, V. (1989). Plasmid-encoded ropiness production in *Lactobacillus casei* ssp. *casei, Biotechnol. Lett., 11*:709–712.

von Wright, A., and Tynkkynen, S. (1987). Construction of *Streptococcus lactis* ssp. *lactis* strains with a single plasmid associated with mucoid phenotype, *Appl. Environ. Microbiol., 53*:1385.

Whitehead, H. R. (1953). Bacteriophage in cheese manufacture, *Bacteriol. Rev., 17*:109.

Wright, C. T., and Klaenhammer, T. R. (1983a). Survival of *Lactobacillus bulgaricus* during freezing and freeze-drying after growth in the presence of calcium, *J. Food Sci., 48*:773.

Wright, C. T., and Klaenhammer, T. R. (1983b). Influence of calcium and manganese on dechaining of *Lactobacillus bulgaricus, Appl. Environ. Microbiol., 4*:785.

Ziv, G. (1980). Availability and usage of new antibacterial drugs in Europe, *J. Am. Vet. Med. Assoc., 176*:1122.

3

Stability of Lactic Acid Bacteria in Fermented Milk

Yuan-Kun Lee
National University of Singapore, Kent Ridge, Singapore

Siew-Fai Wong
Malaysia Dairy Industries, Singapore

I. INTRODUCTION

Lactic acid bacteria are organisms which ferment sugar to yield lactic acid. Some of the lactic acid bacteria associated with fermented milk are listed in Table 1. There is increasing evidence to suggest that some members of the lactic acid bacteria, such as lactobacilli and bifidobacteria, when consumed in sufficiently large numbers exhibit prophylactic and therapeutic properties in both humans and animals (Mitsuoka, 1990; Robinson, 1991; Sandine et al., 1972). It has also been suggested that the consumption of a minimum viable cell number of 1×10^6 to 1×10^9 cells per day is necessary to have any chance of developing beneficial effects in human (Gilliland, 1989; Sellars, 1991; Hawley et al., 1959). Viable lactobacilli and bifidobacteria are currently available commercially either in the form of freeze-dried cells or milk-based products, e.g., yogurt, fermented milk drinks, and nonfermented acidophilus milk. In the former, as long as the water activity in the freeze-dried products is less than 0.25, the viability of cells can be maintained at room temperature for as long as 12 months (Sellars, 1991). The latter however has a limited shelf life, so milk-based products have to be recalled from stores after a display period of two to four weeks on a refrigerated shelf. From the economic point of view, the exercise is costly in terms of wastage and operation. Thus, if the factors determining the shelf life of milk-based products

Table 1 Some of the Major Lactic Acid Bacteria Used in Fermented Milk Products

Bacteria	Products	Remarks
Thermophilic, homofermentative		
Lactobacillus		
Lb. delbrueckii ssp. *bulgaricus*	Yogurt, Swiss and Italian cheeses, Bulgarian buttermilk, yogurt drink, koumiss	Produce acetaldehyde
Lb. delbrueckii ssp. *lactis*		
Lb. delbrueckii ssp. *delbrueckii*		
Lb. acidophilus	Acidophilus milk, yogurt drink, miru-miru, kefir, koumiss	Therapeutic
Lb. helveticus	Kefir, yogurt drink	
Lb. helveticus ssp. *juguri*	Yogurt	
Lb. fermentum		
Mesophilic, heterofermentative		
Lactobacillus		
Lb. casei ssp. *casei*	Yakult, yogurt drink, miru-miru, kefir	Therapeutic
Lb. casei ssp. *pseudoplantarum*		
Lb. casei ssp. *rhamnosus*		
Lb. casei ssp. *tolerans*		
Lb. plantarum		
Lb. brevis	Kefir	
Lb. kefir	Kefir	
Mesophilic *Streptococcus* (*Lactococcus*)		
Str. lactis ssp. *lactis*	Scandinavian fermented milk, cultured buttermilk, cultured cream, kefir	Produce nisin
Str. lactis ssp. *lactis*	Scandinavian fermented milk, cultured buttermilk, cultured cream, kefir	Produce diplococcin
Str. lactis biovar *diacetylactis*	Cultured buttermilk, cultured cream, kefir	Produce diacetyl
Thermophilic *Streptococcus* (*Lactococcus*)		
Str. thermophilus	Yogurt	
Leuconostoc		
Leu. mesenteroides ssp. *mesenteroides*	Kefir	
Leu. mesenteroides ssp. *dextranicum*	Kefir	

Table 1 (Continued)

Bacteria	Products	Remarks
Leu. mesenteroides ssp. *cremoris*	Cottage and cream cheeses, Scandinavian cultured buttermilk	Produce diacetyl
Leu. citrororum	Cultured buttermilk	
Bifidobacterium		
Bif. bifidum	Fermented milk	
Bif. logum	Fermented milk	
Bif. breve	Fermented milk, miru-miru	
Bif. infantis	Fermented milk	

could be understood, the stability of these products could be improved through product development and genetic manipulation of cells.

II. SURVIVAL OF CELLS IN MILK PRODUCTS

A. Kinetics of Cell Death

In general, the viable cell count of lactic acid bacteria in milk cultures decreases rapidly once the cultures have stopped growing (stationary growth phase). The kinetics of loss in viability of the cultures follows the exponential death law (Pirt, 1975):

$$X_t = X_0 \exp(-kt)$$

where

X_t = concentration of cells surviving time after t

X_0 = initial concentration of viable cells

k = specific death rate

The specific death rate of the lactic acid bacteria cultures is strain-dependent and influenced strongly by their physicochemical environment.

B. Effect of Storage Temperature

When stored at room temperature (25°C) the viability of *Lactobacillus casei* in fermented milk decreases by two log-cycles in about two weeks (Table 2), compared to *Lactobacillus plantarum*, *Lactobacillus acidophilus*, and *Bifidobacterium bifidum* which have a higher death rate at the same temperature. The

Table 2 Effect of Storage Temperature on the Time Taken for the Viable Cell Number of Lactic Acid Bacteria to Decrease by 2 Log-Cycle (td) in Fermented Milk (10% milk solid)

Bacteria	Product pH	Storage temperature (°C)	Initial viable cell number (cells/mL)	td (days)
Lactobacillus casei	3.5	25	5×10^9	13
YIT9018	3.8	5	1×10^9	>30
(yakult strain)				
Lactobacillus plantarum	3.4	25	3×10^9	4
MDI133	3.8	5	2×10^9	>30
(Malaysia Dairy strain)				
Lactobacillus acidophilus	4.0	25	3×10^8	3
CH5	3.6	5	1×10^8	15
(Chr. Hansen strain)				
Bifidobacterium bifidum	4.3	25	3×10^9	4
BB12	4.3	5	2×10^9	15
(Chr. Hansen strain)				

viability of lactobacillus cells in fermented milks can, however, be extended when the latter is stored at 5°C (Table 2). There is an additional advantage of storing the final products at low temperature: acid production is arrested and the products will not become overly sour. The kinetics of lactic acid production is partially growth-linked (Aborhey and Williamson, 1977; Hanson and Tsao, 1972; Leudeking and Piret, 1959), with lactic acid continuously being produced at a low rate as long as a carbon source is available, even when cell growth has ceased. A low storage temperature basically lowers the baseline level of acid production.

Casolari (1981) had suggested the following relationship to describe the effect of incubation temperature on the death rate of bacterial cultures:

$$\log X_0 = (1+Mt)\log X_t$$

where M is the frequency of collision between the cells and water molecules having an energy greater than that leading to cell death (E_d). From the Maxwellian distribution

$$M = \exp(103.7293) - (2E_d/RT)$$

The values of E_d for vegetative bacteria cells range from about 130 to 160 kJ/mol. Using these relationships to calculate the predicted death rate of lactobacilli and bifidobacteria at 25°C and 5°C, it is clear that physicochemical factors other than temperature are also involved in causing the much faster cell death at higher storage temperatures.

C. Effect of Product pH

The final product pH appears to play an important role in determining the stability of lactic acid bacteria in fermented milk products. Higher-culture pH values allow a longer survival of *Lactobacillus acidophilus* (Sellars, 1991; Gilliland, 1989). Hence, when the pH of milk cultures was maintained at near neutrality, the viable cell count of the milk cultures was maintained constant for longer than a month at 25°C (Table 3).

D. Effect of Flavor Additive

Some of the lactic acid bacteria, e.g., *Lactobacillus acidophilus*, although having demonstrated therapeutic values, do not produce acetaldehyde which gives the characteristic buttery flavor of regular yogurt (Sellars, 1991). As a result fermented acidophilus milk is tart and plain. In order to improve the flavor of acidophilus milk products, fruit juices are often mixed into the fermented milk in various ratios. Temperate fruits such as strawberry, apple, orange, grape, and tropical fruits such as mango and pineapple have been used. The final products are bottled in clear or translucent containers to show the color and indicate the type of

Table 3 Effect of Product pH on the Time Taken for the Viable Cell Number of Lactic Acid Bacteria to Decrease by 2 Log-Cycle (td) in Fermented Milk Cultures (10% milk solid)

Bacteria	Product pH	Storage temperature (°C)	Initial viable cell number (cells/mL)	td (days)
Lactobacillus casei	3.8	25	5×10^9	13
YIT9018	6.5	25	1×10^9	>30
(yakult strain)				
Lactobacillus plantarum	3.4	25	3×10^9	4
MD1133	6.5	25	3×10^9	>30
(Malaysia Dairy strain)				
Lactobacillus acidophilus	4.0	25	3×10^8	3
CH5	6.6	25	1×10^8	>30
(Chr. Hansen strain)				
Bifidobacterium bifidum	4.3	25	3×10^9	4
BB12	6.6	25	1×10^9	>15
(Chr. Hansen strain)				

Table 4 Effect of Fruit Juices Added on the Time Taken for the Viable Cell Number of *Lactobacillus acidophilus* CH5 (Chr. Hansen strain) to Decrease by 2 Log-Cycle (td), in Fermented Milk (10% milk solid)

Fermented milk	Product pH	Storage temperature (°C)	Initial viable cell number (cells/mL)	td (days)
Milk alone	3.8	5	1×10^9	15
Milk + 3% strawberry	3.8	5	1×10^9	5
Milk + 3–10% orange	3.8	5	8×10^8	15
Milk + 3–10% grape	3.8	5	1×10^9	15
Milk + 3–10% apple	3.8	5	9×10^8	15
Milk + 3–10% mango	3.8	5	1×10^9	15
Milk + 3–10% pineapple	3.8	5	9×10^8	15

fruit juices added. We have consistently observed that the viable cell count of *Lactobacillus acidophilus* in fermented milk containing strawberry juice deteriorates much faster than in fermented milk mixed with other types of fruit juices. Inclusion of as low as 3% (v/v) strawberry juice resulted in the reduction of viable cell number by two log-cycles within five days, while cells in fermented milk containing no fruit juice or mixed with other types of fruit juices of up to 10% (v/v) survived longer (Table 4). We further observed that fast deterioration in the viability of *Lactobacillus acidophilus* cells in strawberry milk occurred only when the product was exposed to daylight. When the product was wrapped in aluminum foil, no fast reduction in viable cell number was observed. The observations suggest that photochemical reactions on fermented strawberry milk induce formation of substances which apparently are toxic to bacterial cells. Redesigning the container for the fermented milks was therefore necessary to prolong the viability of the lactic acid bacterium.

III. BIOCHEMICAL FACTORS IN DETERMINING STABILITY OF CELLS

A. Energy Supply

It was observed that when nongrowing *Lactobacillus delbrueckii* cells were incubated in a phosphate buffer containing glucose alone, temporary stability was conferred on the cells (Rees and Pirt, 1979). The authors argued that active glycolysis and energy generation are prerequisites for maintaining cell integrity and stabilizing intracellular protein levels in the *Lactobacillus* cells. On the other hand, excessive ATP generation resulted in instability in *Lactobacillus* cells. Thomas and Batt (1969) showed that glycolytic activity in resting cells of *Streptococcus lactic* decayed more quickly in the presence of glucose, while arginine, which produces ATP at a rate 7.5 times slower than glucose, conferred more stability on the system. This situation is possibly analogous to the elevated levels of ATP observed in substrate accelerated death (Strange, 1976).

B. Magnesium Ion

The inclusion of Mg ions alone had no effect on the stability of *Lactobacillus delbrueckii* in the absence of glucose but did markedly stabilize glycolysis in the presence of glucose (Rees and Pirt, 1979). The mechanism of this stabilization was not identified by the workers, but all kinases of glycolytic sequence are known to require Mg as a cofactor. The requirement for Mg ion by *Lactobacillus delbrueckii* could not be replaced by manganese ion (Rees and Pirt, 1979).

C. Oxygen Toxicity

The works of Gilliland and Speck (1977) and Hull et al. (1984) suggest that instability of *Lactobacillus acidophilus* cells added to yogurt is caused by hydrogen peroxide produced by the lactobacilli in the yogurt starter culture. They found that milk cultured with *Lactobacillus bulgaricus* was antagonistic to *Lactobacillus acidophilus* cells, causing a rapid loss in their viability. This loss in viability was prevented by the addition of catalase. Hydrogen peroxide could also be produced by the *Lactobacillus acidophulus* inoculum. For instance, exposure of acidophilus inoculum to air caused accumulation of hydrogen peroxide through the action of NADH oxidase (Collins and Aramak, 1980; Seeley and Vandemark, 1951). In this case, the stability of the cells decreased with increasing inoculum size because the production of hydrogen peroxide was proportional to the cell concentration (Seeley and Vandemark, 1951). Through cell growth and induction of NADH peroxidase, hydrogen peroxide was eventually removed (Anders et al., 1970). Other workers found that the inclusion of reducing agents, such as thioglycollate, cysteine, or dithiothreitol, conferred greater stability on the glycolytic activity of nongrowing *Lactobacillus delbrueckii* (Rees and Pirt, 1979). Their protective effect was related to their redox potential. Thus, the toxicity of oxygen can also be mediated via its direct interaction with the –SH group of key metabolic enzymes.

The studies so far suggest that optimum stimulation of resting glycolytic activity and improved stability of nongrowing lactic acid bacteria can be achieved by including Mg ion, and maintaining a low redox potential and a constant availability of energy. These conditions can be easily met in nonfermented acidophilus milk products, e.g., Sweet Acidophilus Milk and Nu-Trish, using appropriate packaging materials. In these products, whole cells of *Lactobacillus acidophilus* are added to buttermilk or pasteurized milk and stored immediately at a low temperature to prevent fermentation.

In fermented milk products, growth of lactic acid bacteria ceases before all available carbon and energy sources in milk are exhausted. Lactose in milk generally decreases by only 30–50% after fermentation (Alm, 1991). Additional energy sources in the forms of added sugar and milk fat did not appear to affect the fermentation kinetics nor the stability of *Lactobacillus acidophilus* during storage (Hull et al., 1984). We had also observed that inclusion of lactate of up to 2% (w/v) in milk did not prevent the growth of *Lactobacillus acidophilus* nor the final attainable cell concentration. The cessation of growth is due to the acidity of the cultures.

IV. MOLECULAR MECHANISMS IN DETERMINING THE ACIDOSTABILITY OF CELLS

As most microorganisms can tolerate a wide range of pH in their culture environment, it is clear that the intracellular pH must be regulated to within narrow limits. In fact, the intracellular pH values of all microorganisms, including the extreme acidophiles (Din et al., 1967; Iwatsuka et al., 1962; Suzuki, 1965), are similar and close to neutrality. The observation that lactobacilli and bifidobacteria survive better at low pH at lower temperature (Table 2) may at first glance suggest that energy-driven proton exclusion (Yamazaki et al., 1973) is not the major mechanism for maintaining pH and viability of these cells in acidic environment.

Lactic acid bacteria employ either homofermentative, heterofermentative, or bifidum pathways for fermentation of carbohydrates to produce a single product or a mixture of lactic acid, acetic acid, ethanol, and CO_2 (Gottschalk, 1979). Almost all of the weak acids produced are extruded from cells. The final concentration of lactic acid in fermented milk is between 0.5 and 1.5% (w/v) (corresponding to a pH of 4.5 to 3.5), depending on the bacteria strain used and the fermentation time. In the formulation of fermented milk beverages, it is usually desirable to have a final product pH of around 3.8 for two reasons. Firstly, it is far below the isoelectric point of milk protein (4.6); secondly, a low pH value prevents contamination and growth of pathogenic bacteria during storage. According to the widely accepted Mitchell hypothesis of energy generation by a proton motive force, a major function of the cell and mitochondrial membranes is to maintain a proton (pH) and charge gradient. For this purpose the cell membrane must have a very limited permeability to protons and charged molecules (Forage et al., 1985) such as ionized weak organic acids at near neutral pH values. Hence, organic acids produced by lactic acid bacteria are practically trapped outside the cells. However, at low pH (near to their p*K* values) a considerable fraction of the organic acids is present in undissociated form. The undissociated lactic acid appears to be the species which stops the fermentation process (Blanch et al., 1984; Viniegra-Gonzalez and Gomez, 1984). Undissociated organic acids are uncharged and lipophilic and thus are allowed to diffuse across the cell membrane into cells. Once inside the cells, where the pH is higher, the acids dissociate to form ions and protons. To maintain a neutral intracellular pH, ATP is hydrolyzed by the proton-translocating ATPase to extrude the protons. Consequently, other vital cellular functions are deprived of ATP, and cell viability cannot be maintained. The cells can also be killed through inactivation of cellular enzymes and other key cellular components. This can happen when the rate of diffusion of acids into cells is faster than the rate of active extrusion of protons, resulting in the lowering of the intracellular pH. The diffusion rate of undissociated organic acids across the cell membrane is determined by the diffusion constant, which is temperature-dependent. Thus, at higher temperatures, e.g., 25°C, the rate of diffusion of acids into the cells is faster and this in turn causes a faster decline in the viability of the lactic acid bacteria (Table 2). In contrast, at 4°C, the acids diffuse into the cells at a slower rate so that cells were able to remain viable for a longer period.

Clearly the acidophilicity and acidostability of lactic acid bacteria depend on the membrane function and its permeability to undissociated weak acids. This notion is further supported by reports that showed the phospholipid composition of the membrane of *Staphylococcus aureus* and its acid tolerance to change with the pH of the incubation medium (Haest et al., 1972) and the absence of peptidoglycan in the cell wall of an acid-tolerant sulfur-oxidizing bacterium (Brock et al., 1972). Biochemical analysis of the membrane composition of lactic acid bacteria possessing various degrees of acidostability would shed light on the mechanisms for determining their stability in fermented milk cultures. It is interesting to note that those lactobacilli which are able to metabolize a wider range of carbohydrates, for example, *Lactobacillus casei*, survive longer in fermented milk products than those with a restricted range of usable carbohydrates, e.g., *Lactobacillus acidophilus*.

V. STRATEGY FOR STRAIN IMPROVEMENT

Since the membrane components in determining the acidostability of lactic acid bacteria have not been identified, the use of gene cloning for strain improvement is not currently possible. However, as acid tolerance is a reflection of their acidostability, it would be impossible to select cells which could survive longer in the acidic environment. A simple selection strategy would be to subculture a mutated lactic acid bacteria population periodically, or to use a continuous-flow culture system (Dykhuizen and Hartl, 1983) to select the more acid-tolerant strains. However, in our attempts to isolate acidostable strains of *Lactobacillus acidophilus*, we have always obtained non-acid-producing strains. This is because acid production was not a selection criterion in the isolation procedure. The improved stability of the cultures could therefore be achieved by the cells losing their acid-producing ability, thus maintaining a favorable pH environment for the cells. In another attempt to isolate phage-resistant strains of lactic acid bacteria through mutagenesis (Lawrence et al., 1976), a similar observation was made. Few useful strains were obtained, as isolated phage-resistant strains usually lose their acid-producing ability. There is now evidence to suggest that the lactose-fermenting ability (*lac*) of lactic acid bacteria is mediated through plasmids (Anderson and Mckay, 1977; Efstathiou and Mckay, 1976) and can be modified through loss of these plasmids.

A selection process which includes acid production as a criterion in isolating acid tolerant lactobacillus strains has been described by Lee and Wong (1992). This procedure for selecting and isolating lactic-acid-producing, acid-tolerant *Lactobacillus* variants is based on two assumptions:

a. The pH of the milk medium will only decrease through the fermentation production of lactic acid from lactose.
b. At low pH, only acid-tolerant variants will continue to grow.

Lactobacillus cells subjected to mutagenesis or somatic hybridization (through protoplast fusion) are introduced into a stirred tank fermentor containing a milk culture medium. The pH of the culture is monitored by a pH indicator/controller, which activates a peristaltic pump when the pH of the culture drops below a preset value, for example, 4.0. The peristaltic pump delivers fresh milk medium into the culture. The decline in pH of the batch culture is due to the fermentative production of lactic acid. Addition of the fresh milk medium restores the pH of the culture due to dilution by a neutral medium.

The volume of the culture in the vessel is controlled by two level-sensing probes, one fixed at the high level and the other at the low level. The signal from the high-level sensor is amplified by an electronic liquid-level controller to activate a harvest pump, which in turn is deactivated by signals from the low-level sensor.

The decrease in pH due to the increased amount of lactic acid through *Lactobacillus* growth in the culture system is restored by the addition of fresh medium. Thus, the growth of lactobacilli can be monitored by measuring the cumulative amount of fresh medium added to the culture system, assuming that the cell yield of lactic acid is a constant.

The principle of the selection procedure is that as acid-tolerant *Lactobacillus* cells grow, more milk medium is fed into the culture system. The dilution rate therefore depends on the overall growth rate of the culture system. While cultures grow, the total volume of the culture increases with time. As the culture level comes into contact with the high-liquid-level sensor inserted in the culture vessel, a harvest pump is switched on. The pump is switched off when the culture level drops and contacts the low-level sensor. Thus, another selection cycle is initiated. The cell concentration remains at about constant level due to the dilution of the culture by a fresh medium as the cell grows.

When the pH value in the culture drops below a critical value, the growth rate of the culture decreases. Therefore an increase in the growth rate of the culture indicates an improvement in acid tolerance of the *Lactobacillus* culture. The higher the growth rate, the faster is fresh medium is added to the culture, and thus the higher is the dilution rate. The selection process is therefore fully automatic. The frequency and pressure of selection is therefore determined by the intrinsic potential of the culture. These factors could be important in the selection of a small population of superior mutants.

Several lactic acid-producing strains of *L. acidophilus, L. casei,* and *L. plantarum* with shelf lives longer than 40 days in fermented milk were isolated by the above screening procedure.

REFERENCES

Aborhey, S., and Williamson, D. (1977). Modeling of lactic acid production by *Streptococcus cremoris* HP, *J. Gen. Appl. Microbiol., 23*:7.

Alm, L. (1991). The therapeutic effects of various cultures, in *Therapeutic Properties of Fermented Milks* (R. K. Robinson, ed.), Elsevier, Essex, p. 45.

Anders, R. F., Hogg, D. M., and Jago, G. R. (1970). Formation of hydrogen peroxide by group N streptococci and its effect on their growth and metabolism, *Appl. Microbiol., 19*:608.

Anderson, D. G., and McKay, L. L. (1977). Plasmids, loss of lactose metabolism, and appearance of partial and full lactose-fermenting revertants in *Streptococcus cremoris* B., *J. Bacteriol., 129*:367.

Blanch, H. W., Vickroy, T. B., and Wilke, C. R. (1984). Growth of prokaryotic cells in hollow-fiber reactors, *Ann. N.Y. Acad. Sci., 434*:373.

Brock, T. D., Brock, K. M., Belly, R. T., and Weiss, R. L. (1972). Sulfolobus: a new genus of sulfur-oxidizing bacteria living at low pH and high temperature, *Arch. Mikrobiol., 84*:54.

Casolari, A. (1981). A model describing microbial inactivation and growth kinetics, *J. Theor. Biol., 88*:1.

Collins, E. D., and Aramaki, K. (1980). Production of hydrogen peroxide by *Lactobacillus acidophilus, J. Dairy Sci., 63*:353.

Din, G. A., Suzuki, I., and Lees, H. (1967). Ferrous iron oxidation by *Ferrobacillus ferrooxisans, Can. J. Biochem., 45*:1523.

Dykhuizen, D. E., and Hartl, D. L. (1983). Selection in chemostats, *Microbiol. Rev., 47*:150.

Efstathiou, J. D., and McKay, L. L. (1976). Plasmids in *Streptococcus lactis*: evidence that lactose metabolism and proteinase activity are plasmid linked, *Appl. Environ. Microbiol., 32*:38.

Forage, R. G., Harris, D. E. F., and Pitt, D. E. (1985). Effect of environment on microbial activity, in *Comprehensive Biotechnology Vol. 1* (A. T. Bull and H. Dalton, eds.), Pergamon Press, Oxford, p. 251.

Gilliland, S. E. (1989). Acidophilus milk products: a review of potential benefits to consumers, *J. Dairy Sci., 72*:2483.

Gilliland, S. E., and Speck, M. L. (1977). Instability of *Lactobacillus acidophilus* in yogurt, *J. Dairy Sci., 80*:1394.

Gottschalk, G. (1979). *Bacterial Metabolism*, Springer-Verlag, New York, p. 173.

Haest, C. W. M., De Gier, J., Op Den Kamp, J. A. F., Bartels, P., and Van Deenen, L. L. M. (1972). Changes in permeability of *Staphylococcus aureus* and derived liposome with varying lipid composition, *Biochim. Biophys. Acta, 255*:720.

Hanson, T. P., and Tsao, G. T. (1972). Kinetic studies of the lactic acid fermentation in batch and continuous cultures, *Biotechnol. Bioeng., 14*:233.

Hawley, H. B., Shepherd, P. A., and Wheater, D. M. (1959). Factors affecting the implantation of lactobacilli in the intestine, *J. Appl. Bacteriol., 22*:360.

Hull, R. R., Roberts, A. V., and Mayes, J. J. (1984). Survival of *Lactobacillus acidophilus* in yoghurt, *J. Dairy Technol., 39*:164.

Iwatsuka, H., Kune, M., and Maruyama, M. (1962). Studies on the metabolism of a sulfur-oxidizing bacterium. II: The system of CO_2-fixation in *Thiobacillus thiooxidans, Plant Cell Physiol., 3*:157.

Lawrence, R. C., Thomas, T. D., and Terzaghi, B. E. (1976). Reviews of the progress of dairy science: cheese starters, *J. Dairy Res., 43*:141.

Lee, Y. K., and Wong, S. F. (1992). A self-regulated screening system for selection and isolation of *Lactobacillus* variants of long shelf-life for the production of fermented milk, U.K. Patent Appl. 9100915-9.

Leudeking, R., and Piret, E. L. (1959). A kinetic study of the lactic acid fermentation, *J. Biochem. Microb. Technol. Eng., 1*:393.

Mitsuoka, T. (1990). Role of intestinal flora in health with special reference to dietary control of intestinal flora, in *Microbiology Applications in Food Biotechnology* (B. H. Nga and Y. K. Lee, eds.), Elsevier, Essex, p. 135.

Pirt, S. J. (1975). *Principles of Microbe and Cell Cultivation*, Blackwell, Oxford.

Rees, J. F., and Pirt, S. J. (1979). The stability of lactic acid production in resting suspensions of *Lactobacillus delbrueckii, J. Chem. Tech. Biotechnol., 29*:591.

Robinson, R. K., ed. (1991). *Therapeutic Properties of Fermented Milks*, Elsevier, Essex.

Sandine, W. E., Muralidhara, K. S., Elliker, P. R., and England, D. C. (1972). Lactic acid bacteria in food and health: A review with special reference to enteropathogenic *Escherichia coli* as well as certain enteric diseases and their treatments with antibiotics and lactobacilli, *J. Milk Food Technol., 35*:691.

Seely, H. W., and Vandemark, P. J. (1951). An adaptive peroxidation by *Streptococcus faecalis, J. Bacteriol., 61*:27.

Sellars, R. L. (1991). Acidophilus products, in *Therapeutic Properties of Fermented Milks* (R. K. Robinson, ed.), Elsevier, Essex, p. 81.

Strange, R. E. (1976). *Microbial Response to Mild Stress*, Meadowfield Press, Sheldon, Co., Durham.

Suzuki, I. (1965). Oxidation of elemental sulfur by an enzyme system of *Thiobacillus thiooxidans*, *Biochim. Biophys. Acta*, *104*:359.

Thomas, T. D., and Batt, R. D. (1969). Metabolism of exogenous arginine and glucose by starved *Streptococcus lactis* in relation to survival, *J. Gen. Microbiol.*, *58*:371.

Viniegra-Gonzalez, G., and Gomez, J. (1984). Lactic acid production by pure and mixed bacterial cultures, in *Bioconversion System* (D. Wise, ed.), CRC Press, Boca Raton, p. 17.

Yamazaki, Y., Koyama, N., and Nosoh, Y. (1973). On the acidostability of an acidophilic thermophilic bacterium, *Biochim. Biophy. Acta*, *314*:257.

4

Lactic Acid Bacteria in Cereal-Based Products

Hannu Salovaara
University of Helsinki, Helsinki, Finland

I. SUMMARY

Cereal-based foods are a major source of economical dietary energy and nutrients worldwide. A considerable proportion of cereal foods is made without any fermentation process, such as boiled or steamed rice, porridge, pasta, cookies, etc. When fermentation is applied it is often the alcoholic fermentation by yeast that prevails, as in bread making and brewing. However, a number of cereal-based foods are fermented by lactic acid bacteria, such as the European sour bread, various Asian fermented flat breads and pancakes, and the numerous types of fermented sour porridges, dumplings, and nonalcoholic beers, which are common in Africa and Asia. Perhaps the most important role of lactic acid bacteria in cereal-based foods with respect to technology and health is the ability of lactic acid bacteria to increase palatability, keeping properties, safety and nutritional value of cereals. In most processes lactic acid bacteria dominate only after suppressing the growth of undesirable or pathogenic bacteria and fungi. The successive growth of microorganisms in fermenting cereals also favors yeast growth, and often a mixed culture of lactic acid bacteria and yeast finally prevail.

Lactic acid fermentation provides some specific benefits in certain cereal processes with respect to technology and nutritional value. The low pH resulting from fermentation suppresses excess activity of alpha-amylase and other enzymes in rye. Lactic acid fermentation contributes to bioavailability of minerals by favoring phytate degradation. As in other industries lactic acid fermentation also gives a natural image to the product.

Most fermented cereal-based foods contain no live lactic acid bacteria, since food preparation is usually completed by a heat treatment which kills the microbes. However, it has been suggested recently that cooked cereal flummeries could be fermented with lactic acid bacteria of the probiotic type, and novel types of physiologically beneficial cereal-based foods could be produced in this way. The present article reviews some aspects of the traditional role of lactic acid bacteria in the production of cereal-based foods. Possibilities for novel fermented cereal foods will be discussed shortly.

II. GENERAL INTRODUCTION: LACTIC ACID FERMENTATION AS AN INTEGRAL PART OF CEREAL FOOD PROCESSES

Dry cereal grains are hard and highly nutritive packages, which can be easily stored for long periods of time. However, they cannot be eaten as such without processing. The techniques used in the preparation of cereal grains for food normally involves some kind of milling or grinding, mixing with water, and a heat treatment. Fermentation either by yeast or lactic acid bacteria or both is often an essential part of the process and may take place in more than one phase during the preparation.

In the production of many cereal foods fermentation by lactic acid bacteria is an integrated process which occurs simultaneously with other required operations. Sometimes lactic acid fermentation may be characterized as secondary or unintentional. For example, acidification, or souring, occurs when grains are soaked prior to wet-milling, and when dough for bread is leavened by sourdough. In some other processes it is the acidification caused by the lactic acid bacteria that is required in particular, and specific starters and other procedures are used to ensure that a proper lactic acid fermentation occurs. The production of nonalcoholic soured beer "mageu" serves as an example.

Certain technological operations give rise to lactic acid fermentation in cereal processes. Table 1 lists some functions and integrated side effects of lactic acid fermentation processes in the context of cereal-based foods. Soaking or steeping of cereal grains in water prior to wet-milling is a customary initial step when corn, sorghum, or millet are ground in traditional food processing. Soaking softens the grain endosperm, and the work input required for grinding is considerably less than in grinding dry kernels. Penetration of water into the interior of the kernels requires hours in an ambient temperature, and simultaneous fermentation obviously occurs as an inevitable but useful integrated side effect. The fermenting

Table 1 Some Functions and Mechanisms of Lactic Acid Fermentations in Processes of Cereal-Based Foods

Operation	Material	Principal purpose and desirable mechanism	Side effect or simultaneous reaction	Example of a typical product
Soaking of grains prior to wet-milling	Whole grains	Softening of grain endosperm	Lactic acid fermentation, control of undesired organisms	Agidi
Slurrying for sour porridge or dumplings	Wet starchy material from wet-milling	Increase in keeping properties	Flavor production, control of undesired organisms	Agidi, kenkey
Slurrying for sour porridge separation of hulls and bran microorganisms	Coarse meal from dry-milling	Separation of hulls, etc., from the starchy endosperm	Lactic acid fermentation, control of undesirable organisms	Kiesa
Doughmaking for bread	Dry flour	Aeration	Acidification, flavor production increase of mold-free time	Sourdough bread
Brewing	Malted or unmalted cereal	Ethanol and flavor production	Acidification by lactic acid bacteria	Country beers lambic beer
Brewing	Boiled corn meal	Lactic acid and flavor production	Control of undesired organisms	Mahew

microorganisms originate from the surface of the kernels and from other sources such as the steeping vessel (Odunfa, 1985). The resulting starchy material from wet-grinding carries the sour flavor, which has become an essential and desired element in the local foods cooked from the fermented slurry (Campbell-Platt, 1987). Typical foods made by the soaking and wet-milling process are the African fermented porridges and dumplings reviewed in more detail by Novellie (1981) and Campbell-Platt (1987).

A more effective lactic fermentation occurs when the flour from dry-milling or the starchy material from wet-grinding is slurried. This procedure is common in the preparation of many African and Asian staple foods. As in other fermentations of the plant material, such as vegetables, a short period of growth by fungi and bacteria belonging to *Enterobacteriaceae* may occur followed by a natural lactic acid bacterial fermentation and alcoholic fermentation by yeast (Odunfa, 1985). The lactic acid bacteria and yeast convert fermentable sugars available into lactic acid, ethanol and other fermentation products, thus lowering the pH of the reaction mixture and suppressing growth of undesirable organisms such as pathogenic bacteria (Wood, 1981).

Slurrying dry-ground cereal grains in water has also been used as a technological measure in the separation of hulls from oats, when traditional sour fermented oat porridges or flummeries were made (Fenton, 1974; Salovaara et al., 1990). When a coarse oatmeal slurry undergoes fermentation, endosperm particles will sediment and the hulls can be removed from the surface.

Doughs for bread making are aerated by carbon dioxide, which is produced by alcoholic fermentation. The main reason for the use of sourdough was probably to achieve leavening caused by spontaneously occurring yeast. However, pure cultures by yeast are difficult to maintain even using today's industrial processes, and lactic acid bacteria emerge originating from the unsterilized ingredients. Therefore, sourdoughs are normally an integrated system of yeasts and lactic acid bacteria (Wood, 1981).

Brewing of cereal grains into beer undoubtedly also was an integrated alcoholic/lactic fermentation process in its original form. Today local low-alcohol beers or cereal beverages fermented with added yeast are made in many societies, especially in Africa. Lactic acid bacteria are a natural flora in these drinks, and they make the beverage not only alcoholic but also sour after a few days of storage (Wood, 1981). The corn-based sour beer "mageu" from South Africa is an example of a different type of beverage, which is fermented by lactic acid bacteria and produced industrially (McGill and Taylor, 1991).

Lactic acid fermentation provides many functions in the production of cereal-based foods. Some of these functions were listed in Table 1. In addition to the technological advantages, enhanced flavor, better keeping properties, and suppression of pathogenic organisms, lactic acid fermentation also improves nutritional value of cereals by increasing bioavailability of minerals through phytate hydrolysis (Reddy et al., 1989).

III. SPECIFIC TOPICS

A. Lactic Acid Bacteria in Bread Making

1. Functions of Sourdough

In bread making, lactic acid bacteria have an essential function when rye plays an important role in the recipe. Acidification or souring of rye dough by the lactic acid bacteria improves bread-making potential of rye and the quality of rye bread by suppressing eventual high activity of alpha-amylase and other enzymes, and by other mechanisms (Spicher and Stephen, 1987). Wheat sourdoughs are also known and utilized in the production of crackers and certain specialties such as the San Francisco French bread and Italian and Spanish specialties (Sugihara, 1978; Spicher and Stephan, 1987; Barber et al., 1989). Sour bread, whether made of wheat or rye, is characterized by a typical flavor and better resistance to mold compared to unsoured bread. General functions of sourdoughs in bread making are listed in Table 2.

Historically sour bread was probably the first type of bread. The aerated dough is easier to bake, and the result is a soft and tasty bread with better keeping properties in comparison to a doughy unflavored and unleavened product, which would soon become stale and would be hard when dry. The aeration of sourdough is caused primarily by fermenting yeasts, whereas the acidification is the consequence of the simultaneous fermentation by lactic acid bacteria. Heterofermentative lactic acid bacteria may also have a significant leavening effect, as has been

Table 2 General Functions of Sourdough in Bread Making

Leavening action by yeast and to a lesser extent by heterofermentative lactic acid bacteria
Control and inhibition of contaminating or spoiling flora
Modification of flour components such as swelling of starch and pentosans
Control of enzymatic activity of flour, especially α-amylase
Accumulation of flavor components such as lactic and acetic acids and other fermentation products
Elongation of mold-free time due to lactic and acetic acids and low pH; prevention of growth of *Bacillus subtilis*, the "rope"-causing organism
Improvement of nutritional value by increasing mineral bioavailability through the degradation of phytate
Characterization of the product by a natural image

shown to be the case, e.g., in "idli," an Indian rice/leguminous-based aerated pancake-type food (Mukherjee et al., 1965).

A good sourdough is of great value, and therefore rebuilding systems based on a previous batch were developed. Repeated rebuilding of sourdough gives rise to selective enrichment of microorganisms. In sourdoughs species belonging to *Lactobacillus* often predominate, although representatives of *Leuconostoc* and *Pediococcus* have also been reported to be present. In addition to lactobacilli, yeasts are also present in a sourdough.

From ancient times until the last century sourdough predominated as the means used for leavening dough for bread. Concentrated live yeast from beer or wine industries may have been available to bakers for thousands of years, but compressed yeast specially made for leavening of dough became available only in the 19th century (Pederson, 1971). Today the sourdough procedure is utilized internationally in spite of the fact that baker's yeast is available worldwide. This is a demonstration of the capacity of lactic acid bacteria to provide flavor and other desired properties to bread.

Most of the scientific literature on lactic acid bacteria in cereal-based foods derives from studies on sour rye bread, eaten in central, northern, and eastern Europe. Much of this work is performed and reviewed by Spicher and Stephan (1987) in their book on sourdough. Lund and Hansen (1987), Salovaara (1988), and Lönner (1988) have also reviewed literature on sourdough and its microbiology. Reports on fermented Asian and African sour breads and pancakes are accumulating. A useful reference with regard to fermented foods in general is the dictionary compiled by Campbell-Platt (1987).

In Arab countries, North Africa, the Middle East, and elsewhere, flat breads made from wheat are the staple food. Dough for the flat bread is often fermented in a process closely similar to that used for sourdough rye bread in Europe. There are numerous local varieties of these breads such as "lavash," "injera," "idli," and others, as described by Steinkraus (1983) and Campbell-Platt (1987).

The need of chemical preservatives such as sorbic and propionic acids is avoided by the acidity of sourdough bread. The mold prevention function of sourdough was a great benefit in earlier times, and it has become very important again because of the negative consumer response against added preservatives.

Lactic acid and other organic acids accumulate in the sourdough primarily due to lactic acid fermentation. Figure 1 shows development of temperature, pH, titratable acidity, and colony counts of lactic acid bacteria and yeast during an industrial rye sourdough bread-making process.

2. *Lactic Acid Bacteria of Sourdoughs*

Whole cereal grains and 100% extraction rye flour may contain 10^4–10^6 CFU of unspecified bacteria per gram. Colony counts of coliform organisms may range

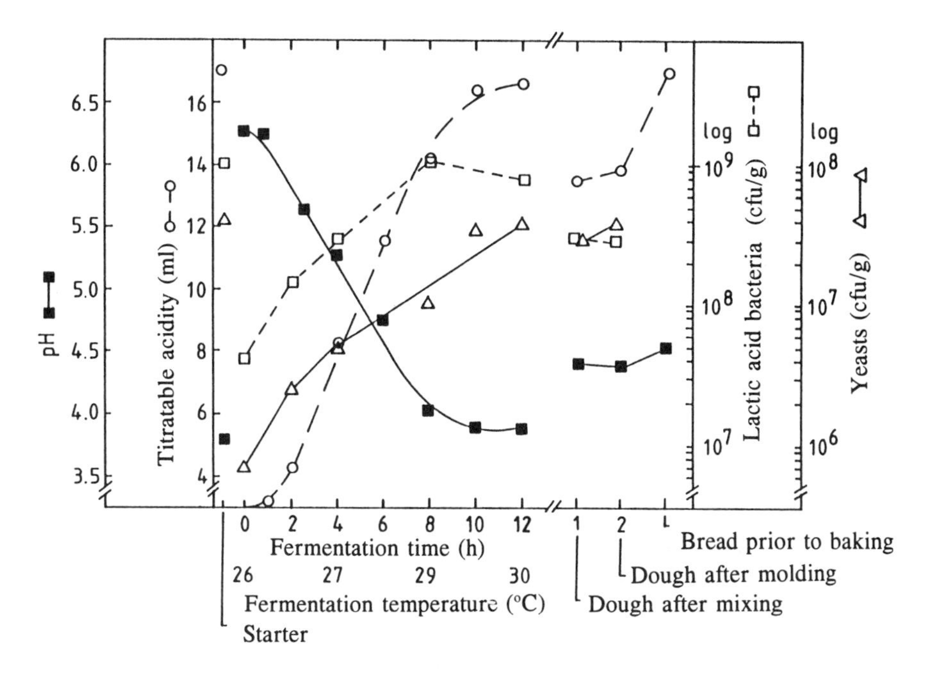

Figure 1 Development of acidity and microbial counts in an industrial rye sourdough process (Salovaara, 1988).

from 10^2 to 10^4 CFU per gram of flour, and those of a spontaneous flora of lactic acid bacteria may be 10^2–10^3 CFU/g. Inoculation of the sourdough with a starter from previous batch increases the number of lactic bacteria to 10^7 to 10^8 CFU/g (Fig. 1). This gives little possibilities for growth to contaminating organisms, including those which are imported by flour.

In a fully fermented sour rye sponge there may be more than 10^9 CFU of lactobacilli per gram. The number of yeast cells may be 10 or 100 times lower (Fig. 1).

Lactobacilli frequently identified in sourdoughs are listed in Table 3. In a given sourdough usually several species of lactobacilli can be detected. Both homofermentative and heterofermentative types of lactobacilli are found. See Lönner (1988) for a more detailed review on the microflora in sourdoughs.

In spite of repeated contamination by microorganisms imported in flour, most sourdough processes seem to be very stable, indicating the presence of well-adapted natural flora. Commercial starter cultures containing certain well-defined

Table 3 Some Identified Species of Lactic Acid Bacteria in Sourdoughs

Species	Fermentation pattern	Reference
Lactobacillus plantarum	homof.	1, 2, 3, 4
L. casei	homof.	1, 2, 3, 4, 5
L. buchneri	homof.	1, 3, 5
L. acidophilus	homof.	2, 3, 4, 5
L. alimentarius	homof.	6, 7
L. farciminis	homof.	3, 4
L. delbrueckii	homof.	1
L. fermentum	heterof.	1, 3, 4, 5
L. brevis	heterof.	1, 3, 8, 9
L. sanfrancisco (*L. brevis* var. *lindneri*)	heterof.	5, 3, 4

References: 1. Kazanskaya et al. (1983), 2. Salovaara and Katunpää (1984), 3. Spicher (1984), 4. Spicher and Lönner (1985), 5. Foramitti and Mar (1982), 6. Lönner et al. (1986), 7. Spicher and Schröder (1978), 8. Azar et al. (1977), 9. Barber et al. (1989).

species or strains of lactic acid bacteria such as those used in many other fermentation industries have not been very successful in the inoculation of sour dough batches so far.

The yeast flora of sourdoughs is more homogenous. The universal sourdough yeast appears to be *Candida millerii* or closely related to it. *Saccharomyces cerevisiae* is also often reported (Salovaara, 1988). A peculiar symbiosis between the *C. millerii* yeast and *L. sanfrancisco* or *L. brevis* var. *lindnerii* has been described for the San Francisco sourdough French bread process. The lactobacilli in the sourdough utilize mainly maltose and leave glucose for the use of yeast which in turn is incapable of assimilating maltose (Kline and Sugihara, 1971).

The final stage of bread making is baking in an oven. The heat treatment stabilizes the product by denaturing proteins and gelatinizing starch. No live lactic acid bacteria or yeast can be found in the baked bread, although spores of *Bacillus subtilis*, the bread-"rope"-inducing bacteria may survive.

With respect to nutritional value of cereal-based foods fermentation enhances bioavailability of minerals from breads made from high-extraction flour. Degradation of phytate during a rye sourdough bread-making process is shown in Fig. 2. The degradation takes place due to the phytase activity present in flour and in yeast (MacKenzie-Parnell and Davies, 1986). A corresponding phytate degradation takes place in other similar fermentation processes which utilize other cereals (Reddy et al., 1989).

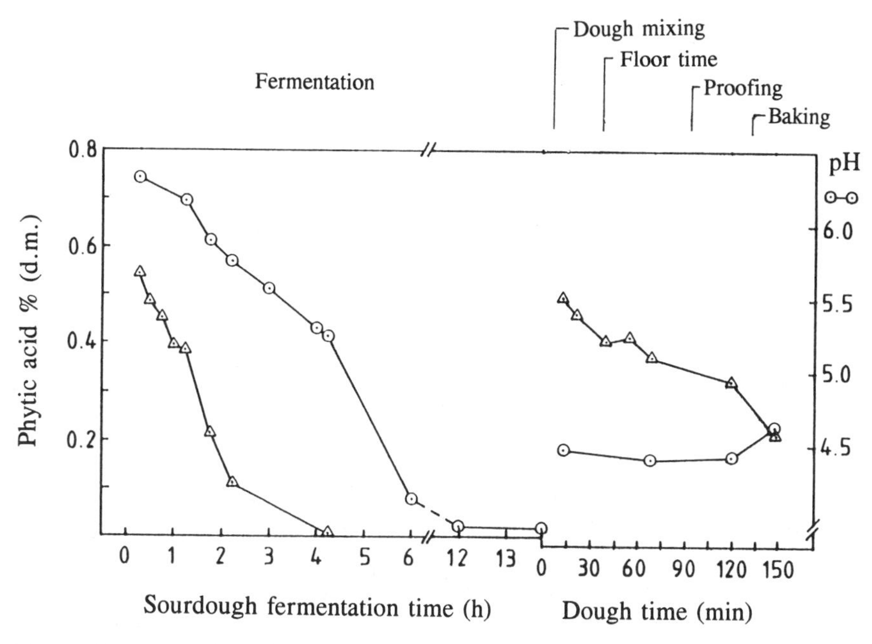

Figure 2 Degradation of phytate during a rye sourdough bread process (Salovaara and Göransson, 1983).

B. Fermented Cereal-Based Soured Porridges, Dumplings, and Beverages

1. Functions of Souring

Sour breads made from various cereals are not the only important cereal-based foods which are made by utilizing lactic acid bacteria. Especially in Africa and in parts of Asia maize, sorghum, millet, and other starchy materials are extensively used for the preparation of various indigenous fermented cereal products other than bread. These foods may be drinkable sour beverages, gruels or porridges eaten with a spoon, or stiff dumplings used in stews and in other foods (Hesseltine, 1979; Campbell-Platt, 1987). Fermented sour porridges have not been unknown in Europe either as indicated by the reports on "sowens," "flummeries," and other similar oat-based products (Fenton, 1974; Campbell-Platt, 1987).

General purposes and functions of lactic acid fermentation in cereal food processes were already discussed (cf. Tables 1 and 2). The preparation of fermented sour cereal foods other than bread often follows the general pattern characterized in Fig. 3. Soaking grains prior to wet-milling gives rise to various fermentation processes. Slurrying in water of the material either from wet- or

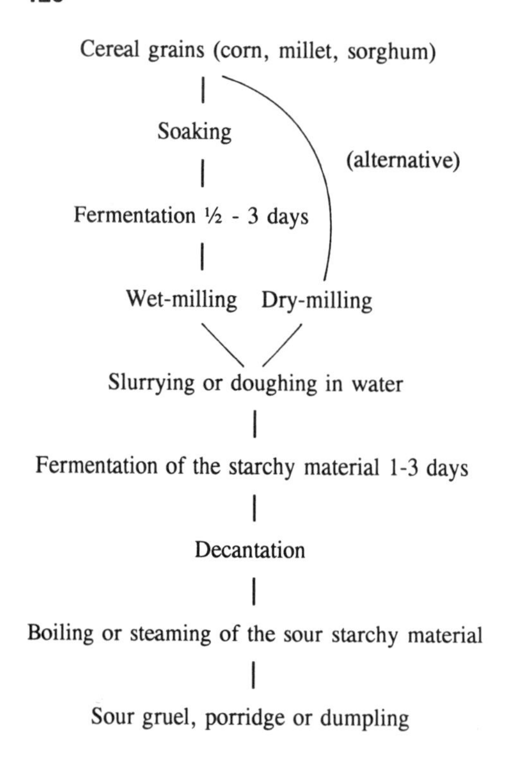

Figure 3 Simplified flow sheet of traditional sour porridge preparation.

dry-milling favors fermentation, which is allowed to take place overnight or longer, usually at ambient temperature. The slurrying or doughing process has many similarities with the sourdough procedures used in European sourdough bread making (Banigo and Muller, 1972). Accumulation of acids occurs much in the same way as in a sourdough for bread as can be seen from the data presented by Akinrele (1970), Abigail Andah and Muller (1973), Mbugua (1981), and others. When the fermentation is completed, the slurry is boiled with a necessary amount of water so that gelatinization of starch occurs and a product of desired consistency is obtained. The final product may be drinkable, spoonable, or stiff dumpling-like.

Examples of local cereal-based products which have been scientifically studied are "ogi" and "agidi" (Nigeria), "koko," "akasa," and "kenkey" (Ghana), "uji" (East Africa), and "mahewu" (Southern Africa). Many of these products are also made commercially for local markets (Odunfa, 1985).

The preparation of "ogi" follows the general pattern shown in Fig. 3. Maize, sorghum, or millet grains are soaked in clean water for one-half day to three days. Mixed fermentations, including some lactic acid fermentation, already take place during the soaking stage (Akinrele, 1970). During soaking the grain softens and becomes easier to wet-mill into a slurry. After grinding, the slurry is sieved and bran particles retained on the sieve are removed. The starchy fraction passes the sieve and is allowed to ferment for two to three days so that metabolites, primarily lactic acid, accumulate. In a fully fermented slurry there may be more than 10^9 CFU/g, or equal cell density to a fermented sourdough (Akinrele, 1970). Lactic acid bacteria identified in soaking and slurrying stages in various processes are listed in Table 4.

The most important microorganisms in the "ogi" souring process were reported to be *Saccharomyces cerevisiae, Aerobacter cloacea*, and *Lactobacillus plantarum* (Akinrele, 1970). "Ogi" is the name for the sour starch sediment, which is a semiproduct and sold locally wrapped in leaves. "Ogi" is prepared for food by boiling it with water to make a porridge or gruel called "agidi." The gel-like "agidi" has the role of a staple food as well as a weaning food in west Africa (Akinrele, 1970; Odunfa, 1985; Adyemi and Oluwamukomi, 1989).

"Koko" and "akasa" from Ghana are reported to be prepared much in the same way as "ogi" (Odunfa, 1985), whereas the preparation of "kenkey" differs in that only part of the fermented dough is boiled, and the boiled fraction is mixed back into uncooked dough and left to ferment for another 3–8 h. The resulting stiff paste is wrapped in leaves, and the dumplings are prepared for food by steaming, boiling, or baking (Amoa and Muller, 1976; Odunfa, 1985; Platt-Campbell, 1987).

Table 4 Identified Species of Lactic Acid Bacteria in Slurries for Fermented Porridges, Dumplings, and Beverages

Species	Fermentation pattern	Substrate reference
Lactobacillus plantarum	homof.	Ogi:1, uji:3 mahew (mageu):4
L. delbruckii	homof.	Mageu:4
Lactobacillus sp.	homof.	Kishk:5
Leuconostoc mesenteroides	heterof.	Koko:2, kenkey:6
Leuconostoc fermenti	heterof.	Koko:2, uji:3, kenkey:6
Pediococcus acidilactici	homof.	Kenkey:6
Streptococcus lactis	homof.	Mahew (mageu):4

References: 1. Odunfa and Adyele (1984), 2. Christian (1970), 3. Mbugua (1981), 4. Hesseltine (1979), 5. Morcos et al. (1973), 6. Amoa and Muller (1976).

"Uji" is a fermented thin gruel or sour beverage from eastern Africa. It is made from a mixture of maize flour and millet, sorghum, or cassava flour. Sucrose is also a regular ingredient. "Uji" is one of the major foods in East African diets. Mbugua (1981) studied its preparation in detail.

2. Cereal-Based Foods Containing Live Lactic Acid Bacteria

Analogous to the final baking stage in bread making, boiling or steaming of the fermented slurry pasteurizes the material at the end of the process and kills the lactic acid bacteria. However, there are a few processes which involve fermentation after the gelatinizing heat treatment, and hence the product contains live lactic acid bacteria.

An example of this type of product is "kishk," a fermented sour milk–wheat mixture widely consumed in Egypt and elsewhere in the Arab world. The process involves actual sour milk fermentation to produce a yogurt-like product, which is mixed with boiled and sun-dried wheat to make a dough. The dough is sun-dried to make dried balls or the "kishk" (Robinson and Cadena, 1978). "Rabadi" is a corresponding fresh type of product from India. It is made from buttermilk and pearl millet (Neerja Dhankher and Chauhan, 1987).

Lactic acid bacteria are also unavoidably present when local country beers are brewed. In contrast, modern beer making relies exclusively on alcoholic fermentation, and lactic acid bacteria are more or less undesirable contaminants. Lactic acid bacteria are in part responsible for the restricted keeping time of local beers (Wood, 1981).

Lactic acid fermentation is utilized in the production of at least two cereal-based sour beverage specialties described in the literature. Representatives of *Lactobacillus* and *Pediococcus* belong to the fermenting flora of "lambic beer" made from barley and wheat, a specialty in Belgium. Typical of the process is a very long fermentation period which takes up to two years or more. The result is a fairly strong, sour alcoholic beer (Verachtert and Dawoud, 1984).

In contrast to the "lambic beer," "mageu" (or mahew) is a nonalcoholic nonyeasted sour maize beverage produced in South Africa (Hesseltine, 1979; McGill and Taylor, 1991). "Magou" is the term used for the commercial counterpart of the traditional "mahew." It is a thin sour maize beverage also produced industrially on a large scale and sold in cartons (McGill and Taylor, 1991). In its original form "mahew" is made by adding maize meal to boiling water and cooking for 10 min. Some wheat flour is added to provide amylolytic activity and a spontaneous inoculum, *Lactococcus lactis* being the main organism. Industrially produced "magou" is made by using starter organisms such as *Lactobacillus plantarum, Lactobacillus delbrueckii*, or *Lactobacillus bulgaricus* (Hesseltine, 1979).

In contrast to many fermented dairy products and pickled vegetables, which contain large quantities of live lactic acid bacteria, cereal-based fermented foods do not generally contain live lactic bacteria except in the few occasions mentioned. On the other hand, interest in the effects of live lactic acid bacteria entering the digestive system has grown enormously in recent years, when the possible beneficial effects of probiotic lactic acid bacteria have been explored.

Cereal-based products containing large quantities of live lactic acid bacteria of the probiotic type are not available. However, research in this area seems to be going on. Shin (1989) reported some experiments with cooked rice fermented with lactic acid bacteria species typical in yogurt fermentation. Recent studies by Molin et al. (1989) and Marklinder (1991) have suggested the potential of a fermented oatmeal soup as a nutritive solution for enteral feeding. Another possibility is to ferment cooked oat bran with probiotic lactic acid bacteria to make a yogurt-type fermented oat bran product (Salovaara et al., 1990; Salovaara and Kurka, 1991). In such a product the postulated physiologically beneficial effects of oat bran and those of probiotic lactic acid bacteria would be combined. However, more research and development are needed in this area.

IV. CONCLUSIONS

Lactic acid bacteria are utilized in the production of cereal-based products in many ways. The lactic acid fermentation contributes beneficially both to processing technology and to quality of the end products in terms of flavor, keeping properties, safety, and overall image of the product. Among cereal foods most scientific research and technological development with respect to lactic acid bacteria has been associated with the rye sourdough process. However, in comparison to dairy processes the research input on lactic acid bacteria in cereal food processes may be regarded as nominal. One obvious reason for this is that the fermentation substrate, the cereal flour, inherently carries a considerable contaminating flora, which restricts possibilities for using specific starters of distinctive properties.

Work remains to be done in many areas. Research on the traditional fermented cereal foods other than bread, i.e., on the soured porridges and dumplings, has been activated in the past two decades or so. However, there is still a considerable lack of information in this area in spite of the significance of these foods for millions of people, especially in Africa and Asia. It is obvious that fermented cereal foods of this category would deserve a major developmental input in order to improve the quality and attractivity of these foods as economical and nutritious staple foods. The studies made so far seem to indicate that proper control of fermentation conditions and application of a starter—possibly from a rebuilding system—would give added value to the local cereal-based foods fermented by lactic acid bacteria. A new possibility would be the use of selected probiotic starters for fermentation of novel cereal-based fermented foods.

REFERENCES

Abigail Andah, and Muller, H. G. (1973). Studies on koko, a Ghanaian fermented maize porridge, *Ghana J. Agric. Sci., 6*:103–108.

Adyemi, I. A., and Oluwamukomi, M. O. (1989). An investigation into the storage stability of agidi, a Nigerian fermented maize gel, *J. Cereal Sci., 10*:239–246.

Akinrele, I. A. (1970). Fermentation studies on maize during the preparation of a traditional African starch-cake food, *J. Sci. Food Agric., 21*:619–625.

Amoa, B., and Muller, H. G. (1976). Studies on kenkey with particular reference to calcium and phytic acid, *Cereal Chem., 53*:365–375.

Azar, M., Ter-Sarkissian, N., Ghavifek, H., Ferguson, T., and Ghassemi, H. (1977). Microbiological aspects of Sangak break, *J. Food Sci. Technol. (India), 14*: 251–254.

Banigo, E. O. I., and Muller, H. G. (1972). Carboxylic acid patterns in ogi fermentation, *J. Sci. Food Agric., 23*:101–110.

Barber, S., Torner, M. J., Martinez-Anaya, M. A., and Benedito de Barber, C. (1989). Microflora of the sour dough of wheat flour bread. IX: Biochemical characteristics and baking performance of wheat doughs elaborated with mixtures of pure microorganisms, *Z. Lebensm. Unters. Forsch., 189*:6–11.

Campbell-Platt, G. (1987). *Fermented Foods of the World. A Dictionary and Guide*, Butterworths, London, p. 291.

Christian, W. F. (1970). Lactic acid bacteria in fermenting maize dough, *Ghana J. Sci., 10*:22–28.

Fenton, A. (1974). "Sowens" of Scotland, *Folk Life—A Journal of Ethnological Studies, 12*:41–47.

Foramitti, A., and Mar, A. (1982). Neue Aspekte und Erkenntnisse zur kontinuerlichen Herstellung von Sauerteig in der Brotindustrie, *Getreide Mehl Brot, 36*: 147–149.

Hesseltine, C. W. (1979). Some important fermented foods in mid-Asia, the Middle-East and Africa, *J. Am. Oil Chemists' Soc., 56*:367–374.

Kazanskaya, L. N., Afanasyeva, O. V., and Patt, V. A. (1983). Microflora of rye sours and some specific features of its accumulation in bread baking plants of the USSR, in *Proc. World Cereal Bread Congress, Cereals '82 Symp.*, Prague, Developments in Food Science, 5B, pp. 759–763.

Kline, L., and Sugihara, T. F. (1971). Microorganisms of the San Francisco sour dough bread process. II: Isolation and characterization of undescribed bacterial species responsible for the souring activity, *Appl. Microbiol., 21*:459–465.

Lund, B., and Hansen, Å. (1987). Fermentation of rye sourdough, *Ugeskrift Jordbruk, Select. Res. Rev., 1987*:47–58.

Lönner, C. (1988). *Starter Cultures for Rye Sour Dough.* Dissertation, Department of Applied Microbiology, Lund University, Lund, Sweden, p. 92 + append.

Lönner, C., Welander, T., Molin, N., Dostalek, M., and Blickstad, E. (1986). The microflora in a sour dough started spontaneously on typical Swedish rye meal, *Food Microbiol., 3*:3–12.

MacKenzie-Parnell, M. M., and Davies, N. T. (1986). Destruction of phytic acid during home breadmaking, *Food Chem., 22*:181–192.

Marklinder, I. (1991). *Fermented Oatmeal Soup—a Nutritive Solution Designed for Enteral Feeding, Fermented by Intestinal Lactobaccilus spp*, Lic. thesis, University of Lund, p. 24 + append.

Mbugua, S. K. (1981). *Microbiological and Biochemical aspects of uji (an East African Sour Cereal Porridge) Fermentation, and Its Enhancement through application of Lactic Acid Bacteria*, dissertation, Cornell University, University Microfilm International, Dissertation Information Service, Ann Arbor, p. 140.

McGill, A. E. J., and Taylor, J. R. N. (1991). Technical developments in African cereal beverages, Paper presented at '91 ICC Symposium, Prague (manuscript).

Molin, G., Albertsson, C-E., Bengmark, S., and Larsson, K. (1989). Nutrient composition and method for the preparation thereof. International Patent Application, PCT, WO 89/08405.

Morcos, S. R., Hegazi, S. M., and El-Damhougi, S. T. (1973). Fermented foods in common use in Egypt, *J. Sci. Food Agric., 24*:1153–1156.

Mukherjee, S. K., Albury, M. N., Pederson, C. S., van Veen, A. G., and Steinkraus, K. H. (1965), Role of *Leuconostoc mesenteroides* in leavening of the batter of idli, a fermented food of India, *Appl. Microbiol., 13*:227–231.

Neerja Dankher, and Chauhan, B. M. (1987). Technical note: Preparation, acceptability and B vitamin content of rabadi—fermented pearl millet food, *Int. J. Food Sci. Technol., 22*:173–176.

Novellie, L. (1981). Fermented porridges, *Proc. Int. Symp. Sorghum grain quality*, Patancheru, India, ICRISAT, pp. 121–128.

Odunfa, S. A. (1985). African fermented foods, in *Microbiology of Fermented Foods*, Vol. 2 (B. J. B. Wood, ed.), Elsevier, New York, pp. 155–192.

Odunfa, S. A., and Adyele, S. (1985). Microbiological changes during the production of ogi-baba, a West African fermented sorghum gruel, *J. Cereal Sci., 3*:175–180.

Pederson, C. S. (1971). Some cereal foods, in *Microbiology of Food Fermentations*, AVI, pp. 173–198.

Reddy, N. R., Pierson, M. D., Sathe, S. K., and Salunkhe, D. K. (1989). *Phytates in Cereals and Legumes*, CRC Press, Boca Raton, FL, p. 152.

Robinson, R. K., and Cadena, M. A. (1978). The potential value of yoghurt-cereal mixtures, *Ecol. Food Nutr., 7*:131–136.

Salovaara, H . (1988). Functions of micro-organisms in sour doughs—An overview of recent research, in *Proc. Int. Symp. Cereal Technology in Sweden*, Ystad, Sweden, pp. 91–101.

Salovaara, H., Bäckström, K., and Mantere-Alhonen, S. (1990). Kiesa," the traditional Karelian fermented oat pudding revisited, in *Proc. 24th Nordic Cereal Congress*, (H. Johansson, ed.), Stockholm, pp. 77–83.

Salovaara, H., and Göransson, M. (1983). Nedbrytning av fytinsyra vid framställning av surt och osyrat rågbröd, *Näringsforskning, 27*:97–101.

Salovaara, H., and Katunpää, H. (1984). An approach to the classification of lactobacilli isolated from Finnish sour rye dough ferments, *Acta Alim. Polon., 10*:231–239.

Salovaara, H., and Kurka, A. (1991). Food product containing dietary fiber and method of making said product, International Patent Application, PCT, WO 91/17672.

Shin, D-H. (1989). A yogurt-like product development from rice by lactic acid bacteria, *Korean J. Food Sci. Technol., 21*:686–690.

Spicher, G. (1984). Die Mikroflora des Sauerteiges. XVII. Mitteilung: Weitere Untersuchungen über die Zusammensetzung und die Variabilität der Mikroflora handelsüblicher Sauerteig-Starter, *Z. Lebensm. Unters. Forsch., 178*:106–109.

Spicher, G., and Lönner, C. (1985). Die Mikroflora des Sauerteiges. XXI. Mitteilung: Die in Sauerteigen schwedischer Bäckereien vorkommenden Lactobacillen, *Z. Lebensm. Unters. Forsch., 181*:9–13.

Spicher, G., and Schröder, R. (1978). Die Mikroflora des Sauerteiges. IV. Untersuchungen über die Art der in 'Reinzuchtsauern' anzutreffenden stäbchenförmigen Milchsäurebakterien (Genus Lactobacillus Beijerinck), *Z. Lebensm. Unters. Forsch., 167*:342–354.

Spicher, G., and Stephan, H. (1987). *Handbuch Sauerteig. Biologie, Biochemie, Technologie*, 3, Aufl., B. Behr's Verlag, Hamburg, p. 397.

Steinkraus, K. H. (ed.) (1983). *Handbook of Indigenous Fermented Foods*, Marcel Dekker, New York, p. 671.

Sugihara, T. F. (1978). Microbiology of the soda cracker process. I: Isolation and identification of microflora, *J. Food Protection, 41*:977–979.

Verachtert, H., and Dawoud, E. (1984). Microbiology of lambic-type beers, *J. Appl. Bact., 57*(3):xi–xii.

Wood, B. J. B. (1981). The yeast/lactobacillus interaction: a study in stability, *Mixed Culture Fermentations* Vol. 5, (M. E. Bushell and J. M. Slater, eds.), Academic Press, London, pp. 137–150.

5

Antimicrobial Components from Lactic Acid Bacteria

P. Michael Davidson
University of Idaho, Moscow, Idaho

Dallas G. Hoover
University of Delaware, Newark, Delaware

I. INTRODUCTION

Lactic acid bacteria of the genera *Lactococcus* (formerly *Streptococcus*), *Lactobacillus, Pediococcus,* and *Leuconostoc* are involved in the production of fermented foods representing virtually all commodity groups including dairy, vegetable, fruit, meat, and cereal products. Historically, the primary role of these microorganisms in fermented foods was one of preservation. With the advent of modern processing and preservation techniques, the importance of that role diminished. While preservation is still important, the lactic acid bacteria are also used to provide variety in the food supply. They carry out this function by altering the flavor, texture, and appearance of raw commodities in a desirable way.

Preservation of fermented foods by lactic acid bacteria is due primarily to sugars being converted to organic acids (lactic, acetic) causing a reduction in pH and removal of carbohydrates as nutrient sources (Gilliland, 1985; Daeschel, 1989). In addition, lactic acid bacteria produce substances including hydrogen peroxide, diacetyl, secondary reaction products, and bacteriocins which have potential to inhibit a variety of other microorganisms (Daeschel, 1989). With the emergence of psychrotrophic foodborne pathogens, the discovery of new food processes and the desire by consumers for "natural" food products, these microorganisms have been recognized as a potential source of biopreservatives for foods. Substantial research on these microorganisms in recent times has focused on such an application.

The objective of this chapter is to acquaint the reader with the various inhibitors produced by lactic acid bacteria. Information on inhibitors includes, where appropriate, chemical or biochemical data, antimicrobial activity spectrum, factors affecting antimicrobial activity, applications or potential applications, mechanism of action, and genetics of production. Detailed information on inhibitors produced by lactic acid bacteria is available in Hurst (1983), Klaenhammer (1988), Daeschel (1989), Delves-Broughton (1990), Hoover (1993), and Hurst and Hoover (1993).

II. NONPEPTIDE INHIBITORS

Acetic (pK_a = 4.75) and lactic (pK_a = 3.08) acids are among the inhibitors produced by lactic acid bacteria (Gilliland, 1985). Acetic acid is generally a more effective inhibitor than lactic acid (Doores, 1990). Acetic acid produced by *Leuconostoc citrovorum* was found to inhibit psychrotrophic bacteria and *Salmonella* (Mather and Babel, 1959; Pinheiro et al., 1968; Sorrels and Speck, 1970).

While organic acids play an important role, they work in concert with other inhibitors to preserve foods. Addition of lactic cultures (*Lactococcus, Lactobacillus, Leuconostoc*, and *Pediococcus*) or their extracts to various food products including milk, meat, eggs, and seafoods has been shown to increase shelf life (Reddy et al., 1970; Juffs and Babel, 1975; Gilliland and Speck, 1975; Raccach et al., 1979; Raccach and Baker, 1979; Martin and Gilliland, 1980; Gilliland and Ewell, 1983; Bremner and Statham, 1983). Inhibition is attributed to a combination of organic acids, hydrogen peroxide and inhibitory substances (Gilliland, 1985). Gilliland and Speck (1972) found that *Salmonella* and *Staphylococcus aureus* were inhibited by lactic streptococci in milk. Inhibition levels were 88.2–93.4% for *Salmonella* and 98.1–98.9% for *S. aureus*. The authors showed that inhibition was due partially to organic acid production and partially to small molecular weight compounds in whey.

Hutton et al. (1991) recently described the use of a combination of *Pediococcus acidilactici* plus dextrose (known as the "Wisconsin process") in chicken salad to protect against the formation of botulinum toxin. At abuse temperatures, *Pediococcus acidilactici* ferments dextrose to produce lactic acid which lowers the pH sufficiently to prevent formation of toxin. This process was shown previously to be effective in reducing the risk of botulinal toxin formation in nitrite-reduced bacon (Tanaka et al., 1980).

Lactobacillus species produce large amounts of hydrogen peroxide (H_2O_2) through pyruvate, L-lactate oxidase, NAD-independent D-lactate dehydrogenase, and NADH oxidase (Daeschel, 1989). The compound accumulates because

Lactobacillus do not produce catalase. Dahiya and Speck (1968) found that *Lactobacillus bulgaricus* (*Lactobacillus delbrueckii* ssp. *bulgaricus*) and *Lactobacillus lactis* (*Lactobacillus delbrueckii* ssp. *lactis*) inhibited the growth of *S. aureus*. Inhibition was due partially to hydrogen peroxide. Storage at low temperature favored hydrogen peroxide formation with the maximum at 5°C and pH 7.0. There was an inverse relationship between acid production and hydrogen peroxide formation. Juffs and Babel (1975) evaluated commercial multistrain cultures containing *Lactococcus lactis* and *Leuconostoc cremoris* added to milk at 0.5% which contained various psychrotrophic microorganisms. Inhibition was demonstrated, but the extent was dependent upon type of starter, initial number and type of psychrotroph, temperature, and time. The antagonist was determined to be hydrogen peroxide since catalase decreased inhibition. As with Dahiya and Speck (1968), storage at low temperature was found to favor hydrogen peroxide formation and an inverse relationship existed between acid production and hydrogen peroxide formation. Price and Lee (1970) studied the effect of added *Lactobacillus* on *Pseudomonas* sp. in oysters. They found that as *Lactobacillus* increased, *Pseudomonas* decreased. They concluded that inhibition was due to hydrogen peroxide. The bacteriostatic concentration for *Pseudomonas* was 2–8 μg/mL at 30°C.

Hydrogen peroxide also functions in the lactoperoxidase system of raw milk (Reiter and Harnulv, 1984; Banks et al., 1986). In this system, lactoperoxidase catalyzes the oxidation of thiocyanate by hydrogen peroxide producing hypothiocyanite ions, higher oxyacids and short-lived oxidation products (Gaya et al., 1991). Potential sources of hydrogen peroxide for this reaction are the lactic acid bacteria.

Diacetyl (2,3-butanedione) is produced by the citrate fermenting lactic acid bacteria, *Leuconostoc cremoris* and *Lactococcus lactis* ssp. *lactis* bv. diacetylactis (formerly *Streptococcus diacetilactis, Streptococcus lactis* ssp. *diacetylactis*, and *L. lactis* ssp. *diacetylactis*). The compound is synthesized from pyruvate. It produces a buttery flavor in fermented dairy products and to foods in which it is added as a flavor additive. The compound has long been known to have antimicrobial activity. Jay (1982) reported that 200 μg/mL diacetyl inhibited yeasts and Gram-negative bacteria, 300 μg/mL inhibited nonlactic Gram-positive bacteria and >350 μg/mL was required to inhibit lactic acid bacteria. Gupta et al. (1973) also reported that diacetyl was most effective against Gram-negatives. Motlagh et al. (1991) recently tested diacetyl at 344 μg/mL against *Listeria monocytogenes, L. ivanovii, L. innocua, Salmonella anatum, S. typhimurium, Yersinia enterocolitica, Escherichia coli*, and *Aeromonas hydrophila*. The compound was only effective against the Gram-negative bacteria causing a slight reduction in viable cells after 24 h at 4°C. The effective concentration for diacetyl is too high for it to play a major role in natural preservation (Gupta et al.,

1973) or as an added biopreservative (Motlagh et al., 1991). Its role in the combination of inhibitors produced by lactic acid bacteria, however, is undoubtedly important.

III. PEPTIDE/PROTEIN INHIBITORS

As was stated above, the lactic acid bacteria produce several inhibitors, one group of which are bacteriocins. Tagg et al. (1976) described bacteriocins as protein-containing macromolecules which exert a bactericidal mode of action on susceptible bacteria. Second, their production and immunity were plasmid-controlled. Klaenhammer (1988) defined bacteriocins as "proteins or protein complexes with bactericidal activity directed against species that are usually closely related to the producer microorganism." In addition, bacteriocins are heterogenous compounds which vary in molecular weight, biochemical properties, activity spectra and mechanism of action. Investigations concerning bacteriocins have focused on identification of novel bacteriocins, characterization of bacteriocin production and mode of action, characterization of plasmids encoding bacteriocin production and immunity, and classification of bacteria based upon bacteriocin production or sensitivity (Klaenhammer, 1988).

Because antimicrobial activity of lactic acid bacteria is due to several compounds, assays for the presence of bacteriocins are designed to minimize the effect of interfering inhibitors. The most used methods are in agar medium and involve direct or deferred methods (Hoover, 1993). In one direct method (Sabine, 1963), the indicator (susceptible) strain is added to molten agar and then to a petri dish. Wells are cut in the agar and then filled with the bacteriocin-producing strain in agar. Sabine (1963) stated that there was no effect of pH with this technique. However, according to Hoover (1993), there is still the possibility that hydrogen peroxide or organic acids may interfere. The deferred method involves some modification of the agar-flip overlay technique of Kekessy and Piguet (1970). In this method, the bacteriocin producer is spot-inoculated onto spread plates of optimal growth and medium and incubated. The agar is flipped into the lid of the plate and the indicator microorganism in soft agar is overlaid. Inhibition is indicated by halo of no growth. This technique is said to minimize the effect of other inhibitors and bacteriophage. Lewus and Montville (1991) recommended the spot-on-the-lawn technique in which TSYE agar (trypticase soy broth, 0.5% yeast extract, and 1.5% agar) is spotted with 2 μL of the bacteriocin producer and incubated overnight at 30°C. Brain heart infusion agar (1.0%) is seeded with log 5.0–6.0 CFU/mL of indicator and ca. 8 mL are added as an overlay to TSYE.

IV. *LACTOCOCCUS*

The inhibitory powers of the *Lactococcus* were probably first recognized in the 1930s. Inhibition of commercial cheese starter cultures by similar dairy bacteria was reported by Whitehead (1933) and Mattick and Hirsch (1944). This inhibition was reported to be caused by inhibitory proteins. Since that time, at least three types of bacteriocins have been described for *Lactococcus*: nisin, diplococcin, and lactostrepcins.

A. Nisin

Nisin was isolated, characterized, and named by Mattick and Hirsch (1947). Since it was produced by group N streptococci, the name was derived from "N inhibitory substance." Nisin is a polypeptide produced by some strains of *Lactococcus lactis* ssp. *lactis* (formerly *Streptococcus lactis*). Commercial production of nisin started in the 1950s and it is currently available under the trade name Nisaplin (Aplin and Barrett, Ltd., England). Jarvis and Farr (1971) described five nisin compounds, A, B, C, D, and E. Nisin A was most active and appears to be the commercial compound (Hurst and Hoover, 1993).

1. *Biochemistry and Stability*

Nisin contains 34 amino acids, 5 thioether bonds, and has a molecular weight of around 3500 daltons (Gross and Morell, 1967; Gross, 1977); however, it usually occurs as a dimer with a molecular weight of 7000 (Jarvis et al., 1968). Nisin contains no aromatic amino acids but does have dehydroalanine and S-amino acids (lanthionine and β-methyllanthionine) which are also common to subtilin (Gross et al., 1969). The hydrochloride of nisin has 5.6% by weight of sulfur (Hall, 1966). Methods for concentration and purification of nisin were described by Cheeseman and Berridge (1957) and Bailey and Hurst (1971). Fukase et al. (1988) were the first to produce nisin synthetically.

The activity of nisin is measured in reading units (RU) or international units (IU) and is based upon the antimicrobial activity of 1 μg of a standard batch of commercial nisin. The approximately activity 1 μg of pure nisin is 40 RU or IU (Tramer and Fowler, 1964).

The solubility of nisin is dependent upon pH. At pH 2.5, the solubility is 12%, at pH 5.0, 4.0%, and it is insoluble at neutral and alkaline pH. In dilute HCl at pH 2.5, nisin is stable to boiling with no marked loss of activity. At pH 7.0, inactivation occurs even at room temperature (Hurst, 1983). Heinemann et al. (1965) studied the effect of processing temperature and pH on stability of nisin in foods. In low-acid (pH 6.1–6.9) or high acid (3.3–4.5) foods, heating for 3 min at 250°F destroyed 25–50% of the nisin. Hurst (1981) reported that nisin was stable after treatment at 115.6°C, pH 2. At the same temperature, 40% and 90% loss of activity occurred at pH 5 and 6.8, respectively. Less inactivation occurred in complex medium (milk, microbiological broth) due to a protective effect. Nisin remains stable for years in the dry form, but in foods nisin activity is gradually lost (Hurst, 1983). Delves-Broughton (1990) determined nisin retention in two brands of pasteurized process cheese at 20, 25, and 30°C. Nisin was added at 250 IU/g to the cheese (pH 5.6–6.0) which had 54–58% moisture. At 20°C, ca 90% of the nisin was retained after 30 weeks while ca. 60% and 40% was retained at 25°C and 30°C, respectively.

Nisin is inactivated by the proteases pancreatin and α-chymotrypsin (Jarvis and Mahoney, 1969). It is resistant to the effects of trypsin, elastase, carboxypeptidase A, pepsin, erepsin, and pronase.

2. Antimicrobial Spectrum

Nisin has a narrow spectrum of inhibitory activity affecting only Gram-positive bacteria including lactic acid bacteria, streptococci, bacilli, and clostridia (Hawley, 1957). It does not generally inhibit Gram-negative bacteria, yeasts, or molds.

The range of concentrations necessary for inhibition of spores of *Bacillus* and *Clostridium* has been studied by a number of researchers (Rayman et al., 1981). Nisin has been reported to be sporicidal. It inhibits germination of spores at the stage of preemergent swelling (Hitchins et al., 1963; Lipinska, 1977). Heat-injured spores have increased sensitivity to nisin. The inhibitory range for *Bacillus* sp. in microbiological medium was from 0.04 to 2.0 μg/mL (1.6–80 IU/mL) (Gould, 1964). *Bacillus stearothermophilus* was sensitive to <0.05 μg/mL (2 IU/mL) while sensitivity of other *Bacillus* sp. was in the range of 0.075–0.325 μg/mL (3–13 IU) (O'Brien et al., 1956; Jarvis, 1967). *Bacillus cereus, B. megaterium*, and *B. polymyxa* had sensitivities of 1.88–2.5 μg/mL (75–100 IU), 0.625–2.5 μg/mL (25–100 IU/mL), and 1.25 μg/mL (50 IU/mL), respectively (Jarvis, 1967; Gupta et al., 1972). Gould and Hurst (1962) found that *Bacillus* sensitivity could be divided into two groups. Spores which ruptured their envelope by mechanical pressure (e.g., *Bacillus subtilis*) were inhibited by 0.125 μg/mL (5 IU/mL). In contrast, spores which outgrew by lytic mechanism (e.g., *Bacillus cereus*) required 2.5 μg/mL (100 IU/mL) nisin for inhibition.

Hirsch and Grinsted (1954) reported that *Clostridium bifermentans, Clostridium sporogenes*, and *Clostridium butyricum* spores were more sensitive than vegetative cells. Inhibition was related to inoculum size. For example, inhibition of log 2.2 *C. butyricum*/mL was achieved with nisin, whereas log 4.2/mL was not. These researchers also reported that the compound was sporicidal to *C. butyricum* but not others. One of the first reports on *Clostridium botulinum* showed that the microorganism was inhibited by 100 μg/mL (4000 IU/mL) nisin (Denny et al., 1961). Later, Scott and Taylor (1981a,b) evaluated *Clostridium botulinum* types

A, B, and E for their nisin susceptibility in brain heart infusion (BHI) broth and cooked meat medium (CMM). Maximum nisin concentrations required for inhibition of the organism in BHI were 200, 80, and 20 μg/mL (8000, 3200, and 800 IU/mL) for types A, B and E respectively. The concentration required to inhibit *C. botulinum* in CMM was not determined as it was beyond the highest concentrations tested for types A (>200 μg/mL) and B (>80 μg/mL). It was theorized that the higher levels required in CMM were due to binding of the nisin by meat particles. *C. sporogenes* was inhibited in a meat system (pork) by 5–75 μg/mL (200–3000 IU/mL) nisin at pH 6.5–6.6 (Rayman et al., 1981). Okereke and Montville (1991) demonstrated that the nisin producer, *Lactococcus lactis* ssp. *lactis* 11454, was capable of inhibiting all 11 strains of type A or B *C. botulinum* spores tested. The minimum number of cells required to inhibit log 4.0 CFU/mL of types A and B *C. botulinum* was log 5.8 CFU/mL in a direct antagonism assay.

A wide variety of vegetative microorganisms are inhibited by nisin. Because it is a bacteriocin, closely related strains such as *L. lactis* ssp. *cremoris* are most sensitive. Other species inhibited include *Corynebacterium, Lactobacillus, Listeria, Micrococcus, Mycobacterium, S. aureus,* and *Streptococcus* (Mattick and Hirsch, 1947; Gowans et al., 1952; Mohamed et al., 1984; Ogden and Tubb, 1985; Daeschel et al., 1988; Benkerroum and Sandine, 1988; Radler, 1990a,b).

A number of excellent studies have been carried out recently investigating the activity of nisin against *Listeria monocytogenes* and other foodborne pathogens. Mohamed et al. (1984) used an initial number of log 5.0 CFU/mL of *L. monocytogenes* in broth and achieved complete inhibition at 32 or 256 IU/mL (ca. 1 or 7 μg/mL), depending on strain. At lower concentrations, viable cells decreased below detectable levels but eventually recovered to a level equivalent to the control. They demonstrated inhibition was more pronounced at pH 5.5 than pH 7.3. Harris et al. (1989) used a deferred antagonism assay and well diffusion assay in BHI to evaluate the effects of *L. lactis* ssp. *lactis* 11454 and *L. lactis* ssp. *lactis* SIK83 (nisin producers) against eight strains of *L. monocytogenes*. Both inhibited all eight strains but only in a deferred antagonism assay. Inhibition was prevented by proteolytic enzymes.

Carminati et al. (1989) tested seven strains of *L. lactis* ssp. *lactis* against six strains of *L. monocytogenes*. They used neutralized, filter-sterilized catalase-treated culture extracts so as to eliminate the effect of organic acids or hydrogen peroxide. Three strains of *L. lactis* ssp. *lactis* (2,16,26) inhibited all *L. monocytogenes* strains. Pure nisin used at 100 IU/mL was also effective. A bactericidal effect was achieved but no lysis occurred. There was a wide sensitivity pattern of the extracts produced by the *L. lactis* ssp. *lactis* strains to proteolytic enzymes and none were the same as nisin. The authors termed these "bacteriocin-like" substances.

Spelhaug and Harlander (1989) used an agar spot assay to determine the effectiveness of *L. lactis* ssp. *lactis* 11454 (nisin producer) against a number of foodborne pathogens. The microorganism inhibited *Bacillus cereus, Clostridium perfringens, L. innocua* (4 strains), *L. ivanovii, L. monocytogenes* (14 strains), *L. seeligeri, L. welshimeri* (2 strains), and *S. aureus.* Based upon zone size, *C. perfringens* was most sensitive. Slight activity was detected against *Aeromonas hydrophila* (2 strains), *E. coli* O157:H7, *Vibrio cholerae*, and *Vibrio parahaemolyticus.* No inhibition of *Campylobacter jejuni, Salmonella enteritidis* (3 serotypes), or *Y. enterocolitica* was demonstrated. This is one of the few reports that indicated inhibition of Gram-negative microorganisms by a nisin producer. The researchers attempted to determine whether inhibition of Gram-negative bacteria was related solely to nisin. They found that pronase and α-chymotrypsin eliminated activity while trypsin had no effect. They concluded that another inhibitor may be causing inhibition.

Motlagh et al. (1991) carried out a similar study using nisin from *L. lactis* ssp. *lactis* DL16 and commercial nisin (Nisaplin). The compounds were tested at 1500 activity units (AU)/ml and 500 IU/mL, respectively. *L. monocytogenes* (9 strains), *L. ivanovii, L. innocua, Salmonella anatum, S. typhimurium, Y. enterocolitica* (2 strains), *E. coli* (2 strains), and *Aeromonas hydrophila* (2 strains) were the indicator organisms in a disk assay. Nisin and Nisaplin inhibited all strains of *Listeria* but none of the Gram-negative bacteria. Nisin and Nisaplin caused an average decrease in viable *L. monocytogenes* cells of 2.1 logs and 3.4 logs, respectively, after 24 h at 4°C. Growth at 4°C of *L. monocytogenes* Ohio$_2$, Scott A and *Y. enterocolitica* in the presence of the inhibitors was evaluated. Strain Ohio$_2$ was the most sensitive to both inhibitors followed by strain Scott A. *Yersinia enterocolitica* was not inhibited by either compound.

Harris et al. (1991) determined the effect of nisin on *L. monocytogenes* Scott A, ATCC 19115 and UAL500 using a direct plating assay. The found that *L. monocytogenes* demonstrated at biphasic survivor curve in BHI agar at pH 6.5. At 0–10 μg/mL, a 5–7 log decrease in survivors occurred. At 10–50 μg/mL, no additional decrease in viability was demonstrated and approximately 100–1000 survivors/mL remained. Colonies of Scott A and ATCC 19115 from 50-μg/mL plates were subcultured and found to have significantly more resistance at 50 μg/mL than the original cultures. Resistance was not due to plasmid DNA. In addition, Harris et al. (1991) investigated the effect of salt addition and pH on nisin inhibition of *L. monocytogenes.* Salt (2.5%) increased the activity of 0–10 μg/mL nisin but had no effect at 10–50 μg/mL. A reduction in pH from 6.5 to 5.5 increased the activity of 10-μg/mL nisin. There was no difference in activity between nisin acidified with HCl or lactate, therefore it was determined not to be associated with type of acid. The authors stated that low pH did not contribute to stability of nisin as some had previously suggested.

3. Factors Affecting Activity

From the above discussion, several factors which affect the activity of nisin can be delineated. In foods, nisin activity decreases with time most likely due to a degradation of the compound. Hirsch (1951) showed that in Swiss-type cheese made with a nisin-producing culture at day 2 there were 270 IU/g while at day 20 nisin was undetectable. Inactivation has been demonstrated in processed Swiss cheese (McClintock et al., 1952), canned mushrooms (Denny et al., 1961), chocolate milk (Fowler and McCann, 1971), simulated cooked ham (Rayman et al., 1981), and pasteurized process cheese (Delves-Broughton, 1990).

Increasing the size of inoculum decreases the effectiveness of nisin (Hirsch and Grinsted, 1954). Ramseier (1960) found that inhibition of *C. botulinum* and *C. butyricum* was correlated with spore load. Increasing the load 10 fold required an increase in nisin of 0.5 log IU (Hurst and Hoover, 1993). The significance of this finding is twofold. First, a food processor could not use nisin to overcome poor sanitation. Second, as was reported by Motlagh et al. (1991) and Harris et al. (1991), a resistant subpopulation of susceptible microorganisms may be selected.

One possible cause for resistance of microorganisms to nisin is the production of the enzyme nisinase. Nisinase has been demonstrated in *Lactobacillus plantarum, Streptococcus thermophilus*, other lactic acid bacteria, and certain *Bacillus* species (Kooy, 1952; Alifax and Chevalier, 1962; Jarvis, 1967). The nisinase from *S. thermophilus* did not inactivate penicillin, aureomycin, bacitracin, polymyxin, or gramicidin. Nisinase from *Bacillus cereus* inactivated subtilin but not bacitracin, polymyxin, or gramicidin (Jarvis, 1967). The nisinase from several *Bacillus* reduced the C-terminal dehydroalanyllysine of nisin to alanyllysine. This suggested that it was a dehydropeptide reductase (Jarvis, 1970; Jarvis and Farr, 1971).

4. Mechanisms of Action

Nisin is a cationic polypeptide which acts as a surface-active cationic detergent (Hall, 1966; Ruhr and Sahl, 1985). Anionic soaps are capable of neutralizing its activity (Ramseier, 1960). The first step in nisin inactivation of a microorganism is adsorption (Hurst, 1983). Ramseier (1960) showed that nisin adsorbed to sensitive vegetative cells of *C. butyricum* while resistant cells did not adsorb the compound. Therefore, the primary site of action of nisin is the cell membrane resulting in disruption of the cytoplasmic membrane and resultant release of cytoplasmic material (Ramseier, 1960; Morris et al., 1984; Ruhr and Sahl, 1985). In spores, the cytoplasmic membrane is apparently destroyed immediately after germination (Lueck, 1980). Gross and Morell (1970) reported that the unsaturated amino acids in the nisin molecule competed with sulfhydryl-containing enzymes, such as coenzyme. A. Linnett and Strominger (1973) found that high concentrations of nisin caused inhibition of peptidoglycan synthesis in *Bacillus stearothermophilus* or *E. coli.*

5. Genetic Control

Kozak et al. (1974) first suggested that plasmid DNA might be involved in production of nisin. In *L. lactis* ssp. *lactis* 11454, curing of 28–30 MDa plasmid caused loss of nisin production, nisin immunity, and sucrose-fermenting ability. In transconjugants however, researchers have been unable to demonstrate presence of appropriate plasmids, which could be due to low copy number (Klaenhammer, 1988). The genes which code for production and immunity reside on large (> 15 MDa) plasmids often with other determinants such as sucrose metabolism in *L. lactis* ssp. *lactis* (LeBlanc et al., 1980; Gasson, 1984; Gonzalez and Kunka, 1985; Steele and McKay, 1986) and bacteriophage resistance in a 40-MDa plasmid of *L. lactis* ssp. *lactis* bv. diacetylactis (McKay and Baldwin, 1984). Steen et al. (1991), using *Lactococcus lactis* ssp. *lactis* 11454, have recently provided evidence that nisin production and resistance are located on the chromosome rather than a plasmid. They suggested that genes for nisin biosynthesis and immunity may be organized as an operon.

The nisin-producing gene from *L. lactis* ssp. *lactis* has been cloned (Buchman et al., 1988; Kaletta and Entian, 1989; Dodd et al., 1990) along with the nisin resistance gene in *L. lactis* ssp. *lactis* bv. diacetylactis (Froseth et al., 1988) and the nisin resistance gene in *L. lactis* ssp. *lactis* (von Wright et al., 1990). Broadbent and Kondo (1991) recently were successful in generating a nisin-producing *L. lactis* ssp. *cremoris* from *L. lactis* ssp. *lactis* ATCC 11454 donor by direct-plate conjugation. Transconjugants retained bacteriophage resistance and ability to produce acid in milk. This accomplishment could allow for development of nisin-producing *L. lactis* ssp. *cremoris* strains for use in making cheese (Hurst and Hoover,1993). Froseth and McKay (1991) were able to sequence the nisin resistance gene of *Lactococcus lactis* ssp. *lactis* bv. diacetylactis and identify the protein encoded by the sequence.

6. Application to Foods

Application of nisin as a food preservative has been studied extensively (Marth, 1966; Lipinska, 1977; Hurst, 1981, 1983; Delves-Broughton, 1990). Nisin-producing starter cultures were first used in foods as a preservative by Hirsch et al. (1951) to prevent "blowing" of Swiss-type cheese caused by *Clostridium tyrobutyricum* and *C. butyricum.* This was followed by a similar application for the preservation of processed Swiss-type cheese (McClintock et al., 1952). Later, Hawley (1957) recommended the addition of a nisin-containing skim milk powder. Tanaka et al. (1986) found that pH, sodium chloride and phosphate influence the amount of nisin required to inhibit *C. botulinum* in pasteurized process cheese. Somers and Taylor (1987) showed that nisin at 500–10,000 IU/g was effective in delaying or preventing growth and toxin production by *C. botulinum* types A and B in pasteurized process cheese spread. Higher levels of nisin were required in cheese with high moisture and low levels of salt and phosphate. Nisin at 30–50 IU/mL was reported to double the shelf life of pasteurized milk stored at chilled, ambient, and elevated temperature (Delves-Broughton, 1990). This could be of benefit in warm climates where shelf life of the product is short. Shehata et al. (1976) used low levels of nisin in sterile whole buffalo milk and chocolate milk and reduced the heat process by 80%. Products were stored successfully for 24 days at 37°C.

Nisin has been recommended for use in canned vegetable products to prevent the outgrowth of *C. botulinum* when less severe sterilization conditions are designed or required (Campbell et al., 1959; Denny et al., 1961). Regulations in the United Kingdom require that nisin-treated low-acid canned foods receive a minimum F_0 (equivalent time at 250°F or 121.1°C) of 3 min to destroy *C. botulinum* spores (Delves-Broughton, 1990). However, heat-resistant thermophilic sporeformers such as *Bacillus stearothermophilus* and *Clostridium thermosaccharolyticum* may still survive a minimum sterilization process. This could result in spoilage of the product if it were stored at elevated temperatures (e.g., in warm climates). With nisin, thermophilic spoilage may be avoided and food processed at lower temperatures. Canned products in which nisin has been shown to be of value include beans in tomato sauce (Gillespy, 1957), tomato juice (Campbell et al., 1959), soups (Bardsley, 1962), evaporated milk (Gregory et al., 1964), mushrooms (Heinemann et al., 1963), cream style corn and chow mein (Wheaton and Hays, 1964), and peas (Kiss et al., 1968). In addition, nisin could be used in high-acid canned foods to control *Clostridium pasteurianum* and *Bacillus* sp. (Delves-Broughton, 1990).

The compound has been shown to have potential benefit in some meat products. Jarvis and Burke (1976) demonstrated that 400 mg/kg nisin, in conjunction with 0.1% sorbic acid and 2.5% (w/w) polyphosphate, retarded spoilage of fresh sausage at 5°C. Nisin (150–200 mg/kg) has been suggested as an adjunct to nitrite in cured meats for the purpose of preventing the growth of clostridia (Caserio et al., 1979; Holley, 1981). Rayman et al. (1981) used nisin as an alternative or adjunct to nitrite in simulated cooked ham against *C. sporogenes* PA3679. The compound at 3000 IU/mL prevented spoilage for 56 days at 37°C. Rayman et al. (1981) stated that nitrite could be reduced from 150 to 40 ppm in the presence of nisin and retain color and safety. Other researchers have demonstrated variable to poor success with nisin in meats (Rayman et al., 1983; Calderon et al., 1985; Taylor and Somers, 1985; Taylor et al., 1985; Bell and DeLacy, 1986, 1987; Henning et al., 1986a). According to Delves-Broughton (1990), only high levels of nisin are effective against *C. botulinum* in cured meats making it less than economically feasible as an additive. Reasons for reduced activity in meats have been reported as binding of the nisin by meat particles, uneven distribution, poor solubility, or interference by phospholipids (Delves-Broughton, 1990).

Nisin has recently been recommended for the preservation of beer, ale, wine, and fruit brandies against spoilage by lactic acid bacteria (Henning et al., 1986b; Ogden et al., 1988; Radler, 1990a,b). The reason it is used successfully is that nisin does not effect yeast used for the fermentations. Daeschel et al. (1991) isolated nisin-resistant mutants of *Leuconostoc oenos* which were then used to carry out a malolactic fermentation in wines treated with 100 IU/mL nisin. Nisin was effective in inhibiting the spoilage bacterium, *Pediococcus damnosus*. Therefore, the potential exists for partial replacement of the sulfur dioxide needed to inhibit spoilage microorganisms in wine.

7. Regulation and Toxicology

Nisin is permitted for use in approximately 47 countries (Delves-Broughton, 1990). Food products for which it is approved include cheeses, processed cheeses, canned vegetables and fruits, confectionery creams, milk, milk products, cooked meats, custard, ice, bakery products, and mayonnaise. It has been approved by the United States Food and Drug Administration (FDA) for use in pasteurized cheese spreads and pasteurized process cheese spread to inhibit the growth of *C. botulinum* at a maximum of 250 ppm (FDA, 1988).

Since it occurs naturally, nisin is generally considered nontoxic. In one study of 251 raw milk samples from nine countries, 109 had nisin-producing *L. lactis* ssp. *lactis* present (Delves-Broughton, 1990). The oral LD_{50} in mice is 6950 mg/kg body weight, which is similar to that of common salt (Hara et al., 1962). Nisin was found to have no effect on animals in studies of subchronic or chronic toxicity, reproduction or sensitization (FDA, 1988; Frazer et al., 1962). There is no evidence of cross-reaction which might affect the therapeutic value of antibiotics (Carlson and Bauer, 1957; Szybalski, 1953; Hossack et al., 1983). The U.S. FDA recently set the acceptable daily intake for nisin at 2.9 mg/ person/day (53 FR 11247). According to the WHO/FAO Experts Committee, the acceptable daily intake is 0–33,000 IU/kg body weight/day.

B. Diplococcin

Diplococcin was first described by Whitehead (1933). It is produced by some strains of *L. lactis* ssp. *cremoris*. Apparently few *L. lactis* ssp. *cremoris* strains produce diplococcin, as Davey and Richardson (1981) tested 150 strains and only 11 were producers. In contrast to nisin, diplococcin lacks the S-containing amino acids, lanthionine, and β-methyllanthionine (Davey and Pearce, 1980; Gross et al., 1969) and dehydroalanine and dehydrobutyrine (Davey and Pearce, 1982).

The compound is unstable when purified but is stabilized in a complex medium. Diplococcin is sensitive to trypsin, pronase, and α-chymotrypsin (Oxford, 1944; Davey, 1981; Davey and Pearce, 1982). It has a molecular weight of 5300 daltons (Davey and Richardson, 1981; Gross and Morell, 1967). The inhibitory spectra of diplococcin is limited to other strains of *L. lactis* ssp. *lactis* and *L. lactis* ssp. *cremoris*. It is ineffective against sporeforming bacteria. Diplococcin causes immediate cessation of DNA and RNA synthesis in sensitive cells but no lysis (Davey, 1981). Production of diplococcin is plasmid associated as was demonstrated by conjugal mating using diplococcin-negative variants as recipients (Davey, 1984). A 54-MDa conjugative plasmid regulates biosynthesis of diplococcin in *Streptococcus cremoris* 346 (Davey, 1984).

Because diplococcin-producing *L. lactis* ssp. *cremoris* will predominate in a cheese culture, nonproducers are desirable. Davey and Pearce (1980) developed a nonproducer strain in which fermentative ability and bacteriophage susceptibility were not affected.

C. Lactostrepcins

Kozak et al. (1978) assayed culture supernatants of 67 non-nisin-producing strains of *L. lactis* for inhibition to streptococci at pH 4.6. All non-nisin-producing *L. lactis* ssp. *lactis* inhibited the indicators. Only 12 of 13 *L. lactis* ssp. *lactis* bv. diacetylactis and 1 of 17 *L. lactis* ssp. *cremoris* produced inhibition. The inhibitors were termed *lactostrepcins*. The spectrum of antimicrobial activity of the compounds includes *Lactococcus* strains, *Streptococcus* groups A, C, and G, *Bacillus cereus*, a few *Leuconostoc* and *Lactobacillus* (Zajdel and Dobrzanski, 1983). Five types of lactostrepcins have been identified based upon range of bactericidal activity (Dobrzanski et al., 1982). *L. lactis* ssp. *lactis* 300 produced two types, Las 2 and Las 3, during different stages of growth (Bardowski et al., 1979). Lactostrepcin 5 (Las 5) was isolated from *Streptococcus cremoris* 202. It had no unusual amino acids, no lipid component, and a molecular weight of 6000–20,000 (Laemmli and Favre, 1973; Davey and Richardson, 1981). Lactostrepcins from *L. lactis* ssp. *lactis* 10 and 300 were heat-resistant (100°C, 10 min (10) or 121°C, 15 min (300)). Optimum pH for activity of the lactostrepcins is 4.6–5.0, with no activity at pH 7.0. They are susceptible to trypsin, α-chymotrypsin, pronase, and phospholipase (Bardowski et al., 1979; Dobrzanski et al., 1982). No plasmid linkage was established for production or immunity of lactostrepcins (Dobrzanski et al., 1982). The mode of action of Las 5 involves immediate inhibition of RNA, DNA, and protein synthesis, however, this is apparently secondary to membrane disruption and loss of intracellular ATP and K^+ ions (Zajdel et al., 1985). While sensitive and insensitive cells adsorb Las 5 equally well, protoplasts of sensitive strains were not affected, indicating that cell wall receptors may be required.

D. Other Inhibitors

Branen et al. (1975) tested *Leuconostoc citrovorum* and *Streptococcus diacetilactis* (*L. lactis* ssp. *lactis* bv. diacetylactis) against *Pseudomonas*. While inhibition by *Leuconostoc* was attributed to organic acid inhibition, *S. diacetilactis* produced a small-molecular-weight, heat-stable protein which was active against *Pseudomonas* at pH 6.0.

Geis et al. (1983) tested a variety of primarily commercial starter strains for bacteriocin activity in agar and broth. Of 54 *L. lactis* ssp. *lactis* tested, 14 were positive in agar and 5 in broth. Of 133 *L. lactis* ssp. *cremoris* strains and 93 *L. lactis* ssp. *lactis* bv. diacetylactis strains, 15 and 36 and 10 and 1 were inhibitory in agar and broth, respectively. The inhibitors produced were classified into eight groups based upon antimicrobial spectra, proteolytic enzyme susceptibility, heat stability, and cross reactivity with other bacteriocin producers. One group of *L. lactis* ssp. *lactis* inhibited a high percentage of *Clostridium* strains tested. There was no effect of the inhibitors against *S. aureus* and little against *Bacillus subtilis*.

Cell-free filtrates of 14 strains of *L. lactis* ssp. *lactis* bv. diacetylactis inhibited the growth of *S. aureus* S6 and *Pseudomonas fragi* (Reddy and Ranganathan, 1983). The extent of inhibition was greater against *P. fragi* than *S. aureus*. *Streptococcus lactis* ssp. *diacetylactis* strain S_1-67/C was inhibitory to a variety of Gram-positive and Gram-negative microorganisms. These results are indicative of inhibition by organic acids.

V. *PEDIOCOCCUS*

A. Pediocin A

In cucumber fermentations with *Pediococcus pentosaceus* FBB-61 and *Lactobacillus plantarum* starters, Etchells et al. (1964) noted inhibition of lactobacilli. This was felt to be more than organic acid inhibition since *Lactobacillus plantarum* is very resistant to acid. Later, bacteriocins were identified from *P. pentosaceus* FBB-61, FBB-63, and L-7230 which were designated pediocin A. Pediocin A is bactericidal, nondialyzable, heat-stable (100°C, 60 min), and sensitive to pronase.

Fleming et al. (1975) assayed 16 strains of *Pediococcus* for inhibition against bacteria and yeasts associated with fermentation of brined cucumbers. *Pediococcus pentosaceus* FBB-61 and L-7230 had similar antimicrobial spectra against the Gram-positive bacteria, *Lactobacillus plantarum, Leuconostoc mesenteroides, Micrococcus luteus, Streptococcus faecalis, S. aureus*, and *Bacillus cereus*. The two *Pediococcus* strains had no effect on yeasts or Gram-negative bacteria. Daeschel and Klaenhammer (1985) showed that pediocin A inhibited other pediococci, *Lactobacillus brevis, Lactobacillus plantarum, L. lactis* ssp. *lactis*

ATCC 11454 (nisin producer), *S. aureus, C. botulinum*, and *C. sporogenes*. Harris et al. (1989) used deferred antagonism and well assays on BHI to test *P. pentosaceus* FBB61-1 and L-7230 (pediocin A producers) and *P. pentosaceus* FBB61-2 (nonbacteriocin producer) against eight strains of *L. monocytogenes*. While the nonbacteriocin producer had no effect, the pediocin A producers inhibited all eight strains in the deferred antagonism assay only. Proteases were found to reduce activity of the bacteriocin. Spelhaug and Harlander (1989) further determined the spectrum of *P. pentosaceus* FBB61 and FBB63-DG2 against foodborne pathogens in an agar spot assay. The microorganisms inhibited *B. cereus, C. perfringens, L. innocua* (4 strains), *L. ivanovii, L. monocytogenes* (14 strains), *L. seeligeri, L. welshimeri* (2 strains), and *S. aureus*. There was no effect on *Aeromonas hydrophila* (two strains), *E. coli* O157:H7, *Vibrio cholerae, Vibrio parahaemolyticus, Campylobacter jejuni, S. enteritidis* (three serotypes), or *Y. enterocolitica*. Okereke and Montville (1991) found that *Pediococcus pentosaceus* 43200 and 43201 were capable of inhibiting all 11 strains of type A or B *C. botulinum* spores in a deferred antagonism assay. The minimum number of cells required to inhibit log 4.0 CFU/mL types A and B *C. botulinum* was log 5.2 CFU/mL and log 6.6 CFU/mL for strains 43200 and 43201, respectively, in a direct antagonism assay. *Pediococcus pentosaceus* 43200 was reported to be the most promising among strains of *Lactococcus, Lactobacillus, Pediococcus*, and *Streptococcus* for bacteriocin-mediated control of *C. botulinum*.

Graham and McKay (1985) first demonstrated that bacteriocin biosynthesis by *Pediococcus cerevisiae* (*pentosaceus*) FBB-63 was linked to a 10.5-MDa plasmid. Bacteriocin production and host cell immunity of *P. pentosaceus* FBB-61 were shown to be controlled by plasmid DNA (Daeschel and Klaenhammer, 1985). A 13.6-MDa plasmid (pMD136) was identified in bacteriocin-producing, bacteriocin-resistant *P. pentosaceus* FBB-61. Loss of pMD136 resulted in loss of bacteriocin production and immunity.

B. Pediocin PA-1

A bacteriocin, designated pediocin PA-1, was found to be produced by *Pediococcus acidilactici* PAC 1.0 (Gonzales and Kunka, 1987). The bacteriocin is proteinaceous, bactericidal, sensitive to pronase, papain, pepsin, and α-chymotrypsin, and heat-stable (100°C, 10 min). Its molecular weight is approximately 16,500. The initial study of antimicrobial activity of this bacteriocin showed that it inhibited other pediococci, some lactobacilli, and strains of *Leuconostoc mesenteroides* ssp. *dextranicum* (Gonzales and Kunka, 1987). There was no activity against lactococci. In a study with similar strains of *P. acidilactici* (PO2, B5627, PC) and *P. pentosaceus* MC-03, Hoover et al. (1988) found that the microorganisms inhibited other pediococci, *S. aureus, B. cereus, Streptococcus faecalis, Leuconostoc mesenteroides*, and four of five *L. monocytogenes* strains. Inhibition of *L. monocytogenes* 19113 by dialyzed growth supernatant of *P. acidilactici* PO2 was found to be reversed after prolonged exposure (Hoover et al., 1989). This phenomenon was similar to that shown for nisin (see above) and was suggested to be selection of resistant subpopulation or binding of the bacteriocin (Hoover et al., 1989). Harris et al. (1989) demonstrated that *P. acidilactici* PAC 1.0 was effective in inhibiting all eight *L. monocytogenes* strains tested in two different ways. Inhibition was prevented by proteolytic enzymes. Pucci et al. (1988) used a powder of *P. acidilactici* PAC 1.0 growth extract to inhibit *L. monocytogenes* LM01 over pH 5.5–7.0 at 4 and 32°C. The extract was effective against 100–1000 *L. monocytogenes* per mL in cottage cheese, half-and-half, and cheddar cheese soup for two weeks. Biosynthesis of pediocin PA-1 by *P. acidilactici* PAC 1.0 was shown to be associated with a 6.2-Mda plasmid (Gonzales and Kunka, 1987). Bacteriocin activity of *P. acidilactici* (PO2, B5627, PC) and *P. pentosaceus* MC-03 was associated with a 5.5-MDa plasmid (Hoover et al., 1988).

C. Pediocin AcH

A bacteriocin produced by *P. acidilactici* H and designated pediocin AcH was characterized by Bhunia et al. (1987, 1988). The compound has a molecular weight of 2700 daltons and is sensitive to trypsin, ficin, papain, chymotrypsin, and proteases IV, XIV, XXIV, and K. Pediocin AcH is resistant to heat (121°C, 15 min) and organic solvents and is reported to be effective at pH 2.5–9.0.

Motlagh et al. (1991) did an extensive study on the antimicrobial activity of pediocin AcH. They first tested 1500 AU/mL of pediocin AcH against *L. monocytogenes* (9 strains), *L. ivanovii, L. innocua, S. anatum, S. typhimurium, Y. enterocolitica* (2 strains), *E. coli* (2 strains), and *Aeromonas hydrophila* (2 strains) in a disk assay. Pediocin AcH inhibited all strains of *Listeria* but none of the Gram-negative bacteria. The compound caused an average decrease in viable *L. monocytogenes* cells of 2.9 logs after 24 h at 4°C. Against *L. monocytogenes* strain $Ohio_2$, a minimum of 150 AU/mL was required for lethality. At 150–450 AU/mL, a 4-log decrease occurred, but, above 450 AU/mL, no further kill resulted. A resistant population of 1000 cells (0.01%) remained. With *L. monocytogenes* strain Scott A, a minimum 600 AU/mL of pediocin AcH was required for lethality, 600–4200 AU/mL caused a 2 log decrease, and above 4200 AU/mL no further kill resulted. Pediocin-AcH lysed cells of *L. monocytogenes* Camp+/Beta- but not strains Scott A, CA, $Ohio_2$, or 117. In growth studies at 4°C, pediocin AcH was most inhibitory to *L. monocytogenes* $Ohio_2$ followed by Scott A. No effect on growth was demonstrated for *Y. enterocolitica*. Overall, the effectiveness of pediocin AcH was greater than another bacteriocin tested, sakacin A, and slightly less or equal to nisin. In contrast to the effectiveness of *Pediococcus acidilactici* H with *Listeria monocytogenes*, Okereke and Montville (1991) found that the organism had no effect on the growth of *C. botulinum* spores.

Bhunia et al. (1991) studied the mechanism of inhibition by pediocin AcH. They found that pediocin AcH binds to specific receptors on cell walls of sensitive strains and destabilizes membrane function, causing cell death. There must be enough pediocin AcH present to reach a saturation point prior to bactericidal effect.

In application studies, *Pediococcus* sp. strains JD1-23 and MP1-08 from commercial suppliers were tested against *L. monocytogenes* Scott A in summer sausage with two different spice blends (Berry et al., 1990). Log 6.0 *L. monocytogenes*/g were reduced 2 logs in presence of bacteriocin-producing *Pediococcus* but only 1 log with bacteriocin-negative strains. Inactivation occurred in the absence of adequate acid production indicating the direct involvement of bacteriocin. Nielsen et al. (1990) evaluated *P. acidilactici* from a commercial meat starter culture on *L. monocytogenes* Scott A in radiation sterilized fresh lean beef muscle. A filter-sterilized growth extract decreased attached *L. monocytogenes* by 0.5–2.2 logs within 2 min. The system was functional for one month at refrigeration temperature. Yousef et al. (1991) determined the activity of *Pediococcus acidilactici* H or pediocin AcH against *Listeria monocytogenes* Scott A, V7, and 101M in wiener exudate stored at 4°C or 25°C. At 4°C, pediocin AcH caused a rapid decrease in viable *Listeria monocytogenes* in the first 2 h,but, thereafter, inactivation was similar to the control. Inactivation by *Pediococcus acidilactici* H was similar to the control. At 25°C, pediocin AcH caused a gradual decrease in viable *Listeria monocytogenes* throughout a seven- to eight-day storage. *Pediococcus acidilactici* H did not affect growth of *Listeria monocytogenes* until 64 h into storage and then it caused a rapid decrease (ca. 6 logs). The authors reported that biopreservatives could potentiate the antilisterial effect in wiener exudate.

VI. *LACTOBACILLUS*

Bacteriocins produced by *Lactobacillus* generally have limited or narrow range, usually with *Lactobacillaceae*. As with previous bacteriocins, they are generally bactericidal and proteinaceous. One of the major differences between *Lactobacillus* and other lactic acid bacteria is the general inability to link bacteriocin biosynthesis with plasmid DNA (Hoover, 1993). One exception may be lactacin F production by *Lactobacillus acidophilus* 88 (Muriana and Klaenhammer, 1987). Because *Lactobacillus* are very strong hydrogen peroxide producers, care must be taken not to confuse this characteristic with bacteriocin production.

There is significant evidence that *Lactobacillus* produce bacteriocins. In 1963, Sabine demonstrated inhibitory activity of *L. acidophilus* against *S. aureus.* De Klerk and Coetzee (1967) screened 121 strains of *Lactobacillus fermenti* for bacteriocin production and found 25 to be positive against *Lactobacillus fermenti* and *L. acidophilus* strains. The inhibitory agents were bacteriocin-like in that they were not affected by pH or catalase, precipitated in the presence of ammonium acetate, and their activity was retained by dialysis membranes. Barefoot and Klaenhammer (1983) found that 42 of 52 strains of *L. acidophilus* produced bacteriocin. Harris et al. (1989) demonstrated that *L. acidophilus* UAL 11 but not *L. acidophilus* C-136 inhibited eight strains of *L. monocytogenes.* Inhibition was prevented by proteolytic enzymes.

A. Broad Spectrum Compounds

Vincent et al. (1959) demonstrated the presence of a bacteriocin-like compound, designated lactocidin, produced by *L. acidophilus* in veal-liver agar cultures. It was nonvolatile, insensitive to catalase, active at neutral pH, and retained within dialysis membranes. The compound reportedly had broad spectrum activity against Gram-positive and Gram-negative bacteria. Because of the broad spectrum activity, lactocidin was later reported to be due to the combined effect of organic acids, H_2O_2, and antibiotic-type substances (Klaenhammer, 1988). Two other bacteriocin-type compounds were produced by strains of *L. acidophilus,* acidolin (Hamdan and Mikolacjik, 1974), and acidophilin (Shahani et al., 1977).

Reuterin is a low-molecular-weight, nonproteinaceous, soluble, pH-neutral compound produced by heterofermentative *Lactobacillus reuterii* (Daeschel, 1989). It has a broad spectrum of inhibitory activity against species of Gram-negative and Gram-positive bacteria, yeasts, fungi, and protozoa. Bacterial species inhibited include *Salmonella, Shigella, Clostridium, Staphylococcus, Listeria, Candida,* and *Trypanosoma.* Dobrogosz (Daeschel, 1989) applied reuterin (50 units/g) to ground beef and found that coliform growth was inhibited for six days and 4°C.

Gilliland and Speck (1977) detected inhibition by *L. acidophilus* of *S. aureus,* enteropathogenic *E. coli, S. typhimurium,* and *C. perfringens.* Inhibition could not be totally related to organic acids or H_2O_2. An inhibitor termed "bulgarican" was produced by *Lactobacillus bulgaricus* DDS14, isolated by methanol-acetone extraction, and purified by silica gel chromatography (Reddy et al., 1983). The spectrum of antimicrobial activity of bulgarican included *Bacillus subtilis, E. coli, Proteus vulgaris, Sarcina lutea, S. aureus, Pseudomonas aeruginosa, P. fluorescens,* and *Serratia marcescens.* No molds were inhibited. The pH for optimum activity was <4.5.

Lactobacillus casei strain GG isolated from normal human feces produced an inhibitor with broad spectrum activity (Silva et al., 1987). The compound was distinct from organic acids and had a molecular weight of <1000. The inhibitor was heat-stable (121°C, 15 min) and resistant to trypsin, proteinase K, α-chymotrypsin, bromelin, carboxypeptidase A, and *Streptomyces griseus* protease. The substance was inhibitory to strains of *E. coli, Pseudomonas, Salmonella, Streptococcus, Bacillus, Clostridium,* and *Bifidobacterium.* It was not inhibitory to *Lactobacillus.* Inhibition occurred at pH 3–5. Silva et al. (1987) suggested that the inhibitor was not a bacteriocin, but rather closely resembled a microcin (low-molecular-weight peptides produced mostly by *Enterobacteriaceae*). *Lactobacillus* GG is currently used to produce a fermented whey drink and yogurt-type product which are reported to have numerous health benefits (Salminen et al., 1991).

B. Lactacin B

Lactacin B is produced by *L. acidophilus* N2 (Barefoot and Klaenhammer, 1983). It is bactericidal, sensitive to proteinase K but not urea, and is heat-stable (100°C, 1 h). The molecular weight was estimated at 6200 daltons, but it can occur in aggregates of 100,000. Activity spectrum for the bacteriocin includes *Lactobacillus bulgaricus, Lactobacillus leichmannii, Lactobacillus helveticus,* and *Lactobacillus lactis.* Harris et al. (1989) found that neither *L. acidophilus* 11759 nor *L. acidophilus* ADH (lactacin B producers) were effective against eight strains of *L. monocytogenes* in deferred antagonism or well assays on BHI agar. *Lactobacillus acidophilus* C-7 (a lactacin M producer) also showed no inhibition of *L. monocytogenes. Lactobacillus acidophilus* N2 did not inhibit the growth of *C. botulinum* spores in a deferred antagonism assay system (Okereke and Montville, 1991). Synthesis of this bacteriocin is chromosomally linked. Adsorption of lactacin B is nonspecific; i.e., sensitive and insensitive cells adsorb the compound.

C. Lactacin F

Lactacin F is produced by *L. acidophilus* 11088 (NCK88). The compound has a molecular weight of 2500 and as many as 56 amino acids (Muriana and Klaenhammer, 1991). It is heat-stable (121°C, 15 min), sensitive to trypsin, proteinase K, ficin, and subtilisin, and production is pH-dependent. Lactacin F has an antimicrobial activity spectrum which includes *Lactobacillus bulgaricus, Lactobacillus leichmannii, Lactobacillus helveticus, Lactobacillus lactis, Lactobacillus fermentum* 1750, and *Enterococcus faecalis* 19433 (Barefoot and Klaenhammer, 1983; Muriana and Klaenhammer, 1987, 1991). The first four strains are the same as those found for lactacin B. Raccach et al. (1989)

demonstrated inhibition of *L. monocytogenes* by *L. acidophilus* NU-A and 88 (lactacin F) in milk. Organic acid inhibition probably contributed since the pH was reduced to 4.7 in 24 h. In contrast, Harris et al. (1989) found no inhibition of eight strains of *L. monocytogenes* by *L. acidophilus* 88 in deferred antagonism or well diffusion assays on BHI. Lactacin F production and immunity in *L. acidophilus* 88 are apparently under the control of transient plasmid determinants (Muriana and Klaenhammer, 1987).

D. Lactocin 27

The source of lactocin 27 is *Lactobacillus helveticus* LP27 (Upreti and Hinsdill, 1973). The compound is a glycoprotein of 12,400 daltons which is sensitive to trypsin and pronase but not ficin. No loss of activity was detected following heating for 1 h at 100°C (Upreti and Hinsdill, 1975). Its antimicrobial spectrum is limited to *L. acidophilus,* and *Lactobacillus helveticus*. It is bacteriostatic to *Lactobacillus helveticus* LS18. There is no evidence for plasmid control of biosynthesis (Upreti and Hinsdill, 1975). The mode of action of lactocin 27 involves termination of protein synthesis, and it has no effect on DNA or RNA synthesis or ATP levels (Upreti and Hinsdill, 1975).

Lactobacillus fermenti 466 was reported to produce a compound similar to lactocin 27 (De Klerk and Smit, 1967; Upreti and Hinsdill, 1973). It is heat-resistant (96°C, 30 min) sensitive to trypsin and pepsin, and resistant to urea and lysozyme.

E. Helveticin J

Joerger and Klaenhammer (1986) characterized helveticin J, a second bacteriocin from *Lactobacillus helveticus*. The compound was sensitive to pronase, trypsin, pepsin, ficin, proteinase K, and subtilisin. Heat (100°C, 30 min) inactivated helveticin J. Purification of the protein revealed a 37,000-dalton peptide which retained inhibitory activity. Helveticin J was bactericidal to *Lactobacillus helveticus* 1846 and 1244, *Lactobacillus bulgaricus* 1373 and 1489, and *Lactobacillus lactis* 970. Harris et al. (1989) screened *Lactobacillus helveticus* 481 (helveticin J producer) against eight strains of *L. monocytogenes*. No inhibition of *L. monocytogenes* was detected. While the helveticin J producer *Lactobacillus helveticus* 481 contained an 8-MDa plasmid, derivatives cured of the plasmid still produced bacteriocin which indicated chromosomally linked biosynthesis (Joerger and Klaenhammer, 1986).

F. Lactolin

The first antimicrobial protein detected in *Lactobacillus plantarum*, an organism important in vegetable fermentations, was designated lactolin (Kodama, 1952). Andersson (1986) characterized an inhibitory protein from *Lactobacillus plantarum* 83 isolated from fermenting carrots. Broth cultures in MRS were incubated at 30°C, 48 h, dialyzed, and filter-sterilized. The filtrate at concentrations of 0–10% (v/v) in broth medium was inhibitory to *S. aureus* and spheroplasts of Gram-negative organisms. Activity was greatest at low pH (5.0). *Staphylococcus aureus* NCTC 6571 used as the indicator adapted to the presence of bacteriocin resulting in selection of a subpopulation immune to effects of the bacteriocin. Characteristics of the inhibitory protein were molecular weight 100,000 (macromolecular protein complex) and inactivation by trypsin, pepsin, and heat (121°C, 15 min). According to Klaenhammer (1988), lactolin may actually be a nisin-like compound produced by *Lactobacillus lactis*.

Another report by West and Warner (1988) described a screening study done with six strains each of *Lactobacillus plantarum* and *Leuconostoc mesenteroides*. Activity was detected in a single strain (1193) of *Lactobacillus plantarum*. The inhibitor was sensitive to proteolytic enzymes and activity was reduced by lipase and α-amylase. The inhibitor was antagonistic to strains of *Lactobacillus plantarum, Leuconostoc mesenteroides* 8015, and *Pediococcus damnosus* 1832. There was no effect on *Leuconostoc cremoris, Lactobacillus casei, Lactobacillus fermentus*, or *L. acidophilus*. Okereke and Montville (1991) screened 3 strains of *Lactobacillus plantarum* strains for inhibitory activity against 11 strains of type A or B *C. botulinum* spores. All three strains (LG592, LB75 and BN) were capable of inhibiting all strains of *C. botulinum* in a deferred antagonism assay. In contrast, none of the *Lactobacillus* strains inhibited the *C. botulinum* in a direct antagonism assay, limiting their usefulness in food products.

G. Plantaricin A

Lactobacillus plantarum C-11 isolated from a cucumber fermentation was found to produce a bacteriocin which was named plantaricin A (Daeschel et al., 1990). The compound was heat-stable (100°C, 30 min) and active at pH 4.0–6.5. The molecular weight was less than 8000. The inhibitory activity spectrum included other *Lactobacillus plantarum, P. pentosaceus*, and *Leuconostoc paramesenteroides*. The compound was bactericidal. Harris et al. (1989) tested *Lactobacillus plantarum* C-11 (plantaricin A producer) against eight strains of *L. monocytogenes* and found no inhibition of the foodborne pathogen. Thus far, there is no evidence for plasmid-controlled biosynthesis of plantaricin A (Daeschel et al., 1990).

H. Sakacin A

Schillinger and Lucke (1989) screened 221 strains of *Lactobacillus* isolated from meats for inhibition. Nineteen strains of *Lactobacillus sake*, three strains of *Lactobacillus plantarum*, and one strain of *Lactobacillus curvatus* inhibited other *Lactobacillus* sp. Antimicrobial spectrum of the supernatants showed them to be different compounds. Further study of *Lactobacillus sake* 706 showed that it produced the bacteriocin, sakacin A. The bacteriocin was shown to be effective against *L. monocytogenes* 8732 and 17A. In a recent study, Motlagh et al. (1991) found that sakacin A from *Lactobacillus sake* 706 at 150 AU/mL was inhibitory to nine strains of *L. monocytogenes, L. ivanovii*, and *L. innocua*. It did not inhibit *Salmonella anatum, S. typhimurium, Yersinia enterocolitica* (2 strains), *E. coli* (2 strains), or *Aeromonas hydrophila* (2 strains). Sakacin A caused a slight decrease in viable *L. monocytogenes* cells in 24 h at 4°C; however, it had no effect on growth of *L. monocytogenes* Ohio$_2$, Scott A, or *Y. enterocolitica* after 8 weeks at 4°C. *Lactobacillus sake* Lb706 and Lb796 also had no effect on the growth of *C. botulinum* in a deferred antagonism assay (Okereke and Montville, 1991). Biosynthesis of sakacin A is mediated by plasmid DNA (Schillinger and Lucke, 1989).

I. Lactocin S

Lactobacillus sake L 45, isolated from dry sausage fermentation, was reported to produce a bacteriocin which was moderately heat-stable and protease-sensitive (Mortvedt and Nes, 1990). Its antimicrobial spectrum included *Lactobacillus, Pediococcus*, and *Leuconostoc*. Biosynthesis and immunity were reportedly plasmid mediated.

VII. *LEUCONOSTOC*

Inhibition by *Leuconostoc* has been attributed primarily to production of lactic acid, acetic acid, and diacetyl (Branen et al., 1975; Devoyod and Poullain, 1988). However, there have been reports of bacteriocin-like compounds produced by the genus. Orberg and Sandine (1984) demonstrated inhibition of *Streptococcus cremoris* U134 but not *Streptococcus lactis* ATCC 11454 (nisin producer) by *Leuconostoc* sp. PO184. Harris et al. (1989) found that a *Leuconostoc* sp. strain UAL14, which produced an unnamed bacteriocin, inhibited all eight strains of *L. monocytogenes* tested using two different assays. Inhibition of *L. monocytogenes* was prevented by proteolytic enzymes.

Tsai and Sandine (1987) transferred a 17.5-MDa plasmid for nisin production and sucrose fermentation from *Streptococcus lactis* 7962 (a nisin producer) to a *Leuconostoc dextranicum* strain by conjugation. The *Leuconostoc* was a significant overproducer of nisin (ca. 1000 times greater than the *Streptococcus*).

VIII. FUTURE OF INHIBITORS PRODUCED BY LACTIC ACID BACTERIA

Much progress has been made in the past 10–15 years to increase our knowledge of antimicrobials produced by lactic acid bacteria. Significant advances have been made in the areas of methods for differentiating the effect of organic acids, hydrogen peroxide, and bacteriocins, expanding the data base on sources, types, and antimicrobial spectra of bacteriocin-producing microorganisms and ascertaining aspects of the genetic control of bacteriocin synthesis and immunity. For antimicrobials produced by lactic acid bacteria to have increased importance as potential biopreservatives in foods, much more information is needed on application of the compounds to foods. As has been shown for other food antimicrobials, in vitro studies on food antimicrobials do not necessarily predict their effectiveness in food products (Davidson and Parish, 1989). Therefore, more studies on the effect of environmental alterations, interaction with food components and potentiation by other food antimicrobials need to be carried out with bacteriocins. These studies should include both foodborne pathogens and spoilage microorganisms. Another concern is the development of bacteriocin-resistant subpopulations of foodborne pathogens. The potential impact of this resistance needs to be evaluated prior to wholesale use of these compounds in foods.

REFERENCES

Alifax, R., and Chevalier, R. (1962). Etude de la nisinase produite par *Streptococcus thermophilus, J. Dairy Res., 29*:233–240.

Andersson, R. E. (1986). Inhibition of *Staphylococcus aureus* and spheroplasts of Gram-negative bacteria by an antagonistic compound produced by a strain of *Lactobacillus plantarum, Int. J. Food Microbiol., 3*:149–160.

Bailey, F. J., and Hurst, A. (1971). Preparation of a highly active form of nisin, *Can. J. Microbiol., 17*:61–67.

Banks, J. G., Board, R. G., and Sparks, N. H. C. (1986). Natural antimicrobial systems and their potential in food preservation of the future, *Biotech. Appl. Biochem., 8*:103.

Bardowski, J., Kozak, W., and Dobrzanski, W. T. (1979). Further characterization of lactostrepcins-acid bacteriocins of lactic streptococci, *Acta Microbiol. Polon., 28*: 93–99.

Bardsley, A. (1962). Antiobiotics in food canning, *Food Technol. Austral., 14*:532–537.

Barefoot, S. F., and Klaenhammer, T. R. (1983). Detection and activity of lactacin B, a bacteriocin produced by *Lactobacillus acidophilus, Appl. Env. Microbiol., 45*: 1808–1815.

Bell, R. G., and DeLacy, K. M. (1986). Factors influencing the determination of nisin in meat products, *J. Food Technol., 21*:1.

Bell, R. G., and DeLacy, K. M. (1987). The efficacy of nisin, sorbic acid, and monolaurin as preservatives in pasteurized cured meat products, *Food Microbiol., 4*:277.

Benkerroum, N., and Sandine, W. E. (1988). Inhibitory action of nisin against *Listeria monocytogenes, J. Dairy Sci., 71*:3237–3245.

Berry, E. D., Liewen, M. B., Mandigo, R. W., and Hutkins, R. W. (1990). Inhibition of *Listeria monocytogenes* by bacteriocin-producing *Pediococcus* during the manufacture of fermented semidry sausage, *J. Food Prot., 53*:194–197.

Bhunia, A. K., Johnson, M. C., and Ray, B. (1987). Direct detection of an antimicrobial peptide of *Pediococcus acidilactici* in sodium dodecyl sulfate polyacrylamide gel electrophoresis, *J. Ind. Microbiol., 2*:319.

Bhunia, A. K., Johnson, M. C., and Ray, B. (1988). Purification, characterization and antimicrobial spectrum of a bacteriocin produced by *Pediococcus acidilactici, J. Appl. Bacteriol., 65*:261–268.

Bhunia, A., Johnson, M. C., Ray, B., and Kalchayanand, N. (1991). Mode of bactericidal action of pediocin AcH, a bacteriocin from *Pediococcus acidilactici, J. Appl. Bacteriol., 70*:25–33.

Branen, A. L., Go, H. C., and Genske, R. P. (1975). Purification and properties of antimicrobial substances produced by *Streptococcus diacetilactis* and *Leuconostoc citrovorum, J. Food Sci., 40*:446.

Bremner, H. A., and Statham, J. A. (1983). Spoilage of vacuum-packed chill-stored scallops with added lactobacilli, *Food Technol. Austral., 35*:284.

Broadbent, J. R., and Kondo, J. K. (1991). Genetic construction of nisin-producing *Lactococcus lactis* subsp. *cremoris* and analysis of a rapid method for conjugation, *Appl. Env. Microbiol., 57*:517.

Buchman, G. W., Banerjee, S., and Hansen, J. N. (1988). Structure, expression and evolution of a gene encoding the precursor of nisin, a small protein antibiotic, *J. Biol. Chem., 263*:16260.

Calderon, C., Collins-Thompson, D. L., and Usborne, W. R. (1985). Shelf-life studies of vacuum-packaged bacon treated with nisin, *J. Food Prot., 48*:330.

Campbell, L. L., Sniff, E. E., and O'Brien, R. T. (1959). Subtilin and nisin as additives that lower the heat process requirements of canned foods. *Food Technol., 13*:462.

Carlson, S., and Bauer, H. M. (1957). Nisin, ein antibakterieller wirkstoff aus *Streptococcus lactis* unter berucksichtigung des resistenzproblems, *Arch. Hyg. Bakteriol., 141*: 445–459.

Carminati, D., Giraffa, G., and Bossi, M. G. (1989). Bacteriocin-like inhibitors of *Streptococcus lactis* against *Listeria monocytogenes, J. Food Prot., 52*:614–617.

Caserio, G., Ciampella, A., Gennari, M., and Barluzzi, A. M. (1979). Utilization of nisin in cooked sausages and other cured meat products, *Industrie Aliment. (Italy), 18*:1.

Cheeseman, G. C., and Berridge, N. J. (1957). An improved method of preparing nisin, *Biochem. J., 65*:603.

Daeschel, M. A. (1989). Antimicrobial substances from lactic acid bacteria for use as food preservatives, *Food Technol., 43*:164.

Daeschel, M. A., Harris, L. J., Stiles, M. E., and Klaenhammer, T. R. (1988). Antimicrobial activity of lactic acid bacteria against *Listeria monocytogenes*, 88th Ann. Mtg. Amer. Soc. Microbiol., Miami, FL.

Daeschel, M. A., Jung, D. S., and Watson, B. T. (1991). Controlling wine malolactic fermentation with nisin and nisin-resistant strains of *Leuconostoc oenos, Appl. Env. Microbiol., 57*:601.

Daeschel, M. A., and Klaenhammer, T. R. (1985). Association of a 13.6-megadalton plasmid in *Pediococcus pentosaceus* with bacteriocin activity, *Appl. Env. Microbiol., 50*:1538–1541.

Daeschel, M. A., McKenney, M. C., and McDonald, L. C. (1990). Bacteriocidal activity of *Lactobacillus plantarum* C-11, *Food Microbiol., 7*:91.

Dahiya, R. S., and Speck, M. L. (1968). Hydrogen peroxide formation by lactobacilli and its effect on *Staphylococcus aureus, J. Dairy Sci., 51*:1568.

Davey, G. P. (1981). Mode of action of diplococcin, a bacteriocin from *Streptococcus cremoris* 346, *N.Z.J. Dairy Sci. Technol., 16*:187.

Davey, G. P. (1984). Plasmid associated with diplococcin production in *Streptococcus cremoris, Appl. Env. Microbiol., 48*:895.

Davey, G. P., and Pearce, L. E. (1980). The use of *Streptococcus cremoris* strains cured of diplococcin production as cheese starters, *N.Z.J. Dairy Sci. Technol., 15*:51.

Davey, G. P., and Pearce, L. E. (1982). Production of diplococcin by *Streptococcus cremoris* and its transfer to nonproducing group N streptococci, in *Microbiology—1982*, p. 221 (D. Schlessinger, ed.), American Society for Microbiology, Washington, DC.

Davey, G. P., and Richardson, B. C. (1981). Purification and some properties of diplococcin from *Streptococcus cremoris* 346, *Appl. Env. Microbiol., 41*:84.

Davidson, P. M., and Parish, M. E. (1989). Methods for testing the efficacy of food antimicrobials, *Food Technol., 43*(1):148–155.

De Klerk, H. C., and Coetzee, J. N. (1967). Bacteriocinogeny in *Lactobacillus fermenti, Nature (London), 214*:609.

De Klerk, H. C., and Smit, J. A. (1967). Properties of a *Lactobacillus fermenti* bacteriocin, *J. Gen. Microbiol., 48*:309.

Delves-Broughton, J. (1990). Nisin and its use as a food preservative, *Food Technol., 44*(11):100–117.

Denny, C. B., Sharpe, L. E., and Bohrer, C. W. (1961). Effects of tylosin and nisin on canned food spoilage bacteria, *Appl. Microbiol., 9*:108.

Devoyod, J. J., and Poullain, F. (1988). Les Leuconostocs properties: leur role en tachnologie laitiere, *Le Lait, 68*:249.

Dobrzanski, W. T., Bardowski, J., Kozak, W., and Zajdel, J. (1982). Lactostrepcins: Bacteriocins of lactic streptococci, in *Microbiology-1982*, p. 225 (D. Schlessinger ed.), American Society for Microbiology, Washington, DC.

Dodd, H. M., Horn, N., and Gasson, M. J. (1990). Analysis of the genetic determinant for production of the peptide antibiotic nisin, *J. Gen. Microbiol., 136*:555.

Doores, S. (1990). pH control agents and acidulants, in *Food Additives*, pp. 477–510 (A. L. Branen, P. M. Davidson, and S. Salminen, eds.), Marcel Dekker, New York.

Etchells, J. L., Costilow, R. N., Anderson, T. E., and Bell, T. A. (1964). Pure culture fermentation of brined cucumbers, *Appl. Microbiol., 12*:523.

Fleming, H. P., Etchells, J. L., and Costilow, R. N. (1975). Microbial inhibition by an isolate of *Pediococcus* from cucumber brines, *Appl. Microbiol., 30*:1040.

Food and Drug Administration (1988). Nisin preparation: affirmation of GRAS status as a direct human food ingredient, *FR 53*:11247–11313 (April 6, 1988).

Fowler, G. G., and McCann, B. (1971). The growing use of nisin in the dairy industry, *Austral. J. Dairy Technol., 26*:44.

Frazer, A. C., Sharratt, M., and Hickman, J. R. (1962). The biological effects of food additives. I: Nisin, *J. Sci. Food Agric., 13*:32.

Froseth, B. R., Herman, R. E., and McKay, L. L. (1988). Cloning of nisin resistance determinant and replication origin on 7.6-kilobase *Eco*RI fragment of pNP40 from *Streptococcus lactis* subsp. *diacetylactis* DRC3, *Appl. Env. Microbiol., 54*:2136–2139.

Froseth, B. R., and McKay, L. L. (1991). Molecular characterization of the nisin resistance region of *Lactococcus lactis* ssp. *lactis* biovar diacetylactis DRC3, *Appl. Env. Microbiol., 57*:804–811.

Fukase, K., Kitzawa, M., Sano, A., Shimbo, K., Fujita, H., Honimoto, S., Wakamiya, T., and Shiba, T. (1988). Total synthesis of the peptide antibiotic nisin, *Tetrahedron Lett., 29*:795.

Gasson, J. M. (1984). Transfer of sucrose fermenting ability, nisin resistance and nisin production in *Streptococcus lactis* 712, *FEMS Microbiol. Lett., 21*:7.

Gaya, P., Medina, M., and Nunez, M. (1991). Effect of the lactoperoxidase system on *Listeria monocytogenes* behavior in raw milk at refrigeration temperatures, *Appl. Env. Microbiol., 57*:3355–3360.

Geis, A., Singh, J., and Teuber, M. (1983). Potential of lactic streptococci to produce bacteriocin, *Appl. Env. Microbiol., 45*:205–211.

Gillespy, T. G. (1957). *Nisin Trials*, Fruit and Vegetable Canning and Quick Freezing Research Association, Leaflet No. 3, Chipping Campden, United Kingdom.

Gilliland, S. E. (1985). Role of starter culture bacteria in food preservation, in *Bacterial Starter Cultures for Food*, p. 175 (S. E. Gilliland, ed.), CRC Press, Boca Raton, FL.

Gilliland, S. E., and Ewell, H. R. (1983). Influence of combinations of *Lactobacillus lactis* and potassium sorbate on growth of psychrotrophs in raw milk, *J. Dairy Sci., 66*:974.

Gilliland, S. E., and Speck, M. L. (1972). Interactions of food starter cultures and food-borne pathogens: lactic streptococci versus staphylococci and salmonellae, *J. Milk Food Technol., 35*:307.

Gilliland, S. E., and Speck, M. L. (1975). Inhibition of psychrotrophic bacteria by lactobacilli and pediococci in nonfermented refrigerated foods, *J. Food Sci., 40*:903.

Gilliland, S. E., and Speck, M. L. (1977). Antagonistic action of *Lactobacillus acidophilus* toward intestinal and foodborne pathogens in associative culture, *J. Food Prot., 40*:820.

Gonzalez, C. F., and Kunka, B. S. (1985). Transfer of sucrose fermenting ability and nisin production phenotype among lactic streptococci, *Appl. Env. Microbiol., 49*:627–633.

Gonzales, C. F., and Kunka, B. S. (1987). Plasmid-associated bacteriocin production and sucrose fermentation in *Pediococcus acidilactici, Appl. Env. Microbiol., 53*: 2534–2538.

Gould, G. W. (1964). Effect of food preservatives on the growth of bacteria from spores, in *Microbial Inhibitors in Food* (N. Molin, ed.), p. 17, Almqvist and Wiksell, Stockholm.

Gould, G. W., and Hurst, A. (1962). Inhibition of *Bacillus* spore development by nisin and subtilin, *8th Int. Congress of Microbiology*, Abstract A2–11.

Gowans, J. L., Smith, N., and Florey, H. W. (1952). Some properties of nisin, *Br. J. Pharmacol., 7*:438.

Graham, D. C., and McKay, L. L. (1985). Plasmid DNA in strains of *Pediococcus cerevisiae* and *Pediococcus pentosaceus, Appl. Env. Microbiol., 50*:532.

Gregory, M. E., Henry, K., and Kon, S. K. (1964). Nutritive properties of freshly prepared and stored evaporated milks manufactured by normal commercial procedure or by reduced thermal processes in the presence of nisin, *J. Dairy Res., 31*:113.

Gross, E. (1977). α, β-Unsaturated and related amino acids in peptides and proteins, in *Protein Cross-Linking-B* (M. Fiedman, ed.), pp. 131–153, Plenum, New York.

Gross, E., and Morell, J. L. (1967). The presence of dehydroalanine in the antibiotic nisin and its relationship to activity, *J. Am. Chem. Soc., 89*:2791.

Gross, E., and Morell, J. L. (1970). Nisin. The assignment of sulfide bridges of β-methyllanthionine to a novel by bicyclic structure of identical ring size, *J. Am. Chem. Soc., 92*:2919.

Gross, E., Morell, J. L., and Craig, L. C. (1969). Dehydroalanyllysine: Identical COOH-terminal structures in the peptide antibiotics nisin and subtilin, *Proc. Nat. Acad. Sci. USA 62*:952.

Gupta, K. G., Chandiok, L., and Bhatnagar, L. (1973). Antibacterial activity of diacetyl and its influence on the keeping quality of milk, *Zentralbl. Bakteriol. Parasitenkd. Infektionskr. Hyg. Abt. 1: Orig. Reihe B., 158*:202.

Gupta, K. G., Sidhu, R., and Yadav, N. K. (1972). Effect of various sugars and their derivatives upon the germination of *Bacillus* spores in the presence of nisin, *J. Food Sci., 37*:971.

Hall, R. H. (1966). Nisin and food preservation, *Proc. Biochem., 1*:461.

Hamdan, I. Y., and Mikolacjik, E. M. (1974). Acidolin: An antibiotic produced by *Lactobacillus acidophilus, J. Antibiot., 27*:631.

Hara, S., Yakazu, K., Nakakawaji, K., Takenchi, T., Kobayashi, T., Sata, M., Imai, Z., and Shibuya, T. (1962). An investigation of toxicity of nisin, *Tokyo Med. Univ. J., 20*:175.

Harris, L. J., Daeschel, M. A., Stiles, M. E., and Klaenhammer, T. R. (1989). Antimicrobial activity of lactic acid bacteria against *Listeria monocytogenes, J. Food Prot., 52*: 384–387.

Harris, L. J., Fleming, H. P., and Klaenhammer, T. R. (1991). Sensitivity and resistance of *Listeria monocytogenes* ATCC, 19115, Scott A, and UAL500 to nisin, *J. Food Prot., 54*:836–840.

Hawley, H. B. (1957). Nisin in food technology, *Food Manuf., 32*:370.

Heinemann, B., Voris, L., and Stumbo, C. R. (1965). Use of nisin in processing food products, *Food Technol., 19*:592.

Henning, S., Metz, R., and Hammes, W. P. (1986a). Studies on the mode of action of nisin, *Int. J. Food Microbiol., 3*:121.

Henning, S., Metz, R., and Hammes, W. P. (1986b). New aspects for the application of nisin to foods based on its mode of action, *Int. J. Food Microbiol., 3*:135.

Hirsch, A. (1951). Growth and nisin production of a strain of *Streptococcus lactis, J. Gen. Microbiol., 5*:208–221.

Hirsch, A., and Grinsted, E. (1954). Methods for the enumeration of anaerobic sporeformers from cheese, with observations on the effect of nisin, *J. Dairy Res., 21*:101.

Hirsch, A., Grinsted, E., Chapman, H. R., and Mattick, A. T. R. (1951). A note on the inhibition of an anaerobic sporeformer in swiss-type cheese by a nisin producing *Streptococcus, J. Dairy Res., 18*:205–206.

Hitchins, A. D., Gould, G. W., and Hurst, A. (1963). The swelling of bacterial spores during germination and outgrowth, *J. Gen. Microbiol., 30*:445.

Holley, R. A. (1981). Review of the potential hazard from botulism in cured meats, *Can. Inst. Food Sci. Technol. J., 14*:183.

Hoover, D. G. (1993). Bacteriocins with potential for use in foods, in *Antimicrobials in Foods*, 2nd ed. (P. M. Davidson and A. L. Branen, eds.), p. 409, Marcel Dekker, New York.

Hoover, D. G., Dishart, K. J., and Hermes, M. A. (1989). Antagonistic effect of *Pediococcus* spp. against *Listeria monocytogenes, Food Biotechnol., 3*:183.

Hoover, D. G., Walsh, P. M., Kolaetis, K. M., and Daly, M. M. (1988). A bacteriocin produced by *Pediococcus* species associated with a 5.5-megadalton plasmid, *J. Food Prot., 51*:29–31.

Hossack, D. J. N., Bird, M. C., and Fowler, G. G. (1983). The effect of nisin on the sensitivity of micro-organisms to antibiotics and other chemotherapeutic agents, in *Antimicrobials and Agriculture* (M. Woodbine, ed.), p. 425, Butterworths, London.

Hurst, A. (1981). Nisin, *Adv. Appl. Microbiol., 27*:85–103.

Hurst, A. (1983). Nisin and other inhibitory substances from lactic acid bacteria, in *Antimicrobials in Foods* (A. L. Branen and P. M. Davidson, ed.), p. 327, Marcel Dekker, New York.

Hurst, A., and Hoover, D. G. (1993). Nisin, in *Antimicrobials in Foods*, 2nd ed. (P. M. Davidson and A. L. Branen, eds.), p. 369, Marcel Dekker, New York.

Hutton, M. T., Chehak, P. A., and Hanlin, J. H. (1991). Inhibition of botulinum toxin production by *Pediococcus acidilactici* in temperature abused refrigerated foods, *J. Food Safety, 11*:255–267.

Jarvis, B. (1967). Resistance to nisin and production of nisin-inactivating enzymes by several *Bacillus* species, *J. Gen. Microbiol., 47*:33.

Jarvis, B. (1970). Enzymic reduction of the C-terminal dehydroalanyllysine sequence in nisin, *Biochem. J., 119*:56P.

Jarvis, B., and Burke, C. S. (1976). Practical and legislative aspects of the chemical preservation of food, in *Inhibition and Inactivation of Vegetative Microbes* (F. A. Skinner and W. B. Hugo, eds.), p. 345, Academic Press, New York.

Jarvis, B., and Farr, J. (1971). Partial purification, specificity and mechanism of action of the nisin-inactivating enzyme from *Bacillus cereus, Biochem. Biophys. Acta, 227*:232–240.

Jarvis, B., Jeffcoat, J., and Cheeseman, G. C. (1968). Molecular weight distribution of nisin, *Biochem. Biophys. Acta, 168*:153.

Jarvis, B., and Mahoney, R. R. (1969). Inactivation of nisin by alpha-chymotrypsin, *J. Dairy Sci., 52*:1448.

Jay, J. M. (1982). Antimicrobial properties of diacetyl, *Appl. Env. Microbiol., 44*: 525.

Joerger, M., and Klaenhammer, T. R. (1986). Characterization and purification of helveticin J and evidence for a chromosomally determined bacteriocin produced by *Lactobacillus helveticus* 481, *J. Bacteriol., 167*:439–446.

Juffs, H. S., and Babel, F. J. (1975). Inhibition of psychrotrophic bacteria by lactic cultures in milk stored at low temperature, *J. Dairy Sci., 58*:1612–1619.

Kaletta, C., and Entian, K. D. (1989). Nisin, a peptide antibiotic: Cloning and sequencing of the *nisA* gene and posttranslational processing of its peptide product, *J. Bacteriol., 171*:1597.

Kekessy, D. A., and Piguet, J. D. (1970). New method for detecting bacteriocin production, *Appl. Microbiol., 20*:282–283.

Kiss, I., Kiss, K. N., Farkas, J., Fabri, I., and Vas, K. (1968). Further data on the application of nisin in pea preservation, *Ellelmiszentudomany (Budapest), 2*:51.

Klaenhammer, T. R. (1988). Bacteriocins of lactic acid bacteria, *Biochimie, 70*:337.

Kodama, R. (1952). Studies on lactic acid bacteria. II: Lactolin, a new antibiotic substance produced by lactic acid bacteria, *J. Antibiot., 5*:72.

Kooy, J. S. (1952). Strains of *Lactobacillus plantarum* which destroy the antibiotic made by *Streptococcus lactis, Neth. Milk Dairy J., 6*:223.

Kozak, W., Bardowski, J., and Dobrzanski, W. T. (1978). Lactostrepcins—acid bacteriocins produced by lactic streptococci, *J. Dairy Res., 45*:247.

Kozak, W., Rajchert-Trzpil, M. and Dobrzanski, W. T. (1974). The effect of proflavin, ethidium bromide and an elevated temperature on the appearance of nisin-negative clones in nisin-producing strains of *Streptococcus lactis, J. Gen. Microbiol., 83*: 295–302.

Laemmli, U. K., and Favre, M. (1973). Maturation of the head of bacteriophage T4. I: DNA packing events, *J. Mol. Biol., 80*:575.

LeBlanc, D. J., Crow, V. L., and Lee, L. N. (1980). Plasmid mediated carbohydrate catabolic enzymes among strains of *Streptococcus lactis*, in *Plasmids and Transposons: Environmental Effects and Maintenance Mechanisms* (C. Stuttand and K. R. Rozee, eds.), pp. 31–41, Academic Press, New York.

Lewus, C. B., and Montville, T. J. (1991). Detection of bacteriocins produced by lactic acid bacteria, *J. Microbiol. Methods, 13*:145.

Linnett, P. E., and Strominger, J. L. (1973). Additional antibiotic inhibitors of peptidoglycan synthesis, *Antimicrob. Agents Chemother., 4*:231.

Lipinska, E. (1977). Nisin and its applications, in *Antibiotics and Antibiosis in Agriculture* (M. Woodbine, ed.), p. 103, Butterworths, London.

Lueck, E. (1980). *Antimicrobial Food Additives*, Springer-Verlag, Berlin.

Marth, E. H. (1966). Antibiotics in foods—naturally occurring, developed and added, *Residue Rev., 12*:5.

Martin, D. R., and Gilliland, S. E. (1980). Inhibition of psychrotrophic bacteria in refrigerated milk by lactobacilli isolated from yogurt, *J. Food Prot., 43*:675.

Mather, D. W., and Babel, F. J. (1959). Inhibition of certain types of bacterial spoilage in creamed cottage cheese by the use of a creaming mixture prepared with *Streptococcus citrovorus, J. Dairy Sci., 42*:1917.

Mattick, A. T. R., and Hirsch, A. (1944). A powerful inhibitory substance produced by group N streptococci, *Nature, 154*:551.

Mattick, A. T. R., and Hirsch, A. (1947). Further observation on an inhibitor (nisin) from lactic streptococci, *Lancet, 2*:5.

McClintock, M., Serres, L., Marzolf, J. J., Hirsch, A., and Mocquot, G. (1952). Action inhibitrice des streptocoques producteurs de nisine sur le developpement des sporules anaerobies dans le fromage de Gruyere fondu, *J. Dairy Res., 19*:187.

McKay, L. L., and Baldwin, K. A. (1984). Conjugative 40-megadalton plasmid in *Streptococcus lactis* subsp. *diacetylactis* DRC3 is associated with resistance to nisin and bacteriophage, *Appl. Env. Microbiol., 47*:68–74.

Mohamed, G. E. E., Seaman, A., and Woodbine, M. (1984). Food antibiotic nisin: Comparative effects on *Erysipelothrix* and *Listeria*, in *Antimicrobials and Agriculture* (M. Woodbine, ed.), pp. 435–442, Butterworths, London.

Morris, S. L., Walsh, R. C., and Hansen, J. N. (1984). Identification and characterization of some bacterial membrane sulfydryl groups which are targets of bacteriostatic and antibiotic actions, *J. Biol. Chem., 259*:13590.

Mortvedt, C. I., and Nes, I. F. (1990). Plasmid-associated bacteriocin production by a *Lactobacillus sake* strain, *J. Gen. Microbiol., 136*:1601.

Motlagh, A. M., Johnson, M. C., and Ray, B. (1991). Viability loss of foodborne pathogens by starter culture metabolites, *J. Food Prot., 54*:873–878.

Muriana, P. M., and Klaenhammer, T. R. (1987). Conjugal transfer of plasmid-encoded determinants for bacteriocin production and immunity in *Lactobacillus acidophilus* 88, *Appl. Env. Microbiol., 53*:553.

Muriana, P. M., and Klaenhammer, T. R. (1991). Purification and partial characterization of lactacin F, a bacteriocin produced by *Lactobacillus acidophilus* 11088, *Appl. Env. Microbiol., 57*:114–121.

Nielsen, J. W., Dickson, J. S., and Crouse, J. D. (1990). Use of a bacteriocin produced by *Pediococcus acidilactici* to inhibit *Listeria monocytogenes* associated with fresh meat, *Appl. Env. Microbiol., 56*:2142–2145.

O'Brien, R. T., Titus, D. S., Devlin, K. A., Stumbo, C. R., and Lewis, J. C. (1956). Antibiotics in food preservation. II: Studies on the influence of subtilin and nisin on the thermal resistance of food spoilage bacteria, *Food Technol., 10*:352.

Ogden, K., and Tubb, R. S. (1985). Inhibition of beer spoilage lactic acid bacteria by nisin, *J. Inst. Brewing, 91*:390.

Ogden, K. M., Waites, J., and Hammond, J. R. M. (1988). Nisin and brewing, *J. Inst. Brew., 94*:233.

Okereke, A., and Montville, T. J. (1991). Bacteriocin inhibition of *Clostridium botulinum* spores by lactic acid bacteria, *J. Food Prot., 54*:349–353.

Orberg, P. K., and Sandine, W. E. (1984). Common occurrence of plasmid DNA and vancomycin resistance in *Leuconostoc* spp, *Appl. Env. Microbiol., 48*:1129.

Oxford, A. E. (1944). Diplococcin, an antibacterial protein elaborated by certain milk streptococci, *Biochem. J., 38*:178.

Pinheiro, A. J. R., Liska, B. J., and Parmelee, C. E. (1968). Properties of substances inhibitory to *Pseudomonas fragi* produced by *Streptococcus citrovorus* and *S. diacetylactis, J. Dairy Sci., 51*:183–187.

Price, R. J., and Lee, J. S. (1970). Inhibition of *Pseudomonas* species by hydrogen peroxide producing lactobacilli, *J. Milk Food Technol., 33*:13.

Pucci, M. J., Vedamuthu, E. R., Kunka, B. S., and Vandenburgh, P. A. (1988). Inhibition of *Listeria monocytogenes* by using bacteriocin PA-1 produced by *Pediococcus acidilactici* PAC 1.0, *Appl. Env. Microbiol. 54*:2349–2353.

Pulusani, S. R., Rao, D. R., and Sunki, G. R. (1979). Antimicrobial activity of lactic cultures: Partial purification and characterization of antimicrobial compound(s) produced by *Streptococcus thermophilus, J. Food Sci., 44*:575.

Raccach, M., and Baker, R. C. (1979). Bacterial repressive action of meat starter culture in pasteurized liquid whole egg, *J. Food Sci., 44*:90.

Raccach, M., Baker, R. C., Regenstein, J. M., and Mulnix, E. J. (1979). Potential application of microbial antagonism to extended storage stability of a flesh type food, *J. Food Sci., 44*:43.

Raccach, M., McGrath, R., and Daftarian, H. (1989). Antibiosis of some lactic acid bacteria including *Lactobacillus acidophilus* toward *Listeria monocytogenes, Int. J. Microbiol., 9*:25.

Radler, F. (1990a). Possible use of nisin in wine-making. I: Action of nisin against lactic acid bacteria and wine yeasts in solid and liquid media, *Am. J. Enol. Viticulture, 41*:1.

Radler, F. (1990b). Possible use of nisin in wine-making. II. Experiments to control lactic acid bacteria in the production of wine, *Am. J. Enol. Viticulture, 41*:7.

Ramseier, H. R. (1960). The action of nisin on *Clostridium butyricum, Arch. Mikrobiol., 37*:57–94.

Rayman, M. K., Aris, B., and Hurst, A. (1981). Nisin: A possible alternative or adjunct to nitrite in the preservation of meats, *Appl. Env. Microbiol., 41*:375.

Rayman, K., Malik, N., and Hurst, A. (1983). Failure of nisin to inhibit outgrowth of *Clostridium botulinum* in a model cured meat system, *Appl. Env. Microbiol. 46*:1450.

Reddy, G. V., Shahani, K. M., Friend, B. A., and Chandan, R. C. (1983). Natural antibiotic activity of *Lactobacillus acidophilus* and *bulgaricus*: III: Production and partial purification of bulgarican from *Lactobacillus bulgaricus, Cult. Dairy Prod. J., 18*:15.

Reddy, N. S. and Ranganathan, B. (1983). Preliminary studies on antimicrobial activity of *Streptococcus lactis* subsp. *diacetylactis, J. Food Prot., 46*:222–225.

Reddy, S. G., Henrickson, R. L., and Olson, H. C. (1970). The influence of lactic cultures on ground beef quality, *J. Food Sci., 35*:787.

Reiter, B., and Harnulv, B. G. (1984). Lactoperoxidase antibacterial system: natural occurrence, biological functions and practical applications, *J. Food Prot., 47*:724–732.

Ruhr, E., and Sahl, H. G. (1985). Mode of action of the peptide antibiotic nisin and influence on the membrane potential of whole cells and on cytoplasmic and artificial membrane vesicles, *Antimicrob. Agents Chemother., 27*:841.

Sabine, D. B. (1963). An antibiotic-like effect of *Lactobacillus acidophilus, Nature, 199*:811.

Salminen, S., Gorbach, S., and Salminen, K. (1991). Fermented whey drink and yogurt-type product manufactured using *Lactobacillus* strain, *Food Technol., 45*:112.

Schillinger, U., and Lucke, F. K. (1989). Antibacterial activity of *Lactobacillus sake* isolated from meat, *Appl. Env. Microbiol., 55*:1901–1906.

Scott, V. N., and Taylor, S. L. (1981a). Effect of nisin on the outgrowth of *Clostridium botulinum* spores, *J. Food Sci., 46*:117.

Scott, V. N., and Taylor, S. L. (1981b). Temperature, pH and spore load effects on the ability of nisin to prevent outgrowth of *Clostridium botulinum* spores, *J. Food Sci., 46*:121.

Shahani, K. M., Vakil, J. R., and Kilara, A. (1977). Natural antibiotic activity of *Lactobacillus acidophilus* and *bulgaricus*. II. Isolation of acidophilin from *L. acidophilus, Cult. Dairy Prod. J., 12*:8.

Shehata, A. E., Khalafalla, S. M., Magdoub, M. N. I., and Hofi, A. A. (1976). The use of nisin in the production of sterilized milk drinks, *Egypt. J. Dairy Sci., 4*:37.

Silva, M., Jacobus, N. V., Deneke, C., and Gorbach, S. L. (1987). Antimicrobial substance from a human *Lactobacillus* strain, *Antimicrob. Agents Chemother., 31*:1231–1233.

Somers, E. B., and Taylor, S. L. (1987). Antibotulinal effectiveness of nisin in pasteurized process cheese spreads, *J. Food Prot., 50*:842.

Sorrels, K. M., and Speck, M. L. (1970). Inhibition of *Salmonella gallinarum* by culture filtrates of *Leuconostoc citrovorum, J. Dairy Sci., 53*:239.

Spelhaug, S. R., and Harlander, S. K. (1989). Inhibition of foodborne bacterial pathogens by bacteriocins from *Lactococcus lactis* and *Pediococcus pentosaceous, J. Food Prot., 52*:856–862.

Steele, J. L., and McKay, L. L. (1986). Partial characterization of the genetic basis for sucrose metabolism and nisin production in *Streptococcus lactis, Appl. Env. Microbiol., 51*:57.

Steen, M. T., Chung, Y. J., and Hansen, J. N. (1991). Characterization of the nisin gene as part of a polycistronic operon in the chromosome of *Lactococcus lactis* ATCC 11454, *Appl. Env. Microbiol., 57*:1181–1188.

Szybalski, W. (1953). Cross resistance of *Micrococcus pyrogenes* var. *aureus* to thirty-four antimicrobial drugs, *Antibiot. Chemother., 3*:1095.

Tagg, J. R., Dajani, A. S., and Wannamaker, L. W. (1976). Bacteriocins of Gram-positive bacteria, *Bacteriol. Rev., 40*:722.

Tanaka, N., Traisman, E., Lee, M. H., Cassens, R. G., and Foster, E. M. (1980). Inhibition of botulinum toxin formation in bacon by acid development, *J. Food Prot., 43*: 450–457.

Tanaka, N., Traisman, E., Plantingua, P., Finn, L., Flom, W., Meske, L., and Guggisberg, J. (1986). Evaluation of factors involved in antibotulinal properties of pasteurized process cheese spreads, *J. Food Prot., 49*:526.

Taylor, S. L., and Somers, E. (1985). Evaluation of the antibotulinal effectiveness of nisin in bacon, *J. Food Prot., 48*:949.

Taylor, S. L., Somers, E., and Krueger, L. (1985). Antibotulinal effectiveness of nisin-nitrite combinations in culture medium and chicken frankfurter emulsions, *J. Food Prot., 48*:234.

Tramer, J., and Fowler, G. G. (1964). Estimation of nisin in foods, *J. Sci. Food Agric., 15*:522.

Tsai, H. J., and Sandine, W. E. (1987). Conjugal transfer of nisin plasmid gene from *Streptococcus lactis* 7962 to *Leuconostoc dextranicum* 181, *Appl. Env. Microbiol., 53*:352.

Upreti, G. C., and Hinsdill, R. D. (1973). Isolation and characterization of a bacteriocin from a homofermentative *Lactobacillus, Antimicrob. Agents Chemother., 4*:487.

Upreti, G. C., and Hinsdill, R. D. (1975). Production and mode of action of lactocin 27: Bacteriocin from homofermentative *Lactobacillus, Antimicrob. Agents Chemother., 7*:139.

Vincent, J. G., Veomett, R. C., and Riley, R. F. (1959). Antibacterial activity associated with *Lactobacillus acidophilus, J. Bacteriol., 78*:477.

von Wright, A., Wessels, S., Tynkynen, S., and Saarela, M. (1990). Isolation of a replicon region of a large lactococcal plasmid and use in cloning of a nisin resistance determinant, *Appl. Env. Microbiol., 56*:2029–2035.

West, C. A., and Warner, P. J. (1988). Plantacin B, a bacteriocin produced by *Lactobacillus plantarum* NCDO 1193, *FEMS Microbiol. Lett., 49*:163.

Wheaton, E., and Hays, G. L. (1964). Antibiotics and control of spoilage in canned foods, *Food Technol., 18*:147.

Whitehead, H. R. (1933). A substance inhibiting bacterial growth produced by certain strains of lactic streptococci, *Biochem. J., 27*:1793.

Yousef, A. E., Luchansky, J. B., Degnan, A. J., and Doyle, M. P. (1991). Behavior of *Listeria monocytogenes* in wiener exudates in the presence of *Pediococcus acidilactici* H or pediocin AcH during storage at 4 or 25°C, *Appl. Env. Microbiol., 57*:1461–1467.

Zajdel, J. K., Ceglowski, P., and Dobrzanski, W. T. (1985). Mechanism of action of lactostrepcin 5, a bacteriocin produced by *Streptococcus cremoris* 202, *Appl. Env. Microbiol., 49*:969.

Zajdel, J. K., and Dobrzanski, W. T. (1983). Isolation and preliminary characterization of *Streptococcus cremoris* strain 202 bacteriocin, *Acta Microbiol. Pol., 32*:119.

6

Genetic Modification of Lactic Acid Bacteria

Atte von Wright
University of Kuopio, Kuopio, Finland; University of Helsinki, Helsinki, Finland; and The Royal Veterinary and Agricultural University, Frederiksberg, Denmark

Mervi Sibakov
Valio Ltd., Helsinki, Finland

I. INTRODUCTION

Serious genetic research on lactic acid bacteria started in the 1970s with the observation of plasmid linkage of many of the essential functional properties of the lactococci (or group N streptococci, as they were known then). Both in vivo and in vitro gene transfer techniques were soon developed for these organisms, and the studies were expanded to other genera of lactic acid bacteria. In the 1980s recombinant DNA technique became possible with many of the important species, and several genes involved in lactose fermentation or coding for proteinases were cloned and analyzed at the molecular level.

During the last few years progress has been increasingly rapid. The structure and function of both plasmid-linked and chromosomal genes of lactic acid bacteria have been further elucidated. Also the potential of these organisms to act as production hosts for heterologous proteins is being evaluated.

In the following sections we describe in more detail the development and present status of genetic research briefly outlined above. The emphasis is on the basic scientific aspects and recent findings, rather than on the potential applications or history of the field. We have chosen this approach in order to give a reader a thorough overview about the methods and genetic techniques available at the moment for this biotechnologically important group of microorganisms.

II. THE PLASMID BIOLOGY OF LACTIC ACID BACTERIA

A. Lactococcal Plasmids

The observations made by the group directed by L. L. McKay at the University of Minnesota about the spontaneous and acriflavine-induced loss of lactose fermentation ability of lactococcal strains suggested involvement of plasmids in this phenotype (McKay et al., 1972). Subsequent demonstration of extrachromosomal DNA in the lactococci (Cords et al., 1974) led soon to the identification of several metabolic plasmids and their functions. Summing up the results of both his own group and those obtained by others, McKay (1983) in his review listed the following phenotypes that apparently are plasmid-linked either in most or in some of the lactococcal strains studied: fermentation of lactose, sucrose, galactose, mannose, xylose, and even glucose, proteinase activity, citrate utilization, phage resistance and DNA restriction and modification, production of bacteriocins, mucoidness, and possibly, resistance to inorganic ions. It is noteworthy that many of the plasmid-linked properties are either essential (lactose fermentation, proteinase activity) or helpful in the dairy applications. In addition, lactococcal strains usually contain many apparently cryptic plasmids ranging in size between 1 and more than 100 kilobase pairs (kbp) (Klaenhammer et al., 1978; Kuhl et al., 1979; Davies et al., 1981; von Wright et al., 1986). Study of plasmid-linked genes has dominated genetic research on lactococci until recent years. Some of the most important findings are described below.

1. *Lactose Fermentation and Proteinase Plasmids*

Both lactose fermentation and proteinase activity were soon found to be associated with relatively large (from about 17 to more than 50 kbp) lactococcal plasmids. The genes coding for these phenotypes can reside either both in the same linkage group, or they may occupy different plasmids (see the review of McKay, 1983).

The lactococci metabolize lactose using the phosphoenolpyruvate (PEP)-dependent phosphotransferase system. Lactose enters the cell as lactose-6-phosphate, which is subsequently split into galactose-6-phosphate and glucose by phospho-β-galactosidase. Glucose is further metabolized by the Embden-Meyerhof-Parnas pathway, while galactose-6-phosphate is first converted to tagatose diphosphate, before it can be split to two triose phosphates and enter the normal glucolytic route (see Thompson, 1987, for review).

A PEP-dependent lactose phosphotransferase system is operating also in *Streptococcus sanguis* Challis, and lactose-negative mutants with genetic defects in different components of this pathway are available. By using these mutants as recipients in transformation and cloning experiments with lactococcal plasmid DNA, and selecting for lactose fermentation, the plasmid linkage of at least phospho-β-galactosidase and two of the proteins involved in lactose transport (Enzyme II^{lac} and Factor II^{lac}) in the lactococci could be confirmed (Harlander and McKay, 1984; Harlander et al., 1984). Evidence of the plasmid linkage of also the genes controlling the tagatose diphosphate part of the pathway has been obtained, and the gene coding for tagatose 1,6-biphosphate aldolase has been isolated from a lactococcal plasmid and cloned in *Escherichia coli* (Limsowtin et al., 1986). For the latest genetical work, see Section VI.B.2.

The proteolytic system of the lactococci has been the subject of intensive biochemical and genetic studies. The plasmid-coded lactococcal proteinases are very closely related to each other, and they represent probably the most thoroughly genetically characterized family of genes in this genus. A more detailed description of their structure and function is therefore included in Section VI.B.1.

One of the best-characterized lactococcal lactose-fermentation and proteinase plasmids is pPL712 from *Lactococcus lactis* ssp. *lactis* NCDO712 (Gasson, 1983; Maeda and Gasson, 1986). The functional map of this plasmid is presented in Fig. 1.

2. *Citrate Utilization Plasmids*

Diacetyl is an important aroma compound in dairy products. Among the lactococci *Lactococcus lactis* ssp. *lactis* biovar diacetylactis strains are able to metabolize citrate to diacetyl and carbon dioxide. Uptake of citrate is controlled in these strains by a small (5.5-MDa or 8.7-kbp) plasmid, which codes for a specific permease (Kempler and McKay, 1979, 1981).

3. *Plasmids Associated with Phage Resistance*

The presence of restriction-modification (R/M) systems in the lactococci was first discovered by Collins (1956), who also observed the instability of lactococcal phage resistance in repeated transfers in milk (Collins, 1958). Several plasmid-linked R/M systems have been characterized during recent years, and also other plasmid-associated phage resistance mechanisms, such as inhibition of phage adsorption or abortive infection have been detected. These three main types of plasmid-mediated phage defense systems are handled separately in the following section.

a. Plasmid Coding for Restriction-Modification. Direct demonstration of a plasmid-encoded R/M system was done by Sanders and Klaenhammer (1981). They could identify in *Lactococcus lactis* ssp. *cremoris* KH a 10-MDa (or 15.8-kbp) plasmid, designated pME100, the absence of which increased the sensitivity of this strain to phage c2 about a thousandfold. There was, however, a considerable residual, apparently chromosomally encoded, R/M activity left in the plasmid-cured variants.

The presence of multiple plasmid-mediated R/M systems with additive effects was demonstrated in *Lactococcus lactis* ssp. *lactis* IL594 by Chopin et al. (1984). Of the nine different plasmids present in this strain two, pIL6 (28 kbp) and pIL7

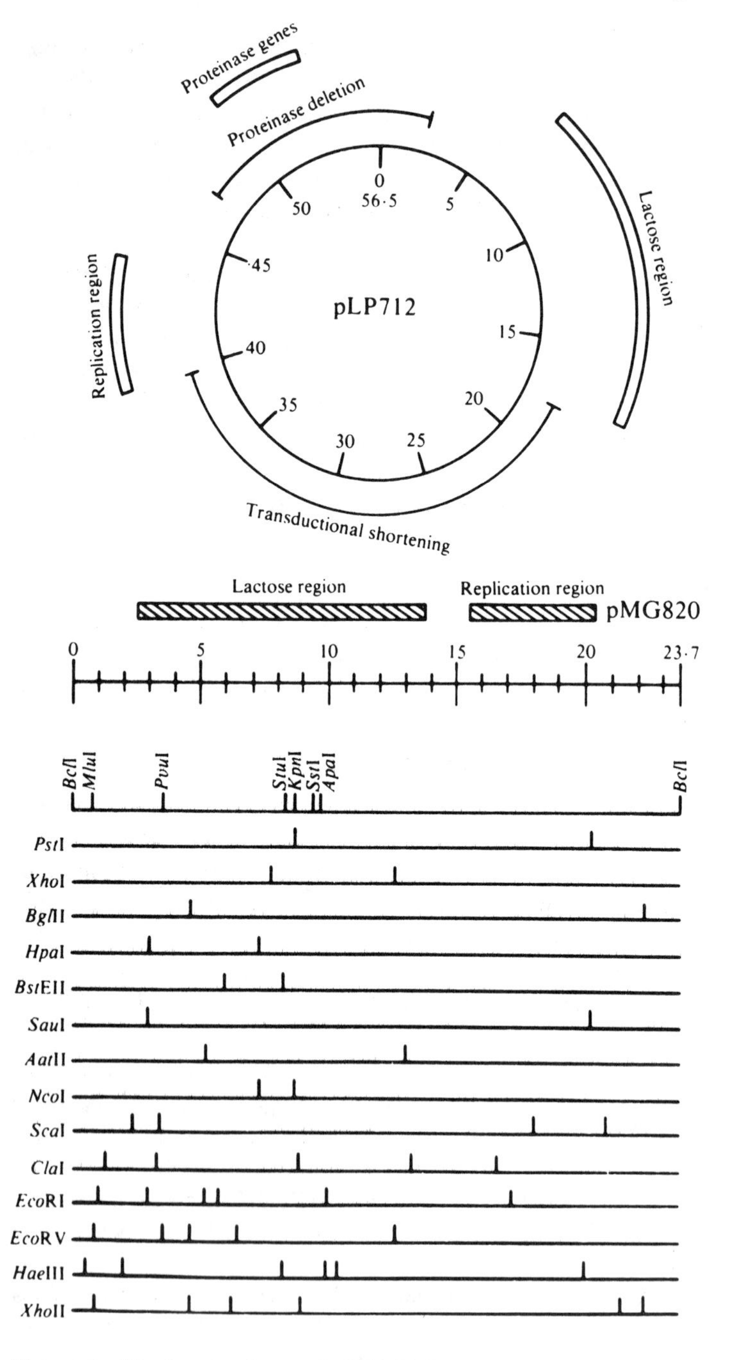

Figure 1 The functional map of pLP712. (From Maeda and Gasson, 1986.)

(31 kbp), coded for separate R/M systems. A similar situation could later be detected in a *Lactococcus lactis* ssp. *cremoris* strain IL964, which likewise has two different R/M plasmids (Gautier and Chopin, 1987). This strain has also a plasmid-mediated abortive infection phenotype.

Since these findings several other R/M plasmids have been detected in different lactococcal strains (see Klaenhammer, 1987, for review). Of particular interest has been pTR2030, a 46.2-kbp plasmid isolated from *Lactococcus lactis* ssp. *lactis* ME2 (Klaenhammer and Sanozky, 1985). This conjugative plasmid encodes both an R/M system and an abortive infection mechanism (Hill, 1989). Consequently this plasmid confers resistance to a wide variety of lytic phages, and one has been able to conjugate it to phage-sensitive lactococcal strains, thereby making them more useful in dairy applications (Sanders et al., 1986; Sing and Klaenhammer, 1986). Another example of a plasmid coding for multiple phage defense mechanisms is pKR223, a 35.5-kbp plasmid from *Lactococcus lactis* ssp. *lactis* biovar *diacetylactis* KR2 (Laible et al., 1987). Also this plasmid is associated both with an R/M system and abortive infection (McKay et al., 1989).

b. Plasmids Conferring Reduced Phage Adsorption. In addition to R/M systems and abortive infection, the lactococcal strain ME2 mentioned above has a third plasmid-encoded defense mechanism against phages. A 30-MDa (47.3-kbp) plasmid, pME0030, causes reduced adsorption of at least four different phages (Sanders and Klaenhammer, 1983).

In *Lactococcus lactis* ssp. *cremoris* SK11 a 35-MDa (55.1-kbp) plasmid, pSK112, prevents completely the adsorption of a certain phage (de Vos, Underwood, and Davies, 1984) forming apparently the main phage defense mechanism of the strain in question.

c. Plasmids Causing Abortive Infection. Abortive infection means reduction of the phage plating efficiency despite normal phage adsorption and lack of R/M activity. One of the first plasmids associated with this phenomenon is pNP40, a 40-MDa (63-kbp) plasmid from *Lactococcus lactis* ssp. *lactis* biovar diacetylactis resistance DRC3 (McKay and Baldwin, 1984). This conjugative plasmid confers resistance against polypeptide antibiotic nisin and protects the strain from a lytic phage c2. Phage resistance is thermosensitive, disappearing at 37°C. However, the phage that had developed at 37°C was still unable to replicate in hosts harboring pNP40 at 32°C. This excludes an ordinary R/M system as an explanation for the reduced phage viability.

The abortive infection system encoded by pTR2030 (see Section II.A.3.a) is also to some degree thermosensitive in *Lactococcus lactis* ssp. *lactis* strains, being ineffective at 40°C (Klaenhammer and Sanozky, 1985). In pTR2030-containing *Lactococcus lactis* ssp. *cremoris* the phage-resistance phenotype is stable even at elevated temperatures (Stevenson and Klaenhammer, 1985), and apparently at least some *Lactococcus lactis* ssp. *lactis* transconjugants retain the pTR2030 mediated phage protection when grown at 39°C (Sanders et al., 1986). The effect of pTR2030 on phage DNA injection, transfection, and release of progeny phage has been studied in detail (Sing and Klaenhammer, 1990). DNA injection was found to be normal, transfection could be achieved, although at a reduced frequency, and the number of phage particles released from infected hosts was decreased. These effects vary somewhat according to the phage tested. For the prolate phage c2 pTR2030 causes formation of small plaques while the majority (92%) of the infected cells die. The plasmid-mediated protection against small isometric phages p2 and ϕ31 is much stronger. Although 72% of the p2-infected cells still die, they do not lyse. With ϕ31 the interaction between the pTR2030-mediated R/M system and abortive infection causes a total protection against the phage.

Several other plasmids associated with abortive infection have been identified. One of them, pKR223, coding for both an R/M system and an abortive infection mechanism, has already been mentioned (Section II.A.3.a). Other examples have been described by both Murphy et al. (1988) and Jarvis et al. (1989).

4. Plasmid-Encoded Bacteriocin Production and Resistance

Bacteriocins are proteinaceous substances secreted by bacteria that inhibit the growth of other bacterial strains or species. Lactococcal strains produce several bacteriocins, some of which are plasmid-encoded. Plasmid linkage was firmly established with diplococcin, a bacteriocin produced by *Lactococcus lactis* ssp. *cremoris* 346 (Davey, 1984) harboring a 54-MDa (85.1-kbp) conjugative plasmid associated with this phenotype. Neve et al. (1984) demonstrated that a 60-kbp plasmid, p9B4-6, in *Lactococcus lactis* ssp. *cremoris* 9B4 is responsible for bactericidal activity that inhibits, in addition to other lactococci, strains of *Clostridium*.

Another example of a bacteriocin plasmid characterized at a molecular level is *Lactococcus lactis* ssp. *lactis* biovar *diacetylactis* WM_4 plasmid pNP2 (131.1 kbp) (Scherwitz-Harmon and McKay, 1987). In cloning experiments the gene(s) for antagonistic activity were located within an 18.4-kbp fragment. As the fragment was cloned in a vector having a higher copy number than pNP2, a gene dose effect was observed in bacteriocin-producing clones, the inhibitory being more prominent that in the original WM_4 strain.

Economically the most important lactococcal bacteriocin is nisin, which is used as a food additive to control the growth of Gram-positive spoilage organisms (Hurst, 1981). Nisin production is often linked with sucrose fermentation, and is both a curable and conjugally transmissible trait (Gasson, 1984; Steele and McKay, 1986; Gonzales and Kunka, 1985). These observations at first suggested involvement of a plasmid in nisin production. Recent findings, however, suggest that nisin production, nisin resistance, and sucrose metabolism are coded by a large transmissible chromosomal gene block (see Section IV.B.6). Nisin resistance genes can exist also extrachromosomally, and while no nisin production plasmids have been detected so far, several examples of large plasmids coding for nisin resistance among other phenotypes have been described (Klaenhammer and Sanozky, 1985; McKay and Baldwin, 1984; von Wright et al., 1990). From two of these plasmids, pNP40 and pSF01, nisin resistance genes have been isolated and cloned (Froseth et al., 1988; von Wright et al., 1990).

5. Other Plasmid-Mediated Phenotypes

Mucoidness is a property of some lactococcal strains that have traditionally been used to give body and texture to certain types of Scandinavian fermented milks (Forsen et al., 1973; Macura and Townsley, 1984). Several groups have identified plasmids ranging in size between 17 and 30 MDa (26.8 and 47.3 kbp) associated with mucoid phenotype (Vedamuthu and Neville, 1986; von Wright and Tynkkynen, 1987; Neve et al., 1988).

Resistance to ultraviolet (UV) irradiation is plasmid-encoded in *Lactococcus lactis* ssp. *lactis* IL594. The relevant plasmid, pIL7, has a size of 33 kbp, and by cloning experiments the UV-resistant determinant has been located in a 5.4-kbp fragment (Chopin et al., 1985, 1986).

No antibiotic resistance plasmids have been yet isolated from the lactotocci. Sinha reported (1986) that a 5.5-MDa (8.7 kbp) plasmid effectively suppresses the formation of streptomycin-resistant mutants in certain strains of *Lactococcus lactis* ssp. *lactis*.

B. Plasmids in Other Groups of Lactic Acid Bacteria

Plasmids are regularly found also in strains of *Lactobacillus, Pediococcus, Leuconostoc*, and *Streptococcus salivarius* ssp. *thermophilus*, although their metabolic role seems to be less prominent in these genera than in the lactococci.

1. Plasmids of the Lactobacilli

Strains of a homofermentative lactobacillus *Lactobacillus casei* harbor plasmids associated with the PEP-dependent lactose phosphotransferase system (Chassy et al., 1978). In *Lactobacillus casei* 64H lactose metablism is controlled by a 35-kbp plasmid, pLZ64, and the relevant gene block has been cloned in *Escherichia coli* (Lee et al., 1982; Chassy, 1987). Another *Lactobacillus helveticus* plasmid, pLY101 (68.2 kbp), contains the gene for 6-P-β-galactosidase, while the rest of the genes for lactose metablism apparently are chromosomal in this particular strain (Shimizu-Kadota, 1987).

In *Lactobacillus helveticus* *N*-acetyl-D-glucosamine utilization appears to be a plasmid-encoded trait, at least in some strains (Smiley and Fryder, 1978). In the same species a 3.5-MDa (5.5-kbp) plasmid coding for proteinase activity has been detected (Morelli et al., 1986).

Correlation of mucoid phenotype and a 4.5-MDa (6.6-kbp) plasmid has been observed in *Lactobacillus casei* (Vescovo et al., 1989). In *Lactobacillus sake* cysteine transport has been associated with a 8.3-kbp plasmid (Shay et al., 1988).

Lactacin F is a bacteriocin produced by *Lactobacillus acidophilus* 88 (Muriana and Klaenhammer, 1987). Although no plasmid linkage of either bacteriocin production or immunity could be detected in strain 88, two plasmids of 68 and 57 MDa (107 and 90 kbp) could be detected in bacteriocin-producing transconjugants obtained in crosses with *Lactobacillus acidophilus* 88 as the donor. Both lactacin F production and immunity appear to be linked with the 107-kbp plasmid (Muriana and Klaenhammer, 1987, 1991).

Vescovo et al. (1982) tested 20 strains of *Lactobacillus acidophilus* and 16 strains of *Lactobacillus reuteri* for antibiotic resistance and presence of plasmids. Complex patterns of antibiotic resistances were observed together with variable plasmid complements. The results of plasmid curing experiments suggest involvement of plasmids in the drug resistance, although it was not possible to link individual plasmids to resistance against certain antibiotics. In *Lactobacillus reuteri* strains isolated from pig intestine erythromycin resistance was found to be coded by a 5.5-MDa (8.7 kbp) or a 6.9-MDa (10.9-kbp) plasmid (Axelsson et al., 1988). The erythromycin resistance region of the latter plasmid, pLUL631, was successfully cloned and expressed both in *Escherichia coli* and *Lactococcus lactis* ssp. *lactis*.

Cryptic plasmids of *Lactobacillus casei* have been restriction endonuclease mapped and cloned in toto both in *Escherichia coli* and *Streptococcus sanguis* as a part of cloning vector construction (Lee-Wickner and Chassy, 1985). The complete nucleotide sequence of a 3.29-kbp cryptic plasmid from *Lactobacillus helveticus* ssp. *jugurti* has also been obtained (Takiguchi et al., 1989). Other examples of cryptic lactobacillus plasmids are presented in Section VII.A.

2. Pediococcal Plasmids

Cryptic plasmids varying in size between 1.2 and 34 MDa (1.9 and 53.6 kbp) have been detected in several pediococcal strains (Gonzalez and Kunka, 1983, 1986; Graham and McKay, 1985). The ability to utilize raffinose, a trisaccharide, was found to be linked with the presence of relatively large plasmids (23–30 MDa, 36.2–47.3 kbp) in three strains of *Pediococcus pentosaceus* (Gonzalez and Kunka, 1986). In a *Pediococcus cerevisiae* strain the production of a bacteriocin-like substance was associated with a 10.5-MDa (16.3-kbp) plasmid (Graham and McKay, 1985).

3. *Plasmids in the Genus Leuconostoc and in Streptococcus Salivarius subsp. Thermophilus*

Cryptic plasmids in the 1–76 MDa (1.6–120 kbp) range were detected in 14 of the 18 *Leuconostoc* sp. strains screened. No phenotypic trait could be associated with any of the plasmids observed (Orberg and Sandine, 1984).

In *Streptococcus salivarius* ssp. *thermophilus* plasmids are not very common, and the majority of them are small (less than 10 kbp). No plasmid linkage of any obvious phenotypic property has been detected in this bacterium, not even with the largest (25.5 kbp) plasmid detected so far (see Mercenier, 1990, for a review). The best-characterized *Streptococcus salivarius* ssp. *thermophilus* plasmid is apparently a 6.9-kbp plasmid pA33 studied by the group of Mercenier (see previous reference). This plasmid apparently can both integrate into the chromosome and be reexcised, the colony morphology and growth characteristics of the host strain apparently being to a degree dependent on whether the plasmid exists as an extrachromosomal element or not. The sequence of pA33 has been determined and found to contain five open reading frames (ORF).

III. GENE TRANSFER IN LACTIC ACID BACTERIA

Natural gene transfer systems observed among lactic acid bacteria include both transduction, or phage-mediated genetic exchange, and conjugation, in which genetic material is transferred from the donor to the recipient by cell-to-cell contact. Physiological transformation, or uptake of naked DNA by the recipient cell, does not seem to function in the species and strains studied so far. In vitro transformation techniques, based either on polyethylene glycol (PEG)–mediated DNA transfer into bacterial protoplasts or on the electroporation of either protoplasts or intact bacteria, have made recombinant DNA techniques feasible in lactic acid bacteria.

A. Gene Transfer In Vivo

1. *Transduction*

First transduction experiments with lactococcal strains were performed using a virulent bacteriophage (Allen et al., 1963). The common occurrence of lysogeny in the lactococci (see the review of Davidson et al., 1990) has since then made the temperate phages an obvious tool for transduction experiments. They have been used to transform genes involved in the lactococcal lactose metabolism (McKay et al., 1973). Subsequently also proteinase genes could be transduced (McKay and Baldwin, 1974). A potentially useful observation was the integration of both lactose fermentation and proteinase genes into the recipient chromosome in some transductants leading to the stabilization of these traits (McKay and Baldwin, 1978; Kempler et al., 1979).

Both chromosomal and plasmid-linked genes can be transduced in the lactococci. With large metabolic plasmids the frequency of transduction often rises significantly, when the transducing lysate is derived from a first generation transductant (McKay et al., 1973, 1976; Gasson, 1983). This is associated with the transductional shortening of the transferred plasmid, apparently to a size optimal for the packaging of DNA into the phage head (Davies and Gasson, 1981).

There are few reports of successful transduction experiments with other lactic acid bacteria than the lactococci. However, transduction of plasmids of both *Streptococcus salivarius* ssp. *thermophilus* (Mercenier et al., 1988) and *Lactobacillus acidophilus* (Raya et al., 1989) has been achieved.

2. Conjugation

The presence of conjugative phage resistance plasmids in lactococcal strains has already been mentioned in Section II.A.3, and conjugation appears to be a useful technique in genetic analysis and strain construction. In the following sections some examples of conjugation systems that either are well characterized at the molecular level or demonstrate the potential of this approach as an alternative to in vitro genetic manipulation are presented.

a. Lactococcal Lactose Plasmid Transfer and Cell Aggregation Phenotype. The first reports of conjugative transfer of lactose fermentation plasmids were done by Gasson and Davies (1980a) and Kempler and McKay (1979). Since then this observation has been repeated with several lactococcal strains and also with *Lactobacillus lactis* (see Fitzgerald and Gasson, 1988, for a review). However, only the conjugation-associated phenomena in closely related *Lactococcus lactis* ssp. *lactis* strains 712 and ML3 appear to be studied in detail.

Although the frequency of lactose-positive transconjugants is usually very low, using 712 donors (Gasson and Davies, 1980a) or ML3 donors (Walsh and McKay, 1981) it is possible to obtain transconjugants having an ability to further transfer the lactose fermentation genes (at a frequency of 10^{-1}–10^{-2} per donor). These transconjugants also have a peculiar cell aggregation or clumping (Clu^+) phenotype. In Clu^+ strains obtained using ML3 as the donor a novel 104-kbp plasmid, subsequently shown to be a cointegrate of a 55-kbp lactose fermentation plasmid (pSK08) and a 48.4-kbp plasmid pRS01, both present in the original donor, could be detected. A cointegrate plasmid of similar size can, however, be detected also in transconjugants displaying neither the high-frequency conjugation nor Clu^+ phenotype. Restriction mapping showed that pSK08 can be integrated into several sites of pRS01 sometimes inactivating either the high-efficiency conjugation genes or genes responsible for the Clu^+ phenotype or both. These observations suggested the presence of an insertion element (see

Section IV. B.1.a) in pSK08 (Anderson an McKay, 1984). Similar observations and conclusions were made with 712 transconjugants having high donor activity and showing cell aggregation, with the exception of the DNA inserted into the lactose plasmid being chromosomal and not representing any known plasmid (Fitzgerald and Gasson, 1988) The insertion element present in pSK08 has been further analyzed (Polzin and Shimizu-Kadota, 1987). It has a size of 808 bp, has 18 bp inverted, repeats at its ends, and contains a single open reading frame for a protein of 226 amino acids. It is designated ISS1S and shows a strong homology with a Gram-negative insertion element IS26.

According to the model developed on the basis of these observations the high frequency of conjugation and cell aggregation, which are separate phenotypes, result from the integration of a sex factor, either a plasmid or a chromosomal gene fragment, into the lactose plasmid. Recently this model was further refined by identifying a new chromosomal genetic determinant designated *agg*, which is needed in addition to the *clu* genes, to produce the cell aggregation phenotype (van der Lelie et al., 1991).

b. Intergeneric Conjugation. Broad host range streptococcal plasmids pAMβ1 (Clewell et al., 1974; Leblanc and Lee, 1984) and pIP501 (Horodniceanu et al., 1976; Evans and Macrina, 1983), the former a 26.5-kbp erythromycin resistance plasmid and the latter a 30-kbp chloramphenicol-erythromycin double-resistance plasmid, have successfully been conjugated to a variety of lactic acid bacteria. Conjugal transfer of pAMβ1 into several lactococcal strains was first demonstrated by Gasson and Davies (1980b). Similar plasmid transfer into *Lactobacillus casei* was reported even earlier (Gibson et al., 1979). Other lactobacilli successfully used as pAMβ1 recipients in conjugation experiments include *Lactobacillus plantarum* (West and Warner, 1985; Shrago et al., 1986), *Lactobacillus acidophilus* and *Lactobacillus reuteri* (Vescovo et al., 1983; Cocconcelli et al., 1985). Subsequently Morelli et al. (1988) demonstrated the transfer of pAMβ1 from *Lactobacillus reuteri* to *Enterococcus faecalis* in the gut of germ-free mice. Outside the lactococci and lactobacilli, pAMβ1 has been introduced into *Leuconostoc* strains (Pucci et al., 1987). With pediococci, the transfer of pAMβ1 proved unsuccessful, but pIP501 was successfully conjugated into *Pediococcus acidilactici* and *Pediococcus pentosaceus* using both *Streptococcus sanguis* and *Lactococcus lactis* ssp. *lactis* as donors (Gonzalez and Kunka, 1983).

Another broad host range plasmid pVA797, a pIP501 derivative, in which the erythromycin resistance region has been replaced by the replication origin of a cryptic streptococcal plasmid (Evans and Macrina, 1983) has been successfully conjugated, in addition of lactococcal strains, to *Streptococcus salivarius* ssp. *thermophilus* in a conjugative mobilization experiment (see the next section) by Romero et al. (1987).

Intergeneric conjugation within lactic acid bacteria using their own plasmids or marker genes has been successful with lactococcal strains as donors and *Leuconostoc* strains as recipients. The lactococcal genetic determinants transferred code for such phenotypes as lactose fermentation, sucrose fermentation, and nisin production (Tsai and Sandine, 1987a,b).

c. Conjugative Mobilization. A conjugative plasmid can enhance the transfer of otherwise non- or poorly conjugative plasmids or chromosomal markers from donor to recipient. Streptococcal plasmid pAMβ1 has been used to mobilize proteinase plasmids in crosses between different lactococcal strains (Hayes et al., 1990). Novel plasmids representing cointegrates between pAMβ1 and lactococcal plasmids could be detected in transconjugants indicating several recombination events associated with conjugation.

With plasmid pVA797 (see previous section) its tendency to form cointegrates with plasmids having homologous regions forms a basis of a mobilization system allowing the transfer of recombinant plasmids carrying cloned genes to nontransformable strains (Smith and Clewell, 1984). This plasmid has two replication regions, one of pIP501 and the other that of a cryptic streptococcal plasmid pVA380-1. When a vector plasmid pVA838 having also the pVA380-1 replication origin is introduced into a strain harboring pVA797, a cointegrate is formed. This cointegrate can be conjugated into other strains. Subsequently it has a tendency to resolve, and the plasmids segregate due to replicon incompatibility. Romero et al. (1987) applied this strategy to develop, in addition to pVA838, other cloning vectors that are suitable to lactic acid bacteria and can be mobilized by pVA797.

An interesting phenomenon was observed with a nonconjugative streptococcal tetracycline resistance plasmid pMV158. It could be mobilized into *Lactococcus lactis* ssp. *lactis* IL1403 in intergeneric conjugation apparently by cotransfer of plasmids pAMβ1 or pIP501. When IL1403 recipients were used as donors in conjugation experiments with other *Lactococcus lactis* ssp. *lactis* strains, the transfer of pMV158 did not require the presence of either pAMβ1 or pIP501. Instead, there was an indication of a lactococcal chromosomal function causing plasmid mobilization (van der Lelie et al., 1990).

Conjugal mobilization of plasmids from Gram-negative donors to lactococcal recipients has also been achieved (Trieu-Cuot et al., 1987) using a shuttle vector consisting of both pAMβ1 and pBR322 replication regions and an antibiotic resistance marker expressed in both Gram-negative and Gram-positive hosts. The vector had a conjugal transfer origin of a P group plasmid, and the mobilizing plasmid belonged also to group P.

B. In Vitro Gene Transfer Techniques

Despite the usefulness of the in vivo methods available for genetic studies, development of different transformation techniques was necessary for the actual gene cloning in lactic acid bacteria. The methods currently available are based either on protoplast transformation or electroporation ("electro-transformation").

1. Protoplast Techniques

Preparation and regeneration of lactococcal protoplasts was achieved by Gasson (1980). He could also demonstrate PEG-induced protoplast fusion and the recombination of both chromosomal and plasmid-linked markers in the regenerated fusion products. Later also intergeneric transfer of several staphylococcal antibiotic resistance plasmids into *Lactococcus lactis* ssp. *lactis* was detected in protoplast fusion with *Bacillus subtilis* (Baigori et al., 1988). Protoplast fusion and consequent plasmid transfer has also been achieved between *Lactococcus lactis* ssp. *lactis* and *Lactobacillus reuteri* (Cocconcelli et al., 1986). Two strains of *Lactobacillus fermentum* with different plasmid compositions have been fused forming a strain containing both parental plasmids (Iwata et al., 1986).

Actual protoplast transformation of *Lactococcus lactis* ssp. *lactis* was reported by Kondo and McKay (1982). First transformations were achieved with lactococcal lactose fermentation plasmids and the transformation frequencies were low (about 8.5 transformants per microgram plasmid DNA). Subsequently the transformation protocols were optimized in different laboratories in respect to protoplast regeneration, PEG concentration and molecular weight, PEG treatment time, protoplast concentration, ionic composition of transformation buffers, and bacterial growth phase before protoplasting (Kondo and McKay, 1984; von Wright et al., 1985; Simon et al., 1986). With optimized methods and using different antibiotic resistance markers transformation frequencies ranging from about 10^4–10^6 per microgram of DNA were achieved allowing shotgun cloning of both plasmid-linked (Kondo and McKay, 1984) and chromosomal lactococcal genes (von Wright et al., 1986) in lactococcal hosts. Sanders and Nicholson (1987) reported high-efficiency transformation of nonprotoplasted *Lactococcus lactis* ssp. *lactis*. As the procedure required PEG treatment and osmotic stabilizer, it appears that the growth conditions applied induced the particular strain used to "self-protoplast" without external enzymatic cell-wall digestion.

There are few reports of successful protoplast transformation of lactic acid bacteria other than lactococci. Low-efficiency protoplast transformation of *Lactobacillus salivarius, Lactobacillus reuteri*, and *Lactobacillus fermentum* has been reported (Skou, 1987): Strains of *Lactobacillus acidophilus* (Lin and Savage,

1986; Morelli et al., 1987) and *Lactobacillus plantarum* (Leer et al., 1987) have been reported transformable. Transformation of *Lactobacillus casei* with plasmid DNA encapsulated in liposomes yields transformants 10^5–10^6 per microgram DNA, and the integration of plasmid into chromosome was observed, while somewhat lower transformation frequencies have been obtained with *Streptococcus salivarius* ssp. *thermophilus* protoplasts and liposome-encapsulated DNA (Mercenier et al., 1988).

2. *Electroporation*

The limitations of protoplast techniques, especially with industrially important lactobacilli, have led to the adaptation of electroporation as a transformation method for lactic acid bacteria. The principle of this method is to permeabilize the cell wall and membrane to DNA by an electric discharge, and it was originally used to transform eukaryotic cells (Neumann et al., 1982). The basic electroporation procedure is relatively straightforward. The cells are suspended in buffer having a low ionic strength and containing high concentrations of sucrose (about 0.3 *M*) or glycerol (10%). They are chilled and exposed either to single or multiple electric pulses with varying field strengths (usually several kV/cm), and the DNA suspended in the electroporation buffer gets transferred into cell through pores generated by the pulse. The optimization necessary for each strain includes determination of suitable growth phase and cell density, growth medium, and optimal electroporation parameters.

Electroporation has proven to be a very useful technique in transforming many of the previously practically nontransformable strains, such as lactobacilli, *Leuconostoc* and *Pediococcus* strains, and, among lactococci, strains of *Lactococcus lactis* ssp. *cremoris*.

First electroporation of a *Lactococcus lactis* ssp. *lactis* strain was reported by Harlander (1987). By using electrodes installed in a chamber of a spinning centrifuge rotor, she was able to obtain transformation frequencies comparable to those achieved by protoplast transformation. High-efficiency electroporation of *Lactobacillus casei* was achieved at the same year (Chassy and Flickinger, 1987), and in 1988 Luchansky et al. published successful results of electroporation experiments involving several Gram-positive genera, including *Lactobacillus, Lactococcus, Leuconostoc*, and *Pediococcus. Streptococcus salivarius* ssp. *thermophilus* was electroporated by Somkuti and Steinberg (1988). Since then electroporation procedures have been further optimized for lactococci (van der Lelie et al., 1988; Powell et al., 1988; McIntyre and Harlander, 1989). *Streptococcus salivarius* ssp. *thermophilus* (Mercenier, 1990), and *Leuconostoc paramesenteroides* (David et al., 1989). Electroporation is also the standard gene transfer method in *Lactobacillus plantarum* (Bates et al., 1989; Scheirlinck et al., 1989).

IV. GENETIC RECOMBINATION IN LACTIC ACID BACTERIA

In the previous section it was noted that genetic rearrangements are often associated with the transfer of genetic material. Recombination can occur in two ways. The usual pathway, when chromosomal DNA is transferred from one strain to another by conjugation or by transduction, is the so-called general recombination. It requires considerable sequence homology between the participating DNAs and also specific bacterial recombination functions. Recombination mediated by transposable genetic elements, on the other hand, is independent of host function, and very short target sequences (about 10 bp or less) are necessary.

A. General Recombination

The mechanisms involved in lactococcal general recombination have mainly been studied using different integration vectors as models (see Section VI A.2). Homologous sequences in these plasmids have been derived either from lactococcal chromosome (Leenhouts et al., 1989) or from a resident prophage (M.-C. Chopin et al., 1989). Amplification of integrated plasmids seems to be a regular phenomenon in these experiments, when selective pressure (the presence of an antibiotic against which the inserted plasmid confers resistance) is applied.

A recombination-deficient *Lactococcus lactis* ssp. *lactis* mutant has been isolated by Anderson and McKay (1983) after mutagenization with methyl methanesulfonate (MMS). Transfer of plasmids into this strain occurs normally, but introduction of chromosomal markers, such as streptomycin resistance, is inhibited.

B. Recombination Mediated by Transposable Elements

Transposable elements can contain either only the functions that are necessary for the transposition (inverted repeats flanking the transposase gene), or they may contain additional genetic information, such as drug–resistant genes (Grindley and Reed, 1985). The first insertion sequence in lactic acid bacteria was identified in *Lactobacillus casei* by Shimizu-Kadota et al. (1985). The sequence, designated ISL1, causes the virulence of an otherwise lysogenic phage possibly by insertional inactivation of phage repressor functions. The role of a lactococcal insertion sequence ISS1S has been described in Section III.A.2.a. Some other well-characterized transposable elements are briefly discussed below.

1. Lactococcal Transposable Elements

a. Insertion Sequences. An observation of cointegrate formation between nonconjugate vector plasmids and conjugative phage-resistant plasmid pTR2030 led to the isolation and characterization of a pTR2030-associated insertion sequence, designated IS946 (Romero and Klaenhammer, 1990). IS946 has about

96% sequence homology with ISS1-type insertion sequences. Other recent examples of plasmid-linked insertion sequences are IS1076L and IS1076R, a pair of nearly identical insertion elements flanking a 3.3-kbp region in lactose fermentation plasmid pUCL22 (Huang et al., 1991). Their sequence shows a strong homology with a lactococcal chromosomal insertion sequence IS904 (see Section IV.B.1.b). Insertional inactivation of the phage-resistant phenotype conferred by plasmid pGBK17 led to the identification of a chromosomal insertion sequence IS981 (McKay et al., 1989; Polzin and McKay, 1990), which apparently is present in several copies in the chromosome of *Lactococcus lactis* ssp. *lactis* LM0230. Lactococcal insertion sequences show significant homology with insertion elements detected in other bacteria, and they apparently are widely distributed among lactococcal plasmids and in the chromosome. These data, which at the time of writing are still partially unpublished, have been reviewed by Gasson (1990).

b. The Nisin-Sucrose Gene Block. As mentioned in Section II.A.4 early genetic experiments suggested plasmid involvement in the production of peptide antibiotic nisin. Failure to isolate the suspected plasmid directed the research to check the possibility of chromosomal location of genes for nisin production. Several groups have isolated and sequenced the chromosomal nisin gene (Buchman et al., 1988; Kaletta and Entian, 1989; Dodd et al., 1990). Dodd, Horn, and Gasson demonstrated the presence of an insertion sequence, designated IS904, upstream of the nisin determinant. IS904 is present in several copies in the chromosome, and a hypothesis was put forward that nisin production and the closely linked genes for nisin resistance and sucrose utilization together with the flanking IS904 sequences would form a transposon-like mobilizable gene block.

2. *Transposons Tn916 and Tn919*

Streptococcal transposons Tn916 and closely related Tn919 (Fitzgerald and Clewell, 1985; Franke and Clewell, 1981) have been introduced by conjugation into several lactococcal strains as well as into lactobacilli and *Leuconostoc* bacteria (for reviews, see Fitzgerald and Gasson, 1988; Gasson, 1990). These transposons carry a selectable tetracycline resistance marker. They can be used to insertionally inactivate genes for subsequent isolation, since inactivated genes can be cloned using restriction enzymes with no recognition sites within the transposon, and selecting the recombinant plasmids with tetracycline resistance. In the absence of tetracycline the transposon is excised with high accuracy restoring the original gene structure and function.

In lactococcal conjugation the transfer of Tn919 occurs at low frequency, but by strains harboring MG600, a highly conjugative derivative of lactose fermentation-protease plasmid pLP712, the transfer frequencies can be raised to the level of 1.25° 10^{-4} per donor (original frequencies are about 10^{-9}) allowing the matings to be done on agar surfaces instead of nitrocellulose filters (Hill et al., 1987). This effect is dependent on the recipient. *Lactococcus lactis* ssp. *cremoris* 17S did not receive Tn919 even in matings with the donor harboring MG600. Interestingly the transfer of MG600 itself is not necessary for efficient conjugation of Tn919, and MG600 can be resident either in the donor or in the recipient without any effect on the enhanced transposon transfer. The insertion of Tn919 is random in several lactococcal strains, but in derivatives of *Lactococcus lactis* ssp. *lactis* 712 there appears to be a hot spot for Tn919 integration (Fitzgerald et al., 1987).

As an example of the use of transposons in genetic analysis, the insertional inactivation by Tn916 of malolactic fermentation together with several other metabolic pathways in *Lactococcus lactis* ssp. *lactis* is worth mentioning (Renault and Heslot, 1987).

V. DNA REPAIR AND MUTAGENESIS

Aspects of DNA repair and mutagenesis have not been extensively studied in lactic acid bacteria. The fact that the repair-deficient lactococcal mutant mentioned in Section IV.A has an increased sensitivity to MMS and UV indicates the existence of recombinational DNA repair pathway in the lactococci (Anderson and McKay, 1983). The presence of UV-resistant plasmids (see Section II.A.5) in some lactococcal strains could be useful in future studies on the DNA-repair phenomena.

Strong alkylating agents such as MMS or *N*-nitro-*N'*-nitrosoguanidine have been used to mutagenize lactic acid bacteria (Mayo et al., 1991; Nardi et al., 1991). There appear to be no studies on the molecular mechanism of mutagenesis, and at the moment it is not known whether inducible SOS functions such as those characterized in *Escherichia coli* (Walker, 1984) are present in any species of lactic acid bacteria. As, however, most lysogenic phages of lactic acid bacteria are inducible by DNA-damaging agents such as UV or mitomycin C (Davidson et al., 1990), a phenomenon associated with inducible SOS mutagenesis, it is likely that this kind of mutagenic pathway is operating also in these organisms.

VI. GENE CLONING IN LACTOCOCCI

A. Cloning Vectors

Since the first report on a successful plasmid transformation in 1982 (Kondo and McKay, 1982) the development of cloning systems have increased exponentially. In order to make recombinant DNA techniques and hence development of cloning vectors feasible, efficient gene transfer methods had to be developed first. It took only a year after the report of a successful gene transfer when the first cloning vector was reported.

1. First-Generation Cloning Vectors

The most commonly applied vectors can be divided into two different classes:

1. Vectors based on large conjugative plasmids
2. Vectors based on small cryptic lactococcal plasmids

Class 1 vectors are based on large conjugative plasmids, which are found in several Gram-positive species and belong to the MLS-resistance group. Deletion derivatives of both pIP501 (Horodniceanu et al., 1976) and pAMβ1 (Clewell et al., 1974) have been obtained by spontaneous or deliberate deletions. Deletion derivatives have lost their conjugative properties, and the plasmid copy number has risen somewhat. The important properties, like ability to replicate and express antibiotic resistance markers, have been preserved.

Class 1 replicons do not appear to produce single-stranded DNA intermediate, a fact that has proven to increase genetic stability in cloning experiments.

Class 2 vectors are based on small cryptic plasmids found in several lactococcal species. Cryptic plasmids are not useful as such, but they have to be tagged with a selectable marker, usually an antibiotic resistance gene. This approach has been used in constructing pGK12 and its derivatives based on a cryptic replicon pWVO1 from *Lactococcus lactis* ssp. *cremoris* (Otto et al., 1982) and also in pCK1 and derivatives based on *L. lactis* ssp. *lactis* replicon pSH71 (Gasson, 1983).

Yet another approach in constructing cloning vectors has been the application of replicon screening techniques. There random fragments from different lactococcal plasmids are joined to DNA fragments bearing selectable markers. Viable constructs showing stable replication activity form useful vectors. An example of this type of vector construction is plasmid pNZ12 bearing a replication region from pSH71 (de Vos, 1986a).

The two most widely used replicons are pWV01 and pSH71, both small plasmids that share other common features as well. Their DNA sequences are nearly identical (de Vos, 1986b), but the few differences result in different copy number is *L. lactis*: low for pWV01 and high for pSH71. A unique and very useful feature is that both plasmids replicate not only in several different Gram-positive bacteria, but also in some strains of *E. coli* (Kok et al., 1984). Replication in *E. coli* has sped up many of the cloning experiments due to more sophisticated methods in that well-known organism. Broad host range replication has also helped in comparative studies of replicon functioning and gene expression in general.

The only drawback of class 2 type cloning vectors appears to be the often encountered instability when foreign DNA is expressed in them. This is most likely due to their mode of replication, which creates a single-stranded intermediate. At present, class 1 type of vectors are increasingly applied, due to their expected better stability.

Since the construction of first-generation cloning vectors, many improvements have been created. These include versatile multiple cloning sites and also vectors for screening regulatory signals such as promoters, terminators, or sequences promoting protein secretion (van der Vossen et al., 1985; Koivula et al., 1991; Sibakov et al., 1991). There are also vectors for translational fusions where regulatory sequences from lactococci can be fused in all three reading frames in front of *E. coli lac*Z gene. By monitoring β-galactosidase activity the functioning of lactococcal gene expression can be studied (Xu et al., 1991).

2. Integration Vectors

Once an efficient gene transfer system has been established for lactococci it is then also possible to integrate foreign genes into their chromosome. This appears to be an especially attractive alternative for dairy starters, since many of the crucial properties are plasmid-coded and hence, prone to instability.

For the integration experiments either Gram-negative, such as pBR322, or Gram-positive plasmids, such as pE194, pSC101, or pTB19 with replication genes not known to function in lactococci, have been used (Leenhouts et al., 1990).

It has been shown that integration occurs generally via a single crossing over (Campbell-type) mechanism. In a few cases, integration via double crossing over has been reported (Venema, personal communication). The latter makes it possible to construct strains with only the needed piece of foreign DNA in the host's genome.

3. Food-Grade Vectors

For industrial applications, the modified organisms must not bear any nonedible DNA. Therefore, e.g., antibiotic resistance markers in the plasmid vectors have to be replaced. Considerable effort has been put and is presently going on in constructing alternative, i.e., food-grade selection systems. One of the candidates for a safe marker is the strain's ability to ferment lactose. Lactose metabolism has been well characterized at the molecular level, and from those studies the gene *lac*F (see Section VI.B.2) of the lactose operon proved suitable for direct selection both in laboratory-scale and in industrial fermentations (de Vos et al., 1990). Plasmids were also stably maintained when lactose was used as the sole carbon source.

Another suitable food-grade marker has been found from the proteolytic system of lactococci, X-prolyl dipeptidyl aminopeptidase (X-PDAP), (Mayo et al., 1991). This peptidase is specific in releasing dipeptides from oligopeptides containing proline at the penultimate position. It is due to this cleavage that proline-containing peptides can be taken into the cell.

A gene-regulating DNA synthesis, thymidylate synthase (*thy*A) has also been successfully applied in lactococcal vectors (Ross et al., 1990a,b). Since *thy*A is a general metabolic regulator, it could have applications in several different species of lactic acid bacteria.

Resistance to endogenous lactococcal bacteriocins has also been studied in concentration with food-grade vector construction. The nisin resistance gene has been cloned by two different groups (Froseth and McKay, 1991; von Wright et al., 1990) and shown to function effectively in direct selection of transformants.

Compared to antibiotic markers, food-grade selection often gives more background colonies and the conditions need to be carefully adjusted for optimal results.

B. Examples of Cloned Genes

Gene cloning in lactococci has started with the careful study of the two most crucial process properties of these bacteria, lactose utilization and proteolytic activity. At present both systems have been cloned and sequenced from several independent strains, and very sophisticated molecular analyses have been carried out. The knowledge obtained from these studies have been further applied in construction of food-grade vectors (see Section VI.A.3), in expression/secretion vector construction, and in attempts to produce heterologous proteins (see Section VI.C). Even protein engineering has been applied.

1. Proteinases

Early observations on instability of lactococcal proteolytic activity led the researchers to analyze the plasmid content of first *L. lactis* ssp. *cremoris* strains. It was from a large proteinase plasmid of *L. cremoris* Wg2 that the first successful cloning experiment was performed, first in *Bacillus subtilis*, then in a proteinase-negative *L. lactis* ssp. *lactis* mutant (Kok et al., 1985). Sequence analysis of proteinases from three different strains (Kok, 1990) have revealed homology to subtilisins, serine proteinases synthesized by *B. subtilis*. Lactococcal proteinases are only much larger, nearly five times the size of subtilisins. Both are synthetized as preproenzymes, and a typical Gram-positive signal sequence of 33 amino acid residues is found at the N-terminus. At the extreme C-terminus a membrane anchorlike sequence is found, which is responsible for the attachment of the protease to the cell envelope. If the anchor is removed the enzyme activity is liberated to the culture medium (Kok et al., 1988).

Upstream of the structural gene *prt*P, a completely conserved gene, *prt*M, is found. It is essential for the maturation step of the protease. It has been shown that *prt*M, which is a lipoprotein, does not perform the cleavage step(s) and the actual mechanism how it brings about or mediates the activation step is still unclear (Haandrikman et al., 1989; Vos et al., 1989). In addition to N-terminal processing of preproteinase, also C-terminal processing appears to be needed to get an active enzyme.

It has also been possible to determine, from the sequences, those amino acids, which are responsible for the caseinolytic specificity of the proteinases. Also, the confusion about the actual number of proteinases present in the lactococcal cells has been clarified. There appears to be only one extremely unstable, large enzyme, whose degradation products are often observed in biochemical experiments.

Genetics have furthermore provided tools to stabilize plasmid-coded proteolytic functions by integration into the lactococcus chromosome via single or double crossing-over event.

2. *The Lactose Utilization System*

Two basically different systems for lactose transport exist in lactic acid bacteria, either phosphoenolpyruvate (PEP)–dependent phosphotransferase system (PTS) or lactose permease system. Mesophilic starter strains, such as all lactococci, have the PTS system. For analysis of lactose permease system, see Sections VI.D, VII.B. The unraveling of the lactococcal lactose utilization pathway began with cloning of the lactose proficient fragment from large metabolic plasmids (Maeda and Gasson, 1986). Since then, detailed molecular analyses of the entire tagatose-6-P pathway and the lactose PTS coded by the 23-kb lactose miniplasmid pMG820 have been completed (de Vos, 1990). Cloning and sequencing of the individual genes have revealed a single operon, whose gene order has been established. Also the function of most of the gene products is known. The transcription of the operon is lactose-inducible. A divergently transcribed monocistronic message from a gene (*lac*R) on the 5′ side of the operon structure has also been characterized. This gene has been shown to function as a repressor in controlling the expression of the operon. The nature of the inducer of the *lac* expression awaits further analysis, but it could be one of the phosphorylated intermediates of the tagatose-6-P pathway (Porter and Chassy, 1988; Alpert and Chassy, 1988; Boizet et al., 1988; de Vos and Gasson, 1989).

3. *Citrate Permease*

The industrially important citrate permease gene from *Lactococcus lactis* has been cloned in *E. coli* and further analyzed by DNA sequencing (David et al., 1990).

4. *Examples of Other Cloned Genes*

A lot of research activity has also been put in the isolation and characterization of bacteriocins produced by lactococci. By far the best characterized is a small peptide antibiotic, nisin (see also Section II.A.4), naturally secreted by several *Lactococcus lactis* ssp. *lactis* strains. The gene coding for nisin has been cloned and sequenced by different groups (Buchman et al., 1988; Kaletta and Entian, 1989). The biosynthesis of this posttranslationally modified peptide is under intensive study. There are hopes to create nisin variants whose antimicrobial spectra and process applications would be much wider than that of the natural peptide.

Lactococci produce other nisin-type as well as non-nisin-type bacteriocins. Earlier work by Neve et al. (1984) (see Section II.A.4) has resulted in a thorough analysis of a bacteriocin of the latter type. In cloning experiments by van Belkum et al. (1989, 1991) two fragments (1.8 kbp and 1.3 kbp), coding for low and high antagonistic activity, could be isolated. The former fragment contained three open reading frames organized in an operon. Two of the reading frames were associated with bacteriocin production, the third with immunity. In the latter fragment two open reading frames, one coding for antagonistic activity and the other for immunity, were detected.

Since lactococcal plasmids code for many of the important metabolic traits, most of the genes characterized thus far of plasmid origin. However, an increasing number of chromosomally coded genes have been analyzed as well. It has been shown that plasmid-coded proteinases alone are not sufficient to render lactococci able to grow in milk, but an additional chromosomally coded peptidase-system is needed. Examples of cloned peptidase genes are X-PDAP (see Section VI.A.3) and a yet not fully identified chromosomal gene block (von Wright et al., 1987; Tynkkynen et al., 1989).

A very interesting chromosomally coded gene characterized by van Asseldonk et al. (1990) is *usp*45. It is a secretory protein found in all tested *Lactococcus* strains. Its DNA sequence has revealed unusual repeats and the physiological role of the protein is still unclear. Construction of secretion vectors based on the *usp*45 signal sequence are under way (van Asseldonk, personal communication).

C. Expression of Heterologous Genes

For the expression of heterologous genes efficient expression vectors need to be constructed first. This is an area not very widely covered in the literature. However, some reports have appeared on the characterization of lactococcal promoters. The classical approach to clone and sequence gene coding for a useful protein has resulted in the elucidation of *prt*P promoter and its application in expressing bovine prochymosin (de Vos et al., 1989) in *Lactococcus lactis* ssp. *lactis*.

Another approach has been to screen *Lactococcus* chromosomal or phage DNA with specific probe vectors (van der Vossen et al., 1985, 1987; Lakshmidevi et al., 1990; Koivula et al., 1991). Usually a promoterless chloramphenicol acetyl-transferase gene (*cat*-86) from *Bacillus subtilis* has been fused to a lactococcal vector, so that ligation of promoter-bearing DNA fragments upstream of it have yielded chloramphenicol resistant colonies. Putative promoter fragments have been sequenced by different groups and been further applied in expression of heterologous genes. Attempts to express hen egg white lysozyme (van de Guchte et al., 1987) and β-lactamase (Sibakov et al., 1991) have been reported.

A similar approach to find sequences promoting protein secretion has also been reported (Sibakov et al., 1991). With the aid of "random" signal sequences found from the chromosome, secretion of β-lactamase and α-amylase have been reported. These randomly screened signal sequences resemble closely their Gram-positive counterparts. The only confirmed signal sequences isolated from lactococci are those of *prt*P, *prt*M and *usp*45 genes (Kok et al., 1988; Haandrikman et al., 1989; van Asseldonk et al., 1990).

Although the attempts above described have demonstrated the potential of lactococci as hosts for production of useful foreign proteins, the yields are still far from commercial applications. Thus, for higher yields, more knowledge on the regulation of gene expression and protein secretion is necessary.

D. Genetics of *Streptococcus salivarius* ssp. *thermophilus*

When the dairy streptococci were renamed lactococci, the thermophilic *Streptococcus thermophilus* retained its old name, and thus its position as a distinct species was emphasized. Officially it is presently called *Streptococcus salivarius* ssp. *thermophilus*. Recently, an excellent review article (Mercenier, 1990) on the molecular genetics of this important starter strain has come out. In this context mainly differences between mesophilic lactococci and *S. thermophilus* are discussed.

The plasmid content of mesophilic lactococci is much more abundant. *S. thermophilus* harbors only one to two if any plasmids per cell. So far only one endogenous plasmid has been fully characterized; plasmid pA33 from a yogurt starter has been sequenced (see Section II.B.3).

Few of the cryptic plasmids of *S. thermophilus* have been used until now to construct shuttle vectors (Mercenier and Lemoine, 1989): pA2 (2 kbp) and pA33 (6.9 kbp) endogenous replicons have been cloned in the pVA891 (em^r) vector. The recombinant plasmids were rapidly lost in the absence of selection pressure. A very peculiar feature is that contrary to endogenous *S. thermophilus* replicons, closely related lactococcal replicons pGK12, pCK17, and pNZ12 appear stably maintained even without antibiotic pressure.

S. thermophilus has also proven to be very difficult to transform with foreign DNA. The reported frequencies have been low, usually only in the range of 10^3–10^4 cfu/μg DNA has been reported.

Gene-cloning studies have begun with the characterization of lactose metabolism. Four genes from the lactose permease pathway have been cloned and sequenced: lactose permease (*lac*S, Poolman et al., 1989), β-galactosidase (*lac*Z, Schroeder et al., 1991), aldose-1-epimerase (mutarotase, *gal*M), and

UDP-glucose-4-epimerase (*galE*) (Poolman et al., 1990). From the data an operon structure has been deduced. The structural gene, β-galactosidase shows considerable homology with its counterpart from other organisms, whereas the regulatory genes are different.

In gene expression studies lactococcal broad host range plasmids have been useful. Chloramphenicol resistance has been the most useful marker gene. At present, there is an obvious lack of reporter genes and food-grade markers. Some of the well-established markers for other Gram-positive bacteria have failed to function in *S. thermophilus* and very little is known about expression of foreign genes. There is only one report on applying *S. thermophilus* expression signals to the production of heterologous proteins (Slos et al., 1991). In that study, although transcriptional functioning was confirmed by mRNA analysis, no protein was detected, suggesting a posttranslational barrier.

In summary, tools are now available to study the genetics and molecular biology of this very important starter species. Excellent pioneering work has been carried out in Dr. Mercenier's group, and this hopefully encourages other researchers to work on this demanding organism.

VII. GENE CLONING IN LACTOBACILLI

Lactobacilli are a very heterogeneous group of bacteria, which has complicated the progress in the study of their genetics and molecular biology. One of the barriers has been inefficient transformation systems that are now, at least, partly overcome by electroporation techniques. Research of lactobacilli is divided according to the intended application: there are several groups working on biological silage fermentation with *Lactobacillus plantarum*, and considerable interest is also focused on *Lactobacillus bulgaricus*, an important yogurt starter.

A. Cloning Vectors

Until now only one comprehensive review on cloning in lactobacilli has appeared (Chassy, 1987). Very intensive research has been pursued during the last years and many new discoveries are presently appearing.

In constructing cloning vectors for lactobacilli basically similar approaches as for lactococci, have been attempted: cryptic small replicons have been isolated, analyzed, and marked with antibiotic resistance markers (Damiani et al., 1987; Skaugen, 1989; Bouia et al., 1989). Some of the cryptic plasmids harbor broad host range replicons naturally; to others, e.g., known *E. coli* replicons have been fused, in order to obtain functional shuttle vectors (Cocconcelli et al., 1990; van Luijk et al., 1990).

Also experiences with lactococcal broad host range replicons have been applied in lactobacilli. Derivatives of pSH71 an pAMβI replicons among others, have been successfully used.

With lactobacilli even more instability problems than with lactococci have been encountered. Integration studies have also been analogous to those in lactococci. The only reports on functional integration are in *Lactobacillus casei* and in *L. plantarum*. Most likely both single and double crossing-over techniques have succeeded (Scheirlinck et al., 1989; Chassy, 1987).

To our knowledge there is very little information as yet, on expression/secretion vectors. Some promoters have been isolated and sequences (Porter and Chassy, 1988; Andrews, 1985; Lerch et al., 1989), but their future application is still to come.

B. Examples of Cloned Genes

Due to its industrial importance, *Lactobacillus bulgaricus* is one of the most studied lactobacilli. Unfortunately it is also one of the most difficult species to transform. Only few groups have succeeded in electroporating foreign DNA into it and all the reported frequencies have been low. It appears that only 1–2 strains out of 10 can, at all, be transformed (Chassy, personal communication). However, it has been possible to clone genes from *Lactobacillus bulgaricus*. These efforts have concentrated in the analysis of lactose metabolism. Genes coding for several components of the lactose permease pathway have been cloned and sequenced by at least two groups (Schmidt et al., 1989; Leong-Morgenthaler et al., 1991). The same has been accomplished also from *Lactobacillus casei* (Porter and Chassy, 1988; de Vos, 1990). Operon structures have been recognized based on these data. Typically the analyses of isolated *Lactobacillus bulgaricus* genes have been performed in *E. coli*, in which mutation and complementation tests are easily available. Similarly, the tryptophan operon of *Lactobacillus casei* has been characterized, with *E. coli* as the host (Natori et al., 1990).

C. Expression of Heterologous Genes in Lactobacilli

There are few two reports on expressing heterologous genes in lactobacilli applying their endogenous expression signals. Ahmad and Stewart (1991) reported on bioluminescent *Lactobacillus casei*, where the lux a/b gene was under a putative *Lactobacillus casei* promoter sequence.

Other studies describe attempts to produce heterologous proteins with nonlactobacillar regulatory signals. Many groups report on expressing α-amylase gene in *Lactobacillus plantarum* (Scheirlinck et al., 1989). In this context also integration has been applied (Scheirlinck et al., 1989). In the same connection, that is in improving silage starters, expression and secretion of clostridial endoglucanase has been reported (Bates et al., 1989). The above-described together with a report on successful secretion of β-lactamase (Sibakov et al., 1991) and staphylococcal lipase (Vogel et al., 1990) confirm the fact that lactobacilli have a good potential for protein secretion, which can hopefully be applied in genetic modification of starter cultures.

VIII. GENE CODING IN OTHER LACTIC ACID BACTERIA

There is hardly any information on recombinant DNA techniques outside lactococci and different lactobacilli. Some preliminary studies such as analysis of plasmid profiles (see Section II.B) and the development of DNA transfer methods have appeared from *Pediococcus* and *Leuconostoc* species.

IX. SUMMARY

The last few years have brought about an enormous increase of knowledge about the genetics and molecular biology of lactic acid bacteria. Although most of the effort has been put into lactococci, information on lactobacilli as well as some minor species is also accumulating.

As to lactococci, all the molecular biological methods are established and very sophisticated analyses have been performed on gene structure and function, unraveling of metabolic pathways, and even genetic modification to improve the process properties of these important mesophilic starters.

The molecular analysis of an important, but difficult thermophile, *Streptococcus thermophilus*, has successfully started, but there are still numerous unanswered questions.

In lactobacilli, although most of the "tools" are available there is still need for, e.g., more efficient gene transfer systems. In addition there is little information on gene expression and regulation and applications still await their breakthrough.

REFERENCES

Allen, L. K., Sandine, W. E., and Elliker, P. R. (1963). Transduction in *Streptococcus lactis, J. Dairy Res., 30*:351.

Alpert, C. A., and Chassy, B. M. (1988). Molecular cloning and nucleotide sequence of the factor III^{lac} gene of *Lactobacillus casei, Gene, 62*:277.

Anderson, D. G., and McKay, L. L. (1983). Isolation of a recombination deficient mutant of *Streptococcus lactis* ML3, *J. Bacteriol., 155*:930.

Anderson, D. G., and McKay, L. L. (1984). Genetic and physical characterization of recombinant plasmids associated with cell aggregation and high frequency conjugal transfer in *Streptococcus lactis* ML3, *J. Bacteriol., 158*:954.

Ahmad, K. A. and Stewart, G. S. A. B. (1991). The production of bioluminescent lactic acid bacteria suitable for the rapid assessment of starter culture activity in milk, *J. Appl. Bacteriol. 70*:113.

Andrews, J., Clore, G. M., Davies, R. W., Gronenborn, A. M., Kalderone, B. et al. (1985). Nucleotide sequence of the dihydrofolate reductase gene of methotrexate resistant *Lactobacillus casei, Gene, 35*:217.

Axelsson, L. T., Ahrné, S. E. I., Anderson, M. C., Ståhl, S. R. (1988). Identification and cloning of a plasmid-encoded erythromycin resistant determinant from *Lactobacillus reuteri, Plasmid, 20*:171.

Baigori, M., Sesma, F., Ruiz Holgado, A. P., and de Mendoza, D. (1988). Transfer of plasmids between *Bacillus subtilis* and *Streptococcus lactis, Appl. Environ. Microbiol., 54*:1309.

Bates, E. E. M., Gilbert, H. J ., Hazlewood, G. P., Huckle, J., Laurie, J. I., and Mann, S. P. (1989). Expression of a *Clostridium thermocellum* endoglucanase gene in *Lactobacillus plantarum, Appl. Environ. Microbiol., 55*:2095.

Boizet, B., Villeval, D., Slos, P., Novel, M., Novel, G., and Mercenier, A. (1988). Isolation and structural analysis of the phospho-β-galactosidase gene from *Streptococcus lactis* Z268, *Gene, 62*:249.

Bouia, A., Bringel, F., Frey, L., Kammerer, B., Belarbi, A., Guyonvardi, A., and Hubert, J-C (1989). Structural organization of pLP1, a cryptic plasmid from *Lactobacillus plantarum* CCM1904, *Plasmid, 22*:185.

Buchman, W. B., Banerjee, S., and Hansen, J. N. (1988). Structure, expression, and evolution of a gene encoding the precursor of nisin, a small protein antibiotic, *J. Biol. Chem., 263*:16260.

Chassy, B. (1987). Prospects for the genetic manipulation of lactobacilli, *FEMS Microbiol. Rev., 46*:297.

Chassy, B. M., and Flickinger, J. L. (1987). Transformation of *Lactobacillus casei* by electroporation, *FEMS Microbiol. Lett., 44*:173.

Chassy, B. M., Gibson, E. M., and Giuffrida, A. (1978). Evidence for plasmid-associated lactose metabolism in *Lactobacillus casei* subsp. *casei, Curr. Microbiol., 1*:141.

Chopin, A., Chopin, M.-C., Moillo-Batt, A., and Langella, P. (1984). Two plasmid-determined restriction and modification systems in *Streptococcus lactis, Plasmid, 11*:206.

Chopin, M.-C., Chopin, A., Rouault, A., and Galleron, N. (1989). Insertion and amplification of foreign genes in the *Lactococcus lactis* subsp. *lactis* chromosome, *Appl. Environ. Microbiol., 55*:1769.

Chopin, M.-C., Chopin, A., Rouault, A., and Simon, D. (1986). Cloning in *Streptococcus lactis* of plasmid mediated UV resistance and effect on prophage stability, *Appl. Environ. Microbiol., 51*:233.

Chopin, M.-C., Moillo-Batt, A., and Rouault, A. (1985). Plasmid-mediated UV protection in *Streptococcus lactis, FEMS Microbiol. Lett., 26*:243.

Clewell, D., Yagi, Y., Dunny, G., and Schultz, S. (1974). Characterization of three plasmid DNA molecules in a strain of *Streptococcus faecalis*. Identification of a plasmid determining erythromycin resistance, *J. Bacteriol., 117*:283.

Cocconcelli, P. S., Morelli, L., and Bottazzi, V. (1990). Single stranded DNA plasmid vector construction, cloning and expression of heterologous genes in *Lactobacillus, FEMS Microbiol. Rev., 46*:11.

Cocconcelli, P. S., Morelli, L., and Vescovo, M. (1985). Conjugal transfer of antibiotic resistances from *Lactobacillus* to *Streptococcus lactis, Microbiologie-Aliments-Nutrition, 3*:130.

Cocconcelli, P. S., Morelli, L., Vescovo, M., and Bottazzi, V. (1986). Intergeneric protoplast fusion in lactic acid bacteria, *FEMS Microbiol. Lett., 35*:211.

Collins, E. B. (1956). Host-controlled variations in bacteriophages active against lactic streptococci, *Virology, 2*:261.

Collins, E. B. (1958). Changes in the bacteriophage sensitivity of lactic streptococci, *J. Dairy Sci., 41*:41.

Cords, R. B., McKay, L. L., and Guerry, P. (1974). Extrachromosomal elements in group N streptococci, *J. Bacteriol., 117*:1149.

Damiani, G., Romagnoli, S., Ferretti, L., Morelli, L., Bottagi, V., and Sgaramella, V. (1987). Sequence and functional analysis of a divergent promoter from a cryptic plasmid of *Lactobacillus acidophilus*., 168 S, *Plasmid, 17*:69.

Davey, G. P. (1984). Plasmid associated with diplococcin production in *Streptococcus cremoris, Appl. Environ. Microbiol., 48*:895.

David, S., Simons, G., and de Vos, W. (1989). Plasmid transformation by electroporation of *Leuconostoc paramesenterioides* and its use in molecular cloning, *Appl. Environ Microbiol., 55*:1483.

David, S., van der Rest, M. E., Driessen, A. G. M., Simons, G., and de Vos, W. M. (1990). Nucleotide sequence and Expression in *Escherichia coli* of the *Lactococcus lactis* citrate permease gene, *Appl. Environ. Microbiol., 172*:5789.

Davidson, B. E., Powell, I. B., and Hillier, A. J. (1990). Temperate bacteriophages and lysogeny in lactic acid bacteria, *FEMS Microbiol. Rev., 87*:79.

Davies, F. L., and Gasson, M. J. (1981). Reviews of the progress of Dairy Science: Genetics of lactic acid bacteria, *J. Dairy Res., 48*:363.

Davies, F. L., Underwood, H. M., and Gasson, M. J. (1981). The value of plasmid profile for strain identification in lactic streptococci and the relationship between *Streptococcus lactis* 712, ML3 and C2, *J. Appl. Bacteriol., 51*:325.

de Vos, W. M., Underwood, H. M., and Davies, F. L. (1984). Plasmid encoded bacteriophage resistance in *Streptococcus cremoris* SK11, *FEMS Microbiol. Lett., 23*:175.

de Vos, W. M. (1986a). Gene cloning in lactic streptococci, *Neth. Milk Dairy 40*:141.

de Vos, W. M. (1986b). Sequence organization and use in molecular cloning of the cryptic lactic streptococcal plasmid pSH71 (paper 223), in *Second ASM Conf. Streptococcal Genetics*, Miami Beach, American Society for Microbiology.

de Vos, W. M. (1990). Disaccharide utilization in lactic acid bacteria, *Proc. 6th Int. Symp. Genetics of Industrial Microorganisms*, Strasbourg, France, pp. 447–457.

de Vos, M. M., Boerrigter, I., van Rooyen, R. J., Reiche, B., and Hengstenberg, W. (1990). Characterization of the lactose-specific enzymes of the phosphotransferase system in *Lactococcus lactis, J. Biol. Chem., 265*:22554.

de Vos, W. M., and Gasson, M. J. (1989). Structure and Expression of the *Lactococcus lactis* gene for phospho-β-galactosidase (*lac*G) in *Escherichia coli* and *L. lactis, J. Gen. Microbiol., 135*:1833.

de Vos, W., Vos, P., Simons, G., and David, S. (1989). Gene organization and expression in mesophilic lactic acid bacteria, *J. Dairy Sci., 72*:3398.

Dodd, H. M., Horn, N., and Gasson, M. J. (1990). Analysis of the genetic determinant for production of the peptide antibiotic nisin, *J. Gen. Microbiol., 136*:555.

Evans, R. P., and Macrina, F. L. (1983). Streptococcal R plasmid p1P501:endonuclease site map, resistance determinant location, and construction of novel derivatives, *J. Bacteriol., 154*:1347.

Fitzgerald, G. F., and Clewell, D. B. (1985). A conjugative transposon (Tn919) in *Streptococcus sanguis, Infect. Immun., 47*:415.

Fitzgerald, G. F., and Gasson, M. J. (1988). *In vivo* gene transfer systems and transposons, *Biochimie, 70*:489.

Fitzgerald, G. F., Hill, C., Vaughan, E., and Daly, C. (1987). Tn919 in lactic streptococci, in *Streptococcal Genetics* (J. J. Ferretti, and R. Curtis, eds.), American Society for Microbiology, Washington, DC, p. 238.

Forsén, R., Nousiainen, R., and Raunio, V. (1973). Studies on slime forming group N *Streptococcus* strains by comparison of the electrophoretic patterns of enzymes, *Acta Univ. Oul. A13 Biochem., 4*:1.

Franke, A. E., and Clewell, D. B. (1981). Evidence for a chromosome-borne resistance transposon (Tn916) in *Streptococcus faecalis* that is capable of "conjugal" transfer in the absence of a conjugative plasmid, *J. Bacteriol., 145*:494.

Froseth, B. R., Herman, R. E., and McKay, L. L. (1988). Cloning of nisin resistance determinant and replication origin on 7,6 kilobase *Eco*RI fragment of pNP40 from *Streptococcus lactis* subsp. *diacetylactis* DRC3, *Appl. Environ. Microbiol., 54*:2136.

Froseth, B. R., and McKay, L. L. (1991). Molecular characterization of the nisin resistance region of *Lactococcus lactis* subsp. *lactis* biovar *diacetylactis* DRC3, *Appl. Environ. Microbiol., 57*:804.

Gasson, M. J. (1980). Production, regeneration and fusion of protoplasts of lactic streptococci, *FEMS Microbiol. Lett., 9*:99.

Gasson, M. J. (1980). Production, regeneration and fusion of protoplasts of lactic streptococci, *FEMS Microbiol. Lett., 9*:99.

Gasson, M. J. (1983). Plasmid complements of *Streptococcus lactis* NCDO 712 and other lactic streptococci after protoplast-induced curing, *J. Bacteriol., 154*:1.

Gasson, M. (1984). Transfer of sucrose fermenting ability, nisin resistance, and nisin production into *Streptococcus lactis* 712, *FEMS. Microbiol, Lett., 21*:7.

Gasson, M. J. (1990). In vivo genetic systems in lactic acid bacteria, *FEMS Microbiol. Rev. 87*:43.

Gasson, M. J., and Davies, F. L. (1980a). High frequency conjugation associated with *Streptococcus lactis* donor cell aggregation, *J. Bacteriol., 143*:1260.

Gasson, M. J., and Davies, F. L. (1980b). Conjugal transfer of the drug resistance plasmid pAMB in the lactic streptococci, *FEMS Microbiol. Lett., 7*:51.

Gautier, M., and Chopin, M.-C. (1987). Plasmid-determined restriction and modification activity and abortive infection in *Streptococcus cremoris, Appl. Environ. Microbiol., 53*:923.

Gibson, E. M., Chace, N. M., London, S. B., and London, J. (1979). Transfer of plasmid-mediated antibiotic resistance from streptococci to lactobacilli, *J. Bacteriol., 137*:614.

Gonzalez, C. F., and Kunka, B. S. (1983). Plasmid transfer in *Pediococcus* spp.:Intergeneric and intrageneric transfer of pIP501, *Appl. Environ. Microbiol., 46*:81.

Gonzalez, C. F., and Kunka, B. S. (1985). Transfer of sucrose-fermenting ability and nisin production phenotype among lactic streptococci, *Appl. Environ. Microbiol., 49*:627.

Gonzalez, C. F., and Kunka, B. S. (1986). Evidence of plasmid linkage of raffinose utilization and associated α-galactosidase and sucrose hydrolase activity in *Pediococcus pentosaceus, Appl. Environ. Microbiol., 51*:105.

Graham, D. C., and McKay, L. L. (1985). Plasmid DNA in strains of *Pediococcus cerevisiae* and *Pediococcus pentosaceus, Appl. Environ. Microbiol., 50*:532.

Grindley, N. D. F., and Reed, R. R. (1985). Transpositional recombination in procaryotes, *Ann. Rev. Biochem., 54*:863.

Haandrikman, A. J., Kok, J., Laan, H., Soemitro, S., Ledeboer, A. M., Konings, W. N., and Venema, G. (1989). Identification of a gene required for the maturation of an extracellular serine proteinase, *J. Bacteriol., 171*:2789.

Harlander, S. K. (1987). Transformation of *Streptococcus lactis* by electroporation, *Streptococcal genetics* (J. J. Ferretti and R. Curtis III, eds.), American Society for Microbiology, Washington, DC, p. 229.

Harlander, S. K., and McKay, L. L. (1984). Transformation of *Streptococcus sanguis* Challis with *Streptococcus lactis* plasmid DNA, *Appl. Environ. Microbiol., 48*:342.

Harlander, S. K., McKay, L. L., and Schachtele, C. F. (1984). Molecular cloning of the lactose. Metabolizing genes from *Streptococcus lactis, Appl. Environ. Microbiol., 48*:3347.

Hayes, F., Caplice, E., McSweeney, A., Fitzgerald, G. F., and Daly, C. (1990). pAMβ1-associated mobilization of proteinase plasmids from *Lactococcus lactis* subsp. *cremoris* UC205, *Appl. Environ. Microbiol., 56*:195.

Hill, L., Daly, C., and Fitzgerald, G. F. (1987). Development of high-frequency delivery system for transposon Tn919 in lactic streptococci: random insertion in *Streptococcus lactis* subsp. *diacetylactis* 18-16, *Appl. Environ. Microbiol., 53*:74.

Hill, C., Pierce, K., and Klaenhammer, T. R. (1989). The Conjugative plasmid pTR2030 encodes two bacteriophage defence mechanisms in lactococci, restriction modification (R^+/M^+) and abortive infection (Hsp^+), *Appl. Environ. Microbiol., 55*:2416.

Horodniceanu, T., Bouanchand, D. H., Bieth, G., Chabbert, Y. A. (1976). R-plasmids in *Streptococcus agalactiae* (group B), *Antimicrob. Ag. Chemother., 10*:795.

Huang, D. C., Novel, M., and Novel, G. (1991). A transposon-like element on the lactose plasmid of *Lactococcus lactis* subsp. *lactis* Z270, *FEMS Microbiol. Lett., 77*:101.

Hurst, A. (1981). Nisin, *Adv. Appl. Microbiol., 27*:85.

Iwata, M., Mada, M., and Ishiwa, H. (1986). Protoplast fusion of *Lactobacillus fermentum, Appl. Environ. Microbiol., 52*:392.

Jarvis, A. W., Heap, H. A., and Limsowtin, G. K. Y. (1989). Resistance against industrial bacteriophages conferred on lactococci by plasmid pAJ1106 and related plasmids, *Appl. Environ. Microbiol., 55*:1537.

Kaletta, C., and Entian, K.-D. (1989). Nisin, a peptide antibiotic: cloning and sequencing of the *nis*A gene and post-translational processing of its peptide product, *J. Bacteriol., 171*:1597.

Kempler, G. M., Baldwin, K. A., McKay, L. L., Morris, H. A., Halambeck, S., and Thorse, G. (1979). Use of genetic alterations to improve *Streptococcus lactis* C2, *J. Dairy Sci., 62:Suppl.*1. 42.

Kempler, G. M., and McKay, L. L. (1979a). Characterization of plasmid DNA in *Streptococcus lactis* subsp. *diacetylactis*: evidence for plasmid-linked citrate utilization, *Appl. Environ. Microbiol., 37*:316.

Kempler, G. M., and McKay, L. L. (1979b). Genetic evidence for plasmid-linked lactose metabolism in *Streptococcus lactis* subsp. *diacetylactis, Appl. Environ. Microbiol., 37*:1041.

Kempler, G. M., and McKay, L. L. (1981). Biochemistry and genetics of citrate utilization in *Streptococcus lactis* ssp. *diacetylactis, J. Dairy Sci., 64*:1527.

Klaenhammer, T. R. (1987). Plasmid-directed mechanisms for phage defence in lactic streptococci, *FEMS Microbiol. Rev. 46*:313.

Klaenhammer, T. R., McKay, L. L., and Baldwin, K. A. (1978). Improved lysis of group N streptococci for isolation and rapid characterization of plasmid DNA, *Appl. Environ. Microbiol., 35*:592.

Klaenhammer, T. R., and Sanozky, R. B. (1985). Conjugal transfer from *Streptococcus lactis* ME2 of plasmids encoding phage resistance and lactose fermenting ability: evidence for a high-frequency conjugative plasmid responsible for abortive infection of virulent bacteriophage, *J. Gen. Microbiol., 131*:1531.

Koivula, T., Sibakov, M., and Palva, I. (1991). Isolation and characterization of *Lactococcus lactis* subsp. lactis promoters, *Appl. Environ. Microbiol., 57*:333.

Kok, J. (1990). Genetics of the proteolytic system of lactic acid bacteria, *FEMS Microbiol. Rev., 87*:15.

Kok, J., Leenhouts, K. J., Haandrikman, A. J., Ledeboer, A. M., and Venema, G. (1988). Nucleotide sequence of the cell wall proteinase gene of *Streptococcus cremoris* Wg 2, *Appl. Environ. Microbiol., 54*:231.

Kok, J., van der Vossen, J. M. B. M., and Venema, G. (1984). Construction of plasmid cloning vectors for lactic streptococci which also replicate in *Bacillus subtilis* and *Escherichia coli, Appl. Environ. Microbiol., 48*:726.

Kok, J., van Dijl, J. M., van der Vossen, J. M. B. M., and Venema, G. (1985). Cloning and expression of a *Streptococcus cremoris* proteinase in *Bacillus subtilis* and *Streptococcus lactis, Appl. Environ. Microbiol., 50*:94.

Kondo, J. K., and McKay, L. L. (1982). Transformation of *Streptococcus lactis* protoplasts by plasmid DNA, *Appl. Environ. Microbiol., 43*:1213.

Kondo, J. K., and McKay, L. L. (1984). Plasmid transformation of *Streptococcus lactis* protoplasts: Optimization and use in molecular cloning, *Appl. Environ. Microbiol., 48*:252.

Kuhl, S. A., Larsen, L. D., and McKay, L. L. (1979). Plasmid profiles of lactose-negative and proteinase-deficient mutants of *Streptococcus lactis* C10, ML3, and M18, *Appl. Environ. Microbiol., 37*:1193.

Laible, N. J., Rule, P. L., Harlander, S. K., and McKay, L. L. (1987). Identification and cloning of plasmid deoxyribonucleic acid coding for abortive phage infection from *Streptococcus lactis* ssp. *diacetylactis* KR2, *J. Dairy Sci., 70*:2211.

Lakshmidevi, G., Davidson, B. E., and Hillier A. J. (1990). Molecular characterization of promoters of the *Lactococcus lactis* subsp. *cremoris* temperate bacteriophage BKT-5 and identification of a phage gene implicated in the regulation of promoter activity, *Appl. Environ. Microbiol., 56*:934.

Leblanc, D. J., and Lee, L. N. (1984). Physical and Genetic analyses of streptococcal plasmid pAMβ1 and cloning of its replication region, *J. Bacteriol., 157*:445.

Leenhouts, K. J., Kok, J., and Venema, G. (1989). Campbell-like integration of heterologous plasmid DNA into the chromosome of *Lactococcus lactis* subsp. lactis, *Appl. Environ. Microbiol., 55*:394.

Leenhouts, K. J., Kok, J., and Venema, G. (1990). Stability of integrated plasmids in the chromosome of *Lactococcus lactis, Appl. Environ. Microbiol., 56*:2726.

Lee, L.-J., Hansen, J. B., Jagusztyn-Krynicka, E. K., and Chassy, B. M. (1982). Cloning and expression of the β-D-phosphogalactoside galactohydrolase gene of *Lactobacillus casei* in *Escherichia coli, J. Bacteriol., 152*:1138.

Lee-Wickner, L.-J., and Chassy, B. M. (1985). Molecular cloning and characterization of cryptic plasmids isolated from *Lactobacillus casei, Appl. Environ. Microbiol., 49*:1154.

Leer, R. J., Posno, M., van Rijn, J. M. M., Lokman, B. C., and Pouwels, P. H. (1987), *FEMS Microbiol. Rev., 46*:P20.

Leong-Morgenthaler, P., Zwahlen, M. C., Hottinger, H. (1991). Lactose metabolism in *Lactobacillus bulgaricus*: Analysis of the primary structure and expression of the genes involved, *J. Bacteriol., 173*:1951.

Lerch, H.-P., Blöcker, H., Kallwass, H., Hoppe, J., Tsai, H., and Collins, J. (1989). Cloning, sequencing and expression in *Escherichia coli* of the D-2-hydroxyisocaproate dehydrogenase gene of *Lactobacillus casei, Gene, 78*:47.

Limsowtin, G. K. Y., Crow, V. L., and Pearce, L. E. (1986). Molecular cloning and expression of the *Streptococcus lactis* tagatose 1,6-biphosphate aldolase gene in *Escherichia coli, FEMS Microbiol. Lett., 33*:79.

Lin, J. H.-C., and Savage, D. C. (1986). Genetic transformation of rifampicin resistance in *Lactobacillus acidophilus, J. Gen. Microbiol., 132*:2107.

Luchansky, J. B., Muriana, P. M., and Klaenhammer, T. R. (1988). Application of electroporation for transfer of plasmid DNA to *Lactobacillus, Lactococcus, Leuconostoc, Listeria, Pediococcus, Bacillus, Staphylococcus, Enterococcus, and Propionibacterium. Molecular Microbiol., 2*:637.

Macura, D., and Toensley, P. M. (1984). Scandinavian ropy milk-identification and characterization of endogneous ropy lactic streptococci and their extracellular excretion, *J. Dairy Sci., 67*:735.

Maeda, S., and Gasson, M. J. (1986). Cloning, expression and location of the *Streptococci lactis* gene for phospho-β-D-galactosidase, *J. Gen. Microbiol., 132*:331.

Mayo, B., Kok, J., Venema, K., Bockelmann, W., Teuber, M., Reinke, H., and Venema, G. (1991). Molecular cloning and sequence analysis of the X-prolyl dipeptidyl aminopeptidase gene from *Lactococcus lactis* subsp. *cremoris, Appl. Environ. Microbiol., 57*:38.

McIntyre, D. A., and Harlander, S. K. (1989). Improved electroporation efficiency of intact *Lactococcus lactis* subsp. *lactis* cells grown in defined media, *Appl. Environ. Microbiol., 55*:2621.

McKay, L. L. (1983). Functional properties of plasmids in lactic streptococci, *Antonie van Leeuwenhoek, 49*:259.

McKay, L. L., and Baldwin, K. A. (1974). Simultaneous loss of proteinase and lactose-utilizing enzyme activities in *Streptococcus lactis* and reversal of loss by transduction, *Appl. Microbiol., 28*:342.

McKay, L. L., and Baldwin, K. A. (1978). Stabilization of lactose metabolism in *Streptococcus lactis* C2, *Appl. Environ. Microbiol., 36*:360.

McKay, L. L. and Baldwin, K. A. (1984). Conjugative 40-megadalton plasmid in *Streptococcus lactis* subsp. *diacetylactis* DRC3 is associated with resistance to nisin and bacteriophage, *Appl. Environ. Microbiol., 47*:68.

McKay, L. L., Baldwin, K. A., and Efstathiou, J. D. (1976). Transductional evidence for plasmid linkage of lactose metabolism in *Streptococcal lactis* C2, *Appl. Environ. Microbiol., 32*:45.

McKay, L. L., Baldwin, K. A., and Zottola, E. A. (1972). Loss of lactose metabolism in *Streptococcus lactis, Appl. Environ. Microbiol., 23*:1090.

McKay, L. L., Bohanon, M. J., Polzin, K. M., Rule, P. L., and Baldwin, K. A. (1989). Localization of separate genetic loci for reduced sensitivity towards small isometric-headed bacteriophage sk1 and prolateheaded bacteriophage c2 on p6K17 from *Lactococcus lactis* subsp. *lactis* KR2, *Appl. Environ. Microbiol., 55*:2702.

McKay, L. L., Cords, B. R., and Baldwin, K. A. (1973). Transduction of lactose metabolism in *Streptococcus lactis, J. Bacteriol., 115*:810.

Mercenier, A. (1990). Molecular genetics of *Streptococcus thermophilus, FEMS Microbiol. Rev., 87*:61.

Mercenier, A., and Lemoine, Y. (1989). Genetics of *Streptococcus thermophilus*: a review, *J. Dairy Sci., 72*:3444.

Mercenier, A., Robert, A., Romero, D. A., Castellino, I., Slos, P., and Lemoine, Y. (1988). Development of an efficient spheroplast transformation procedure for *S. thermophilus*: the use of transfection to define a regeneration medium, *Biochimie, 70*:567.

Mercenier, A., Slos, P., Fralen, M., and Lecocq, J. P. (1988). Plasmid transduction in *Streptococcus thermophilus, Mol. Gen. Genet., 212*:386.

Morelli, L., Cocconcelli, P. D., Bottazzi, V., Darniani, G., Feretti, L., and Sqaramella, V. (1987). *Lactobacillus* protoplast transformation, *Plasmid, 17*:73.

Morelli, L., Sarra, P. G., and Bottazzi, V. (1988). *In vivo* transfer of pAMβ1 from *Lactobacillus reuteri* to *Enterococcus faecalis, J. Appl. Bacteriol., 65*:371.

Morelli, L., Vescovo, M., Cocconcelli, P., and Bottazzi, V. (1986). Fast and slow milk coagulating variants of *Lactobacillus helveticus* HLM-1, *Can. J. Microbiol., 32*:758.

Muriana, P. M., and Klaenhammer, T. R. (1987). Conjugal transfer of plasmid-encoded determinants for bacteriocin production and immunity in *Lactobacillus acidophilus* 88, *Appl. Environ. Microbiol. 53*:553.

Muriana, P. M., and Klaenhammer, T. R. (1991). Purification and partial characterization of lactacin F, a bacteriocin produced by *Lactobacillus acidophilus* 11088, *Appl. Environ. Microbiol., 57*:114.

Murphy, M. C., Steele, J. L., Daly, C., and McKay, L. L. (1988). Concomitant conjugal transfer of reduced bacteriophage sensitivity mechanisms with lactose- and sucrose-fermenting ability in lactic streptococci, *Appl. Environ. Microbiol., 54*: 1951.

Nardi, M., Chopin, M.-C., Chopin, A., Cals, M. M., and Gripon, J.-C. (1991). Cloning and DNA sequence analysis of an X-prolyl dipeptidyl aminopeptidase from *lactococcus lactis* subsp. *lactis* NCDO763, *Appl. Environ. Microbiol., 57*:45.

Natori, Y., Kano, Y., and Imamoto, F. (1990). Nucleotide sequences and genomic constitution of five tryptophan genes of *Lactobacillus casei, J. Biochem. 107*:245.

Neumann, E., Schaeffer-Ridder, M., Wang, Y., and Hofschneider, P. H. (1982). Gene transfer into mouse lymphoma cells by electroporation in high electric fields, *EMBO J., 1*:841.

Neve, H., Geis, A., Teuber, M. (1984). Conjugal transfer and characterization of bacteriocin plasmids in group N (lactic acid) streptococci, *J. Bacteriol., 157*:833.

Neve, H., Geis, A., and Teuber, M. (1988). Plasmide encoded functions of ropy lactic strains from Scandinavian fermented milk, *Biochimie, 70*:437.

Orberg, P. K., and Sandine, W. E. (1984). Common occurrence of plasmid DNA and vancomycin resistance in *Leuconostoc* spp., *Appl. Enocron. Microbiol., 48*: 1129.

Otto, R., de Vos, W. M., and Gavrieli, J. (1982). Plasmid DNA in *Streptococcus cremoris* Wg2: influence of pH on selection in chemostats of a variant lacking a protease plasmid, *Appl. Environ. Microbiol., 43*:1272.

Polzin, K. M., and McKay, L. L. (1990). Identification, DNA sequence and distribution of IS981, a new, widespread high-copy number insertion sequence in the lactococci, related to the IS2/IS3 family of IS elements. Abstracts of the Third International ASM Conference on Streptococcal Genetics, A63.

Polzin, K. A., and Shimizu-Kadota, M. (1987). Identification of a new insertion element, similar to gram-negative IS26, on the lactose plasmid of *Streptococcus lactis* ML3, *J. Bacteriol., 169*:5481.

Poolman, B., Royer, T. J., Mainzer, S. E., and Schmidt, B. F. (1989). The lactose transport system of *Streptococcus thermophilus*: a hybrid protein with homology to the melibiose carrier and enzyme III of phosphoenolpyruvate-dependent phosphotransferase systems, *J. Bacteriol., 171*:244.

Poolman, B., Royer, T. J., Mainzer, S. E., and Schmidt, B. F. (1990). Carbohydrate utilization in *Streptococcus thermophilus*: characterization of the genes for aldose-1-epimerase (mutarotase) and UDP glucose-4-epimerase, *J. Bacteriol., 172*: 4037.

Porter, E. V., and Chassy, B. M. (1988). Nucleotide sequence of the β-D-phosphogalactoside galactohydrolase gene of *Lactobacillus casei*: comparison to analogous pbg genes of other Gram-positive organisms, *Gene, 62*:263–276.

Powell, I. V., Achen, M. G., Hillier, A., and Davidson, B. E. (1988). A simple and rapid method for genetic transformation of lactic streptococci by electroporation, *Appl. Environ. Microbiol., 54*:655.

Pucci, M. J., Monteschio, M. E., and Vedamuthu, E. R. (1987). Conjugal transfer in *Leuconostoc* ssp: intergeneric and intrageneric transfer of plasmid-encoded antibiotic resistance determinants, *FEMS Microbiol. Lett., 46*:7

Raya, R. R., Kleeman, E. G., Luchansky, J. B., and Klaenhammer, T. R. (1989). Characterization of the temperate bacteriophage ϕadh and plasmid transduction in *Lactobacillus acidophilus* ADH, *Appl. Environ. Microbiol., 55*:2206.

Renault, P. P., and Heslot, H. (1987). Selection of *Streptococcus lactis* mutants defective in malolactic fermentation, *Appl. Environ. Microbiol., 53*:320.

Romero, D. A., and Klaenhammer, T. R. (1990). Characterization of insertion sequence IS946, an Iso-ISS1 element, isolated from the conjugative lactococcal plasmid pTR2030, *J. Bacteriol., 172*:4151.

Romero, D. A., Slos, P., Robert, C., Castellino, I., and Mercenier, A. (1987). Conjugative mobilization as an alternative vector delivery system for lactic streptococci, *Appl. Environ. Microbiol., 53*:2405.

Ross, P., O'Gara, F., and Condon, S. (1990). Cloning and characterization of the thymidylate synthase gene from *Lactococcus lactis* subsp. *lactis, Appl. Environ. Microbiol., 56*:2156.

Sanders, M. E., and Klaenhammer, T. R. (1981). Evidence of plasmid linkage of restriction and modification in *Streptococcus cremoris* KH, *Appl. Environ. Microbiol., 42*:944.

Sanders, M. E., and Klaenhammer, T. R. (1983). Characterization of phage-insensitive mutants from a phage-sensitive strain of *Streptococcus lactis*: evidence for a plasmid determinant that prevents phage adsorption, *Appl. Environ. Microbiol., 46*:1125.

Sanders, M. E., Leonhard, P. J., Sing, W. D., and Klaenhammer, T. R. (1986). Conjugal strategy for construction of fast acid-producing, bacteriophage resistant lactic streptococci for use in Dairy fermentations, *Appl. Environ. Microbiol., 52*:1001.

Sanders, M. E., and Nicholson, M. A. (1987). A method for genetic transformation of nonprotoplasted *Streptococcus lactis, Appl. Environ. Microbiol., 53*:1730.

Scheirlink, T., Mahillon, J., Joos, H., Dhaese, P., and Michiels, F. (1989). Integration and expression of α-amylase and endoglucanase genes in the *Lactobacillus plantarum* chromosome, *Appl. Environ. Microbiol., 55*:2130.

Scherwitz-Harmon, K., and McKay, L. L. (1987). Restriction enzyme analysis of lactose and bacteriocin plasmids from *Streptococcus lactis* subsp. *diacetylactis* WM_4 and cloning of *BCl*I fragments coding for bacteriocin production, *Appl. Environ. Microbiol., 53*:1171.

Schmidt, B. F., Adams, R. M., Requad, C., Power, S., and Mainzer, S. E. (1989). Expression and nucleotide sequence of *Lactobacillus bulgaricus* β-galactosidase gene cloned in *Escherichia coli, J. Bacteriol. 171*:625.

Schroeder, C. J., Robert, C., Lenzen, G., McKay, L. L., and Mercenier, A. (1991). Analysis of the *lac*Z sequences from two *Streptococcus thermophilus* strains: comparison with the *Escherichia coli* and *Lactobacillus bulgarius* β-galactosidase sequences, *J. Gen. Microbiol., 137*:369.

Shay, B. J., Egan, A. F., Wright, M., and Rogers, P. J. (1988). Cysteine metabolism in an isolate of *Lactobacillus sake*: plasmid composition and cysteine transport, *FEMS Microbiol. Lett., 56*:183.

Shimizu-Kadota, M. (1987). Properties of lactose plasmid pLY101 in *Lactobacillus casei., Appl. Environ. Microbiol., 53*:2987.

Shimizu-Kadota, M., Kviaki, M., Hirokawa, H., and Tsuchida, N. (1985). ISL1: a new transposable element in *Lactobacillus casei, Mol. Gen. Genet., 200*:193.

Shrago, A. W., Chassy, B. M., and Dobrogosz, W. J. (1986). Conjugal plasmid transfer (pAMβ1) in *Lactobacillus plantarum, Appl. Environ. Microbiol., 52*:574.

Sibakov, M., Koivula, T., v. Wright, A., and Plava, I. (1991). Secretion of TEM β-lactamase with signal sequences isolated form the chromosome of *Lactococcus lactis* subsp. *lactis, Appl. Environ. Microbiol., 57*:341.

Simon, D., Rouault, A., and Chopin, M.-C. (1986). High-efficiency transformation of *Streptococcus lactis* protoplasts by plasmid DNA, *Appl. Environ. Microbiol., 52*:394.

Sing, W. D., and Klaenhammer, T. R. (1986). Conjugal transfer of bacteriophage resistance determinants on pTR2030 into *Streptococcus cremoris* strains, *Appl. Environ. Microbiol., 51*:1264.

Sing, W. D., and Klaenhammer, T. R. (1990). Characteristics of phage abortion in lactococci by the conjugal plasmid pTR2030, *J. Gen. Microbiol., 136*:1807.

Sinha, R. P. (1986). Development of high-level streptomycin resistance affected by a plasmid in lactic streptococci, *Appl. Environ. Microbiol., 52*:255.

Skaugen, M. (1989). The complete nucleotide sequence of a small cryptic plasmid from *Lactobacillus plantarum, Plasmid, 22*:175.

Skou, B. (1987). Transformation of *Lactobacillus* spp. with *Lactobacillus fermentum* DNA, *FEMS Microbiol. Rev., 46*:P21.

Slos, P., Bourquin, J.-C., Lemoine, Y., and Mercenier, A. (1991). Isolation and characterization of chromosomal promoters of *Streptococcus salivarius* subsp. *thermophilus, Appl. Environ. Microbiol., 57*:1333.

Smiley, M. B., and Fryder, V. (1978). Plasmids, lactic acid production and *N*-acetyl-D-glucosamine fermentation in *Lactobacillus helveticus* subsp. *jugurti, Appl. Environ. Microbiol., 35*:777.

Smith, M. D., and Clewell, D. B. (1984). Return of *Streptotoccus faecalis* DNA cloned in *Escherichia coli* to its original host via transformation of *Streptococcus sanguis* followed by conjugative mobilization, *J. Bacteriol., 160*:1109.

Somkuti, G. A., and Steinberg, D. H. (1988). Genetic transformation of *Streptococcus thermophilus* by electroporation, *Biochimie, 70*:579.

Steele, J. L., and McKay, L. L. (1986). Partial characterization of the genetic basis for sucrose metablism and nisin production in *Streptococcus lactis, Appl. Environ. Microbiol., 51*:57.

Stevenson, L. R., and Klaenhammer, T. R. (1985). *Streptococcus cremoris* M12R transconjugants carrying the conjugal plasmid pTR2030 are insensitive to attack by lytic bacteriophages, *Appl. Environ. Microbiol., 50*:851.

Takiguchi, R., Hashiba, H., Ayoama, K., and Ishii, S. (1989). Complete nucleotide sequence and characterization of a cryptic plasmid from *Lactobacillus helveticus* subsp. *jugurti, Appl. Environ. Microbiol., 55*:1653.

Thompson, J. (1987). Regulation of sugar transport and metabolism in lactic acid bacteria, *FEMS Microbiol. Rev., 46*:221.

Trieu-Cuot, P., Curlier, C., Martin, P., and Courvalin, P. (1987). Plasmid transfer by conjugation from *Escherichia coli* to gram-positive bacteria, *FEMS Microbiol. Lett., 48*:289.

Tsai, H. J., and Sandine, W. E. (1987a). Conjugal transfer of nisin plasmid genes from *Streptococcus lactis* 7962 to *Leuconostoc dextranicum* 181, *Appl. Environ. Microbiol., 53*:352.

Tsai, H.-J., and Sandine, W. E. (1987b). Conjugal transfer of lactose-fermenting ability from *Streptococcus lactis* C2 to *Leuconostoc cremoris* CAF7 yields *Leuconostoc* that ferment lactose and produce diacetyl, *J. Ind. Microbiol., 2*:25.

Tynkkynen, S., von Wright, A., and Syväoja, E.-L. (1989). Peptide utilization encoded by *Lactococcus lactis* subsp. *lactis* SSL135 chromosomal DNA, *Appl. Environ. Microbiol., 55*:2690–2695.

van Asseldonk, M., Rutten, G., Oteman, M., Siezen, R. J., de Vos, W. M., and Simons, G. (1990). Cloning of *usp45*, a gene encoding a secreted protein from *lactococcus lactis* subsp. *lactis* MG1363, *Gene, 95*:155.

van Belkum, M. J., Hayema, B. J., Geis, A., Kok, J., and Venema, G. (1989). Cloning of two bacteriocin genes from a lactococcal bacteriocin plasmid, *Appl. Environ. Microbiol., 55*:1187.

van Belkum, M. J., Hayema, B. J., Jeeninga, R. E., Kok, J., and Venema, G. (1991). Organization and nucleotide sequences of two lactococcal bacteriocin operons, *Appl. Environ. Microbiol., 57*:492.

van de Guchte, M., Kodde, J., van der Vossen, J. M. B. M., Kok, J., and Venema, G. (1990). Heterologous gene expression in *Lactococcus lactis* subsp. *lactis*: synthesis, secretion, and processing of the *Bacillus subtilis* neutral protease, *Appl. Environ. Microbiol., 56*:2606.

van der Lelie, D., Chavarri, F., Venema, G., and Gasson, M. J. (1991). Identification of a new genetic determinant for cell aggregation associated with lactose plasmid transfer in *Lactococcus lactis, Appl. Environ. Microbiol. 57*:201.

van der Lelie, D., van der Vossen, J. M. B. M., and Venema, G. (1988). Effect of plasmid incompatibility on DNA transfer to *Streptococcus cremoris, Appl. Environ. Microbiol., 54*:865.

van der Lelie, D., Wösten, H. A. B., Bron, S., Oskam, L., and Venema, G. (1990). Conjugal mobilization of streptococcal plasmid pMV158 between strains of *Lactococcus lactis* subsp. *lactis, J. Bacteriol., 172*:47.

van der Vossen, J. M. B. M., Kok, J., and Venema, G. (1985). Construction of cloning, promoter-screening and terminator-screening shuttle vectors for *Bacillus subtilis* and *Streptococcus lactis, Appl. Environ. Microbiol., 50*:540.

van der Vossen, J. M. B. M., van der Lelie, D., and Venema, G. (1987). Isolation and characterization of *Streptococcus cremoris* Wg2-specific promoters, *Appl. Environ. Microbiol., 53*:2452.

van Luijk, N., Leer, R. J., Posno, M., Lokman, B. C., van Giegen, M. J. F. et al. (1990). Development of a host-vector system for lactobacillus: analysis of the stability of lactobacillus plasmid DNA vectors, *FEMS Microbiol. Rev., 46*:P12.

Vedamuthu, E. R., and Neville, J. (1986). Involvement of a plasmid in production of ropiness (mucoidness) in milk cultures by *Streptococcus cremoris, Appl. Environ. Microbiol., 29*:807.

Vescovo, M., Morelli, L., and Bottazzi, V. (1982). Drug resistance plasmids in *Lactobacillus acidophilus* and *Lactobacillus reuteri, Appl. Environ. Microbiol., 43*:50.

Vescovo, M., Morelli, L., Bottazzi, V., and Gasson, M. I. (1983). Conjugal transfer of broad host-range plasmid pAMβ1 into enteric species of lactic acid bacteria, *Appl. Environ. Microbiol., 46*:753.

Vescovo, M., Scolari, G. L., and Bottazzi, V. (1989). Plasmid encoded ropiness production in *Lactobacillus casei* ssp. *casei, Biotechnol. Lett., 11*:709.

Vogel, R. F., Gaier, E., and Hammes, W. P. (1990). Expression of the lipase gene from *Staphylococcus hyicus* in *Lactobacillus curvatus* 1c2-c, *FEMS Microbiol. Lett., 69*:289.

von Wright, A., Suominen, M., and Sivelä, S. (1986). Identification of lactose fermentation plasmids of streptococcal dairy starter strains by southern hybridization, *Letters in Appl. Microbiol., 2*:73.

von Wright, A., Taimisto, A.-M., and Sivelä, S. (1985). Effect of Ca^{2+} ions on plasmid transformation of *Streptococcus lactis* protoplasts, *Appl. Environ. Microbiol., 50*:1100.

von Wright, A., and Tynkkynen, S. (1987). Construction of *Streptococcus lactis* subsp. *lactis* strains with a single plasmid associated with mucoid phenotype, *Appl. Environ. Microbiol., 53*:1385.

von Wright, A., Tynkkynen, S., and Suominen, M. (1987). Cloning of a *Streptococcus lactis* subsp. *lactis* chromosomal fragment associated with the ability to grow in milk, *Appl. Environ. Microbiol., 53*:1584.

von Wright, A., Wessels, S., Tynkkynen, S., and Saarela, M. (1990). Isolation of a replication region of a large lactococcal plasmid and use in cloning of a nisin resistance determinant, *Appl. Environ. Microbiol., 56*:2029.

Vos, P., van Asseldonk, M., van Jeveren, F., Siezen, R., Simons, G., and de Vos, W. M. (1989). A maturation protein is essential for the production of active forms of *Lactococcus lactis* SK11 serine proteinase located in or secreted from the cell envelope, *J. Bacteriol., 171*:2795.

Walker, G. C. (1984). Mutagenesis and inducible responses to deoxyribonucleic acid damage in *Escherichia coli*, *Microbiol. Rev., 48*:60.

Walsh, P. M., and McKay, L. L. (1981). Recombinant plasmid associated with cell aggregation and high frequency conjugation of *Streptococcus lactis* ML3, *J. Bacteriol., 146*:937.

West, C. A., and Warner, P. J. (1985). Plasmid profiles and transfer of plasmid-encoded antibiotic resistance in *Lactobacillus plantarum*, *Appl. Environ. Microbiol., 50*:1319.

Xu, F., Pearce, L. E., and Yu, P.-L. (1991). Construction of a family of lactococcal vectors for gene cloning and translational fusion, *FEMS Microbiol. Lett., 77*:55.

7

Lactic Acid Bacteria in Health and Disease

Seppo Salminen
Valio Ltd., Helsinki, and University of Helsinki, Helsinki, Finland

Margaret Deighton
Royal Melbourne Institute of Technology, Melbourne, Victoria, Australia

Sherwood Gorbach
Tufts University School of Medicine, Boston, Massachusetts

I. INTRODUCTION

Lactic acid bacteria are a group of Gram-positive rods and cocci with physiological and ecological characteristics in common. These bacteria use carbohydrates as an energy source and produce lactic acid either as the sole product of metabolism (homolactic fermentation) or as the major end product (heterolactic fermentation). Lactic acid bacteria are commonly found on the mucous membranes of humans and animals, in dairy products and naturally on some plant surfaces. Several species of lactic acid bacteria are used commercially for the production of fermented milks, yogurt, other dairy and meat products, as well as other foods.

Lactic acid bacteria include members of the genera *Streptococcus, Enterococcus, Lactobacillus, Leuconostoc, Lactococcus*, and *Pediococcus*. Food microbiologists have found it convenient to group these organisms together according to their metabolism and natural habitat. Recently, the rationale for this approach was confirmed by studies on isoenzymes (London et al., 1976) and on 16S rRNA relatedness (Stackebrandt et al., 1983), which demonstrate a close relationship between members of the group. The general properties of these bacteria are described by Axelsson (Chapter 1) and Mikelsaar (Chapter 9).

This chapter emphasizes the species of lactic acid bacteria that have been used as human probiotics or in fermented dairy products claimed to promote health.

A. Normal Flora of Human Gastrointestinal Tract

The human colonic flora consists of at least 400 different species of bacteria, with total numbers reaching approximately 10^{12}. These organisms comprise a highly developed ecosystem, and are in dynamic equilibrium with each other and with their host (Simon and Gorbach, 1986; Fuller, 1989). Although both anaerobes and facultative anaerobes are present in gut flora, anaerobes outnumber the facultative anaerobes by a factor of 1000 to 1. Among the facultative flora are several species of lactic acid bacteria including members of the genera *Streptococcus, Enterococcus*, and *Lactobacillus*.

The normal intestinal flora of man performs an important protective function. Any bacterial strain administered as a probiotic must compete with the established flora for nutrients, atmospheric requirements, and attachment sites. Attachment is a highly specific process involving adhesion on the bacterial surface and complementary host receptors. It follows that colonization of human gut flora is most likely to be successful with probiotic strains derived from human sources rather than from animal sources or traditional dairy products.

B. Lactobacilli as Members of the Human Gut Flora

Lactic acid bacteria are normal residents of the gastrointestinal tract of human beings. Most studies of lactic acid bacteria have involved lactobacilli. Lactobacilli, although not predominant organisms in the intestinal flora, are isolated throughout the gastrointestinal tract of healthy humans (Finegold et al., 1977). There is, however, host specificity in colonisation by individual species. In studies of the fecal flora of Americans, 14 of 15 omnivores and 11 of 13 vegetarians had lactobacilli in their feces (Finegold et al., 1977). There was a relatively limited number of species cultivated. Among the omnivores, the most common species were *L. acidophilus, L. fermentum*, and *L. plantarum* with mean viable counts in feces of 10^6, 10^4, and 10^7 per gram of dry weight, respectively. Lower numbers of *L. brevis, L. casei, L. leichmanii*, or *L. minutus* were found in 1 to 3 of the 14 omnivores. Notably absent from this list is *L. bulgaricus*, the most common organism along with *S. thermophilus*, used to produce yogurt.

II. PROBIOTIC BACTERIA

A. Introduction

Probiotics generally refer to viable bacteria, cultured dairy products or food supplements containing viable lactic acid bacteria. Probiotic products are thought to have beneficial health effects in the host by improving the integrity of the intestinal microflora. Probiotic products can include freeze dried bacteria in tablet or capsule form and more recently, fermented dairy products such as yogurt, fermented milks, and sweet acidophilus milks.

The most important property for probiotics is clinically proven efficacy. Any probiotic product should be backed by well-designed, randomized double-blind clinical trials that ensure its health effects and benefits to the host. Moreover, strain characteristics should be carefully defined. Since most probiotic products are backed by a few, if any clinical and microbiological studies, more work is needed to clarify their clinical benefits and define the characteristics of the strains used in these products.

B. Requirements of Probiotic Strains

Several basic properties are required for an effective probiotic strain of lactic acid bacteria. Among the most important properties is the ability to survive passage through the mouth, stomach, small intestine, and large intestine. Thus, the strains must be stable in gastric conditions and resistant to acid. In addition, stability to bile acids is required.

In order to be able to influence functions of the human intestinal tract, probiotic strains must have the capacity to adhere to intestinal mucosal cells and to grow in intestinal conditions (Table 1). Several types of tests can be conducted to observe the adherence of lactic acid bacteria to human intestinal cells. Such methods are described by Goldin et al. (1992) and Conway and co-workers (1987). Various human cell lines have also been utilized. Table 2 summarizes the results of an experiment utilizing a human Caco-2 cell line in testing the adhesion of some lactic acid bacteria and bifidobacteria (Elo et al., 1991). Also other researchers have later verified the adhesion properties of this strain using similar test systems and indicating that normal dairy strains do not show adherence properties (Coconnier et al., 1992).

Table 1 Some Properties of a Good Probiotic Strain

Acid stability (especially gastric acid)
Bile stability
Human origin (species-specific properties)
Adherence to human intestinal cells
Colonization of the human intestinal tract
Production of antimicrobial substances
Antagonism against pathogenic bacteria
Antagonism against cariogenic bacteria
Good growth in vitro
Safety in human use

Table 2 Results of In Vitro Adhesion Tests Conducted on Dairy and Reference Organisms and the Human Caco-2 Cell Line

Tested strain	Degree of adhesions[c]
Lactobacillus GG (Original)	+++
Lactobacillus GG[a]	++
Lactobacillus GG lot 3689	+++
Lactobacillus GG lot 950719	+++
Lactobacillus GG lot 1490	+++(+)
Lactobacillus GG lot 1690	++++
Lactobacillus GG lot 1790	+++(+)
Lactobacillus GG lot 4788	++(+)
Lactobacillus GG lot 1789	+++
Lactobacillus GG (Gefilus/Gefilac fermented whey drink)	+++
Lactobacillus GG col 1[b]	+++
Lactobacillus GG col 2[b]	+++
Lactobacillus GG col 3[b]	+++
Lactobacillus bulgaricus	—
Lactobacillus acidophilus	—
Bifidobacterium 12-1	—, (+)
Bifidobacterium 2572	—, (+)
Bifidobacterium 2919	—, (+)
Bifidobacterium 2394	—, (+)
E. coli (B44)	—

Each value is a summary of at least five determinations except in case of Bifidobacteria (2 determinations).
[a]Maintained in MRS broth by weekly transfer for three to five years.
[b]Isolated from stool samples of persons consuming Gefilac products.
[c]No adhesion, + slight adhesion, ++ some adhesion, +++ good adhesion, ++++ very good adhesion.
Source: According to Elo et al. (1991).

Other requirements of an effective probiotic strain include antagonism against pathogenic bacteria, production of antimicrobial substances, and proven safety in human use. In addition, any proposed probiotic strain needs to be suitable for culturing in common dairy processes to enable product development. If the strain is easy to grow and can be freeze-dried, other types of probiotic products may be processed.

B. Mechanisms of Action of Lactic Acid Bacteria as Probiotics

Data from various studies on probiotic strains of lactic acid bacteria have produced information on the probable mechanisms of action. For some strains, several mechanisms of action have been proposed, but for others, the mechanism of action is unknown. Closely related strains may differ significantly in their probiotic properties, and some may lack the full complement of properties required for activity. Proposed mechanisms include the suppression of harmful bacteria and viruses, stimulation of local and systemic immunity (Kaila et al., 1991b, 1992), and alteration of gut microbial metabolic activity (Goldin and Gorbach, 1977; Goldin et al., 1980; Marteau et al., 1990). The ability of probiotic bacteria to suppress the growth of pathogens has been attributed to the production of antibacterial substances such as lactic acid (Choi et al., 1984; Gibbs, 1987; Bhatia et al., 1989), peroxide (Collins and Aranki, 1980), bacteriocins (Barefoot et al., 1983; McCormick et al., 1983), and bacteriocin-like inhibitory substances (Silva et al., 1987; McCormick et al., 1983). Other proposed methods of suppression include competition for attachment sites, promotion of intestinal mucus production, and competition for nutrients.

D. Species of Lactic Acid Bacteria Used as Probiotics

Various strains of *Lactobacillus acidophilus* have been used in studies on the use of probiotics in humans. Included among the successful organisms are two strains of human origin *L. acidophilus* (NFCO 1748) and *Lactobacillus casei* strain GG.

Lactobacillus GG colonizes the human intestinal tract and adheres more strongly to human intestinal and mucosal buccal cells than other strains of *Lactobacillus* or *Streptococcus* used as starter cultures in the dairy industry. Adhesion of the *Lactobacillus* GG has also been indicated in Caco-2 cells in different studies (Elo et al., 1991; Coconnier et al., 1992). When individuals consume the GG strain, either in a lyophilized form or as a fermented dairy milk product, it is consistently isolated from feces in concentrations of 10^6 cfu per gram (Gorbach, 1990; Goldin et al., 1992). *Lactobacillus* GG strain elaborates an antimicrobial substance which has a broad spectrum of activity against a range of bacteria (Silva et al., 1987). Furthermore, *Lactobacillus* GG has a distinctive colonial morphology on *Lactobacillus* specific (LBS) tomato juice agar and on DeMan-Rogosa-Sharpe (MRS) agar, which facilitates its recognition in fecal specimens (Goldin et al., 1992; Saxelin et al., 1991).

Other studies have investigated the use of preparations containing *Enterococcus faecium* strain 68 (SF68) for the treatment of acute enteritis and other gut disorders. *E. faecium* strain 68 has several attributes which makes it attractive as a probiotic: it has short generation time (20 min), is more stable to acid pH than *L. acidophilus* (Friis-Møller and Hey, 1983), and appears to have antibacterial activity against some enteropathogens (Lewenstein et al., 1979).

It is not clear whether strains of *E. faecium* used as probiotics are the same as these found in the feces of healthy adults. One recent observation, that some so-called *E. faecium* strains used as probiotics are not *E. faecium* but another unidentified enterococcus (Fuller, 1989), suggests that strain differences exist.

In Japan, considerable research interest has been given to *Lactobacillus casei* Shirota. Reportedly this organism has been isolated from human feces, but very little information has been published in English (Nishizato et al., 1985).

E. Colonization of the Human Intestinal Tract with Administered Lactic Acid Bacteria

It has been difficult to study intestinal implantation of administered lactic acid bacteria, and especially *Lactobacillus* strains, because the colonial morphology of these organisms is virtually identical to that of lactobacilli normally resident in the bowel (Goldin et al., 1992). Hence, the administered strains cannot be distinguished, in a fecal culture, from endogenous lactobacilli which are abundant in the normal flora. The presence of the administered strains can only be inferred by an increase in fecal lactobacilli counts over baseline values.

Using this method Robins and co-workers (1981) fed freeze-dried commercial preparations of *L. acidophilus* and *L. bulgaricus* (cfu/day) to volunteers and then cultured aspirates of gastric and intestinal juice. An increase in the number of lactobacilli was detected in the stomach and small intestine after approximately 3 h in fasting subjects and 6 h in nonfasting subjects. However, the viable counts of lactobacilli returned to baseline levels after this period. In another study Clements and co-workers (1983) used strains of *L. acidophilus, L. bulgaricus*, and *Bifidobacterium infantiis*. These organisms could only be recovered from the small intestine for up to 3 h.

Pettersson and co-workers (1983) studied patients with ileostomy. They administered buttermilk, yogurt (both containing 10^8 cfu/mL viable lactobacilli), or pure cultures 10^8 cfu/mL of *L. bulgaricus* or *L. acidophilus* NDCO 1748. The levels of lactobacilli in ileostomy effluents remained constant throughout the study period for patients receiving buttermilk, yogurt and *L. bulgaricus*. Those receiving pure cultures of *L. acidophilus* had transient increases in lactobacilli counts (1–3 logs) in ileostomy effluent. The organisms could be identified in the effluent 1.5 h after feeding and peak concentrations occurred at 3.5 h. By 4 h, 99.5% of the administered organisms had passed out of the small intestine (Petersson et al., 1983).

In a study by Paul and Hoskins (1972), utilizing a powdered preparation of *L. acidophilus*, only half of the human volunteers had moderately elevated fecal counts of lactobacilli during the feeding phase and these levels rapidly returned to baseline values.

Lidbeck et al. (1987) studied the effect of a commercial fermented milk containing *L. acidophilus* NDCO 1748 by giving 500 mL ($10^8 - 4 \times 10^9$ cfu) of the product daily to human volunteers. Significantly increased counts of lactobacilli were observed in the fecal flora on the fourth day after feeding, and the counts remained high until nine days after administration was stopped. The administered organism was not differentiated from endogenous lactobacilli due to the similar colonial morphology. It was noted however that administering 500 mL of fermented milk daily may also increase the levels of endogenous lactobacilli by providing substrate (Lidbeck et al., 1987).

In a more recent study (Lidbeck et al., 1991) the same dose of a fermented milk containing *Lactobacillus acidophilus* NCFB 1748 was given twice daily (10^9 cfu/day) for six weeks to 14 patients with colonic cancer. A significant increase in fecal lactobacilli was observed in 10 patients, but again no identification of the fecal strains was conducted.

In human volunteer studies using *Lactobacillus* GG given as a freeze-dried culture (4×10^9 cfu/day) or in fermented dairy products (1.6×10^{11} cfu/day), the organism could be recovered from the feces of all subjects (Goldin et al., 1992). After discontinuing feeding, it was still present in 81% of the subjects after four days, and 33% after seven days. *Lactobacillus* GG persisted longer in the human gastrointestinal tract than *E. coli* K-12 strains (Smith, 1969; Formal and Hornick, 1978). Moreover distinctive colonial morphology enabled *Lactobacillus* GG to be distinguished from endogenous flora.

In another study of human volunteers, freeze-dried *Lactobacillus* GG was given to healthy subjects for periods of seven days and colonization was measured by performing fecal counts of the administered strain. Colonization of some volunteers occurred at dose levels of 10^9 cfu/day. When 10^{10} cfu was fed daily, all subjects were colonized (Saxelin et al., 1991).

Colonization with *Lactobacillus* GG is more efficient in infants and children than adults. In a study of infants with rotavirus diarrhea, feeding 10^9 cfu/day *Lactobacillus* GG is a freeze-dried powder resulted in colonization of all babies (Kaila et al., 1991b). Considerably lower doses are required to colonize healthy infants (Kaila et al., personal communication).

III. STUDIES ON LACTIC ACID BACTERIA AND HEALTH

A. Reduction of Gastrointestinal Symptoms in Lactose-Intolerant Individuals

Lactose intolerance is caused by a deficiency in the intestinal brush-border enzyme β-galactosidase (lactase), resulting in the inability to digest the disaccharide, lactose. Acquired lactase deficiency may be the result of pelvic radiotherapy, or it may follow an intestinal infection such as rotavirus infection, with associated destruction of lactase-producing brush-border cells (Christensen,

1989). The prevalence of congenital lactose intolerance varies depending on the ethnic origin. It is common in Japan, China, Africa, and among Australian Aborigines but less common in North European and North American countries.

Several authors have reported studies in which lactose-intolerant individuals were given equivalent amounts of lactose in milk, yogurt, or other fermented dairy products. In most studies, hydrogen excretion in breath was used as a marker of the presence of unabsorbed lactose in the colon. In some studies levels of β-galactosidase and viable counts were performed on duodenal aspirates.

Savaiano et al. (1984) reported that viable yogurt is more effective than pasteurized yogurt in enhancing the digestion of lactose in lactose-intolerant subjects. Since pasteurization reduces bacterial counts as well as bacterial lactase levels, this result is not surprising. Buttermilk (a fermented product containing live lactic acid bacteria) was also less effective than yogurt in enhancing digestion of lactose and reducing gastrointestinal symptoms, although the levels of malabsorbed lactose were similar in both groups. This is an unexpected finding, but may relate to different bacterial strains in the yogurt and buttermilk preparations used.

Similar results were obtained from a case-control study reported by Kolars et al. (1984), in which lactose-intolerant subjects were given an equivalent amount of lactose in either yogurt, milk or water. Lactose levels in the colon were reduced by about one third in the yogurt group compared with all other control groups. The yogurt group also suffered significantly fewer gastrointestinal symptoms. Aspiration of duodenal contents demonstrated significant lactase activity for at least 1 h after ingestion of yogurt. The authors suggested that lactase released from organisms present in the yogurt contributed to the enhanced absorption of lactose.

This conclusion is supported by the work of McDonough and co-workers (1987), who compared milk, yogurt, yogurt heated to inactive bacterial lactase, heated yogurt plus lactose, (i.e., lactose added to bring the concentration to the prefermentation level), heated yogurt plus lactase, sweet acidophilus milk (unfermented milk with *L. acidophilus* culture), and sweet acidophilus milk sonicated to release bacterial intracellular lactase. The highest breath hydrogen levels were seen in subjects given milk or sweet acidophilus milk. Intermediate levels were observed with heated yogurt, yogurt plus lactose, and sonicated sweet acidophilus milk. The lowest breath hydrogen levels were found for viable yogurt and heat-treated yogurt with added lactase. These results suggest that bacterial lactase contributes to the improved degradation of lactose in lactose-intolerant individuals. It is important to note, however, that lactose levels are reduced by the fermentation of milk.

Further support for the role of bacterial lactase in improving the digestion of lactose in intolerant individuals was provided by a carefully controlled study by Dewitt et al. (1988). Viable yogurt resulted in less undigested lactose than either heated yogurt or milk.

A group of French investigators have provided evidence that viable yogurt reaches the duodenum and contains β-galactosidase activity. Such activity may be transient, however due to the buffering capacity of yogurt, which reduces duodenal pH, thus preventing yogurt-derived β-galactosidase from hydrolyzing lactose (Pochart et al., 1989).

Brand and Holt (1991) studied the relative effectiveness of milks with reduced amounts of lactose in alleviating lactose intolerance. They utilized reduced lactose milks commercially available in Australia and evaluated their effects in six subjects with lactose intolerance. The differences observed between the lactose-reduced milks and whole milk were statistically significant ($p < .05$). Only one of the six subjects reacted to 50% or 80% lactose-reduced milk, while none reacted to 95% lactose-reduced milk. The mean maximum breath hydrogen level rises were 31, 7, 5, 5, and 8 ppm respectively for 0%, 50%, 80%, and 95% lactose-reduced milk. It has been indicated by Siigur and co-workers that fermentation of the unabsorbed lactose by bacteria colonizing the small intestine may act as a complementary protective mechanisms reducing the osmotic load of lactose and preventing diarrhea (Siigur et al., 1990). Thus, complete avoidance of lactose in the diet may result in more severe symptoms.

Martini et al. (1991) indicated that lactic acid bacteria (*Streptococcus thermophilus, Lactobacillus bulgaricus, Lactobacillus acidophilus*) in yogurts dramatically improved lactose digestion regardless of the observed lactase activity. Fermented milks, on the other hand, varied in response. *Bifidobacterium bifidum* milk caused practically no increase in lactose digestion, while *Lactobacillus bulgaricus* milk improved lactose digestion in healthy volunteers with lactose intolerance (Martini et al., 1991). The authors suggest that the use of novel lactic acid bacteria in fermented milks may improve lactose digestion and reduce lactose-induced symptoms in intolerant persons.

B. Constipation

Constipation is a common problem especially in the elderly. Intestinal microflora has often been connected to constipation and meteorism. There is some indication that the presence of lactobacilli in the intestinal tract may reduce intestinal gas formation. It was suggested by Kochar et al. (1989) that lactobacilli may alter the preexisting normal intestinal flora and thereby alter intestinal gas production and composition. Chronic constipation has been treated using lactobacilli and fermented milk products. Rajala et al. (1988) showed that a *Lactobacillus acidophilus* yogurt supplemented with fiber and lactitol significantly reduced symptoms of constipation in elderly hospitalized patients. During the study 51 patients were followed and 33 patients qualified for the final analysis. All the patients were followed for a week prior to the study and the yogurts were served for two weeks. Afterwards, there was a follow-up of one week. The mean fecal output increased 1.6-fold while in the control yogurt group the output increased only 1.2-fold. It was concluded that fiber and lactitol supplemented fermented milk may offer one dietary means of treating constipation. A later study utilizing a *Lactobacillus* GG fermented whey drink in nursing home patients indicated that the ingestion of *Lactobacillus* GG did not alter the frequency of bowel movements. However, when using radioopaque markers the fermented whey drink was calculated to decrease intestinal transit time from an average of 4.5 to 3.8 days (Mykkänen, personal communication).

C. Lactic Acid Bacteria and Enteric Infections

Several investigators have used fermented dairy products containing lactic acid bacteria or the organisms themselves as dietary supplements for the treatment of gastrointestinal infections in humans and animals. It is difficult to compare the results of these studies because they differ in the strains of lactic acid bacteria used, the type of preparation (yogurt, fermented milk, bacterial cultures in milk or water), and the study group. Moreover, the etiological agent was not always defined.

D. Treatment of *Salmonella* Infections

Gilliland and Speck (1977) showed that *Lactobacillus acidophilus* exerted antagonistic action on the growth of *Salmonella typhimurium* in vitro. The effect was produced under anaerobic conditions in a prereduced medium and the degree of antagonism varied among different strains of *L. acidophilus*.

In an animal study, Fichera and co-workers (1987) showed an increased resistance in mice against lethal infection by *Salmonella enteriditis*, due to the administration of *Bifidobacterium bifidum* and *Lactobacillus acidophilus*. It was observed that the index of *Salmonella* survival in the intestinal contents decreased if it was administered with both of the previously mentioned strains. The authors suggested that a competitive anti-infective action may exist in the intestine between *Salmonella* and the *Bifidobacterium/Lactobacillus* combination. One explanation may also be the decreased pH value in the intestinal contents (Fichera et al., 1987).

Studies of weanling rats, challenged by intragastric inoculation of *Salmonella enteriditis*, suggested that yogurt might be useful for treatment of salmonellosis (Hitchins et al., 1985). Rats fed yogurt after *Salmonella* challenge had significantly lower morbidity and improved weight gain, compared with controls on a diet of milk.

In humans, daily consumption of very large numbers of *L. acidophilus* (500 mL of milk containing 6×10^9 cfu/mL) shortens the period of carriage in patients infected with *Salmonella* (Alm, 1983).

Choi and co-workers (1984) have reported an in vitro study on growth inhibitory effects of culture supernatants of *Lactobacillus casei* isolated from yakult on *Salmonella* indicating inhibition by whole cultures but not culture supernatants. However, Silva and co-workers (1987) reported in vitro inhibition by *Lactobacillus* GG using 10-fold concentrates of culture supernatants.

E. Treatment of Infantile Diarrhea

Many different diseases may induce diarrhea in infants and young children. Diarrhea and diarrheal diseases remain a major public health problem of infants and young children especially in developing countries.

In search of more physiologic methods of treating infantile diarrhea the early concepts introduced by Metchnikoff (1901), Kopeloff (1924), and Kleeberg (1927) have been tested in several studies. However, many of the studies have not defined the strains studied or the numbers of viable bacteria given to patients. Therefore, there is great variability in the published results. We attempt to review some of the recent and more important studies in this area.

Hall et al. (1990) determined the composition of fecal microflora in 46 preterm infants and 52 full-term infants. Lactobacilli were found in high counts in the feces of full-term infants by 30 days of age. Bifidobacteria were not as frequent. In preterm infants the authors found a selective deficiency of lactobacilli when compared to coliforms. In another study (Reuman et al., 1986) the effect of *Lactobacillus acidophilus* treatment on colonization by antibiotic-resistant organisms was studied. The infants were treated with *Lactobacillus* formula or a regular formula and no differences were observed in case of aminoglycoside resistant bacteria.

In one study, Isolauri and co-workers (1991) investigated the effect of *Lactobacillus* GG on recovery of infants from acute diarrhea, mainly due to rotavirus. After oral rehydration, 71 hospitalized infants aged between 4 and 45 months were randomly divided into three groups. Group 1 received *Lactobacillus* GG fermented milk product (10^{10}–10^{11} cfu), group 2 received the same number of organisms as a freeze-dried powder, and group 3 received pasteurized yogurt containing no viable lactic acid bacteria. All treatments were given twice daily for five days in addition to the normal diet, which was free of fermented dairy products. The mean (SD) number of days of diarrhea after commencing therapy, was significantly shorter in group 1 [1.4 (0.8) days], and in group 2 [1.4 (0.8) days], than in group 3 [2.4 (1.1) days]; $F = 8.70$, $p < .001$. The results demonstrated a significant reduction in the duration of rotaviral diarrhea by administration of human *Lactobacillus* GG. The authors suggested that the *Lactobacillus* GG fermented milk product is beneficial for the recovery from rotavirus diarrhea and that the product can be safely administered during an episode of acute diarrhea in patients previously tolerant to cow's milk proteins (Isolauri et al., 1991).

Later, the study was confirmed by Kaila and co-workers (1992). They also showed that the reduction in duration of diarrhea was associated with potentiation of the intestinal immune response and an increase in rotavirus-specific antibodies. This result was confirmed in a report on mucosal barrier function in patients with rotavirus enteritis and rat pups with experimental rotavirus infection (Isolauri, 1992). It was concluded that intestinal implantation of lactic acid bacteria may reverse mucosal dysfunction caused by enteral infection.

Kaila and co-workers (1991c) found that *Lactobacillus* GG survived the passage through the gut during rotavirus diarrhea in infants. All patients who received *Lactobacillus* GG became colonised with the strain as measured by fecal *Lactobacillus* GG counts. They suggested that *Lactobacillus* GG promotes the establishment of colonization resistance even during acute gastroenteritis (Kaila et al., 1991c). It was also suggested that *Lactobacillus* GG may prove to be a beneficial supplement in the treatment and prevention of clinical disorders associated with increased intestinal permeability. In healthy newborns similar colonization by *Lactobacillus* GG was observed and it was concluded that the composition of the newborn's intestinal microflora can be influenced by early administration of *Lactobacillus* GG resulting in colonization of most newborns (Mikelsaar et al., 1992).

An earlier report using conventional yogurt (Niv et al., 1963) had similar findings. Forty-five infants hospitalized for diarrhea were given either yogurt as treatment or a neomycin/kaolin/pectin mixture. The duration of disease was significantly less in the yogurt group compared with the other treatment group. In another early study, 39 children with diarrhea were treated with either *E. faecium* or an inert control preparation. The group treated with the probiotic preparation appeared to recover more quickly than the control group, but the composition of the placebo was not stated (D'Apuzzo and Salzberg, 1962).

In a double-blind controlled study (Bellomo et al., 1982), children with diarrhea were given either *E. faecium* (3.75×10^7 cfu) or a mixture containing 5×10^8 *L. acidophilus*, 5×10^8 *L. bulgaricus*, and 4×10^9 *Lactococcus lactis* in a lyophilized form. Treatment was given throughout the illness and was continued for a few days after clinical recovery (total duration 3–10 days). Duration of diarrhea was less with the *E. faecium* preparation compared with the reference mixture.

In two studies conducted in Algeria (Boudraa et al., 1990; Touhami et al., 1989), milk, yogurt, and diluted milk were compared for the treatment of persistent diarrhea in children. Feeding yogurt rather than milk or diluted milk was of significant benefit to these children (Touhami et al., 1989). The clinical failure rate (defined as more than 5% loss of body weight per day or the persistence of diarrhea after five days) was 14% in the yogurt group compared with 42% in the milk group (Boudraa et al., 1990). The beneficial effect of yogurt in these patients was attributed to intraluminal activity of bacterial β-galactosidase (Boudraa et al., 1990).

F. Prevention of Travelers' Diarrhea

Diarrhea is usually the most frequent health problem of travelers to developing countries. It was estimated that at least 6 million tourists to developing countries in 1985 would develop travelers' diarrhea (Anonymous, 1985). Travelers' diarrhea is a syndrome characterized by a twofold or greater increase in the frequency of unformed bowel movements. Associated symptoms include abdominal cramps, nausea, urgency, and fever.

Some forms of travelers' diarrhea appear to be preventable when viable lactic acid bacteria are administered during the period of risk. Oksanen and co-workers (Oksanen et al., 1990) reported a placebo-controlled double-blind study on the efficacy of *Lactobacillus* GG in preventing travelers' diarrhea. Eight hundred and twenty travelers from Finland to two different destinations in southern Turkey were randomized to receive either freeze-dried *Lactobacillus* GG or placebo (ethyl cellulose). Symptoms of gastrointestinal disease were carefully monitored by physicians who accompanied the travelers. On the return flight, participants were asked to complete a questionnaire on gastrointestinal symptoms experienced during travel. The overall incidence of diarrhea, defined as more than three unformed stools in 24 h or at least one watery stool during the trip, was 43.8%. *Lactobacillus* GG appeared to be effective in reducing travelers' diarrhea at one of the two destinations with a protection rate of 39.5% (Table 3).

In another double-blind study (Black et al., 1989) (Table 3), a lyophilized mixture (90% *L. acidophilus* and *Bifidobacterium bifidum*; 10% *L. bulgaricus* and *S. thermophilus*) was given prophylactically to tourists on a two-week trip to Egypt. A dose of 3×10^9 cfu was commenced two days prior to travel and continued throughout the trip. Travelers' diarrhea was defined as more than two watery stools in 24 h or one watery stool with at least one of the following symptoms; fever, nausea, vomiting, abdominal pain. The probiotic preparation had no effect on duration of diarrhea but significantly reduced the frequency from 71% to 43%. Enterotoxigenic *Escherichia coli* strains were isolated from 5 of 16 (31%) and 10 of 20 (50%) treatment and placebo groups respectively, but stool specimens were collected only from patients with symptoms. No treatment-related side effects were observed.

Clements and co-workers (1981) also tested the efficacy of a *Lactobacillus* preparation in preventing travelers' diarrhea caused by a challenge with

Table 3 The Efficacy of Prophylactic Lactobacillus and Enterococcus Preparations in Preventing Traveler's Diarrhea (TD)

Preparation	Number of cases studied	TD present (number of cases)	TD absent (number of cases)	Protection rate[a]	Year and type of publication
L. acidophilus[b]					1990
2×10^9	154	82	72	n.s.	Conference
Placebo	165	78	87	n.s.	proceedings
E. faecium[b]					1990
$1{,}5 \times 10^7$	672	234	438	n.s.	Conference
Placebo	652	248	404	n.s.	proceedings
E. faecium[b]					1990
1×10^9	401	188	213	n.s.	Conference
Placebo	419	210	209	n.s.	proceedings
Lactobacillus GG[c]					1990
10^9 cfu/day	153	68	85	8.0	Peer review
Placebo	178	74	104	n.s.	journal
Lactobacillus GG[d]					1990
10^9 cfu/day	71	17	54	39.5	Peer review
Placebo	75	30	45	(p<.05)	journal
Lactic acid bacteria					1989
mix[e,f]				n.s.	Peer review
10^9 cfu/day	40	17	23		journal
Placebo	41	29	12		
L. acidophilus					1978
L. bulgaricus mix[g]					Peer review
10^9 cfu/day	17	7	10	n.s.	journal
Placebo	14	2	12	n.s.	

[a]Protection rate = 100 × (% diarrhea in placebo group – % diarrhea in treatment group)/% diarrhea in placebo group; n.s. = not significant.
[b]Austrian tourists (Kollaritsch and Widermann, 1990).
[c]Turkey/1 week (Oksanen et al., 1990).
[d]Turkey/1 week (Oksanen et al., 1990).
[e]Egypt (Black et al., 1989).
[f]Mixture of *L. acidophilus, B. bifidum, S. thermophilus,* and *L. bulgaricus.*
[g]Mexico (Pozo-Olano et al., 1978).
Source: Adapted from Salminen and Deighton (1991) and other references.

enterotoxigenic *E. coli.* No significant differences were observed between *Lactobacillus* treated and placebo-treated groups.

In contrast to these findings, a series of randomized double-blind studies (Kollaritsch and Wiedermann, 1990) of travelers' diarrhea among Austrian tourists showed that neither of two probiotic preparations containing

L. acidophilus and *E. faecium* strain SF68 respectively offered significant protection against travelers' diarrhea (Table 3).

In a recent study the therapeutic efficacy of *Streptococcus faecium* SF68 was also tested in acute watery diarrhea caused by enterotoxigenic *E. coli* and *Vibrio cholerae*. The study included 41 *E. coli* patients and 114 *Vibrio cholerae* patients who received either placebo or *Streptococcus faecium* preparation (1×10^9 cfu three times a day). No demonstrable antidiarrheal property was observed in *Streptococcus faecium* treated patients in either group. The authors did not support the idea of using *Streptococcus faecium* as an agent in acute cholera or in acute *E. coli* associated diarrhea.

G. Activity against *Helicobacter*

In a study by Bhatia and co-workers (1989), in vitro inhibition of *Helicobacter pylori* by *L. acidophilus* was reported. *H. pylori* has been implicated as a cause of antral gastritis. Because lactobacilli are acid tolerant and able to persist in the stomach longer than other bacteria, it was suggested that *L. acidophilus* preparations may be useful for the treatment of gastritis.

H. Treatment of Irritable Bowel Syndrome

E. faecium preparations have been evaluated for treatment of patients with irritable bowel syndrome whose symptoms had been present for an average of seven years. Patient-recorded symptoms did not differ significantly in the placebo and *E. faecium* groups, but it was claimed that the physician's subjective clinical evaluation of symptoms revealed a significant improvement in the treated group (Gade and Thorn, 1989). More carefully designed and controlled studies are needed to confirm the effects of lactic acid bacteria in patients with irritable bowel syndrome.

A preliminary report from Cambridge indicates that probiotic therapy with *Lactobacillus casei* strain GG has some promise in the treatment of irritable bowel syndrome and Crohn's disease and may prove to be of help in controlling these disease states (Coley et al., 1992).

I. Antibiotic-Associated Diarrhea

Diarrhea and other gastrointestinal disturbances are frequent complications of treatment with antimicrobial agents. About one third of cases of antibiotic-associated diarrhea and virtually all cases of pseudomembraneous colitis are related to overgrowth of toxin-producing strains of *C. difficile* in the large intestine. Probiotic preparations containing lactic acid bacteria have been proposed as prophylaxis for antibiotic-associated diarrhea (Gotz et al., 1979; Lidbeck et al., 1988).

In an early study, conducted with 200 patients with pulmonary tuberculosis treated with antitubercular drugs, an *E. faecium* preparation appeared to reduce the incidence of diarrhea and to have some normalizing effect on biochemical parameters (Borgia et al., 1982). The authors defined diarrhea as more than two watery bowel movements per day for at least two days, but gave no clear indication on how the episodes of diarrhea and other parameters were recorded.

In 1983, Clements and co-workers reported that one of two batches of a lyophilized *Lactobacillus* preparation significantly reduced the volume and number of diarrheal stools in patients receiving neomycin. In contrast, a second batch had no effect of diarrhea (Clements et al., 1983). This indicates that there may be batch to batch variation even in commercial lactic acid preparations.

In a recent study, the incidence of antibiotic-associated diarrhea was less in healthy human volunteers receiving erythromycin if they took *Lactobacillus* GG yogurt, compared with a control group taking pasteurized yogurt (Siitonen et al., 1990). Other side effects of erythromycin, such as abdominal distress, stomach cramps, and flatulence were also less common in the *Lactobacillus* group than in the group taking pasteurized yogurt. Moreover, fecal counts of *Lactobacillus* GG showed that colonization the bowel occurred in spite of erythromycin treatment.

In another recent double-blind controlled study, Lactinex (*L. acidophilus* and *L. bulgaricus*) did not appear to consistently prevent amoxicillin-induced diarrhea in children. Patients were given the probiotic (5×10^8 cfu/day) or placebo for five days. Standard doses of amoxycillin were given during the same period. It was suggested that the patient's age (mean 27 ± 17 months), diet, and parental definition of diarrhea may have been factors influencing the outcome (Tankanow et al., 1990).

J. Lactic Acid Bacteria and the Bioavailability of Oral Antibiotics

It is also important to observe the effects of fermented milks and lactic acid bacteria on the absorption and bioavailability of orally administered antibiotics. In two studies it has been examined whether *Lactobacillus* preparations interfere with the absorption of antibiotics, as food does with some antimicrobial agents. Yost and Gotz (1985) studied the influence of Lactinex on the bioavailability of ampicillin in 12 healthy volunteers. Ampicillin was either given alone or with *Lactobacillus* preparation. Blood was sampled over 6 h. No differences in maximum concentration of ampicillin in plasma or any other parameters were observed in any group and it was concluded that the *Lactobacillus* preparation did not interfere with the bioavailability of oral ampicillin.

In another study, the effects of a *Lactobacillus* GG fermented milk were observed on erythromycin acistrate absorption. The erythromycin preparation was given orally to healthy male volunteers either with pasteurized yogurt or with *Lactobacillus* GG yogurt (10^8 cfu/mL). No differences in the peak or mean plasma erythromycin concentrations were observed between the groups. In general, the concentrations were similar to those observed in other erythromycin studies. Therefore, it was concluded that the *Lactobacillus* GG fermented milk or pasteurized normal yogurt did not influence erythromycin absorption in healthy human volunteers (Siitonen et al., 1991).

K. Treatment of Recurrent *Clostridium difficile* Diarrhea

Relapses are common after oral vancomycin treatment of *C. difficile* colitis. Treatment of relapsing *C. difficile* colitis with a freeze-dried concentrate of *L. acidophilus* (8×10^7 cfu, three times daily for 10 days) had no clinical effect and did not result in a return to normal fecal flora (Aronson et al., 1987). This lack of efficacy could be strain-dependent, since the *L. acidophilus* used was not an adherent strain.

Lactobacillus GG, on the other hand, may be efficacious in terminating relapsing colitis due to *C. difficile*. This strain was given to 11 patients as treatment for relapsing *C. difficile* colitis, which followed antibiotic treatment (Gorbach et al., 1987). A daily dose of 10^{10} viable lactobacilli in skim milk was given for 7–10 days. The results are summarized in Table 4. Eight of the patients had no resolution of diarrhea, no further relapses, and negative, or very low, toxin titers in the stool. Three patients showed initial improvement, but relapsed. After a second course of treatment, however, no further relapses occurred.

In an uncontrolled study, Bennett and co-workers (Bennett et al., 1990) administered high doses of *Lactobacillus* GG (10^9 cfu twice daily for 7 to 14 days) to nine symptomatic patients with recurrent *C. difficile* diarrhea. All nine patients improved and no relapses occurred for 60 days of follow-up. Four patients, however, had late recurrences beginning from 60 to 180 days after commencement of treatment. At a lower dose, *Lactobacillus* GG was not effective in eradicating *Clostridium difficile* carriage.

In an in vitro study, Kochar and co-workers (1990) showed that *Lactobacillus* GG had antagonistic activity against most strains of *C. difficile* tested. Further work is required to verify these results, however, the previous human studies appear to support these findings.

Table 4 Details of Outcome of Patients with Relapsing *C. difficile* Colitis Treated with *Lactobacillus* GG

			Toxin titers	
Patient	No. of relapses before treatment	Duration of illness (months before treatment)	Before treatment	After treatment
1	2	2	1:1250	0
2	3	4	1:1250	0
3	3	3	1:1250	0
4	4	6	1:1250	0
5	5	10	Pos	0
6	4	4	Pos	0
7	5	4	1:50	0
8	4	5	Pos	0
9	3	2	Pos	0
10	3	2	1:50	0
11	5	4	Pos	0

Lactobacillus GG was given as a concentrate in 5 mL of skim milk at a daily dose of 10^{10} viable bacteria for 7–10 days.
Source: Adapted from Goldin et al. (1987).

IV. INFLUENCE OF YOGURT-CONTAINING PREPARATIONS ON THE SIDE EFFECTS OF RADIOTHERAPY

Side effects of pelvic radiotherapy are usually treated by medication or by interrupting the therapy. No effective means have been found to prevent acute side effects although the use of bacterial preparations or intensive nutrition may decrease symptoms (Haller and Kräubig, 1960; Neumeister and Schmidt, 1963; Mettler et al., 1973; Donaldsson, 1983). Since dysbiosis of the intestinal flora is a promoting factor in radiotherapy-related colitis (Friberg, 1980; Bounous, 1983), treatment with lactobacilli, in an attempt to restore a more balanced microflora, might be beneficial.

A pilot study was conducted to assess the influence of a test yogurt containing *L. acidophilus* (NDCO 1478) on the side effects of radiotherapy (Salminen et al., 1988). Twenty-four female patients suffering from gynecological malignancies and scheduled for internal and external irradiation of the pelvic area (pelvic dose 5000 cGy) were randomized into two groups. Both groups received dietary counselling recommending a low-fat and low-residue diet during radiotherapy. In addition to dietary counselling the test group received 2×10^9 viable

L. acidophilus daily, in a product prepared from pasteurized fat-free milk. The product also contained lactulose to support the growth of *Lactobacillus* in the large intestine and lactase to hydrolyse lactose. Treatment commenced five days prior to radiotherapy, continued throughout the radiotherapy period including the interval, and for 10 days thereafter. In the treatment group, gastrointestinal side effects were less frequent and less severe than in the control group. The test product appeared to prevent radiotherapy-associated diarrhea, but flatulence was increased, probably due to the lactulose (Salminen et al., 1988).

In a study on mice, Dong and co-workers (Dong et al., 1988) observed that feeding *Lactobacillus* GG prior to lethal irradiation reduced early mortality of treated mice due to lower incidence of *Pseudomonas* bacteremia.

A. Hepatic Encephalopathy

In healthy individuals, the ammonia which is produced in the intestine by the action bacterial urease is absorbed and detoxified in the liver. Detoxification is impaired in patients with liver failure, resulting in elevated blood levels of ammonia leading to encephalopathy. Alteration of the bacterial flora, using lactic acid bacteria to displace organisms with strong urease activity, might be a useful prophylactic measure in these patients.

Two studies by Macbeth et al. (1965) and Read et al. (1966) demonstrated a decrease in fecal urease, a lowering of blood ammonia, and associated clinical improvement in patients with hepatic encephalopathy treated with a *L. acidophilus* preparation.

In a more recent study (Loguercio et al., 1987), patients with nonadvanced hepatic encephalopathy were treated for 10 days with an oral preparation of *E. faecium* SF68 or a control preparation which is widely used for the treatment of hepatic encephalopathy (lactulose). The *Enterococcus* preparation was as effective as lactulose in lowering blood ammonia, and in improving mental state and psychometric performance. Moreover, the effects of *Enterococcus* persisted longer after completion of treatment compared to those of lactulose. Some patients reported diarrhea and abdominal pain with lactulose, but no major adverse side effects were reported for the bacterial preparation.

B. Lactic Acid Bacteria and Carcinogens

Several experimental studies have shown that preparations containing lactic acid bacteria inhibit the growth of tumor cells in experimental animals.

Studies on male Swiss mice implanted with Ehrlich ascites tumor cells, demonstrated that feeding yogurt (prepared using *L. bulgaricus* and *S. thermophilus*) for seven days postimplantation inhibited tumor proliferation. Measurement of the total cell count and DNA content demonstrated that the number of tumor cells was reduced by 23–35% in treated animals compared with controls (Reddy et al., 1983; Friend et al., 1982). In an earlier study, the antitumor activity of *L. bulgaricus*, was shown to be mediated by three glycopeptides with in vitro activity against sarcoma-180 and solid Ehrlich ascites tumor cells (Bogdakov et al., 1978).

Support for the antitumor effect of lactic acid bacteria, in humans as well as in animals comes from a series of studies conducted by Goldin and Gorbach (Goldin and Gorbach, 1984, 1977; Goldin et al., 1980). Oral supplementation of the diet of humans or rats with viable *L. acidophilus* of human origin caused a significant decline in the fecal levels of bacterial β-glucuronidase, azoreductase, and nitroreductase. These enzymes are believed to contribute to the pathogenesis of bowel cancer by converting procarcinogens to proximate carcinogens. In rats fed with carcinogen precursors, fecal levels of carcinogenic amines were also reduced if diets were supplemented with *L. acidophilus* (Goldin et al., 1980). In a more recent study Ling (1992) observed that *Lactobacillus* GG decreased fecal β-glucuronidase, azoreductase, glycocholic acid hydrolase, and urease in elderly nursing home patients, in young healthy female volunteers, and in children (Ling, 1992). Similarly, Goldin and co-workers observed that *Lactobacillus* acidophilus or yogurt reduced fecal bacterial enzymes in the elderly (Pedrosa et al., 1990). These studies suggest that lactobacilli have the capacity to suppress the metabolic activity of the colonic microflora and may reduce the formation of carcinogens in the intestine.

Further studies by the same group (Goldin and Gorbach, 1980) were performed using an animal model in which colon cancer was induced by 1,2-dimethylhydrazine (DMH). This chemical carcinogen is activated in the large intestine by the bacterial enzyme β-glucuronidase. It was postulated that suppression of enzyme activity might reduce DMH activation and suppress subsequent tumor formation. Animals were given DMH to induce tumor formation and treated with *L. acidophilus* in powdered form or left untreated. At 20 weeks, 40% of the *L. acidophilus*–treated animals but 77% of controls had colonic tumors ($p < .02$). After 36 weeks, however, there was no significant difference in the incidence of tumors in treatment and control groups (73% versus 83%). The addition of at least one strain of *Lactobacillus* to the diet may delay tumor formation in rats.

In a study by Marteau and co-workers (Marteau et al., 1990), human volunteers were fed a fermented dairy product containing, human strains of *L. acidophilus* and *B. bifidum*, for three weeks. Although a significant reduction in the fecal nitroreductase activity occurred, the study failed to detect changes in the levels of β-glucuronidase or azoreductase. Differences between the metabolic modifications of the colonic flora in this and other studies could be explained by differences in the host-specificity of the strains of lactobacilli used.

In a recent study by Goldin and co-workers a dimethylhydrazine-induced intestinal tumor model was used on rats. It was indicated that *Lactobacillus* GG inhibits the initiation phase of colonic tumor formation in rats fed a high fat diet (Goldin et al., 1992).

V. CONCLUSIONS

Lactic acid bacteria have documented benefits for the prevention and treatment of some forms of diarrhea and related conditions. Their main role appears to be preservation of the integrity of the normal intestinal flora and promoting colonization resistance. The precise mechanisms of action are unknown, but specific adherence to intestinal mucosal cells and production of antibacterial substances appear to be important. Since marked strain differences exist, it is important to define strains used in human and animal studies. Further controlled clinical studies are needed to verify the described effects of lactic acid bacteria and to elucidate mechanisms of activity.

REFERENCES

Alm, L. (1983). The effect of *Lactobacillus acidophilus* administration upon the survival of *Salmonella* in randomly selected human carriers, *Prog. Food. Nutr. Sci., 7*:13–17.

Anonymous. (1985). Travelers' diarrhea—consensus conference, *JAMA, 253*:2700–2704.

Aronsson, B., Barany, P., Nord, C., et al. (1987). *Clostridium difficile*-associated diarrhoea in uremic patients, *Eur. J. Clin. Microbiol., 6*:352–356.

Barefoot, S., and Klaenhammer, T. (1983). Detection and activity of lactacin B, a bacteriocin produced by *Lactobacillus acidophilus, Appl. Envir. Microbiol., 45*:1808–1811.

Bartlett, J. G., Chang, T. W., Gurwith, M., et al. (1978). Antibiotic-associated pseudomembraneous colitis due to toxin-producing clostridia, *N. Engl. J. Med. 293*:531–534.

Bellomo, G., Mangiagle, A., Nicastro, L., et al. (1982). A controlled double-blind study of SF68 strain as a new biological preparation for the treatment of diarrhoea in pediatrics, *Cur. Ther. Res., 28*:927–936.

Bennet, R. G., Laughon, B., Lindsay, J. et al. (1990). *Lacrobacillus* GG treatment of *Clostridium difficile* infection in nursing home patients, Abstract, 3rd Int. Conf. Nosocomial Infections, Atlanta.

Bhatia, S. J., Kochar, N., Abraham, P. et al. (1989). *Lactobacillus acidophilus* inhibits growth of *Campylobacter pylori* in vitro, *J. Clin. Microbiol., 27*:2328–2330.

Black, F. T., Andersen, P. L., Ørskov, F., et al. (1989). Prophylactic efficacy of lactobacilli on travelers' diarrhoea, *Travel Medicine, 8*:333–335.

Bogdanov, I. G., Velichkov, V. T., Gurvich, A. L. (1978). Antitumor action of glycopeptide from the wall of *Lactobacillus bulgaricus, Bull. Exptl. Biol. Med., 84*:1750–1753.

Borgia, M., Sepe, N., Brancato, V., and Borgia, R. (1982). A controlled clinical study on *Streptococcus faecium* preparation for the prevention of side reactions during long term antibiotic therapy, *Cur. Ther. Res., 31*:265–271.

Boudraa, G., Touhami, M., Pochart, P. et al. (1990). Effect of feeding yogurt versus milk in children with persistent diarrhoea, *J. Pediatr. Gastro. Nutr., 11*:509–512.

Bounous, G. (1983). The use of elemental diets during cancer therapy, *Anticancer Res., 3*:299–304.

Brand, J. C., and Holt, S. (1991). Relative effectiveness of milks with reduced amounts of lactose in alleviating milk intolerance, *Am. J. Clin. Nutr., 54*:148–151.

Choi, C. S., Chung, J. B., Chung, S. I., Yang, Y. T. (1984). Inhibition of pathogenic enterobacteria by *Lactobacillus casei* isolated from yakult, *Korean J. Veter. Public Health, 8*:49–58.

Christensen, M. L. (1989). Human viral gastroenteritis, *Clin. Microbiol. Rev., 2*:51–89.

Clements, M. L., Levine, M. M., Black, R. E., Robins-Browne, R. M., Cisneros, L. A., Drusano, G. L., Lanata, C. F., and Saah, A. J. (1981). *Lactobacillus* prophylaxis for diarrhea due to enterotoxigenic *Escherichia coli, Antimicrob. Agents Chemother., 20*:104–108.

Clements, M. L., Levine, M. M., Ristaino, P. A., et al. (1983). Exogenous lactobacilli fed to man—their fate and ability to prevent diarrhoeal disease, *Prog. Food. Nutr. Sci., 7*:29–37.

Coconnier, M-H, Klaenhammer, T. R., Kerneis, S., Bernet, M-F., and Servin, A. L. (1992). Protein mediated adhesion of *Lactobacillus acidophilus* BG2FO4 on human enterocyte and mucus secreting cell lines, *Appl. Environ. Micr., 58*: 2034–2039.

Coley, A. M., Lee, A. J., and Hunter, J. O. (1992). *Lactobacillus casei* GG in the treatment of colonic disorders, *Microbiol. Ecol. Health Dis., 5*:iv–v.

Collins, E. B., and Aramaki, K. (1980). Production of hydrogen peroxide by *Lactobacillus acidophilus, J. Dairy Sci., 63*:353–357.

Conway, P., Goldin, B. R., and Gorbach, S. L. (1987). Survival of lactic acid bacteria in the human stomach and adhesion to intestinal cells, *J. Dairy Sci., 70*:1–12.

Coudron, P. E., Mayhall, C. G., Facklam, R. R. et al. (1984). *Streptococcus faecium* outbreak in a neonatal intensive care unit, *J. Clin. Microbiol., 20*:1044–1048.

D'Apuzzo, V. and Salzberg, R. (1962). Die Behandlung der akuten Diarrhö in der Pädiatrie mit *Streptococcus faecium*: Resultate einer Doppelblindstudie, *Ther. Umschau, 39*: 1033–1039.

Dewit, O., Pochart, P., and Desjeux, J. F. (1988). Breath hydrogen concentration and plasma glucose, insulin and free fatty acid levels after lactose, milk, fresh or heated yogurt ingestion by healthy young adults with or without lactose malabsorption, *Nutrition, 4*:131–135.

Donaldsson, S. (1983). Nutritional support as an adjunct to radiation therapy, *J. Parent Enter. Nutr., 8*:302–310.

Dong, M. Y., Chang, T. W., and Gorbach, S. L. (1987). Effects of feeding *Lactobacillus* GG on lethal irradiation in mice, *Diagn. Microbiol. Infect. Dis., 7*:1–7.

Drasar, B. S., and Hill, M. J. (1974). *Human Intestinal Flora*, Academic Press, New York, 1974, pp. 36–43.

Elo, S., Saxelin, M., and Salminen, S. (1991). Attachment of *Lactobacillus casei* strain GG to human colon carcinoma cell line Caco-2: comparison with other dairy strains, *Lett. Appl. Microbiol., 13*:154–156.

Facklam, R. R., and Collins, M. D. (1989). Identification of *Enterococcus* species isolated from human infections by a conventional test scheme, *J. Clin. Microbiol., 27*:731–734.

Finegold Sutter, V. L., Sugihara, P. T., Elder, H. A., Lehmann, S. M., and Phillips, R. L. (1977). Fecal microflora in Seventh Day Adventist populations and control subjects, *Am. J. Clin. Nutr., 30*:1781–1792.

Formal, S. B., and Hornick, R. B. (1978). Invasive *Escherichia coli, J. Infect. Dis., 137*:641–644.

Friberg, L. (1980). Effects of irradiation on the small intestine of the rat: a SEM study, Ph.D. thesis, University of Lund, Sweden.

Friend, B. A., Farmer, R. E., and Shahani, K. M. (1982). Effect of feeding and intraperitoneal implantation of yogurt culture cells on Ehrlich ascites tumour, *Milchwiss, 37*:708–710.

Friis-Møller, A., Hey, H. (1983). Colonization of the intestinal canal with a *Streptococcus faecium* preparation (Paraghurt), *Curr. Ther. Res., 33*:807–815.

Fuller, R. (1989). Probiotics in man and animals, *J. Appl. Bacteriol., 66*:365–378.

Gade, J., and Thorn, P. (1989). Paraghurt for patients with irritable bowel syndrome. A controlled clinical investigation from general practice, *Scand. J. Prim. Health Care, 7*:23–26.

Gibbs, P. A. (1987). Novel uses of lactic acid fermentation in food preservation, *J. Appl. Bacteriol., 63*:615–625.

Gilliland, S. E., and Speck, M. L. (1977). Antagonistic action of *Lactobacillus acidophilus* toward intestinal and foodborne pathogens in associative cultures, *J. Food Prot., 40*:820–823.

Goldin, B., and Gorbach, S. L. (1977). Alterations in fecal microflora enzymes related to diet, age, *Lactobacillus* supplements and dimethylhydrazine, *Cancer, 40*:2421–2426.

Goldin, B. R., and Gorbach, S. L. (1980). Effect of *Lactobacillus acidophilus* dietary supplements on 1,2-dimethylhydrazine dihydrochloride induced intestinal cancer in rats, *JNCI, 84*:283–285.

Goldin, B. R., and Gorbach, S. L. (1984). The effect of milk and *Lactobacillus* feeding on human intestinal bacterial enzyme activity, *Am. J. Clin. Nutr., 39*:756–761.

Goldin, B. R., Gorbach, S. L., Saxelin, M. L., Barakat, S., Gualtieri, L., and Salminen, S. (1992). Survival of *Lactobacillus* species (strain GG) in human gastrointestinal tract, *Dig. Dis. Sci., 37*:121–128.

Goldin, B. R., Gualteri, L., Barakat, S., Moore, R., and Gorbach, S. L. (1992). The effect of *Lactobacillus* species (strain GG) on DMH induced rat intestinal tumors and on human immunological function, *XIV Annual Meeting Intestinal Microecology*, Helsinki, Abstracts p. 6.

Goldin, B. R., Swenson, L., Dwyer, J., Sexton, M., and Gorbach, S. L. (1980). Effect of diet and *Lactobacillus acidophilus* supplements on human fecal bacterial enzymes, *JNCI, 64*:255–260.

Gorbach, S. (1990). Lactic acid bacteria and human health, *Ann. Med., 22*:37–41.

Gorbach, S. L., Chang, T., and Goldin, B. (1987). Successful treatment of relapsing *Clostridium difficile* colitis with *Lactobacillus* GG, *Lancet, 2*:1519.

Gotz, V., Romankiewics, J. A., Mose, J., et al. (1979). Prophylaxis against ampicillin-associated diarrhoea with a lactobacillus preparation, *Am. J. Hosp. Pharm., 36*:754–757.

Hall, M. A., Cole, S. L., Smith, S. L., Fuller, R., and Rolles, C. J. (1990). Factors influencing the presence of faecal lactobacilli in early infancy, *Arc. Dis. Childh., 65*:185–188.

Haller, J., and Kraübig, H. (1960). Zur Beinflussung der Strahlenreaktionen des Darmes durch eine Kombination lebender Azidophilus-, Bidifus-, und Kolibakterien, *Strahlenterapie, 113*:272–280.

Hitchins, A. D., Wells, P., McDonough, F. E., et al. (1985). Amelioration of the adverse effects of gastrointestinal challenge with *Salmonella enteriditis* on weanling rats by a yogurt diet, *Am. J. Clin. Nutr., 41*:92–100.

Isolauri, E. (1992). Mucosal barrier and lactic acid bacteria. XIV Annual Meeting of the Society of Intestinal Microecology, Helsinki, Abstracts p. 3–4.

Isolauri, E., Juntunen, M., and Rautanen, T. (1991). A human *Lactobacillus* strain (*Lactobacillus* GG) promotes recovery from acute diarrhoea in children, *Pediatrics, 88*:90–97.

Kaila, M., Isolauri, E, Salminen, S., and Mikelsaar, M. (1991b). Successful intestinal colonization by human *Lactobacillus casei* strain GG during rotavirus diarrhoea. SOMED Meeting, Mountain Lake, VA.

Kaila, M., Isolauri, E., Virtanen, E., Laine, S., Soppi, E., and Arvilommi, H. (1992). Enhancement of circulating antibody response in human diarrhea by a human *Lactobacillus* strain, *Ped. Res., 32*:141–144.

Kleeberg, J. (1927). Die therapeutische Bedeutung von Yoghurt und Kefir in der inneren Medizin, *Deutsche. med. Wchschr., 53*:1093–1095.

Kochar, N., Mehta, A., Abraham, P., and Bhatt, R. (1989). Effect of lactobacilli on intestinal anaerobic flora and intestinal gas, *Microecol. Ther., 19*:119–120.

Kochar, N., Mehta, A., Abraham, P., and Bhatt, R. M. (1990). In vitro effect of *Lactobacillus* on *Clostridium* strains, *Micr. Ecol. Health Dis.*, 250.

Kolars, J. C., Levitt, M. D., Aouji, M., et al. (1984). Yogurt—an autodigesting source of lactose, *N. Engl. J. Med., 3110*:1–3.

Kollaritsch, H., and Wiedermann, G. (1990). Travelers' diarrhoea among Austrian tourists: epidemiology, clinical features and attempts at nonantibiotic drug prophylaxis, in *Proc. Second Int. Conf. Tourist Health* (W. Pasini, ed.), WHO, Rimini, pp. 74–82.

Kopeloff, N. (1924). Clinical results obtained with *Bacillus acidophilus, Arch. Int. Med., 33*:47–54.

Lewenstein, A., Frigerio, G., and Moroni, M. (1979). Biological properties of SF 68, a new approach for the treatment of diarrhoeal diseases, *Cur. Ther. Res., 26*:967–981.

Lidbeck, A. (1991). Studies on the impact of *Lactobacillus acidophilus* on human microflora and some cancer-related intestinal ecological variables. Ph.D. thesis, Karolinska Institute, Stockholm, Sweden.

Lidbeck, A., Edlund, C., Gustafsson, J.-Å. et al. (1988). Impact of *Lactobacillus acidophilus* on the normal intestinal microflora after administration of two antimicrobial agents, *Infection, 16*:329–336.

Lidbeck, A., Geltner Allinger, U., Orrhage, K., Ottova, L., Brismar, B., Gustafsson, J-Å., and Nord, C. E. (1991). Impact of *Lactobacillus acidophilus* supplements on the intestinal microflora and soluble faecal bile acids in colon cancer patients, *Microb. Ecol. Health Dis., 4*:81–88.

Lidbeck, A., Gustafsson, J-Å., and Nord, C. (1987). Impact of *Lactobacillus acidophilus* supplements on the human oropharyngeal and intestinal microflora, *Scand. J. Infec. Dis., 19*:531–537.

Lidbeck, A., Øvervik, E., Rafter, J., Nord, C. E., and Gustafsson, J-Å. (1991). Effect of *Lactobacillus acidophilus* supplements on mutagen excretion in faece and urine in humans, *Microb. Ecol. Health Dis.*, in press.

Ling, W-H. (1992). Effect of lactobacilli containing vegan diet and Lactobacillus GG on colonic chemical loading in man, Ph.D. thesis, Kuopio University Medical Publications, University of Kuopio, Finland.

Loguercio, C., Del Vecchio, B., and Coltori, M. (1987). *Enterococcus* lactic acid bacteria strain SF68 and lactulose in hepatic encephalopathy: a controlled study, *J. Internat. Med. Res., 15*:335–343.

London, J., and Chase, N. M. (1976). Aldolases of the lactic acid bacteria. Demonstration of immunological relationships among eight genera of gram positive bacteria using an anti-pediococcal aldolase serum, *Arch. Microbiol., 110*:121–128.

Macbeth, W. A., Kass, E. H., and McDermott, W. V. (1965) Treatment of hepatic encephalopathy by alteration of intestinal flora with *Lactobacillus acidophilus, Lancet, 1*: 399–403.

Majamaa, H., Arvola, T., and Isolauri, E. (1992). *Lactobacillus* GG normalizes cow's milk induced enhancement of intestinal permeability in suckling rats, *XIV Annual Meeting Soc. Intestinal Microecology*, Helsinki, Abstracts p. 8.

Marteau, P., Pochart, P., Flourie, B., et al. (1990). Effect of chronic ingestion of a fermented dairy product containing *Lactobacillus acidophilus* and *Bifidobacterium bifidum* on metabolic activities of the colonic flora in humans, *Am. J. Clin. Nutr., 52*:685–688.

Martini, M. C., Lerebours, E., Lin, W-J., Harlander, S. K., Berrada, N. M., Antoine, J. M., and Savaiano, D. A. (1991). Strains and species of lactic acid bacteria in fermented milks (yogurts): effect on in vivo lactose digestion, *Am. J. Clin. Nutr., 54*:1041–1046.

McCormick, E. L., and Savage, D. C. (1983). Characterization of *Lactobacillus sp.* strain 100-37 from the murine gastrointestinal tract: ecology, plasmid content and antagonistic activity toward *Clostridium* ramosum H7, *Appl. Envir. Microbiol., 46*:1103–1112.

McDonough, E., Hitchins, A. D., Wong, N. P., et al. (1987). Modification of sweet acidophilus milk to improve utilization by lactose-intolerant persons, *Am. J. Clin. Nutr., 45*:570–574.

Metchnikoff, I. (1901). *Mem. Proc. Manchester Lit. Phil. Soc., 45*:1–38.

Mettler, L., Romeyke, A., and Brieler, G. (1973). Zur Beeinflussung der paraund post-radiologischen Dysbakterie und Strahlenreaktion des Darmes durch Bakterium-Bifidum-Substitutionstherapie, *Strahlenterapie, 145*:588–591.

Mikelsaar, M., Sepp, E., Kaila, M., Isolauri, E., and Salminen, S. (1992). *Lactobacillus casei* strain GG colonizes healthy newborns, *XIV Annual Meeting Soc. Intestinal Microecology*, Helsinki, Abstracts p. 9.

Mitra, A. K., and Rabbani, G. M. (1990). A double-blind controlled trial of Bioflorin (*Streptococcus faecium* SF68) in adults with acute diarrhea due to *Vibrio cholerae* and enterotoxigenic *Escherichia coli, Gastroenterology, 99*:1148–1152.

Neumeister von, K., and Schmidt, W. (1963). Die behandlung der intestinalen Strahlenreaktionen, *Med. Klin, 58*:842–844.

Nishizato, Y., Shimizu, K., and Miyaoka, M. (1985). Clinical trials with the *Lactobacillus casei* preparation Biolactis on patients with irregular bowel movements, *Jap. Pharmacol. Ther., 13*:2423–2428.

Niv, M., Levy, W., and Greenstein, N. M. (1963). Yogurt in the treatment of infantile diarrhea, *Clin. Ped., 2*:407–411.

Oksanen, P., Salminen, S., Saxelin, M., et al. (1990). Prevention of travelers' diarrhoea by *Lactobacillus* GG, *Ann. Med., 22*:53–56.

Paul, D., and Hoskins, L. C. (1972). Effects of oral lactobacillus counts, *Am. J. Clin. Nutr., 25*:763–765.

Pedrosa, M. C., Golner, B., Goldin, B., Barakat, S., Dallal, G., and Russel, R. M. (1990). Effect of *Lactobacillus acidophilus* or yogurt feeding on bacterial fecal enzymes in the elderly, *Gastroenterology, A425*.

Petersson, L., Graf, W., and Sevelin, U. (1983). Survival of Lactobacillus acidophilus in the human gastrointestinal tract, in *XV Symp. Swedish Nutrition Foundation*, (Hallgren, ed.), Almquist. Wiksell, Uppsala, Sweden, pp. 127–130.

Pochart, P., Dewit, O., Desjeux, J., and Bourlioux, P. (1989). Viable starter culture, β-galactosidase activity and lactose in duodenum after yogurt ingestion in lactase-deficient humans, *Am. J. Clin. Nutr., 48*:828–831.

Rajala, S. A., Salminen, S. J., Seppänen, J. H., and Vapaatalo, H. (1988). Treatment of chronic constipation with lactitol sweetened yoghurt supplemented with guar gum and wheat bran in elderly hospital in-patients, *Comp. Gerontol., 2*:83–86.

Read, A. E., McCarthy, C. F., Heaton, K. W., et al. (1966). *Lactobacillus acidophilus* (Enpac) in treatment of hepatic encephalopathy, *Br. Med. J., 1*:1267–1269.

Reddy, G. V., Friend, B. A., and Shahani, K. M. (1983). Antitumor activity of yoghurt components, *J. Food. Prot., 46*:8–11.

Reuman, P. D., Duckworth, D. H., Smith, K. L., Kagan, R., Bucciarelli, R. L., and Ayoub, E. M. (1986). Lack of effect of *Lactobacillus* on gastroenteritis bacterial colonization in premature infants, *Pediatr. Infect. Dis., 5*:663–668.

Robins-Browne, R. M., and Levine, M. (1981). The fate of ingested lactobacilli in the proximal small intestine, *Am. J. Clin. Nutr., 34*:514–519.

Roswitt, B., Malsky, S., and Reid, C. (1972). Severe radiation injury of the stomach small intestine, colon and rectum, *Am. J. Roentgenol., 114*:460–475.

Salminen, E., Elomaa, I., Minkkinen, J., Vapaatalo, H. et al. (1988). Preservation of intestinal integrity using live *Lactobacillus acidophilus* cultures, *Clin. Rad., 39*: 435–437.

Savaiano, D. A., Abouelanaour, A., Smith, D. E. et al. (1984). Lactose malabsorption from yogurt, pasteurized yogurt, sweet acidophilus milk, and cultured milk in lactase-deficient individuals, *Am. J. Clin. Nutr., 40*:1219–1223.

Saxelin, M., Elo, S., and Salminen, S. (1991). Dose response colonization of faeces after oral administration of *Lactobacillus casei* strain GG, *Microb. Ecol. Health Dis., 4*:209–214.

Siigur, U., Tamm, A., and Tammur, R. (1991). The faecal SCFAs and lactose tolerance in lactose malabsorbers, *Eur. J. Gastr. Hepatol., 3*:321–324.

Siitonen, S., Vapaatalo, H., Salminen, S. et al. (1990). Effect of *Lactobacillus* GG yogurt in prevention of antibiotic associated diarrhoea, *Ann. Med., 22*:57–59.

Silva, M., Jacobus, N. V., Deneke, C. et al. (1987). Antimicrobial substance from a human *Lactobacillus* strain, *Antimicrob. Agents Chemother., 31*:1231–1233.
Simon, S. L., and Gorbach, S. L. (1986). The human intestinal microflora, *Dig. Dis. Sci., 31*:147S–162S.
Smith, H. W. (1969). Transfer of antibiotic resistance from animal and human strains of *E. coli* in the alimentary tract of man, *Lancet, 1*:1174–1176.
Stackebrandt, E., Fowler, V. J., and Woese, C. R. (1983). A phylogenetic analysis of Lactobacilli, *Pediococcus pentosaceus* and *Leuconostoc mesenteroides, Syst. Appl. Microbiol., 4*:326–337.
Tankanow, R. M., Ross, M. B., Ertel, J., Dickinson, D. G., McGormick, L. S., and Garfinkel, J. (1990). A double-blind, placebo-controlled study of the efficacy of Lactinex in the prophylaxis of amoxicillin-induced diarrhea, *DICP, Ann. Pharmacother., 24*:382–384.
Touhami, M., Boudraa, G., Adllaoui, M. et al. (1989). La dilution du lait est-elle indispensable dans les diarrhées aigues benignes du nourrisson eutrophique, *Arch. Fr. Pediatr., 46*:25–30.
Yost, R. L., and Gotz, V. (1985). Effects of a *Lactobacillus* preparation on the absorption of oral ampicillin, *Antimicr. Agents Chemother., 28*:727–729.

8

Lactic Acid Bacteria and Intestinal Drug and Cholesterol Metabolism

Alice H. Lichtenstein
USDA Human Nutrition Research Center on Aging at Tufts University, Boston, Massachusetts

Barry R. Goldin
Tufts University School of Medicine, Boston, Massachusetts

I. INTRODUCTION

The intestinal microflora of humans is a complex ecosystem of metabolically active microorganisms in close proximity to an absorptive mucosal surface. Substrates for bacterial transformation can reach the intestinal flora through direct oral ingestion, by biliary secretion into the upper bowel, or by secretion across the mucosa. This chapter reviews the metabolic activity of the microorganisms that reside in the intestine and reviews the current knowledge regarding the role of *Lactobacillus* in this environment.

II. THE BACTERIAL COMPOSITION OF THE GASTROINTESTINAL TRACT

The gastrointestinal tract is inhabited by a diverse bacterial population that constitute a complex ecosystem. More than 400 different bacterial species have been isolated and identified in feces (Finegold et al., 1974; Moore and Holderman, 1974). Strict anaerobic bacteria are the most common organisms in the intestinal tract outnumbering facultative bacteria by a factor of 10^2 to 10^4. In Table 1 the most prevalent microorganisms found at various locations in the human gastrointestinal tract are shown. In healthy individuals the stomach and upper small intestine have relatively low numbers of microorganisms.

Table 1 Distribution and Composition of the Intestinal Flora

Site	Composition[a]	Total number of organisms per mL contents
Stomach	*Streptococcus*	10^1–10^2
	Lactobacillus	
Duodenum and jejunum	Similar to stomach	10^2–10^4
Ileal-cecal	*Bacteroides*	10^6–10^8
	Clostridium	
	Streptococci	
	Lactobacilli	
Colon	*Bacteroides* (10^{10}–10^{11})	$10^{11.5}$–10^{12}
	Clostridium (10^{10})	
	Eubacterium (10^{10})	
	Peptococcus (10^{10})	
	Bifidobacterium (10^9–10^{10})	
	Streptococcus (10^{10})	
	Fusobacterium (10^9–10^{10})	

[a]Organisms listed represent only the major species isolated from the different sites.

The lower small intestine is a transition zone between the sparely populated upper gastrointestinal tract and the heavily bacterially populated colon. In the lower ileum the number of bacteria increases to between 10^6 and 10^7 organisms per milliliter of contents.

In the colon the bacterial concentration increases to between 10^{11} and 10^{12} organism per milliliter of fecal material (Table 1). To illustrate the density of bacteria in the colon, one third of the fecal dry weight consists of viable bacteria.

III. BACTERIAL COLONIZATION OF THE GASTROINTESTINAL TRACT

Colonization of the gastrointestinal tract in humans occurs within a few days after birth (Haenel, 1970). The course of colonization is influenced by gestational age, type of delivery, and dietary constituents. The initial phase of colonization occurs over approximately a two-week period. During this period the bacterial colonization is similar for breast- and formula-fed infants. Almost always *E. coli* and *Streptococcus* are the first organisms detected in the feces at concentrations between 10^8 and 10^{10} organisms per gram of feces (Gruette et al., 1965). This is often followed by the appearance of anaerobic organisms namely *Clostridium, Bifidobacterium*, and *Bacteriodes*. In breast-fed infants there follows a period in which there is a significant reduction in the populations of *E. coli* and *Streptococcus* and a partial or complete disappearance of *Clostridium* and *Bacteroides.* This decrease in bacterial populations results in the predominance of *Bifidobacterium* in the intestine of breast-fed infants. In formula-fed infants the bacterial reductions and disappearances do not occur, resulting in a more complex intestinal microflora (Haenel and Bendig, 1975; Ellis-Pegler et al., 1975; Rotimi and Druerden, 1981; Stark and Lee, 1982). The relatively simple flora of the breast-fed baby continues until other foods are included in the diet. After the introduction of other foods there is a return of *E. coli, Clostridium,* and *Streptococcus* to the intestinal tract of the breast-fed infant as witnessed by isolation of these organisms from the feces. The intestinal flora of the breast-fed infant now resembles that of the formula-fed baby. There is then a period of transition which continues into the second year of life at which time the composition of the intestinal microflora evolve to resemble the bacterial composition found in the adult.

IV. LACTOBACILLI RESIDING IN THE HUMAN GASTROINTESTINAL TRACT

The indigenous intestinal microflora of most healthy individuals harbor representatives of *Lactobacillus* genera. Finegold et al. (1974) reported finding in feces, *L. acidophilus, L. fermentum, L. plantarum, L. minutus*, and *Lactobacillus* species (not identified). The concentration of viable organisms varied between 10^8 and 10^9. Holderman and Moore (1974) also analyzed fecal specimens and found similar results. They also detected *L. leichmannii* and *L. ragosae* in the fecal flora. It is interesting to note that *L. bulgaricus* the organism commonly used in the production of yogurt was not routinely isolated from the fecal cultures. It is clear that *Lactobacillus* is a component of the normal intestinal flora, however, only certain species are normally present in the intestinal tract.

V. INTESTINAL REACTIONS IN WHICH *LACTOBACILLUS* HAS BEEN IDENTIFIED AS A PARTICIPANT

Scheline (1972) tested a number of different substrates and intestinal microorganisms to determine the type of reactions specific *Lactobacillus* species catalyze. In the study cited above, these investigators found that a *Lactobacillus* species could reduce the double bond in hydroxycinnamic acid, reduce the nitro group of 4-nitrobenzoic acid and reduce the azo bonds found in methyl red and acid yellow. In more recent publication Pradham and Majumdar (1986) reported that *Lactobacillus acidophilus* cleaves the azo bond of sulfasalazine, also known as azulfidine, a drug used to treat patients with ulcerative colitis. These investigators also found that *Lactobacillus acidophilus* degraded 17.6% of the antimicrobial agent phthalylsulfathiazole and 8% of the antibiotic chloramphenicol palmitate. These investigators also confirmed the previous study and demonstrated that *Lactobacillus acidophilus* could rapidly hydrolyze the azo bond of tartrazine and methyl red. *Lactobacillus helveticus* and *Lactobacillus salivarius* also had similar but lower activity when compared to *Lactobacillus acidophilus*. The most rapid reaction performed by *Lactobacillus* was the reductive hydrolysis of the azo bond followed by hydrolysis of the amide bond and the least active of the reactions studied was the hydrolysis of the ester bond of chloramphenicol palmitate. Gilliland and Speck (1977) studied the ability of *Lactobacillus acidophilus* and *Lactobacillus casei* to hydrolyze conjugates of bile acids. They found all six strains of *Lactobacillus acidophilus* deconjugated taurocholate, but only one of six deconjugated glycocholate. None of 13 strains of *Lactobacillus casei* hydrolyzed glycocholate. Lundeen and Savage (1990) reported that lactobacilli were responsible for 86% of the hydrolysis of bile acids in the ileum and about 74% in the cecum of mice. Another reaction that *Lactobacillus* has been shown to carry out is dehydroxylation (Peppercorn and Goldman, 1972). The dehydroxylation product *meta*-hydroxy phenylproponic was isolated in the urine of gnotobiotic rat fed caffeic acid and which had been coinfected with two strains of lactobacilli, a strain of bacteriocides, and a group N streptococci. The combination of organisms was required for the dehydroxylation, since none of the individual bacteria carried out the reaction. Lactobacilli have also been shown to reduce the double bond of 3-hydroxy cinnamic acid (Soleim and Scheline, 1972) and cinnamic acid (Whiting and Carr, 1970) to produce respectively 3-hydroxy phenylproponic acid and phenylproponic acid. A *Lactobacillus* species has been shown to be capable of decarboxylating amino acids (Melnoykowycz and Johansson, 1955), and a *Lactobacillus acidophilus* isolate from the stomach of rat was shown to exhibit histidine decarboxylase activity (Horakova et al., 1971).

VI. INTESTINAL CHOLESTEROL METABOLISM

The intestine has a profound effect on cholesterol metabolism, many areas of which have recently been reviewed (Field et al., 1990; Lutton, 1976). The intestine is the site of cholesterol synthesis and absorption. A unique aspect of this organ is the presence of a large and diverse bacterial population, which impacts greatly on this process. Complicating the understanding of those processes the extreme variation in the normal intestinal processing of cholesterol by different population groups and among individuals with diseases of the intestinal tract. Additionally, there is variability with respect to the effect of dietary supplements and probiotics on plasma cholesterol concentrations. For these reasons, the understanding of the complex fate of cholesterol in the intestinal tissue and lumen of the intestine is required to understand sterol balance in the body.

VII. SOURCES OF INTESTINAL CHOLESTEROL

A major source of intestinal cholesterol derives from the de novo synthesis of this sterol. In many animal species mucosal cells secrete cholesterol directly into the lumen (Lutton, 1976); however, this does not appear to occur in humans (Spritz et al., 1965). In humans cholesterol synthesized by the intestinal cells is introduced into the small intestine via the exfoliation of intestinal cells. Additional sources of intestinal cholesterol come from secreted bile and the diet Connor et al. (1969) suggested that bacterial metabolism of cholesterol was highest in persons consuming a mixed Western diet. Subsequent work has indicated that there is not a homogeneous group of converters, and that the percent conversion may be lower than first thought.

VIII. ASSOCIATION WITH BACTERIAL CHOLESTEROL AND COLONIC CANCER

Considerable interest has focused on determining whether bacterial metabolism of cholesterol influences the development of certain disease states of the colon, especially in light of the findings that there exist certain individuals who have markedly different rates of bacterial metabolism of cholesterol. Aries et al. (1969) have suggested that the composition of the intestinal microflora was dependent on diet and that variations in dietary intakes influenced intestinal secretions and the substrates available to the bacteria for metabolism. Surveys of different population groups with different risks of colon cancer support this hypothesis (Hill and Aries, 1971; Reddy et al., 1977). Reddy et al. (1977) found that switching a person from a high-colon-cancer-risk (high meat) diet to a low-colon-cancer-risk (nonmeat) diet resulted in shifts in the composition of the intestinal microflora and neutral sterols.

Reddy et al. (1977) have reported that patients with diagnosed colon cancer who were consuming what was considered to be a mixed Western diet had a higher total sterols output than controls. This increased excretion of sterols was contributed to by increased amounts of both cholesterol and coprostanol per gram dry weight of feces. A previous study assessing the enzymatic activity of cholesterol dehydrogenase in the feces of patients with colon cancer and controls showed a higher level of activity in the patient group, possibly explaining the higher concentration of coprostanol previously observed (Mastromarino et al., 1976).

Reddy et al. (1977) next investigated the rate of bacterial metabolism of cholesterol in patients with ulcerative colitis since this group of persons is at high risk for developing colon cancer. They compared fecal sterol concentration to a classic control group, relatives of the patients and a group of persons with other digestive diseases. They found that in those persons with ulcerative colitis total neutral fecal sterol output was significantly greater than that in any of the control groups. This increase was contributed to by increases in the concentration of cholesterol (four-fold) and coprostanol (two-fold). When the data was expressed as a ratio of cholesterol to its major metabolites, those patients with ulcerative colitis had significantly higher ratio than any of the control groups, indicating a lower level of bacterial metabolism.

IX. INFLUENCE OF DAIRY PRODUCTS ON INTESTINAL CHOLESTEROL METABOLISM

Interest has been focused on the effect of specific foods on the bacterial metabolism of cholesterol and going a step further, the implications of this on plasma cholesterol levels. Given that elevated plasma cholesterol has been identified as a major risk factor for coronary heart disease and that nonpharmacological approaches to normalizing the levels are the treatment of choice, the progression is logical.

A number of studies, in both animals and humans, have looked at the effect of a variety of fermented and nonfermented dairy products on plasma cholesterol. Mann (1977) reported that the consumption of both skim and full-fat yogurt significantly decreased plasma cholesterol after the administration of radiolabeled acetate, he attributed the effect to an inhibitor of hydroxymethylglutaryl CoA reductase.

Hepner et al. (1979) fed pasteurized and nonpasteurized yogurt and 2% butterfat milk to humans for varying periods of time. They reported that both yogurts significantly lowered plasma cholesterol 5–10% by one week and that this was maintained for four weeks, whereas the buttermilk had no effect.

Roussouw et al. (1981) tested the hypothesis that the "milk factor" proposed by Mann (1977) could lower plasma cholesterol in young males. The subjects were fed skim milk, 1.8% fat yogurt, or 3.3% fat milk for five weeks. Only the skim milk resulted in a sustained decrease in plasma cholesterol, which the authors attributed to the decreased consumption of fat and cholesterol.

Considering two rat studies, Grunewald (1982) reported that feeding fermented acidophilus skim milk, but not unfermented skim milk, for four weeks resulted in a decrease in plasma cholesterol. Pulusani and Rao (1983) fed skim milk, whole milk, 2% fat buttermilk, yogurt, buttermilk, or sweet acidophilus milk to humans for three weeks. No significant differences in plasma cholesterol were observed during that period. Lin et al. (1989) conducted a double blind study to determine the effect of Lactinex, a commercially available tablet containing *Lactobacillus acidophilus* and *Lactobacillus bulgaricus*. The viable bacterial count in the preparation was 2×10^6. There was a second group receiving a placebo. There was a total of 354 subjects entered into the study. There was no difference detected between the treatment and placebo for total plasma cholesterol or any of the lipoprotein fractions. The subjects received the Lactinex for two six-week periods separated by a three-week washout. These results are not surprising given the very low dose of organisms ingested by the subjects. This dose is several orders of magnitude lower than would be ingested by an individual eating 6-oz. container of yogurt daily.

In an interesting study, Gilliland et al. (1985) isolated a strain of *Lactobacillus acidophilus* selected for its ability to grow in the presence of bile and assimilate cholesterol. Administration of this culture for 10 days to pigs partially prevented a dietary induced elevation in serum cholesterol. Strains which grow in the presence of bile, but did not assimilate cholesterol, served as a negative control. As a result of these studies Gilliland and Walker (1990) looked for a human strain with similar cholesterol assimilating properties. They reported that an *L. acidophilus* designated strain NCFM had an appreciable ability to assimilate cholesterol, although not as high as the pig strain, and may be useful for lowering plasma cholesterol in humans.

Bacterial metabolism of cholesterol can be influenced by diet as evidenced by significant variations among different population groups with different dietary habits. Altered patterns of intestinal bacterial metabolism of cholesterol may place persons at a higher risk of developing the disorder. Additionally there appears to be a relationship between the intake of certain dairy products and plasma cholesterol, although that relationship is far from being defined. More work needs to be done to clarify the relationship between dietary intake bacterial metabolism and plasma cholesterol levels.

X. CONCLUSION

The large number of microorganisms (approximately 10^{14}) that occupy the normal human intestinal tract constitute an ecosystem capable of metabolizing a large number of exogenous and endogenous compounds. Among this population the lactobacilli form an active component and participate in many of these reactions. The fate and pharmacokinetics of drugs, procarcinogens, dietary components, and endogenous compounds, such as bile acids, are influenced by the intestinal microflora.

The metabolic events occurring in the intestine have a central role in the fate and regulation of cholesterol in the body. As a consequence of cholesterol's central importance in normal physiology and disease the role of the intestine in cholesterol metabolism has great significance. A large body of information has been acquired regarding intestinal cholesterol metabolism; however, many important details and questions have not been determined or answered, and this area is a fertile ground for future research.

REFERENCES

Aries, V., et al. (1969). Bacteria and the aetiology of cancer of the large bowel, *Gut, 10*:334–335.

Connor, W. E., et al. (1969). Cholesterol balance and fecal neutral steroid and bile acid excretion in normal men fed dietary fats of different fatty acid composition, *J. Clin. Invest., 48*:1363–1375.

Ellis-Pegler, R. B., Crabtree, C., and Lampert, H. P. (1975). The faecal flora of children in the United Kingdom, *J. Hyg., 75*:135–142.

Field, F. J., Kam, N. T. P., and Mathur, S. N. (1990). Regulation of cholesterol metabolism in the intestine, *Gastro., 99*:539–551.

Finegold, S. M., Attebery, H. R., and Sutter, V. L. (1974). Effect of diet on human fecal flora: comparison of Japanese and American diets, *Am. J. Clin. Nutr., 27*:1546–1569.

Gilliland, S. E., Nelson, C. R., and Maxwell, C. (1985). Assimilation of cholesterol by *Lactobacillus acidophilus, Appl. Environ. Microbiol., 49*:377–381.

Gilliland, S. E., and Speck, M. L. (1977). Deconjugation of bile acids by intestinal lactobacilli, *Appl. Environ. Microbiol., 33*, 15–18.

Gilliland, S. E., and Walker, D. K. (1990). Factors to consider when selecting a culture of *Lactobacillus acidophilus* as a dietary adjunct to produce a hypocholesterolemic effect in humans, *J. Dairy Sci., 73*:905–911.

Gruette, F. K., Horn, R., and Haenel, H. (1965). Ernahrung unf biochemisch Mikrookologische Vorgange in enddarm Von Sauglingen, *A. Kindesheilkd, 93*:28–39.

Grunewald, K. K. (1982). Serum cholesterol levels in rats fed skim milk fermented by *Lactobacillus acidophilus, J. Food Sci., 47*:2078–2079.

Haenel, H., and Bendig, J. (1975). Intestinal flora in health and disease, *Prog. Food Nutr. Sci., 1*:21–64.

Hepner, G., Fried, R., St. Jeor, S., Fusetti, L., and Morin, R. (1979). Hypocholesterolemic effect of yogurt and milk, *Am. J. Clin. Nutr., 32*:19–24.

Hill, M.-J., and Aries, V. C. (1971). Fecal steroid composition and its relationship to cancer of the large bowel, *J. Pathol., 104*:129–139.

Horakova, Z., Zeirdt, C. H., and Zeaven, M. A. (1971). Identification of *Lactobacillus* as the source of bacterial histidine decarboxylase in rat stomach, *Eur. J. Pharmacol., 16*:67–77.

Lin, S. Y., Ayres, J. W., Winkler, W., and Sandine, W. (1989). *Lactobacillus* effects on cholesterol: in vitro and in vivo results, *J. Dairy Sci., 72*:2885–2899.

Lundeen, S. G., and Savage, D. C. (1990). Characterization and purification of bile salt hydrolase from *Lactobacillus* sp strain 100-100, *J. Bacteriol., 172*:4171–4177.

Lutton, C. (1976). The role of digestive tract in cholesterol metabolism, *Digestion, 14*: 342–356.

Mann, G. (1977). A factor in yogurt which lowers cholesteremia in man, *Atherosclerosis, 26*:335–340.

Mastromarino, A., Reddy, B. S., and Wynder, E. L. (1976). Metabolic epidemiology of colon cancer: enzymatic activity of fecal flora, *Am. J. Clin. Nutr., 29*:1455–1460.

Melnoykowycz, J., and Johansson, K. R. (1955). Formation of amines by intestinal microorganisms and the influence of chlortetracycline, *J. Exp. Med., 101*:507–517.

Moore, W. E. C., and Holderman, L. V. (1974). Human fecal flora: the normal flora of 20 Japanese-Hawaiians, *Appl. Microbiol., 27*:961–979.

Peppercorn, M. A., and Goldman, P. (1972). Caffeic acid metabolism by gnotobiotic rats and their intestinal bacteria, *Proc. Natl. Acad. Sci., 69*:1413–1415.

Pradham, A., and Majumdar, M. K. (1986). Metabolism of some drugs by intestinal lactobacilli and their toxicological considerations, *Acta Pharmacol. Toxicol., 58*:11–15.

Pulusani, S. R., and Rao, D. R. (1983). Whole body liver and plasma cholesterol levels in rats fed thermophilus, bulgaricus and acidophilus milks, *J. Food Sci., 48*:280–281.

Reddy, B. S., Mastromarini, A., and Wynder, E. (1977). Diet and metabolism: large-bowel cancer, *Cancer, 39*:1815–1819.

Rossouw, J. E., Burger, E. M., Van Ne Vyver, P., and Ferreira, J. J. (1981). The effect of skim milk, yogurt, and full cream milk on human serum lipids, *Am. J. Clin. Nutr., 34*:351–356.

Rotimi, V. O., and Duergen, B. I. (1981). The development of the bacterial flora in normal neonates, *J. Med. Microbiol., 14*:51–62.

Scheline, H. A. (1972). Metabolism of xenobioties by strains of intestinal bacteria, *Acta. Pharmacol. Toxicol., 31*:471–480.

Spritz, et al. (1965). Sterol balance in man as plasma cholesterol concentrations are altered by exchanges of dietary fats, *J. Clin. Invest., 44*:1482–1493.

Stark, P. L., and Lee, A. (1982). The microbial ecology of the large bowel of breast-fed and formula-fed infants during the first year of life. *J. Med. Microbiol., 15*:189–203.

Whitting, G. C., and Carr, J. G. (1959). Metabolism of cinnomic acid and hydroxycinnomic acids by *Lactobacillus pastroranius* var. *quinicus, Nature, 184*:1427–1428.

9

Development of Individual Lactic Acid Microflora in the Human Microbial Ecosystem

Marika Mikelsaar and Reet Mändar
University of Tartu and Tartu University Hospital, Tartu, Estonia

I. GENERAL INTRODUCTION

The very first days of our existence pass pleasantly in the deep warm safety of our mother's womb, and it must be a shock of supreme power for a neonate to enter this world with its cosmic numbers of microbes. However, "wise Nature" could not have planned the first contamination of a neonate to happen at random. It is the task of biological sciences to find out the natural precautions. One can suggest it to be the normal microflora, inhabiting the genital tract of a healthy mother, on which the newborn's defense mechanism relies. The predominant microorganisms of the vagina are microaerophilic and anaerobic lactobacilli, streptococci and some other lactic acid producing microbes (Larsen and Galask, 1980; Cook et al., 1984;).

Lactobacilli, one of the most frequent Gram-positive bacteria in the human microflora, usually inhabit various organs as innocuous commensals. Since the turn of the century human lactobacilli have been considered contributors to human health (Metschnikoff, 1908). The interest in their existence in different areas of the body and their role in the host and microflora-related physiochemical conditions in the organism has become one of the subjects of a comparatively new research field—microbial ecology (Haenel, 1957a,b, 1980;

Reuter, 1965, 1975; Dubos, 1966; Gilliland et al., 1975; Goldin and Gorbach, 1984; Gorbach, 1990).

In this chapter an attempt is made to summarize and interpret the current state of our knowledge of the composition, maintenance and formation of human individual lactoflora.

The recognition of lactobacilli as an important part of the human microbial ecosystem and the understanding of various interconnected influences of that system is our starting point. Several researchers have underscored (Reuter, 1965; Lencner, 1973, 1984, 1987; Mitsuoka et al., 1975) the large individual differences of lactoflora, expressed in numbers and species' composition. Population studies and clinical investigations are the main sources of information about the effect of the genetic background of hosts, and their age and health on the persistence of lactobacilli in different persons (Lencner, 1973; Mikelsaar et al., 1982, 1984; Hanson et al., 1989, 1990). The data of the transmission of lactobacilli from mother to neonate during the birth, mode of delivery, hospital conditions, and the effect of feeding upon the establishment of lactoflora in economically differently developed countries contribute to the understanding of the influence of various environmental factors for neonatal ecology of lactic acid bacteria (Raibaud et al., 1980; Bennet, 1987; Hall et al., 1990).

The factors determining the development of individual lactic acid microflora are not completely known yet, and the question about which strains become residential in various human organs remains an object of discussion (Tannock et al., 1990b). The in vitro studies determining the adherence capacities of lactobacilli to eukaryotic cells and other microorganisms are useful for understanding the selective adherence of lactobacilli to a particular host (Fuller, 1975; Brilis, 1983; Dalin and Fisch, 1985; Wadström and Aleljung, 1989). The relationship between the colonizing properties and the in vitro determined antimicrobial substances of particular strains of lactobacilli has not been solved yet.

Direct assertion of the mechanisms by which lactobacilli colonize various areas of a human body require advanced experimental techniques (Savage, 1977, 1989). Here, the selection of any appropriate animal model and particular bacterial strains may be crucial. One point of importance seems to be that the lactobacilli should originate from the animal species studied (Tannock, 1983).

The practical goal of the investigations of lactoflora formation is to develop anti-infectious treatments such as controlled colonization of neonates with biologically highly active strains of lactobacilli (Perdigon et al., 1986; Goldmann, 1988; Moshchich et al., 1989). The revised theoretical basis of formation of lactoflora may improve the development of the infants so treated.

II. LACTOBACILLI IN THE HUMAN MICROBIAL ECOSYSTEM

A. Components of the Microbial Ecosystem

On their surfaces and organs humans harbor a normal, or "indigenous" microflora (Savage, 1977, 1989) whose quantity (10^{13}–10^{14}) exceeds the total number of body cells (Luckey, 1977). These populations of microorganisms form microbial ecosystems of humans which are considered to be a mode of dynamic, relatively stable interactions between the normal microflora of various locations and the host (Starr et al., 1981; Mackowiak, 1982). The gastrointestinal, genital, and skin microbial ecosystems are the most important ones harboring the highest populations of microorganisms in humans (Gustafsson and Norin, 1977).

The main unit of a microbial ecosystem is the *microbiocenosis*: the association of qualitatively different groups of microorganisms having dynamically fluctuating but somehow relatively stable quantitative characteristics (Haenel, 1980; Freter et al., 1983; Atlas and Bartha, 1987; Redondo-Lopez, 1990). Different microbial ecosystems always contain several microbiocenoses located in certain microbial biotopes of the host.

A *microbial biotope* (microbiotope) is an area or place suitable for the survival and growth of particular microorganisms (Haenel, 1980). The most striking and important scientific achievements in the microbial ecology studies have been the discovery of the association of microflora with definite habitats and the division of microflora of various organs into luminal and mucosal (Dubos and Schaedler, 1962; Savage et al., 1968, 1987). Nowadays the so-called horizontal distribution of microorganisms in the gastrointestine (GI) distinguishes among the localization of microbes in the lumen content, in the mucus layer covering the epithelium, and on the epithelial cells of the mucosa, localized for instance on the tips of villi or deep inside Lieberkuhn's crypts (Costerton et al., 1983). Beside the latter division there exists a vertical distribution of microorganisms in the intestine, differentiating between gastric, duodenal, jejunal, ileal, cecal, colonic, and rectal microfloras. Analogous divisions could also be found in other microbial ecosystems, for example, the urogenital one.

The microbial biotope determines not only the anatomical localization for the microbiocenosis, but also the wide complex of host- and microflora-derived physiochemical conditions, including pH, redox potential, nutrient availability, peristalsis, and transit time (Midtvedt, 1989a). Thus in the microbiotope various factors, both endo- and exogenous, monitor the composition and the functional state of microbiocenosis.

B. Lactobacilli in Various Biotopes

Human microflora consist of a large number of different groups and species of microorganisms. The human body is inhabited by more than 500 different species and among them are also lactobacilli (Tannock, 1983; Simon and Gorbach, 1984). Interest in their occurrence, numbers, and role within various microbial ecosystems of humans has survived a whole century (Metshnikoff, 1908; Rettger, 1935; Haenel, 1957a,b; Geimberg, 1957; Reuter, 1965; Lencner, 1973; Gorbach, 1990). Lactobacilli have been found on the skin, in the nasal and conjunctival secretions, in the ear, breast milk, and sperma (Haenel and Bendig, 1975).

There is some confusion about the group definition of lactic acid bacteria, on the basis of their type of metabolism. Most species of lactobacilli are considered to be microaerophilic which are able to replicate in the presence of reduced quantities of oxygen. But there are some species or particular strains which are obligate anaerobes. According to Redondo-Lopez (1990) the main reason for lack of unanimity seems to be that the methods used for cultivating lactic acid bacteria are not standardized. Investigators may use 3–10% CO_2 in the air or even anaerobic condition (84% N_2, 10% CO_2, 5% H_2). Lactobacilli can be generally identified by means of gas-liquid chromatography of fatty acids released during fermentation of glucose (Holdeman et al., 1977). Yet, species like *Lactobacillus reuteri* are still ascertained only on the basis of their fermentation patterns (Kandler and Weiss, 1986).

Sufficient data are available about the numbers of lactobacilli in the proximal and distal parts of the intestine, but the knowledge of the lactobacillar content of the ileum, caecum and colon of healthy subjects is quite inadequate (Fig. 1). The same can be said about the mucosal lactoflora, and the main reasons for this are the methodological limitations of getting biopsy samples from these areas (Bernhardt and Knoke, 1980; Knoke and Bernhardt, 1985). In fecal microflora the *Lactobacillus* counts (10^{10} cfu/g) are outnumbered only by obligate anaerobes (Simon and Gorbach, 1984).

The lactoflora of the human GI consists of various species, subspecies, and biotypes of homo- and heterofermentative lactic acid bacteria. The most frequent lactobacilli belong to six species: *Lactobacillus acidophilus, L. salivarius, L. casei, L. plantarum, L. fermentum*, and *L. brevis* in various combinations (Lerche, Reuter, 1962; Lencner, 1973; Mitsuoka et al., 1975; Bernhardt and Knoke, 1980; Mikelsaar et al., 1982; Lencner et al., 1984; Conway, 1989).

In various GI biotypes the species of lactoflora are not the same (Knoke and Bernhardt, 1985). Comparing the lactoflora of 20 healthy young men (Table 1), differences in the composition of *Lactobacillus* species in their saliva and feces were observed (Lencner et al., 1987). The authors regarded as lactobacilli catalase negative Gram-positive rods isolated in microaerophilic conditions with 10% CO_2

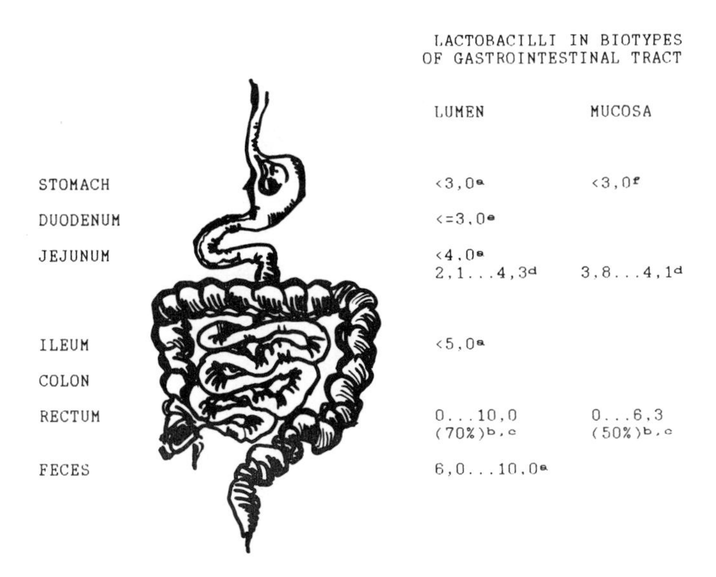

Figure 1 Schematic representation of lactobacilli numbers (log dfu/g) in a healthy man's gastrointestinal tract. Data are adapted from references: a: Simon and Gorbach (1984); b: Mikelsaar et al. (1984); c: Mikelsaar et al. (1986) (in Russian); d: Bhat (1980); e: Bernhardt and Knoke (1984); f: Mikelsaar et al. (1990).

on a somewhat modified MRS medium (Rogosa and Sharp, 1960; Lencner, 1973) and expressing particular biochemical activity (Lencner et al., 1984). The comparison of salivary and fecal lactoflora showed that the latter was usually richer in the number of species and biovariants. Yet the numbers of different biovariants detected in the saliva and feces of any one investigated person coincided to some extent ($p < .05$, Fisher's exact test). The *Lactobacillus acidophilus* I biovariant occurred most frequently both in the saliva and feces (14 cases), *L. fermentum* II and III biovariants were detected in both the examined materials of five and four persons, respectively. At the same time, these data also showed wide interindividual variations of lactoflora of different biotopes.

Lactobacilli are the predominant microorganisms isolated from the vagina of healthy women, their number reaching 10^8–10^9 cfu/mL (Stahl and Hill, 1986). However, data about the predominant lactobacilli species isolated from the vagina are controversial. In early studies *L. acidophilus* and *L. fermentum* were considered to be the most prevalent ones (Reuter, 1965; Lencner, 1973). In contrast, Giorgi et al. (1987) using more precise DNA-DNA homology studies, suggested

Table 1 Comparison of Species and Biotypes of Salivary and Fecal Lactobacilli in Healthy Young Men

NN/ Biotope		Lactobacillus species and biotypes																		
		I[a]	Ia	aI	aII	sI	sII	ccI	ccII	cr	ca	pI	pII	fI	fII	fIII	fIV	ce	brI	brII
1	sal			+											+	+	+			
	fec			+					+			+		+	+	+				+
2	sal			+																
	fec			+				+		+									+	
3	sal			+		+			+						+					
	fec			+																
4	sal			+			+			+		+	+		+					
	fec											+								
5	sal											+								
	fec								+			+							+	
6	sal			+						+	+					+				
	fec			+								+								
7	sal			+												+				
	fec			+		+			+		+	+	+	+					+	
8	sal			+																
	fec	+						+		+									+	
9	sal			+					+						+		+			
	fec									+				+	+					
10	sal			+																
	fec			+										+						
11	sal			+													+			
	fec			+					+	+		+							+	
12	sal	+		+	+	+	+													
	fec			+		+			+			+								
13	sal			+					+	+	+			+	+		+			
	fec	+		+		+	+		+	+		+		+	+					
14	sal			+		+								+	+	+			+	
	fec	+						+	+								+			
15	sal															+	+			
	fec			+	+									+		+			+	

Table 1 (Continued)

NN/	Lactobacillus species and biotypes																		
Biotope	I[a]	Ia	aI	aII	sI	sII	ccI	ccII	cr	ca	pI	pII	fI	fII	fIII	fIV	ce	brI	brII
16 sal	+		+											+	+				
fec			+						+					+	+	+			
17 sal			+	+	+			+	+				+		+	+		+	
fec			+		+			+					+		+	+		+	
18 sal			+	+					+					+			+		
fec			+	+															
19 sal			+	+											+				+
fec			+																
20 sal			+					+	+			+	+		+				
fec			+	+					+			+	+	+	+			+	

[a]Biovariants by A. Lenzner (1973).
Number of different lactobacilli: I = *L. leichmannii*; Ia = *L. lactis*; aI and aII = *L. acidophilus* I and II; sI and sII = *L. salivarius* I and II; ccI and ccII = *L. casei* ssp. *casei* I and II; cr = *L. casei* ssp. *rhamnosus*; ca = *L. casei* ssp. *alactosus*; pI and pII = *L. plantarum* I and II; fl, fII, fIII and fIV = *L. fermentum* I, II, III and IV; c = *L. cellobiosus*; brI and brII = *L. brevis* I and II.
Source: From Lenzner et al. (1987).

L. crispatus, L. jensenii, L. fermentum, and *L gasseri* to be the predominant lactobacilli in the vagina. The main reason for the lack of reliable knowledge seems the fact that the above-mentioned studies are expensive and labor consuming.

Redondo-Lopez (1990) has summarized the data on the species composition of lactoflora of different individuals and found such studies to be quite rare (Bartlett et al., 1977; Eschenbach et al., 1989). According to the unpublished data of Cook (Redondo-Lopez, 1990) there were up to four lactobacillar species in healthy nonpregnant women and each woman having an individual type of species composition.

The discovery of lactobacilli in unusual biotopes also creates some controversy. Lactobacilli are mostly regarded as nonpathogenic microorganisms, but some authors connect their presence in the blood stream with bacteraemia and endocarditis (Stille, 1971; Shinar et al., 1984; Thangkhiew et al., 1987). We have found lactobacillar bacteremia immediately after extraction of tooth roots in chronic periodontitis and consider it transitory, caused by the translocation of microbes after disruption of the mucosal barrier of the mouth (Türi et al., 1982; Mikelsaar and Türi, 1990).

The main species of lactobacilli suspected to have some pathogenic potential are *L. plantarum* (Struve et al., 1988) and *L. casei* (Allison and Galloway, 1988). Türi et al. (1980) have shown the proliferative tissue reaction to the inoculation of *L. casei* into the testes of the guinea pigs. However, the simple fact of isolating lactobacilli from an empyema of the gallbladder (Allison and Galloway, 1988) or from the amniotic fluid (Cox et al., 1986) cannot prove their pathogenicity because some other less recognizable infectious agents may have been overlooked.

C. Stability of Human Microflora

1. *Individuality of Microflora*

The individual differences in the species composition and the numbers of microorganisms of the skin (Lynch and Poole, 1979; Kearney et al., 1984; Leeming et al., 1984), vagina (Reuter, 1965, 1975; Lencner, 1973; Solovyeva, 1986, 1987; Kasesalu et al., 1990), and intestine (Holdeman et al., 1976; Mitsuoka and Ohno, 1977; Finegold et al., 1983) are documented by many investigators, who have studied the microbial ecosystems of adult organisms. Controversial data have been obtained about the stability of individual microflora of various biotopes.

The numbers of microbes in the fecal microflora of a person vary greatly for period of some months (Gorbach et al., 1967) or even a few days (Meijer-Severs, 1986). However, Holdeman et al. (1976) found the quantitative composition of fecal microflora to be very specific for a particular host. The stability of the quantitative composition of the fecal microflora of a particular host means mostly the persistence of stable quantitative relations between the most frequent and predominant groups of microorganisms (Zubrzycki and Spaulding, 1962; Mitsuoka and Ohno, 1977). Thus, after a whole year's study of the fecal microflora of 10 healthy volunteers, stable relationships between the numbers of different aerobic and anaerobic groups of microorganisms were ascertained (Mikelsaar, 1969).

Only few species of microorganisms inhabiting the GI have been followed for their persistence in the same person during long periods. The stable occurrence of the same biotypes of bifidobacteria (Gossling and Slack, 1974), bacteroides (Johnson, 1980; Moore et al., 1979) and certain phenotypes of *E. coli* (Kuhn et al., 1986) has been proved. The latter microbes are suggested to be especially adapted for colonizing the human intestine. It is obvious that the host- and microflora-derived physiochemical conditions of microbial biotopes cannot be too similar for different persons and in that sense microbial ecosystems are always deeply individual, having specific interindividual peculiarities (Haenel and Bendig, 1975).

The microbial ecosystem as a whole is successfully characterized by biochemical studies determining the metabolites of the microorganisms excreted from the human body. The very specific and personally stable composition of various bacterial metabolites excreted by urine or feces has been revealed by several researchers (Hoverstad et al., 1984; Weaver et al., 1989; Midtvedt, 1989a,b, 1990; Siigur et al., 1991). These data also confirm the occurrence of individually specific microflora in humans.

2. *Individual Stability of Lactoflora*

a. Feces. The number of lactobacilli in fecal samples of different persons appears to be quite different (Mitsuoka and Ohno, 1977; Knoke and Bernhardt, 1985). In a survey over a long period (Mikelsaar, 1969; Mikelsaar et al., 1982, 1984) of the number of lactobacilli in the faeces of 10 healthy volunteers quite stable and characteristic quantities were revealed (Table 2). The range of the number of lactobacilli in the feces of four persons (nos. 1, 7, 9, 10) was low. Three of them maintained this situation during the study, although they grew older, had several failures of their health, and used some medicines. The variation of different estimations of the number of lactobacilli of the investigated persons (except no. 7) did not significantly exceed (<1 log cfu/g, the normal error of quantification) the first examination range. It is important to mention that all 10 volunteers are sill in full health even now, when more than 25 years have passed.

Table 2 Quantity of Lactobacilli in the Feces of Healthy Persons During 15 Years Follow-Up

	Quantity of lactobacilli (log cfu/g)		
	I investigation[a]	II investigation	
No. of persons	Range of 6–8 samples		III investigation
1	5.2–7.0	5.6–6.0	5.9
2	5.9–7.5	6.3–7.5	6.0
3	5.8–7.4	6.8–7.5	7.8
4	5.7–7.7	6.9–7.3	7.6
5	5.7–7.7	7.9–8.3	7.3
6	6.2–7.6		7.0
7	5.5–6.7	6.7–8.3	7.3
8	6.3–8.1	6.6	7.3
9	5.5–6.7		7.1
10	6.2–7.0	7.0	

[a]At the I investigation the age of patients was 23–44, at the II, 30–51, and at III, 38–57 years.

In addition to the stability of the number of lactobacilli in feces, the stable persistence of their fecal lactobacillar species (Table 3) was revealed (Mikelsaar et al., 1975, 1982). Among two to four species of lactobacilli isolated from every adult volunteer, one or two species occurred repeatedly and in four persons we could even observe the persistence of the same biotype. Similarly, in a study of the seasonal variation of fecal lactoflora, the investigation of feces of children aged 7–12 revealed that the same biotypes of lactobacilli were detected in 8 out of 11 children (Golyanova, 1972).

The prolonged biological isolation of healthy persons during special training or space flight periods of astronauts caused some shift to opportunistic microorganisms (Shilov et al., 1972). Yet the close physical contact could not eliminate the individual specificity of their lactoflora in terms of species or bacterial number (Lencner et al., 1973, 1981, 1984).

b. Vagina. The number and composition of vaginal lactobacilli are significantly influenced by the hormonal status of the host (Levin, 1968; Brilene and Brilis, 1986; Redondo-Lopez, 1990). The flat epithelium of the vagina contains much glycogen, which is more readily converted into lactic acid during pregnancy, thus creating a more beneficial environment for colonization by lactobacilli, nonhemolytic streptococci, and other lactic acid bacteria (Larsen and Galask, 1980).

Unfortunately, few studies of the vaginal microflora of pregnant women have included a control group of nonpregnant women (Redondo-Lopez, 1990; Tashjian et al., 1976). In that sense our unpublished data (Mändar et al.) comparing the lactoflora of healthy pregnant women with those with a threatened abortion seem to be very informative. It was demonstrated that the relative amount of lactobacilli in 24 women out of 42 increased as gestation advanced (Fig. 2). This tendency started in women with a threatened abortion somewhat earlier (in 24–26 weeks of pregnancy) than in the control group (in 32–34 weeks). It could be explained by the influence of hormonal preparations administered to support the gravidity in cases of threatened abortion. The predominance of lactobacilli may be considered a preventive mechanism offering protection to the fetus and neonate just before labor. After delivery, such a bacterial status disappears (Bartlett et al., 1977; Stahl and Hill, 1986).

The problem of individual stability of human microflora is closely connected to its formation, mainly during vaginal delivery. However, the factors determining which bacterial strains colonize and persist in the infant from the very first days of its life are not known yet (Tannock et al., 1990b). The main reason for that may be the complexity of interactions in a human's microbial ecosystem.

Table 3 Persistence of Lactoflora in the Feces of Healthy Persons During 15 Years Follow-up

	Species and biotypes of lactobacilli																	
No. of persons	a I	a II	s I	s II	cc I	cc II	cr	ca	p I	p II	f I	f II	f III	f IV	br I	br II	c	bu
1 a[a]	+				+								+					
b	+									+	+							
c	+	+				+							+		+			
2 a	+																	
b	+	+				+	+		+						+	+		
c	+					+			+	+				+	+			
3 a	+	+	+														+	+
b	+	+					+				+							
c	+	+																
4 a	+	+	+										+					
b	not determined																	
c	+	+				+								+				
5 a				+														
b	not determined																	
c						+		+					+				+	
6 a						+		+	+			+						
b	not determined																	
c		+							+									
7 a	+							+										
b	+						+		+		+							
c	+	+					+				+							
8 a	+	+	+							+								
b			+	+		+	+				+							
c	+		+							+								
9 a	+								+									
b	not determined																	
c		+	+				+											

[a] a = the year 1965; b = the year 1971; c = the year 1979.
The abbreviations for lactobacilli species and biotypes are the same as in Table 1.

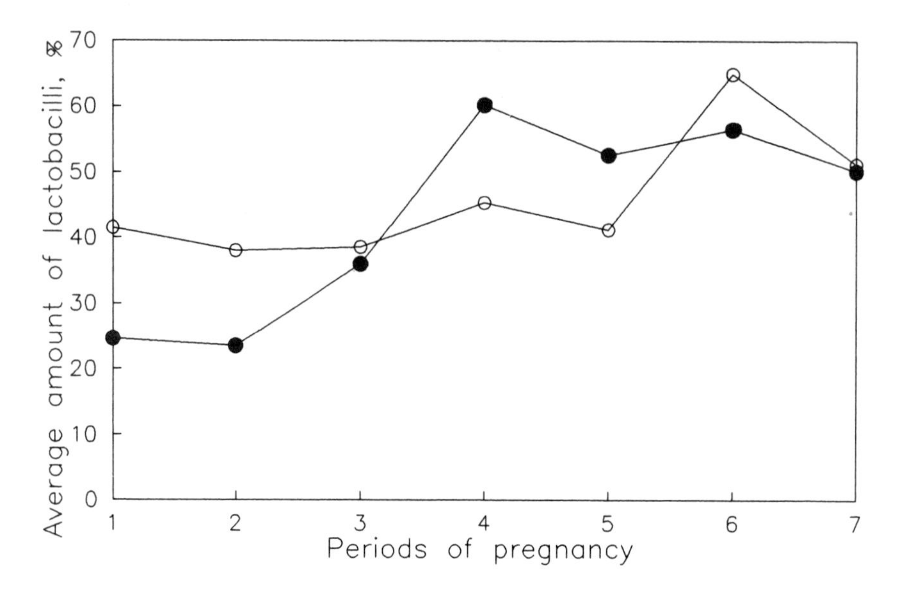

Figure 2 Average relative amounts of lactobacilli in vaginal microflora in women harboring them. 1–10 weeks of gestation age; 2—11–16 w; 3—17–22 w; 4—24–26 w; 5—28–30 w; 6—32–34 w; 7—36–38 w. o———o control group; • – – – • women with threatened abortion.

D. Interactions Between the Components of the GI Microbial Ecosystem

The influences of various factors of host and microbial origin, participating in the formation of the individual GI microbial ecosystem are mainly interconnected. That creates great methodological limitations for research. In order to better understand the complex problem of lactoflora formation, the interactions between the components of a microbial ecosystem may be divided into three different stages (Fig. 3).

1. First Stage

In the first stage there exist specific different interactions between various groups of microorganisms performing a particular microbiocenosis. The species composition and number of microorganisms is mostly influenced by bacterial physicochemical activities. The widely known specific substances excreted by some bacteria to inhibit the growth of others are low- or high-molecular proteins, for example, bacteriocins (Conway, 1989). In addition to that some metabolic products of microbes such as organic acids, short-chain fatty acids (SCFA), H_2O_2,

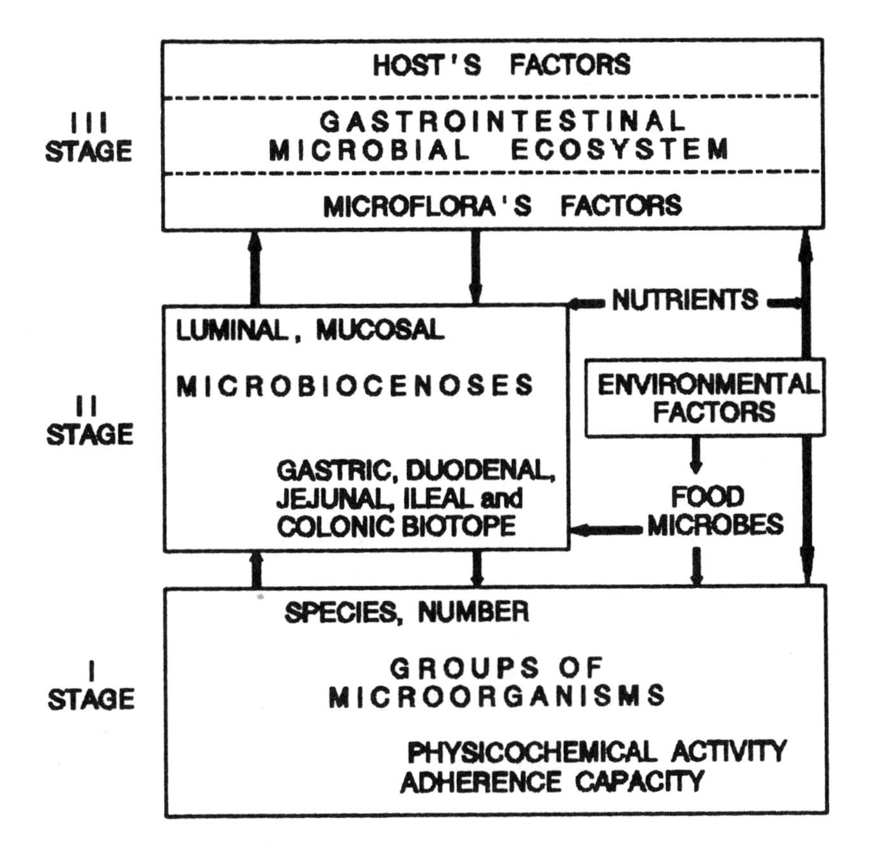

Figure 3 Various factors influencing the human GI ecosystem.

etc., are also very important. The influence of these substances has been shown by in vitro experiments and continuous culture studies where the nutrient supply is constant (Freter et al., 1983; Ushima and Ozaki, 1988). Yet some other factors may also influence the interaction of two different microorganisms.

The colonization of the biotope is mostly influenced by the adhesion phenomenon. The data according to which the less active of two strains with different metabolic activities has to adhere, sometimes even only to the surface of the continuous culture system (Freter et al., 1983) seem very relevant.

In a continuous culture model of a fecal microflora, it was recently shown by Bernhardt et al. (1990) that the relative proportion of lactobacilli in the total count of microorganisms was not influenced by admitting into the system biologically active strains of other lactobacilli. Thus even the isolated colonic content appears to be a well-balanced self-regulating system the mathematical modeling of which is possible (McKay and Speekenbrink, 1984; Freter, 1990).

The particular aspects of interactions of lactobacilli with other microorganisms in the process of colonization will be described in the section of this review.

2. Second Stage

Studying the second stage of interactions, it is obvious that the composition of either luminal or mucosal microbiocenoses depends on the local biotopes' conditions (Savage, 1977, 1987). The main variables of the biotope that influence the microbial population level seem to be the gas concentration and composition (oxygen, carbon dioxide, methane, hydrogen), the pH and *Eh* (reduction-oxidation potential), and the mucosal secretions, such as lysozyme, hormones, immunoglobulins (Knoke and Bernhardt, 1985). What part do these variables play in different microbiocenoses of the GI is not easy to determine and the conclusions are mostly drawn on the basis of in vitro experiments.

Very few data are available about interactions between different microbiocenoses of the GI of a particular host. It is not known how much the luminal content reflects the state of microflora on the mucosal surfaces. The main reason for that is obviously the difficulty in getting samples from various parts of both the lumen and the mucosa of healthy individuals. Several data have been gathered about the diversity of luminal and mucosal microfloras (Gorbach et al., 1967; Bergogne-Berezin et al., 1973; Bhat et al., 1980) or the similarity of them (Dolby et al., 1984; Kanareikina et al., 1986).

We tried to apply mathematical analysis (unpublished data) for the data of Bhat et al. (1980), who have stated the nonresemblance of luminal and mucosal jejunal microfloras just by comparing the presence of concordant groups of microbes in these biotopes. We could reveal a close correlation between the luminal and mucosal counts of lactobacilli and coliforms of the jejunum in healthy test persons (respectively $r = 0.774$, $n = 10$, $p < .05$ and $r = 0.580$, $n = 10$, $p < .05$). In the patients, suffering from sprue, even several other groups of microbes (aerobes and anaerobes) were closely interconnected. The resemblance between the microbial groups predominant in the mucosa and those found in the lumen was detected in 40% of cases in healthy subjects and in 54% of cases in patients. This example shows the possibility of reaching quite different conclusions by using different methods on the same data.

The fecal samples have been for practical reasons, the most widespread means of studying the human intestinal microflora. In interpretating the results, however, it is often difficult to say if a given isolated species is a true member of the normal GI microflora—an indigenous organism—or a transient colonizer, ingested with food or washed down from some proximal habitat (Savage, 1987).

Some data about the similarities and dissimilarities of mucosal and luminal microflora have been collected in animal research (Savage, 1977). The mucosal microflora of pigs was shown to be quite different from that of lumen (Pollmann et al., 1980; Robinson et al., 1984). However, the functional groups of microorganisms of the lumen and the mucosa of the colon were quite similar in horses (Mackie and Wilkins, 1988). Controversial data may result from the different methods used: microscopic and bacteriological comparison may be quite different in terms of their sensitivity.

In a survey on the colonization of piglets of various ages by lactobacilli (Mikelsaar et al., 1987) we observed identical frequency of lysozyme-resistant (15–50 mg/mL) strains of *L. acidophilus* and *L. fermentum* in the small intestine's mucosa (63%) and in the lumen (64%). That confirms the close interconnection between these floras. However, the luminal strains may certainly have their origin in the more proximal parts of the GI, as the ileal lysozyme-resistant strains of 120-day-old piglets could originate from the jejunal mucosa (Fig. 4). No lysozyme-resistant strains of lactobacilli could be isolated from fecal samples (Mikelsaar et al., 1987). The latest investigation on plasmid profiles of lactobacilli present in different parts of GI also revealed a remarkable difference between the inhabitants of pigs' proximal digestive tract and rectal contents (Tannock et al., 1990a).

In several experiments with Wistar rats (starvation, creation of a self-filling jejunal loop, administration of kanamycin and metronidazole) we compared bacteriologically the mucus-associated and the luminal microflora of different parts of the GI (Mikelsaar et al., 1987, 1989). It was found that the influence of changes in microflora of one biotope to another depends on their GI location, and is more frequent in the proximity of the intestine. Also, the effect was quite different for various predominant groups of microbes (Table 4). An overgrowth by coliforms and streptococci of the jejunal mucosa could be detected in 84% and 33% of cases, respectively, in the luminal content, and in 46% and 33% of cases in feces. In the jejunal juice the mucosal overgrowth by bacteroides can be detected rarely (in 33% of cases) but never in the bacteriological examination of feces.

In the experimental rats we found a close correlation between the jejunal luminal and mucosal *Lactobacillus* counts and their numbers in feces (respectively $r = 0.4087$, $n = 70$, $p < .01$ and $r = 0.3539$, $n = 73$, $p < .01$). Yet the predominance of lactobacilli on the mucosa of the small intestine (found in 6 samples out of 28) was never reflected in the fecal samples.

Consequently, there exist some interconnections between the mucosal and luminal microfloras, but the fecal microflora investigation may give only some approximate data about the microbial ecology of the GI. The occurrence of a particular strain of lactobacilli in the fecal samples only proves its persistence in a

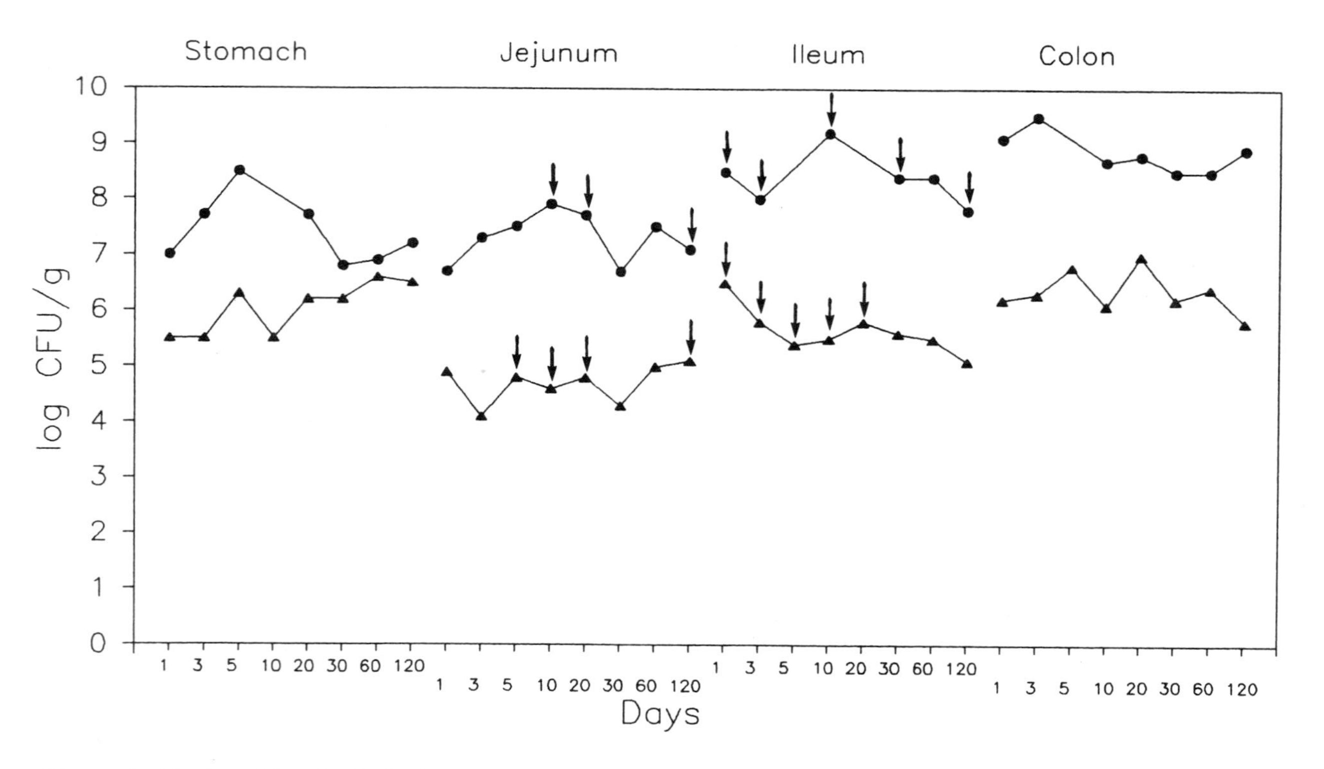

Figure 4 Number of lactobacilli in the gastrointestinal tract of piglets various ages: ●——●lumen; ▲——▲ mucosa; ↓ lysosyme-resistant strains (Mikelsaar et al., 1988a).

Table 4 Identical Changes of Microflora in Different Biotopes of the GI after Experimental Intervention of Rats[a]

Biotope with changes (>1 log cfu/g of control)	Biotope compared	Frequency of occurrence (%) of similar changes		
		Coliforms	Streptococci	Bacteroides
Jejunum	Jejunum			
Mucosa	Content	84.6	33.3	33.3
	Large intestine			
	Feces	46.2	33.3	0
	Mucosa	44.4	12.5	0
Jejunum				
Content	Feces	37.5	75.0	0
Large intestine				
Mucosa	Feces	38.0	0	0

[a] $n = 73$; see the text.

certain part of the GI but does not indicate its number or relative proportion either in the lumen or on the mucosa of the proximal parts of the GI.

3. *Third Stage*

The attention of medical investigators has been focused on the third stage of interactions and particularly on the influence of the microflora on the functional state of the host. The role of microflora in the morphogenesis of the organism and in the immunological and metabolical functions in health and disease has been proven (Gracey, 1986; Borriello et al., 1987; Lizko, 1987; Smolyanskaya, 1987; Shenderov, 1987; Midtvedt, 1989a; Nord et al., 1989; Norhagen et al., 1990).

The lactoflora has been shown to play an important part in the health of humans (Gorbach, 1990). Mainly, the lactoflora is involved in the prevention of various GI infections (Rettger, 1935; Knoke and Bernhardt, 1985) by forming a colonization barrier (Van der Waaij, 1973) against the establishment of microbes with pathogenic potential (Lencner et al., 1987).

The reported beneficial effects of lactobacilli have been objectively summed up by Tuschy (1986). They include, beside colonization resistance, the relieving of bacterial diarrhea, increase in weight, stimulation of the immune system, reduction of the serum cholesterol, deconjugation of the bile acids, improved lactose tolerance, decreased risk of developing a colon cancer, etc. However, the results often prove to be statistically insignificant when attempts are made to confirm the conclusions made by different authors (Conway, 1989). The main reason for this may be that the microbial ecosystems are individually different even in cases of similar diseases.

In clinical microecology, there still are many problems, the main one of which is how to find valuable criteria for evaluating the status of individually different composition of microflora (both at the species level and in cell numbers) (Borriello, 1983, 1990). Moreover, it has not been solved whether the absence of lactobacilli in fecal samples or their high quantities on the mucosa of the small intestine could be considered an abnormality.

Lactobacilli are described as sensitive to various exo- and endogenous factors such as diets (reviewed by Hentges, 1983), antibacterial preparations (Heimdahl and Nord, 1979; Lencner et al., 1981; Bennet et al., 1982), and emotional stress before and during space flights (Lizko et al., 1984).

Lerche and Reuter (1962) could not reveal lactobacilli in the feces of elderly people, but in our investigations the counts and frequency of occurrence of fecal lactobacilli in the aged (65–89 years old) were similar to those of young (21–44 years old) healthy persons (Table 5). Even the number of different lactobacilli species isolated from one sample coincided (one to four species) in both groups (Mikelsaar, 1969).

Some interesting data were obtained from studying the *Lactobacillus* content in patients suffering from some noninfectious GI diseases. Voronina (1968) has revealed that the occurrence of lactobacilli in the gastric content increases in patients with chronic gastritis and gastric cancer associated with anacidity. Alongside with rising interest towards gastric mucosa colonization with *Helicobacter pylori* in cases of chronic gastritis it was found that lactobacilli do not persist in the normal gastric mucosa and rarely occur in association with *H. pylori* in the heavily colonized mucosa of the antrum (Mikelsaar et al., 1990). It is probable that only the colonization of transient nonindigenous food-deprived lactobacilli may cause, in cases of anacidity, the overgrowth syndrome-like situations in the upper part of the intestine. Our experience (Mikelsaar, 1969) revealing increased quantities of fecal lactobacilli counts in cases of anacidity without changes of peristalsis of the gut (Table 5) suggests it. At the same time, in anacidic patients with a dumping syndrome (which causes intensive emptying of the intestine), the number of lactobacilli was not increased. Yet, these mostly descriptive studies are regrettably limited by the lack of real understanding of the function of lactoflora of a particular biotope.

Many efforts are taken to influence various aspects of human health by administering *Lactobacillus* supplements (Alm, 1983; Gorbach, 1990; Oksanen et al., 1990). These supplements obviously affect the microbial ecosystem of the GI of a particular host as environmental factors. As can be seen from Fig. 3, they may exercise their influence as food microbes or nutrients on different stages of the

Table 5 Lactobacilli in the Feces of Patients with Some Noninfectious Gastrointestinal Diseases

	Group of persons	*n*	Lactobacilli: Frequency of occurrence (%)	Number (log cfu/g): Range	Median	Mean
I	Young healthy persons 21–44 y old	10	98	5.5–8.1	6.6	6.7 ± 0.9
II	Old healthy persons 65–89 y old	13	100	6.0–8.0	7.0	7.2 ± 0.9
III	Chronic gastritis, anacidity 21–69 y old	10	100	7.0–9.0	8.0	8.3 ± 0.7[a]
IV	Carcinoma ventriculi 50–74 y old					
	anacidity	8	100	7.8–10.0[b]	8.8	8.8 ± 0.9[a]
	hyperacidity	6	100	5.7–8.0	7.3	7.1 ± 0.8
V	Peptic ulcer 22–68 y old					
	normacidity	3	100	6.1–7.0	7.0	6.8 ± 0.5
	anacidity	5	100	7.1–9.8[b]	7.8	7.7 ± 1.2
	hyperacidity	14	100	5.0–8.0[b]	5.8	6.6 ± 1.1
VI	Status post resectionem ventriculi without complaints 35–62 y old					
	anacidity	6	100	8.7–10.0[b]	9.6	9.5 ± 0.6[a]
	hyperacidity	4	100	7.7–8.7	8.3	8.2 ± 0.4
	with complaints (dumping syndrome) 24–64 y old					
	anacidity	23	100	5.8–10.6	7.5	7.6 ± 0.9

[a] $p < .01$ Student's *t*-test.
[b] $p < .05$ Mann-Whitney test.
Source: From Mikelsaar (1969).

microbial ecosystem, which makes the investigation of their role in the GI a very complicated task.

III. ESTABLISHMENT OF NEONATAL MICROFLORA

The pathways, manners, and various factors causing the contamination of the sterile fetus during labor and of the newborn in the first hours and days of its life and also determining its microflora formation, have been drawing scientists' attention for nearly 100 years (Tissier, 1905; Cooperstock and Zedo, 1983; Goncharova et al., 1987; Hall et al., 1990).

In the formation of normal human microflora two stages can be distinguished: the acquisition of microorganisms by contamination with mother's microorganisms or with other environmental microbes and the successive colonization of different habitats of the neonate (Savage, 1977).

The real colonization means the persistence of microorganisms in the biotope even up to 14 days after their first appearance (Van der Waaij, 1973, 1988).

A. Transmission of Microorganisms from Mother to Infant

Several authors suggest that the neonate is sterile during the period of intrauterine life (Haenel and Bendig, 1975; Davies and Grothefors, 1984). Contamination with commensal bacteria, derived from the microflora of mother's vagina, intestine, and skin and from the environment occurs soon after birth (Bullen, 1977; Ross and Needham, 1980; Rotimi and Duerden, 1981; Bennet and Nord, 1987; Ekwempu et al., 1982; Enhtuja, 1984). Many of these microbes are unable to colonize habitats in the neonate and disappear soon after birth, whereas other microorganisms remain or may support the successive colonization during the early life period to form climax communities in the adult (Savage, 1977, 1989).

1. Infectious Microorganisms

The interest of investigators in the above field has focused on two directions. The widest trend of investigations has been prompted by the practical goal of preventing ante- and perinatally derived infections, such as listeriosis, toxoplasmosis, chlamydial, and various viral infections (Iwasaka et al., 1986; Maciejewski et al., 1987). These infections are generated by pathogenic microorganisms the transmission of which from mother to her neonate has been documented (Davies and Gothefors, 1984).

In the last years, no less attention has been paid to the transmission of the potentially pathogenic (opportunistic) microorganisms like *Staphylococcus aureus, Clostridium difficile*, and *Group B Streptococci* (GBS) from the mother to her neonate. These bacteria may or may not cause various early postnatal infections (Manso et al., 1986; Rudigoz, 1988; Tullus et al., 1989; Kay et al., 1990).

Lately, it has been suggested that they have a certain role even in preterm labor connected with extra- and intra-amniotic infections (Iams et al., 1987; McGregor et al., 1988, 1990).

It has become known that these opportunistic microorganisms are kept in mutual or commensal interactions with the host by various factors characteristic to the macroorganism and other microorganisms of the indigenous microflora (Hungate, 1984; Savage, 1989). It has been speculated that the resident strains of the indigenous microflora, through various mechanisms, protect the baby from the very beginning of its life from the randomly acquired opportunistic strains (Haenel and Bendig, 1975; Davies and Gothefors, 1984).

Thus the second direction of the investigations of microflora transmission from mothers to their newborns is concentrated on the problem of how the colonization controlling indigenous microflora is formed. In this respect there has been a resurgence of interest towards the colonization of the neonate by lactobacilli (Hall et al., 1990; Tannock et al., 1990b).

2. *Indigenous Microflora*

Several studies have described the early skin contamination of the newborn, comparing the neonate's and its mother's microfloras. The investigation by Graham (1975) has shown that only 25% of healthy neonates harbor microorganisms on their skin immediately after birth and the latter are similar to those of the vagina. Ivanov (1985) has shown that the contamination of healthy neonates with microbes varies in different clinics from 7% to 30%, the most frequent contaminants being epidermic staphylococci.

It may be possible that the results of the investigations depend on the location in the baby from which the material was taken. So Bakuleva et al. (1984) claim that the gastric aspirates of neonates were bacterioscopically and bacteriologically uncontaminated. Deshchetkina et al. (1990) described sterile samples of meconium in 83% of the investigated newborns. On the other hand, Sycheva et al. (1986) found microorganisms on the conjunctiva of 83% of the neonates and the microbes resembled those detected on mother's skin. However, the conflicting results may also depend on the different media and methods used for the search of various groups of microorganisms. Very important data are gained when comparing, on the one hand, the mother's vaginal microflora with that of the neonate after vaginal delivery or, on the other hand, mother's amniotic fluid with the newborn's microflora after a caesarean section.

Cesarean section thoroughly alters the colonization patterns in newborn infants. Anaerobic colonization is delayed and there appears an overgrowth by enterobacteria (Rotimi et al., 1985). The colonization of newborns delivered by cesarean section occurs during the first days of life by bacteria provided by the outer environment (Neut et al., 1987). While Lennox-King et al. (1976) find the most common sources for *Escherichia coli* colonization to be the nurses' hands and the contaminated air, Bezirtzoglou and Romond (1990) deny the role of hospital environment and the type of feeding in Enterobacteriaceae colonization.

Recently Torres-Alipi et al. (1990) could reveal no correlation between the microorganisms isolated from the amniotic fluid and the neonate's oral cavity after a cesarean section. In case of the neonates obtained by vaginal delivery there was, however, a correlation between the microorganisms of the oral cavity and the maternal vaginal flora.

Rotimi and Duerden (1981) in their now classical study found that immediately after birth in 52% of cases the lactobacilli were present both in childrens' mouths and in mothers' vaginas. Only in rare cases lactobacilli were isolated either from the mother (14%) or from her neonate (9%).

We compared the microflora of 21 healthy mothers during delivery with that of their newborns also immediately after birth. We examined the skin microflora of mother's perineum and her baby's external ear canal where the amniotic fluid may possibly persist and which remains untouched by the hands of the medical personnel. Aerobic and anaerobic microorganisms were present both in perineal and ear samples (Mikelsaar et al., 1989). In mothers and in their neonates 13 and 11 groups of microorganisms were detected, respectively (Figs. 5, 6). The most frequent microorganisms were the same both in mothers and their newborns: lactobacilli, epidermic staphylococci, and nonhemolytic streptococci. Various combinations of microbes were found mostly dissimilar in the investigated mothers. The maternal perineum was heavily colonized with microorganisms in 75% and the infants' ear skin in 24% of the cases (more than 100 cfu/tampon) and there was no real accordance in the quantitative characteristics of microbial densities between mothers and their infants. Yet, if we studied, using the same methods, the maternal vaginal and neonatal ear microfloras (19 mother–newborn pairs) we could see a close correlation between the colonization rates of the above-mentioned biotopes ($p < .01$; Kasesalu et al., 1991). We found similar predominant groups of microbes in 12 pairs out of 19 (63%). Of the remaining seven pairs, in four babies we could not detect any of the microorganisms searched for. Thus, it has been demonstrated that maternal vaginal microflora species and its quantitative characteristics directly influence the initial contamination of the newborn.

The assumption that the source of the bacteria which finally colonize the infant is the maternal vagina, has not been completely proved yet. Lately discrimination tests permitting the comparison of bacterial strains isolated from maternal and infant sources have been developed. Plasmid profiling has proved a useful technique for that (Davies et al., 1981; Farrar, 1983).

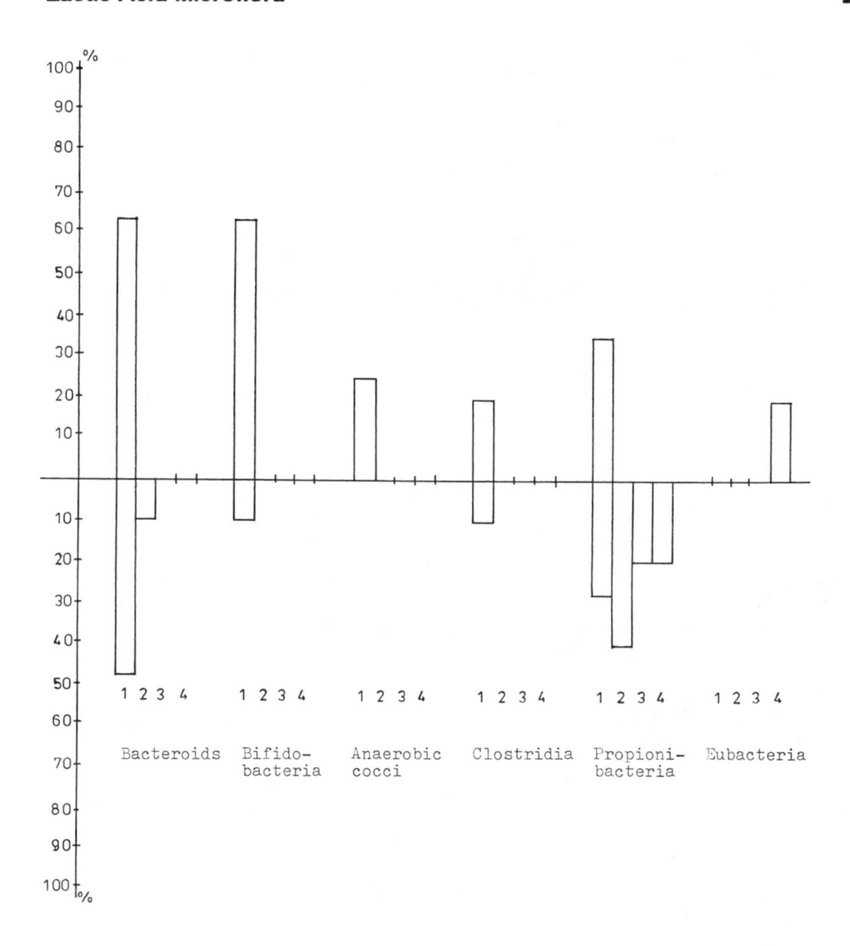

Figure 5 Frequency of occurrence of anaerobic microorganisms on the skin of mothers during labor and their newborns immediately after birth: 1—in labor; 2—at the age of 5–6 days; 3—at the age of 6 months; 4—at the age of 1 year (Mikelsaar et al., 1989a).

Tannock et al. (1990b) compared the plasmid profiles of the microbes of the family Enterobacteriaceae, lactobacilli and bifidobacteria cultured from the vaginal, oral, and rectal swabs of birth-giving mothers with the strains detected in the feces of their infants 10 and 30 days after birth. Lactobacilli inhabiting the mothers' vaginas did not appear to colonize the infants' digestive tracts, but the authors got evidence of the transmission of fecal isolates of enterobacteria and bifidobacteria from mothers to their infants.

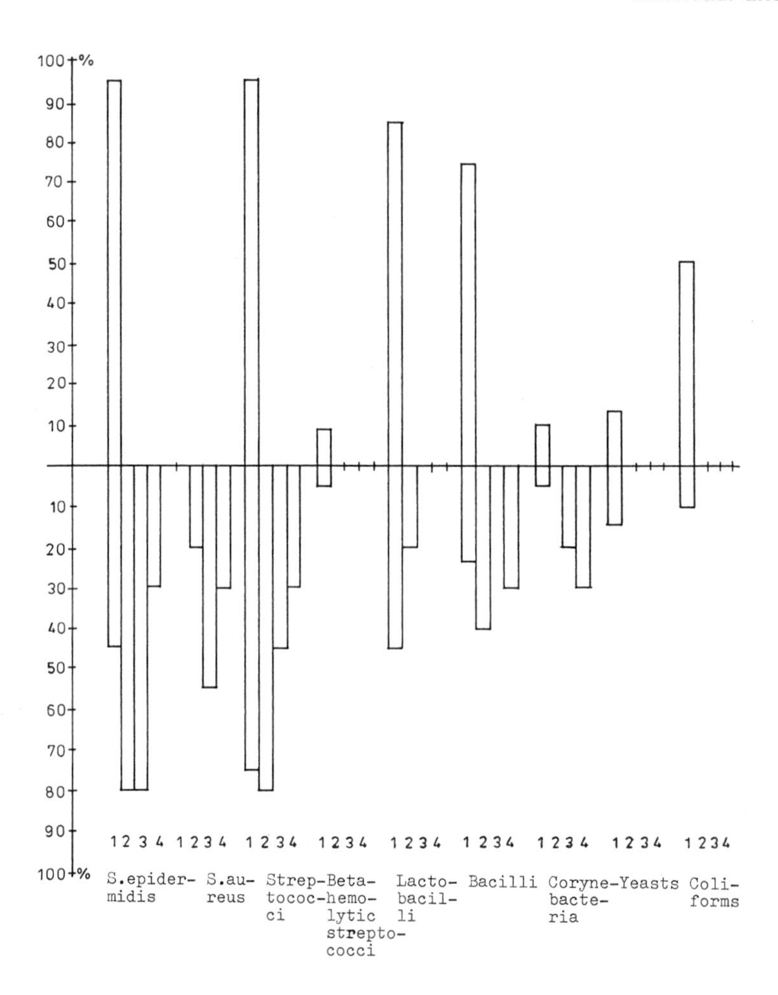

Figure 6 Frequency of occurrence of aerobic microorganisms on the skin of mothers during labor and their neonates immediately after birth. Legend is the same as Fig. 5 (Mikelsaar et al., 1989a).

B. Development of Infants' Individual Microflora

1. *Skin*

Investigations studying the development of normal skin microflora of infants beginning from the moment of birth and lasting through the first year of life are comparatively rare. As a rule, the authors tend to follow the development of skin microflora during the first week of life in the maternity clinic (Rotimi, Duerden, 1981; Ivanov et al., 1985).

It has been shown that during the first one to two days the skin microflora is similar to that of the birth canal and it is only on the third day that the indigenous skin microflora starts to develop (Ivanov et al., 1985). Most authors confine themselves to aerobic microorganisms. By the sixth day the quantitative and species composition is told to be maximally developed (Rotimi, Duerden, 1981). Yet the data of Bochkov et al. (1988) show that on the fifth day the microbiocenosis of skin was not developed yet and many bacteria could be detected (coliforms on the mucosa of the throat, nose and conjunctiva) which are not characteristic of the biotope. However, the average data about the frequency of isolation or quantities of microbes prove to be of little assistance in understanding the formation of individual microflora (Rotimi and Duerden, 1981; Ivanov et al., 1985; Bochkov et al., 1988).

We studied the development of skin microflora in the external ear canal of 10 infants during the first year of life in connection with their first contamination. On group level, mostly ordinary alterations were registered: the disappearance of the lactobacilli predominant immediately after birth and the emergence of *S. aureus*. Out of the anaerobic microorganisms detected immediately after birth, only propionibacteria persisted on the skin of the six months and one year old infants. According to the average data (Fig. 7a) it appears that from the end of the first week of life up to the end of the first year mostly the same microorganisms (*S. epidermidis*) prevail. However, observing the distribution of microorganisms of five infants with high colonization rates (>100 cfu/per tampon), individual differences of microflora development could be noticed (Fig. 7b). So, at the end of the first week in some infants lactobacilli or streptococci formed the predominant microflora, while in others epidermic staphylococci showed a high colonization rate. The tempo and colonization rate of epidemic staphylococci does not seem to affect the occurrence of *S. aureus* on the skin of the ear. In no infants these microbes predominated over other kinds of microflora and no infections with *S. aureus* were documented either. Thus, the mere fact of isolating *S. aureus* from the skin of a neonate even in the case of a high colonization rate of the biotope does not predict the possibility of infection.

2. *Intestine*

We have found no studies following the formation of gastric or small intestinal microflora during infancy. All the available investigations concern the fecal microflora development, mostly comparing its development in breast-fed infants and in babies fed with various formulas (Stark and Lee, 1982; Benno et al., 1984; Goncharova et al., 1987; Baquero et al., 1988; Bennet and Nord, 1987; Kay et al., 1990). Unfortunately, bacteriological investigation of fecal samples cannot give

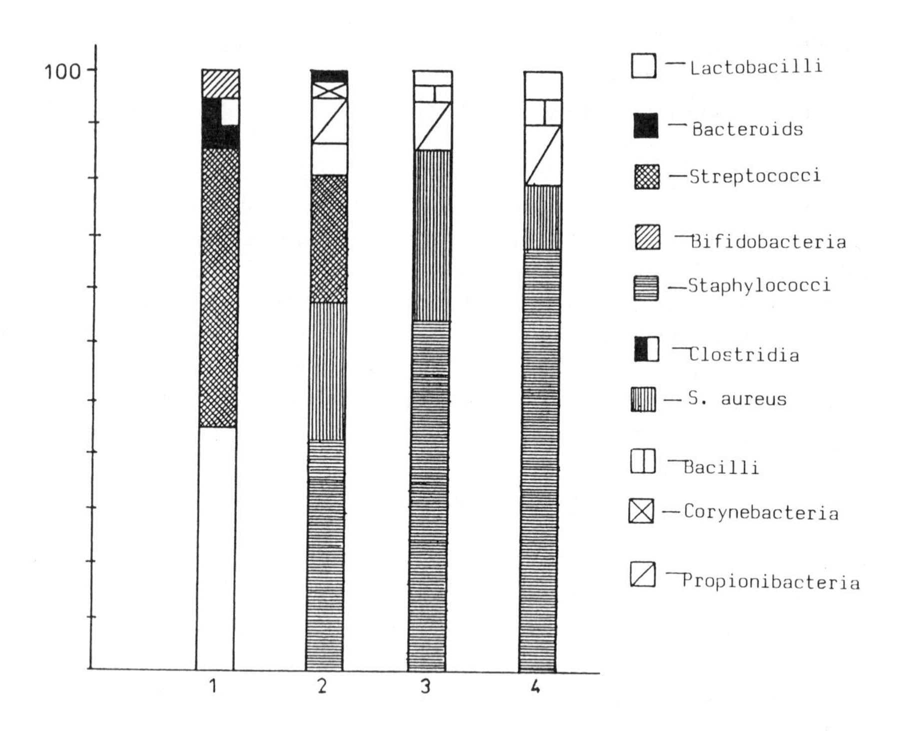

Figure 7a Distribution of microorganisms of infants in the cases of high colonization rate of skin. Mean data: 1—in labor; 2—5–6 days; 3—6 months; 4—1 year.

any information about either the predominance or subordinance of lactobacilli in the contents or the mucosa of the upper parts of the intestine as was shown in Section D.2. It is well known that the first intestinal colonizers are coliforms, streptococci, and staphylococci (Lundequist et al., 1985). From the third day on up to the end of the first week of life anaerobes—bacteroides, bifidobacteria, and eubacteria—gain high population levels (Bullen, 1977; Yoshioka, 1984; Lundequist, 1985).

The frequency of occurrence of lactobacilli in the first days of life is variable, according to the data provided by different authors (15–100%), and it is higher with formula-feeding (Lundequist et al., 1985; Deshchetkina et al., 1990). Recently Hall et al. (1990) reported of having found high counts of lactobacilli but not of bifidobacteria in the stools of most 30-day-old infants born at full term. These data contradict the earlier well-known suggestion that with breast-feeding

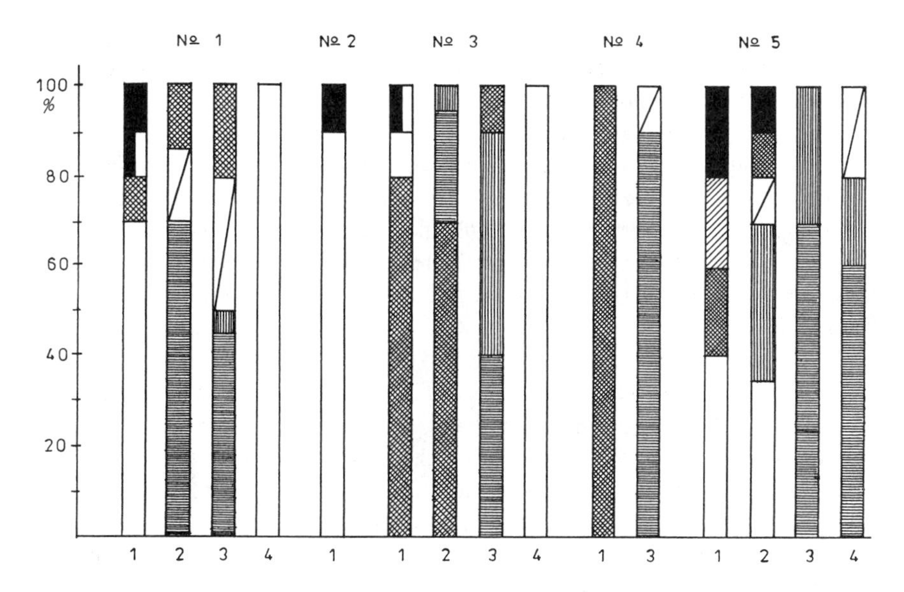

Figure 7b Distribution of microorganisms of infants in the cases of high colonization rate of skin. Individual development in cases: 1–5. Legend is the same as Fig. 7a.

bifidobacteria predominate from the end of the first week of life on (Mata and Urrutia, 1971; Mitsuoka and Kaneuchi, 1977; Lundequist et al., 1985; Goncharova et al., 1987). Alternatively the investigations of Simhon et al. (1982) and Gothefors (1989) showed that there was no difference between breast- and formula-fed English infants' fecal microflora, as in both groups bacteroides were the predominant microorganisms. At the same time, in Nigerian children bifidobacteria were observed to predominate in fecal flora (Simhon et al., 1982). Lundequist et al. (1985) explain it with the ecological imbalance of women's vaginal microflora in the developed countries and the excessively high hygiene at birth giving which eliminates the normal microflora otherwise acquired from the mother.

When the child starts to get solid food (mostly from the fifth to sixth month), high quantities of bacteroides, anaerobic cocci and clostridia in the feces of infants are detected (Hentges, 1983). If breast-feeding is totally stopped, the infants' intestinal microflora succession is broken (Cooperstock and Zedo, 1983).

The microbial metabolite investigators have also proved the long step-by-step formation of the GI microflora of neonates and infants. Thus the microbial SCFA composition in the feces of neonates is described as quite specific with a high concentration of acetic acid, connected with aerobic metabolism. As they grow older, the composition changes, becoming more complex and displaying wide interindividual variations (Rasmussen et al., 1988).

The mucin degrading capacity is one of the microflora-associated characteristics (MAC) of the host which serves as a good indicator for the study of microbial ecology of the intestine (Midtvedt, 1989b). Norin et al. (1988) followed the mucin degradation process from birth and found that mucin breakdown starts only in the second year of life. Yet the figures offered by the authors suggest that mucin degradation may occur in some children already at the age of 2–11 months. Thus, the metabolite studies confirm the fact that the process of microflora formation of infants is individually different.

We studied the predominant (over 10% of the total count in cfu/g) populations of 18 infants at different times during a year (Fig. 8). The wide interindividual variations were observed just as in the biochemical studies of fecal SCFA of the same infants. In several infants we could detect the same predominant groups of microbes during the investigation period. So in six children out of nine (67%) in three observations, and in seven infants out of nine (78%) in two observations the same predominant microorganisms were found.

Yet lactobacilli were never among the predominant fecal microflora of the infants investigated by us (Mikelsaar, 1989a). However, what role they might play in the upper parts of the intestine, cannot be deduced on the basis of fecal lactobacilli counts.

The question about how the stable population level and the same species of lactobacilli are retained in the intestine of individuals cannot be answered without previous animal modeling.

Very important are the data about the acquisition of the vaginal microflora by infants (Redondo-Lopez, 1990). The microorganisms inhabiting the vagina of a newborn female are similar to those found in her mother because the oestrogen level in the newborn's vaginal epithelium remains high even two to three weeks after birth. Later changes can be detected both in the vaginal epithelium and in its microflora, owing to the disappearance of the hitherto predominant lactobacilli. All these facts suggest the presence of the complex microbial ecosystems in various organs, each having a specific microflora formation process of its own.

C. Factors Contributing to the Development of Individual Lactoflora

As noted, individual lactoflora means the persistence of individually different lactobacilli strains in relatively stable quantities in the particular biotope of the host. The selective colonization of a biotope by lactobacilli is a function of the host as well as of the microorganism.

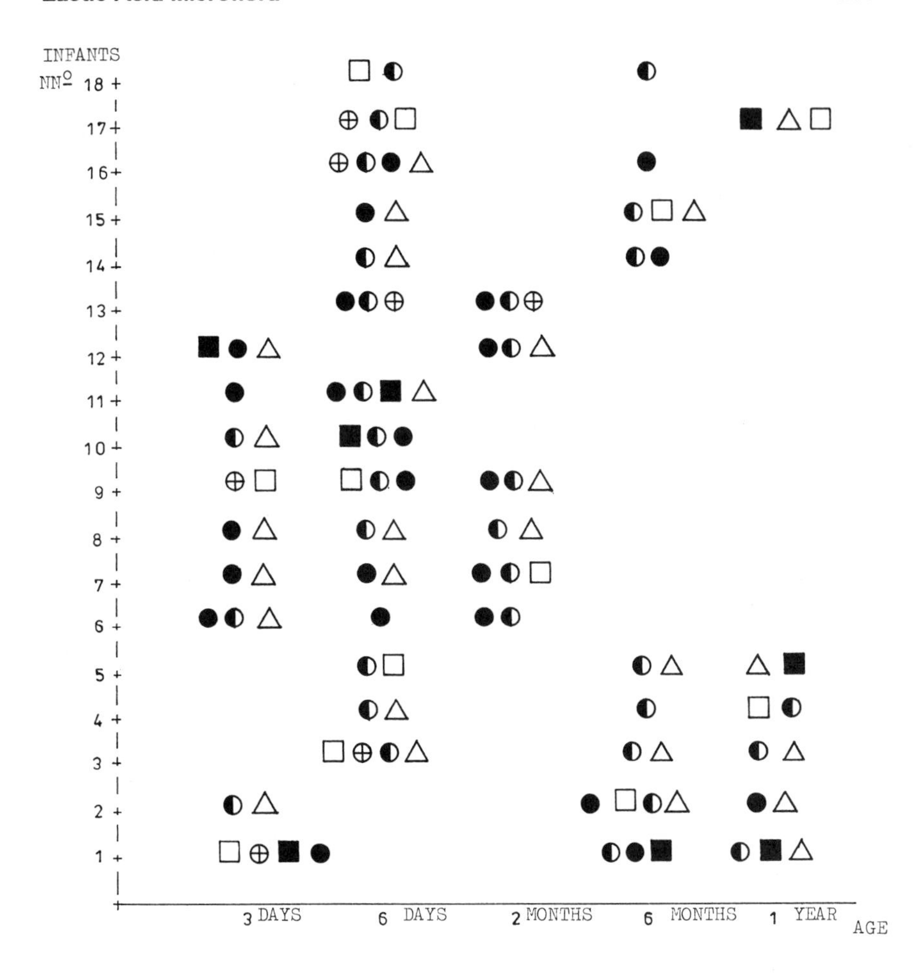

Figure 8 Predominant microorganisms of fecal microflora of infants during a year. □ Coliforms, ■ Eubacteria, Δ Bacteroides, ● Anaerobic cocci, ⊕ Streptococci, ◐ Bifidobacteria.

Colonization of the biotope by particular strains of lactobacilli requires some important characteristics of microorganisms: an ability to survive in the environment and in the secretions of the host, resistance against the antagonistic activity of other microorganisms and an ability to depress their growth to some extent, adherence to host structures (epithelial cells) or secretions (the mucin layer of the mucosa), and degradation of the endogenous nutrients of the host for achieving stable population levels (Hungate, 1984; Savage, 1987). All of these properties may simultaneously influence the formation of individually specific lactoflora, which makes investigations into the microbial ecosystem of different organs a highly complicated task.

1. Survival in the Secretions of the Host

We have found no studies on the bacteriostatic or bacteriocidal influence of host secretions of a particular biotope upon the lactobacilli isolated from that biotope. There may be some specific characteristics of both variables—secretion and microbes—which determine the survival of a specific strain. The survival of mainly commercial lactobacilli in the human stomach and in transition through the intestine has been investigated both in vivo and in vitro (Bernhardt, 1980; Robins-Browne et al., 1981; Conway et al., 1987; Saxelin et al., 1991). The idea of Reuter (1975) and Knoke and Bernhardt (1985) that the extreme conditions (acidity) of the stomach select the biotope-specific lactobacillar strains has not been completely validated yet. However, it seems to be of real value.

One of the host secretions that can influence the composition of microflora is lysozyme excreted by the cells of the host (Knoke and Bernhardt, 1985). It is known that in plasma the mean concentration of lysozyme is 30 μg/mL (Korvjakova and Gorohov, 1979). Practically all of the lactobacilli isolated from the normal microflora of humans are resistant to high concentrations of lysozyme (50 mg/mL; Lencner and Lencner, 1982). Yet the strains of lactobacilli detected in the environment are susceptible to lysozyme (15 mg/mL). Lysozyme resistance was not confined to particular species, there were strains with different susceptibility in various species of lactobacilli. For example, all the strains of *L. fermentum* isolated from the mucosa of the intestine of mice were resistant to high concentrations of lysozyme, but there was a variable susceptibility among the strains of *L. acidophilus* (Kolts et al., 1986). Consequently, the difference of susceptibility/resistance patterns to lysozyme may be one of the factors determining the individuality of lactoflora.

The influence of hormones on the number and species of lactobacilli has been successfully followed in vaginal microflora studies. A high estrogen level of the fetus generates predominance of lactobacilli in the vagina. And the reverse is also true—a decrease of estrogen in postmenopausal women causes the disappearance of lactobacilli from their genital tract. Estrogen therapy can restore the earlier situation (Larsen, 1982).

The importance of hormones for colonization by lactobacilli is vividly proved by the data revealing that progesterone and estradiol were able in vitro to intervene the adhesion process of different lactobacilli (Brilene and Brilis, 1986). Whether these processes are connected with the individuality of lactoflora is not clear yet.

2. *Antibacterial Substances*

In vitro experiments have shown the ability of lactobacilli to inhibit the growth of pathogenes, while their primary metabolites like lactic and acetic acid as well as hydrogen peroxide are well-recognized antibacterial agents (Speck, 1976; Redondo-Lopez, 1990).

Lactobacilli are also known to either produce a variety of inhibitors (viz., bacteriocins) with a narrow spectrum (McCormick and Savage, 1983) or have a broad antibacterial activity (Goldin and Gorbach, 1987; Dobrogosz et al., 1989). One of the broad-spectrum antimicrobial substances, termed reuterin (Chung et al., 1989) is synthesized in vitro at the pH, temperature and the relatively anaerobic conditions similar to those believed to exist in the parts of the pig GI ecosystem inhabited by *L. reuteri.* Reuterin appears to be an antibacterial, antifungal, antiprotozoan and antiviral agent. With its low molecular weight and nonprotein origin it is a neutral end product of glycerol fermentation (Dobrogosz et al., 1989).

Another low-weight antimicrobial substance of lactobacilli is described in the *L. casei* GG strain (Silva et al., 1987). The GG strain was isolated from the feces of a normal human. It produces a substance having a potent inhibitory effect on a wide range of bacterial species. The substance differed from lactic and acetic acid and resembled microcin, previously associated only with the members of the Enterobacteriaceae family. A very valuable property of the aforementioned substance is that it is not inhibitory against other lactobacilli. That the *Lactobacillus* GG strain is able to colonize human subjects after the administration at a dose of 10^{10}–10^{11} cfu per day has been shown recently by Saxelin et al. (1991). A positive effect on relapsing *Clostridium difficile* colitis was reported but there are no data available about the impact of the *Lactobacillus* GG strain colonization on the composition of microbiocenosis of the gut.

Yet, a high antibacterial activity determined in vitro cannot always predict the behavior of a particular strain in the GI microbial ecosystem. Thus, in rats monoassociated with *Bacillus subtilis* the *Lactobacillus* strains with a high bacteriocidal effect upon this bacterium could not exclude the *B. subtilis* from the mucosal microflora of the ileum (Lencner et al., 1980). These data seem to show that in the formation of a particular microbiocenosis the main role is played by the ecological niche.

It is thought that in microbiocenosis, lactobacilli may exclude pathogens or potentially pathogenic microorganisms by either occupying the receptors of epithelial cells or exhausting the nutrients (Tannock and Archibald, 1984; Savage, 1987). Conway (1989) suggests competitive colonization which includes inhibition of growth and colonization. Both aspects seem to be important for the genesis of individually specific lactoflora.

3. Adhesion Properties

Attachment (adherence, adhesion) to solid substrates is a property characteristic of many pathogenic microorganisms and also of the indigenous ones (Baddour et al., 1990). By attaching, microorganisms avoid being swept along and eliminated by the normal flow of body fluids (Reid and Sobel, 1987). Although human cells associated with bacteria can be shed, too, this is a way to eliminate the microbes in spite of adherence (Savage, 1987). Attachment is considered as a necessary first step in the colonization of host mucosal surfaces. The attachment of micro-organisms to epithelial cells of the mucosa is the result of a specific binding process involving surface adhesins on the bacteria and mucosal receptors in cell membrane (Wadström, 1988).

Very many studies have been performed on the adhesion characteristics of Gram-negative microorganisms, especially of *E. coli* strains to the cells of the urinary tract in case of urinary tract infections (UTI). It has been shown that bacterial surface adhesins are expressed as fimbria (pili), filaments (like "fuzzy layer"), nonfimbrial cell-surface lectins (hemagglutinins) and intracellular lectins which are able to recognize specific sugars. They all express high surface hydrophobicity (Wadström, 1988).

Lactobacilli, originating from human intestinal microflora also show high surface hydrophobicity (Wadström et al., 1987). This indicates that members of normal microflora can probably use the same mechanisms for colonizing mucosal surfaces as pathogens do.

At the turn of the 1970s the adhesion of lactobacilli to the nonsecreting epithelium of the stomach or crop in some experimental animals (chicken, rodents, pigs) was thought to be associated to the microbial cell wall filaments, containing acidic mucopolysaccharides (Fuller, 1975; Lorenz et al., 1982). Nevertheless, it was inexplicable how pepsin could intervene into that process (Fuller, 1975). Nowadays it is known that lactobacilli may adhere to the epithelial cells due to their teichoic and lipoteichoic acid-containing structures, polyvalently linked to the peptidoglycans of the cell membrane (Wicken, 1982; Sherman and Savage, 1986; Courtney, 1986). Adhesion to the stomach epithelium is also mediated by the proteinaceous structure of the microbial cell wall showing high hydrophobicity (Savage, 1987; Conway, 1989). Especially in the ileum, hydrophobicity is concluded to be of utmost importance for lactobacilli adhesion and the polysaccharides may not be involved at all (Wadström, 1987). So lactobacilli own specific structures on their cell wall like the "fuzzy layer" consisting of proteins and polysaccharides on the two external covers of the cell wall (Dalin and Fish, 1985; Cook et al., 1988). These structures should be complementary to the receptors of epithelial cells and the mucus layer of the intestine for determining their adherence specificity.

It is known that the adhesion of lactobacilli is host specific because only lactobacilli originating from a particular host species can colonize the epithelium in germ-free animals (Tannock et al., 1982).

However, with these described mechanisms it is hard to explain the individual specificity of lactoflora. To elucidate that problem, we have studied the biochemical basis of the in vivo occurring adhesion of lactobacilli to the GI epithelium of experimental rodents (Kolts et al., 1986). For that purpose we tested the impact of enzymes, hydrolyzing the glycosidic linkages in polysaccharides (lysozyme, hyaluronidase, neuraminidase) on the in vivo attachment of lactobacilli to the gastric and colonic mucosa of conventional rats and mice. Before and after the exposure of slices of the mucosa to the enzymes the counts of lactobacilli were ascertained by cultivation in the MRS medium. Lysozyme (egg, Olaine, Latvia, 15 mg/mL) released lactobacilli from the gastric mucosa of rats but not of mice. In another biotope—the colon—the lactobacilli of mice were released, whereas those of rats remained attached to the mucosa (Fig. 9).

Additionally, we investigated the susceptibility of lactobacilli, originating from mice and rats, to different concentrations of lysozyme. It was revealed that all the rats' strains were highly resistant to lysozyme (50 mg/mL). The susceptibility of mice's lactobacilli to lysozyme was to some extent dependent on their species: all the strains of *L. fermentum* were resistant but only 5 out of 28 of *L. acidophilus*-line strains did not show any growth even at the lowest concentration of lysozyme (15 mg/mL), while 13 strains showed high resistance (50 mg/mL).

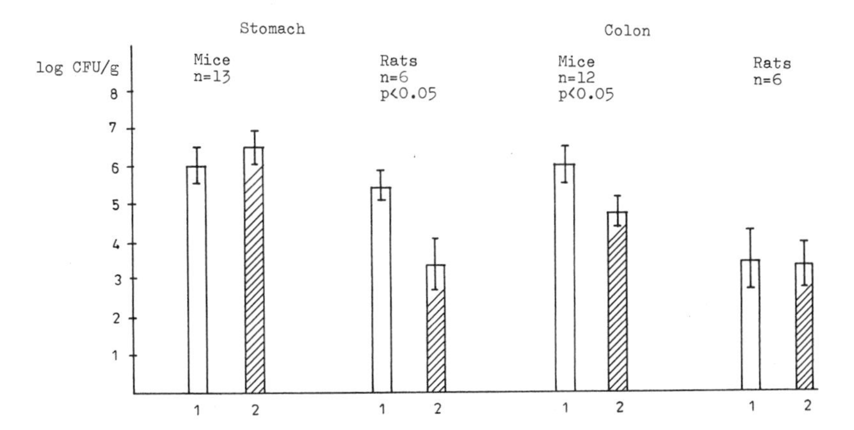

Figure 9 Release of lactobacilli from the gastric and colonic mucosa of mice and rats by lysozyme: 1—control; 2—lysozyme (15 mg/mL).

Lysozyme hydrolyses the glycosidic bonds between glucosamine and *N*-acetyl-muraminic acid in the compounds of neutral mucopolysaccharides and peptidoglycans of the cell wall of microorganisms (Barnard and Holt, 1985). But the detected differences in releasing the lactobacilli of mice and rats cannot be explained by the differences of their peptidoglycans since the lysozyme-susceptibility of all the strains originating from mice and rats was not the same. Our experience seems to demonstrate the particular polysaccharidic specificity of adhesin-receptor bonds of both the stomach and the colon. In the latter, lactobacilli adhere to the mucus layer of the intestinal epithelium. The biochemical mechanisms of adhesion proved to be different in case of even such close species as rats and mice. Thus, adhesion seems to be the basis for residence of lactobacilli in specific biotopes and for particular species of macroorganisms.

Hyaluronidase (testes, Reanal, 15 mg/mL) did not release lactobacilli either from the stomach or from the colon of either mice or rats. Consequently, acidic polysaccharide bonds cannot account for the specificity of adhesion of lactobacilli. These experiments are complementary to the data of Fuller (1981) who, by treating the lactobacilli strains or epithelial cells with hyaluronidase before attachment in vitro, could not find any differences in the adhesion of lactobacilli to the cells.

Supposing that the adhesion of lactobacilli to the epithelium or mucus may be brought about by the terminal residues of glycoproteins, glycoseaminoglycans, or glycolipids, containing sialic acid, we tested their release by neuraminidase. Neuraminidase (type 5, sigma 0.002, 0.005, 0.006, and 0.05 U/mL) released lactobacilli after different exposition times and in different concentrations from the epithelium of the stomach but not from the colon of rats. The impact of neuraminidase on the *Lactobacillus* counts in the colon varied: in some animals the counts increased, in others it decreased after the exposure to the enzyme (Fig. 10). The decrease of the number of lactobacilli in the gastric mucosa of rats and the colon of some animals may be caused by the breakdown of the α-glycosidic bond between sialic acid and the other part of polysaccharide determining the adhesion of lactobacilli. On the other hand, the increase of adhesion of lactobacilli to the colonic mucosa may be connected with the ability of neuraminidase to release and expose the β-galactosidic residues of epithelial cells to which lactobacilli originating from the wash water of the mucosa slices may adhere. These processes were not the same in different animals, confirming the idea of individual specificity of the adherence of lactobacilli.

It cannot be said whether the individual specificity is due to differences between *Lactobacillus* strains or the mucins. Both of them contain, in their polysaccharide sialic acid or its derivate, neuraminic acid. The wide phenotypic variations in heteropolysaccharides of *Lactobacillus* strains have been documented (Wicken, 1982). The mucins of individuals may also differ in the content of terminal sialic

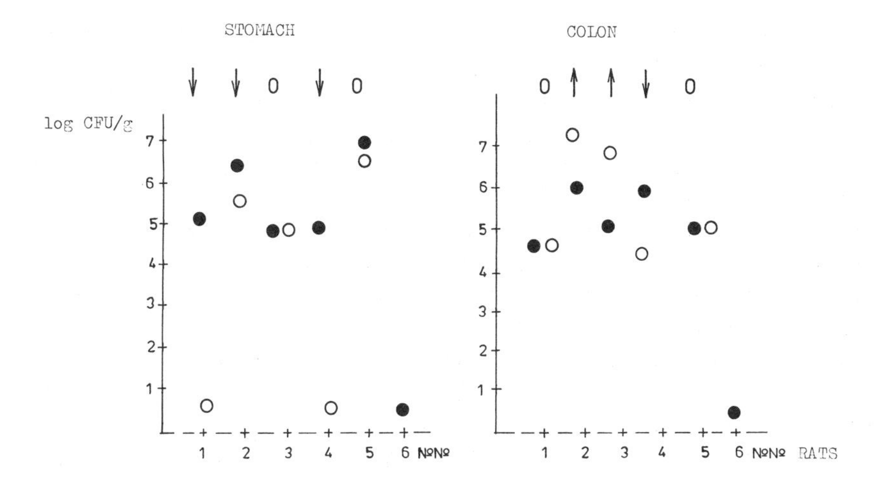

Figure 10 Influence of neuraminidase on the lactobacilli counts of gastric and colonic mucosa of rats: ●—control; ○—neuraminidase (0.005 U/mL; ↓—decrease; ↑—increase; 0—no change).

acid, which plays the receptor function for specific adhesins (Wadström and Aleljung, 1989; Carlsson et al., 1989). Hopefully, more precise methods of investigation will provide the answer. However, the practical aspect of these studies is how to make use of the ability of lactobacilli to interfere in vivo with the adherence of pathogens to the mucosal cells of various organs. Competition for the epithelial cell surface receptors has been an important mechanism for uroinfection prevention by lactobacilli administered to rodents (Reid et al., 1987). It has been shown, too, that adhesion of lactobacilli to vaginal epithelial cells reduces their colonization with *Candida albicans* (Knoke and Bernhardt, 1985). The complexity of that process is obvious, because various factors like the pH and hormonal level are involved (Brilene and Brilis, 1986; von Neumann, 1988). The possibility to intervene in the adhesion phenomenon by changing the chemotaxis of microorganisms with different complement compounds and bacterial products once more demonstrated the complex microecological system present in a particular biotope. Several aspects concerning the functioning of this system are still unclear.

4. Degradation of Endogenous Nutrients

Beside adherence, colonization of a biotope requires availability of suitable nutrients for the microorganisms (Hungate, 1984; Tannock, 1984). In that sense colonization of a particular biotope by certain lactobacilli is a function of the host as well as of the microorganism.

The mucus and its main glycoproteinic compound, mucin, cover the epithelium of the GI system like a viscous substance. Wide variations in the thickness (up to 400 μm in rats and twice as thick in humans) of the gel adhering to the epithelial surface have been described (Hentges, 1983; Carlstedt-Duke, 1989). The intestinal mucin is degraded by members of the normal intestinal microflora. It has been shown that complete degradation of the mucin requires a sequential action of several bacteria. Miller and Hoskins (1981) showed that 1% of all fecal anaerobes are able to degrade the mucus, but Bayliss and Houston (1984) claim a higher mucus-fermenting organisms density in normal human feces.

Lactobacillus strains seem to be able to ferment the monosaccharides like L-fucose or compounds like D-glycoseamine which are components of many types of mucin (Salyers et al., 1977). That lactobacilli survive during long periods in mucin was shown by Sims (1964) many years ago.

In the degradation of mucins the highly substrate specific α- and β-glycosidases (Hoskins, 1981; Salyers and Leedle, 1983) and the microbial endoproteases (Variyan, 1981; Wadström and Aleljung, 1989) participate. Some Gram-negative pathogenic and apathogenic microorganisms have adhesins for the protein compound of mucin (Cohen et al., 1983; Dinari et al., 1986). At the beginning of the 1990s Wadström et al. (1989, 1990) recovered from *Helicobacter pylori*, (inhabiting the mucus of the antral part of the stomach) a lectin (glycoprotein) with a specificity for sialic acid. Thus, the receptors of mucin may be either proteins or sugars.

We have investigated the mucosal microflora of the intestine of piglets of various ages (1–80 days) in different experimental series (Kaarma et al., 1974; Mikelsaar et al., 1982, 1987). Lactobacilli were detected in the mucosa of the whole intestine, whereas the microbiocenosis of each individual piglet had its own quantitative counts and proportions of the investigated microbial groups. The colonization by lactobacilli started from the very first day and a high level of population was immediately achieved (Fig. 4). The high density of *Lactobacillus* colonization of the intestine of pigs has also been described by other investigators (Robinson et al., 1984; Graham and Aman, 1987).

The data from our experiments seem to be very important, showing that the population level of lactobacilli did not change during aging and weaning. At the same time, the other microorganisms of the intestinal microbiocenosis showed significant changes in their counts. In up to 14-day-old piglets the counts of coliforms exceeded those of lactobacilli in the mucosa of the jejunum and the ileum and were equal to each other in the content of these parts of the intestine. However, in 40–80-day-old piglets the predominance of lactobacilli in the content of the upper part of the intestine was obvious. At the same time the range and median counts of lactobacilli in the content of the jejunum during the first weeks up to the age of 80 days did not change (ranges 5.1–10.0 and 4.9–9.9 log cycles;

medians 7.8 and 7.5, consequently). Thus, the structure of microbiocenosis was changed completely during the aging of the piglets.

These experiments showed that lactobacilli may utilize the endogenous nutrients to keep a stable population level. The changes in the composition of mucins, observed during aging, cannot be as drastic (Carlstedt-Duke, 1989) as the change from milk diet to solid food during weaning. The latter causes the decrease of coliforms (Miller, 1981) and increase of anaerobic cellulolytic bacteria (Lee and Gemmell, 1972; Varel et al., 1982). Our experiments are in full accordance with the data of Kuritza et al. (1986) and Hwa and Salyers (1989), who showed that a mucin-utilizing strain did not always prevail in the microbiocenosis. These data also confirm the results of the continuous culture experiments of Freter (1983, 1990) according to which the structure of microbiocenosis is determined by the competition for the substrate where the metabolically more active strains with suitable enzymes, lactobacilli in our case, degrade the mucin, and thus keep a stable population level. On the other hand, the close connections between lactobacillus counts and mucin can be proved in our experiments by starvation of rats (Mikelsaar, 1987). The five-day-long starvation of rats led to the reduction of *Lactobacillus* counts of their stomach and intestine, which could be explained by the loss of mucin from the intestine during starvation, already described earlier by Dene et al. (1988).

Thus, the specificity of lactoflora of the various parts of the intestine cannot be explained solely on the receptor-adhesin interactions but also by the presence of the suitable endogenous nutrients. At the same time, the individually specific mucin composition is obviously the basis which integrates the different microbiocenoses of particular individuals into one complete microbial ecosystem.

In germ-free mice we have observed the same tendency: despite the use of the particular *L. fermentum* and *E. coli* strains, originating from the intestine of Balb-c mice, their quantitative relations in various biotopes of the GI were individually different. So even pheno- and genotypically identical strains develop an individually different structure of microbiocenosis, depending, it seems, on the host's biotope characteristics. There is also the possibility that special genes of the host are involved. We have not found any data concerning the indigenous microflora. However, there are data suggesting that the individuals differ in harboring special genes controlling their resistance toward pathogenic microorganisms (Skamene, 1983).

5. *Host's Genetic Influences*

Concerning pathogenic microorganisms, for example, viruses, it has been shown that the mother's genotype is an important factor in determining the fetal outcome of murine cytomegalovirus infection (Fitzgerald and Shellam, 1991). On the other hand, van der Merwe et al. (1983) has shown that in the cases of Crohn's disease the quantitative composition of normal fecal microflora is also genetically determined. We have observed that the quantitative composition of fecal microflora of adult monozygotic twins has the same degree of similarity as have the paired samples of a single young healthy person (Mikelsaar et al., 1984). Monozygotic twins reveal identity of many genetic markers, which are important for the selective colonization of the indigenous microflora (Lenz, 1976). So both the antigenic structure of somatic cells and secretions of macroorganisms, as well as the immune reaction are determined by the genotype (Warner, 1988). In these pairs of twins we have also found the correlation ($r = 0.711$; $n = 29$; $p < .001$) of the biochemical activity of microflora (Fig. 11). Consequently, the finding of similar microflora and its biochemical activity of the monozygotic twins proves the genetical determination of microflora.

The species composition of the monozygotic twins' lactoflora was individually different. Yet, in 6 out of 10 pairs of monozygotic twins we succeeded in isolating strains with similar biochemical activity of *L. acidophilus* I, *L. casei* ssp. *casei* II, and *L. brevis* I. From two pairs we could even isolate two similar strains of lactobacilli: the combination of *L. acidophilus* II with *L. brevis* I and *L. plantarum* II with *L. brevis* II (Mikelsaar and Lencner, 1982; Lencner et al., 1982). Consequently, in the GI of adult monozygotic twins identical strains of lactobacilli could persist. This is probably due to the identical specificity of adhesin-receptors.

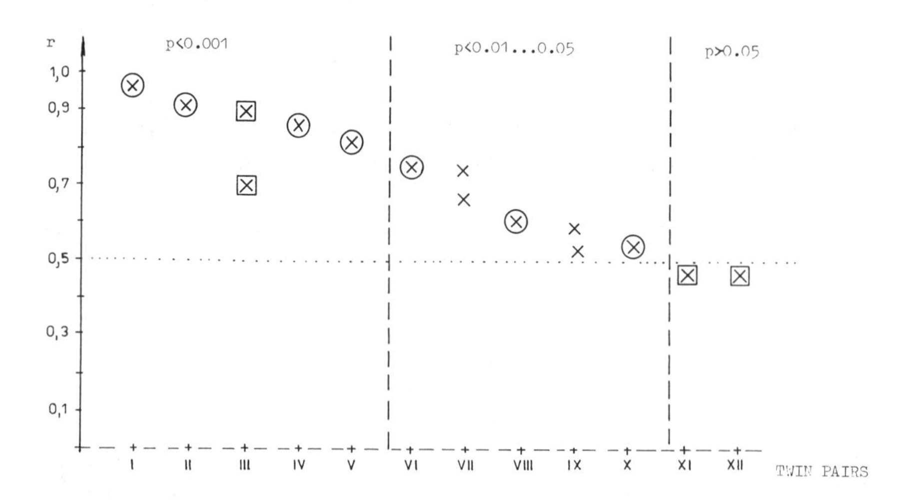

Figure 11 Similarity of quantitative composition of fecal microflora with the biochemical activity (urinary phenols) of microflora in pairs of monozygotic twins: X—the correlation coefficients (r) values of microflora; ⊗—identical biochemical activity; ⊠—nonidentical biochemical activity.

However, we could also isolate *Lactobacillus* strains with identical properties from the GI microflora of piglets of litters of three sows (Lencner et al., 1982). From the piglets of one litter there were isolated five to seven biovariants of lactobacilli. In piglets of two litters we managed to spot identical strains at the age of 1–5 days and 60–120 days. These strains colonized the epithelium of stomach and jejunum. These experiments prove at the same time the primary importance of first colonization and that of the genetic markers on the formation of individual lactoflora. To elucidate the problem we propose our view or rather "speculations" about how the individual lactoflora develops.

D. Some Speculations on Individual Lactoflora Formation

Hereby we propose for discussion the idea of antenatal development of individually specific lactoflora.

We suppose that the normal lactoflora of genital tract of healthy women offers protection for the fetus. In the study of the transfer of microflora from mother to her baby we observed some opportunistic bacteria—clostridia, beta-hemolytic streptococci, and bacteroides on the skin immediately after birth. This demonstrates the possibility of infection at birth (Davies and Gothefors, 1984). However, the infants had no symptoms of infection. This can be explained by the antenatally formed microecological relations where lactic acid microflora predominates over the opportunistic pathogens, preventing them from realizing their pathogenic potential (Fig. 7a).

The fact that not all the microorganisms present on the mother's perinea during delivery contaminate the neonates may support the idea that the baby is intrauterally selectively colonized with microbes. The reason could also be the characteristic features of the microorganisms or on the other hand, the resistance of the infant (Haenel and Bendig, 1975). It is well known that the mucosal cells of neonates are a selective environment for different streptococci (Long and Swenson, 1977) and differ individually from one infant to another in the number of reception sites for adhesion of microorganisms (Broughton and Baker, 1983). However, in the acquisition of microorganisms the adherence due to different characteristics of microorganisms themselves seem to be of primary importance. So it is possible that the fetus gets contaminated with its mother's Gram-positive microflora already during pregnancy.

We suggest that these microbes cause the production of IgG antibodies by the fetus and at the same time induce tolerance towards themselves in order to later become the indigenous microflora with a low immunogenic capacity (Freter, 1983). The idea of tolerance induction was also postulated by van der Waaij (1985) as an inclination of the neonate's immune system to induce tolerance toward the first contaminants. But if there exists a general tolerance induction without selection of strains suitable for adherence and survival (van der Waaij, 1985) then it would be difficult to explain the specific composition of the normal microflora of particular biotope.

In 1977 we succeeded in show that a particular strain of *L. fermentum* with streptomycin resistance which was administered to mice together with food in the last days of pregnancy did colonize their offspring. We could detect this strain in the offspring during four generations, whereas it quickly disappeared from mother's GI (Mikelsaar et al., 1977). Promising results have now been obtained in the very early controlled colonization of human babies (Borderon et al., 1981; Lodinova-Zadnikova et al., 1985).

In support of this idea (Mikelsaar, 1986, 1989a) there are also the data about the increase of the relative amount of lactobacilli in the vagina of pregnant women and some data about the selective ability of some microorganisms to attach to and penetrate chorioamniotic membranes (Galask et al., 1984). In several intranatal infection studies there have been revealed lactobacilli or lactic acid in the amniotic fluid by amniocentesis in cases of intact membranes (Gibbs et al., 1982, 1987; Maye et al., 1983; Wahbeh et al., 1984; Iams et al., 1987). The early formation of GI lactoflora of piglets also supports the idea of peri/intra/antenatal colonization of the fetus by lactobacilli.

The claim that mother–fetus interaction plays an important role in the indigenous microflora formation of the neonate has recently been supported by other investigators also (van der Waaij, 1985; Anufrijeva and Shenderov, 1989; Hanson et al., 1990). Yet their explanations are different. Mellander (1985) for example, correlates the IgG and IgM antibodies of neonate with the transfer of anti-idiotypic antibodies via the placenta to the fetus. But this hypothesis offers no explanation for the presence of such antibodies in the fetus in cases of agammaglobulinemia of mothers (Hanson, 1990).

Hence, we suggest that the selective contamination of the fetus during pregnancy by its mother's lactoflora, the ability of these strains to survive the bacteriocidity of the amniotic fluid, their capacity to adhere to and colonize particular mucosal surfaces postnatally, and the induction of IgG production and tolerance are the mechanisms instrumental in individual lactoflora formation.

IV. CONCLUSIONS

Humans harbor an individually specific lactic acid microflora in various organ systems, especially in the gastrointestinal and genital tracts. This lactoflora consists of several species combinations typical for each individual and its quantity is relatively stable during long survey periods. Unfortunately, the direct investigation of the lactoflora of various biotopes, for example, GI, is often impossible and in making conclusions about the genetics and relative proportions of microbes on the mucosa of the upper part of GI tract, one ought to know how changes in one biotope are reflected in another. In that sense the experimental animal studies appear to be highly valuable.

The formation of individually specific lactoflora starts in the perinatal period due to the selective microbial colonization of the baby with its mother's lactobacilli. The selective colonization of the biotope is a characteristic of both the host and the microorganisms. This conclusion is based, firstly, on the demonstration of the importance of the genetic background of the host and, secondly, on the specific microbial properties (including the ability to survive in the host's secretions, to withstand the antagonistic activity of other microorganisms, to adhere via specific adhesins to the epithelial cells and mucin of the host and degrade the endogenous nutrients of the host to achieve a stable population level) involved in the process of individual lactoflora formation. However, more detailed attention to the precise mechanisms connected with the transfer of indigenous microflora from the mother to its neonate is needed to clarify the role of specificity in lactoflora formation.

The precise purpose of the above studies is to learn how to create by means of particular lactobacilli antenatally the high colonization resistance of neonates. The ecological imbalance of the mother's microflora, often caused by wide-spectrum antibacterial treatment, on one hand, and the use of big maternity clinics with their strict antiseptic procedures at birthgiving, on the other, have distorted the natural source of the normal microflora. To overcome these problems, early close contact of the neonate with its mother immediately after birth has been suggested.

This decade can probably see the development of practical prophylactic interventions such as controlled colonization of babies by new probiotics via their mothers or by early administration of them to control neonatal infection. Fulfillment of this promise will require the application of the results of basic scientific investigations by clinicians. However, according to our present knowledge the clinicians should be very careful in their attempts and not forget the individuality of lactoflora. The latter is the main reason why only some individuals get colonized by various commercial strains of lactobacilli.

In the search of new probiotic strains for the colonization of neonates the idea of using in one probiotic a mixture of strains of lactobacilli isolated from a particular biotope of different persons may prove useful. From that mixture the individual selective conditions of each host can enrich the strains suitable for the colonization in order to develop a stable high population level of lactobacilli in the particular biotope, where possible protection is needed.

Thus, although a great deal has been learned regarding individual lactoflora formation, much remains speculative, yet the practical application of our knowledge for the prevention and treatment of neonatal infections has only just begun. The use of preparations of lactic acid bacteria for treating various infections of adults also needs more theoretical substantiation.

V. SUMMARY

The present review summarizes the current state of knowledge regarding the composition and persistence of the individually specific lactoflora of various human biotopes. The effect of the genetic background of hosts, their age, and their health on the persistence of lactoflora in different persons is demonstrated. Three different stages of interactions between the components of the GI microbial ecosystem, influencing the formation and stability of individual lactoflora, are described.

Data are presented about transmission of lactobacilli from mothers to their neonates during birth giving, and about how the mode of delivery, hospital care conditions, and feeding affect further lactoflora development. The role of the main factors contributing to the individual lactoflora development (the ability to survive in the host's secretions, to withstand the antagonistic activity of other microorganisms, to adhere to host structures or secretions and degrade the endogenous nutrients of the host) are discussed. A hypothesis is put forward about the basis for individual lactoflora formation being founded perinatally together with some supporting data from experimental animal studies.

REFERENCES

1. Aleljung, P., Paulsson, M., Emodi, L., Andersson, M., Naidu, A. S., and Wadström, T. (1991). Collagen binding by lactobacilli, *Current Microbiol., 25*: 33.
2. Allison, D., and Galloway, A. (1988). Empyema of the gallbladder due to *Lactobacillus casei* [letter], *J. Infect., 17*: 191.
3. Alm, L. (1983). The effect of *Lactobacillus acidophilus* administration upon the survival of Salmonella in randomly selected human carriers, *Progr. Food Nutr., 7*: 13.
4. Anufrijeva, R. G., and Shenderov, B. A. (1989). Effect of sisomycine on the composition of intestinal microflora of rats and the colonization resistance of their offspring, *Antibiotiki, 8*: 596 (in Russian).
5. Atlas, R. M., and Bartha, R. (1987). *Microbial Ecology, Fundamentals and Applications*, Benjamin Cummings, California.
6. Baddour, L. M., Christensen, G. B., Simpson, W. A., and Beachey, E. H. (1990). Microbial adherence, in *Principles and Practice of Infectious Diseases*, 3rd ed. (G. L. Mandell, R. G. Douglas, and J. E. Bennett, eds.), Churchill Livingstone, New York, pp. 9–25.
7. Bakuleva, L. P., Nesterova, A. A., Surskaya, O. A., Jakutina, M. F., Mironova, I. I., and Kuklina, M. A. (1984). Microscopical and bacteriological indicators of gastric aspirate in neonates with risk for development of septic infection, *Voprosy ohrany materinstva i detstva, 29*: 69 (in Russian).
8. Baquero, F., Fernandez-Jorge, A., Vicente, M. F., Alos, J. I., and Reig, M. (1988). Diversity analysis of the human intestinal flora: a simple method based on bacterial morphotypes, *Microb. Ecol. Health Disease, 1*:101.

9. Barnard, M. R., and Holt, S. C. (1985). Isolation and characterization of the peptidoglycans from selected Gram-positive and Gram-negative periodontal pathogens, *Can. J. Microbiol., 31*:54.
10. Bartlett, J. G., Onderdonk, A. B., Drude, E., Goldstein, C., Anderka, M., Alpert, S., and McCormack, W. M. (1977). Quantitative bacteriology of the vaginal flora, *J. Infect. Dis., 136*:271.
11. Bayliss, C. E., and Houston, A. P. (1984). Characterization of plant polysaccharide- and mucin-fermenting anaerobic bacteria from human feces, *Appl. Environ. Microbiol., 48*:626.
12. Bennet, R. (1987). *The faecal microflora of newborn infants during intensive care management, and its relationship to neonatal septicaemia*, Stockholm, 1987.
13. Bennet, R., Eriksson, M., Nord, C. E., and Zetterstrom, R. (1982). Suppression of aerobic and anaerobic faecal flora in newborns receiving parenteral gentamicin and ampicillin, *Acta Paediatr. Scand., 71*:559.
14. Bennet, R., and Nord, C. E. (1987). Development of the faecal anaerobic microflora after caesarean section and treatment with antibiotics in newborn infants, *Infection, 15*:332.
15. Benno, Y., Sawada, K., and Mitsuoka, T. (1984). The intestinal microflora of infants: composition of fecal flora in breast-fed and bottle-fed infants, *Microbiol. Immunol., 28*:975.
16. Bergogne-Berezin, E., Zechovsky, N., Cerf, M., Pappo, M. E., and Debray, Ch. (1973). Etude critique de 100 tubages jejunaux. Essai de confrontation des donnees cliniques et bacteriologiques, *Pathol.-Biol., 21*:505.
17. Bernhardt, H., and Knoke, M. (1980). I Bd. Episomatische Biotope, *Handbuch der Antiseptik*, Veb Verlag Volk und Gesundheit, Berlin, pp. 556–580.
18. Bernhardt, H., Knoke, M., Zimmermann, K., and Warm, V. (1990). Different influences on faecal microflora—a study in a continuous flow culture, in *Current Problems of Medical Gnotobiology and Orogastrointestinal Microflora*, Rostock, p. 63.
19. Bezirtzoglou, E., and Romond, C. (1990). Occurrence of Enterobacteriaceae and *E. coli* in the intestine of the newborn delivered by caesarean section, *Acta Microbiol. Bulg., 25*:76.
20. Bhat, P., Albert, M. J., and Rajan, D. (1980). Bacterial flora of the jejunum: a comparison of luminal aspirate and mucosal biopsy, *J. Med. Microbiol., 13*:247.
21. Bochkov, I. A., Semina, N. A., Darbeeva, O. S., Cherkasskaya, R. S., Trofimova, O. D., Shevchuk, M. S., and Pavlova, E. V. (1988). Symbiotic microflora of newborns, in *Antibiotics and Microecology of Man and Animals*, Moscow, pp. 8–13 (in Russian).
22. Borderon, J. C., Gold, F., and Laugier, J. (1981). Enterobacteria of the neonate. Normal colonization and antibiotic-induced selection, *Biol. Neonate, 39*:1.
23. Borriello, P. (1983). Ecological perspective of enteropathogenicity, in *Human Intestinal Microflora in Health and Disease* (J. Hentges, ed.), Academic Press, New York, pp. 143–167.
24. Borriello, S. P. (1990). The influence of the normal flora on *Clostridium difficile* colonization of the gut. Intestinal microflora and Health, *Ann. Medicine, 22*:61.
25. Borriello, S. P., and Hardie, J. M. (1987). Recent Advances in Anaerobic Bacteriology, in *New Perspectives in Clinical Microbiology*, Dordrecht.

26. Brilene, T. A., and Brilis, V. I. (1986). Effect of progesterone and oestradiole on the adhesion of vaginal microorganisms, in *Advances of Medical Science*, Tartu, pp. 47–49 (in Russian).
27. Brilis, V. I. (1983). *Adhesion properties of lactobacilli*. Dissertation, Moscow, p. 19 (in Russian).
28. Brilis, V. I., Mikelsaar, M. E., Kasesalu, R. H., Türi, E. I., Utt, M. M., and Lencner, A. A. (1989). Effect of lectin *Arachis hypogaeae* (PNA) on cytadhesion and hemagglutination of lactobacilli and coliforms, in *Investigation and Usage of Lectins, T.2.*, Tartu, pp. 230–236 (in Russian).
29. Broughton, R. A., and Baker, C. J. (1983). Role of adherence in the pathogenesis of neonatal group B streptococcal infection, *Infect. Immun., 39*: 837.
30. Bruce, A. W., and Reid, G. (1988). Intravaginal instillation of lactobacilli for prevention of recurrent urinary tract infections, *Canad. J. Microbiol., 34*: 339.
31. Bullen, C. L. (1977). The effect of "humanised" milks and supplemented breast feeding on the faecal flora of infants, *J. Med. Microbiol., 10*: 403.
32. Carlsson, A., Emody, L., Ljungh, A., and Wadström, T. (1989). Carbohydrate receptor specificity of haemagglutinins of Campylobacter pylori, in *Campylobacter Pathology and Campylobacter pylori* (S. Megraud and J. Lamoullialle, eds.), Elsevier, New York.
33. Carlstedt-Duke, B. (1989). The normal microflora and mucin, in *The Regulatory and Protective Role of the Normal Microflora* (R. Grubb, T. Midtvedt, and E. Norin, eds.), M. Stockton Press, Stockholm, pp. 109–129.
34. Chung, T. C., Axelsson, L., Lindgren, S. E., and Dobrogosz, W. In vitro studies on reuterin synthesis by *Lactobacillus reuteri, Microb. Ecol. Health Disease, 2*: 137.
35. Cohen, P. S., Rossoll, R., Cabelli, V. J., Yang, S.-L., and Laux, D. C. (1983). Relationship between the mouse colonizing ability of a human fecal Escherichia coli strain and its ability to bind a specific mouse colonic mucous gel protein, *Infect. Immun., 40*: 62.
36. Conway, P. (1989). Lactobacilli: fact and fiction, in *The Regulatory and Protective Role of the Normal Microflora* (R. Grubb, T. Midtvedt, and E. Norin, eds.), M. Stockton Press, Stockholm, pp. 263–283.
37. Conway, P. L., Gorbach, S. L., and Goldin, B. R. (1987). Survival of lactic acid bacteria in the human stomach and adhesion to intestinal cells, *J. Dairy Sci., 70*: 1.
38. Cook, R. L., Harris, R. J., and Reid, G. (1988). Effect of culture media and growth phase on the morphology of lactobacilli and their ability to adherence to epithelial cells, *Current Microbiol., 17*:159.
39. Cook, R. L., Tannock, G. W., and Meech, R. J. (1984). The normal microflora of the vagina, Proc. Univ. Otago Medical School, vol. 62, pp. 72–74.
40. Cooperstock, M. S., and Zedo, A. J. (1983). Intestinal flora of infants, in *Human Intestinal Microflora in Health and Disease* (D. J. Hentges, ed.), Academic Press, New York, pp. 73–80.
41. Costerton, J., Rozee, K., and Cheng, K. J. (1983). Colonization of particulates, mucous and intestinal tissues, *Progr. Fed. Nutr. Sci., 7*: 91.
42. Courtney, H. S., Ofek, I., Simpson, W. A., Hasty, D. L., and Beachey, E. H. (1986). Binding of *Streptococcus pyogenes* to soluble and insoluble fibronectin, *Infect. Immun., 53*: 454.

43. Cox, S. M., Philips, L. E., Mercer, L. J., Stager, C. E., Waller, S., and Faro, S. (1986). Lactobacillemia of amniotic fluid origin, *Obstet. Gynecol., 68*: 134.
44. Dalin, M. V., and Fish, N. G. (1985). Adhesins of microorganisms, in *Advances of Science and Technics. Series of Microbiology, T. 16*, Moscow (in Russian).
45. Davies, F. L., Underwood, H. M., and Gasson, M. J. (1981). The value of plasmid profiles for strain identification in lactic streptococci and the relationship between *Streptococcus lactis* 712, ML3, and C2, *J. Appl. Bacteriol., 51*: 325.
46. Davies, P., and Gothefors, L. A. (1984). *Bacterial Infections in the Fetus and Newborn Infant*, W. B. Saunders, Philadelphia.
47. Dene, M., Clamp, J. R., and Roberts, C. J. (1988). Effects of fasting and refeeding on somatostatin concentration and binding to cytosol from rabbit gastric mucosa, *Gut, 28*: 642.
48. Deshchetkina, M. F., Korshunov, V. M., Demin, V. F., Kholodova, I. N., and Chernova, N. D. (1990). Study of the formation of intestinal microflora in newborn infants staying with or separated from their mothers, *Pediatria, 1*: 13 (in Russian).
49. Dinari, G., Hale, T. L., Washington, O., and Formal, S. B. (1986). Effect of guinea pig or monkey colonic mucus on *Shigella* aggregation and invasion of HeLa cells by *Shigella flexneri* 1b and 2a, *Infect. Immun., 51*: 975.
50. Dobrogosz, W. J., Casas, I. A., Pagano, G. A., Talarico, T. L., Sjoberg, B.-M., and Karlsson, M. (1989). *Lactobacillus reuteri* and the enteric microbiota, in *The Regulatory and Protective Role of the Normal Microflora* (R. Grubb, T. Midtvedt, and E. Norin, eds.), M. Stockton Press, Stockholm, pp. 283–293.

50a. Dolby, J. M., Webster, A. D. B., Borriello, S. P., Barclay, F. E., Bartholomew, B. A., and Hill, M. Y. (1984). Bacterial colonization and nitrite concentration in the achlorhydric stomachs of patients with primary hypogammaglobulinaemia or classical pernicious anaemia, *Scand. J. Gastroent., 19*: 105–110.

51. Dubos, R. (1966). The microbiota of the gastrointestinal tract, *Gastroenterology, 51*: 868.
52. Dubos, R., and Schaedler, R. W. (1962). Some biological effects of the digestive flora, *Trans. Assoc. Am. Phys., 75*: 160.
53. Ekwempu, C., Lawande, R., and Egler, L. (1982). Bacterial colonization of various sites at birth of babies born in Zaira, *J. Infect., 5*: 177.
54. Enhtuja, R. (1984). *Formation of intestinal microflora in healthy newborns during the first week of life on different feeding*, Dissertation, Moscow (in Russian).
55. Eschenbach, D. A., Davick, P. R., Williams, B. L., Klebanoff, S. J., Young-Smith, K., Critchlow, C. M., and Holmes, K. K. (1989). Prevalence of hydrogen peroxide-producing *Lactobacillus* species in normal women and women with bacterial vaginosis, *J. Clin. Microbiol., 27*: 251.
56. Farrar, W. E. (1983). Molecular analysis of plasmids in epidemiologic investigation, *J. Infect. Dis., 148*: 1.
57. Finegold, S. M., Mathiesen, G. E., and George, W. L. (1983). Consequences of the upsetting the ecology of the intestine. Changes in human intestinal microflora related to the administration of antimicrobial agents, in *Human Intestinal Microflora in Health and Disease* (D. J. Hentges, ed.), Academic Press, New York, pp. 351–500.

58. Fitzgerald, N. A., and Shellam, G. R. (1991). Host genetic influences on foetal susceptibility to murine Cytomegalovirus after maternal or fetal infection, *J. Infect. Dis., 163*: 276.
59. Freter, R. (1983). Mechanisms that control the microflora in the large intestine, in *Human Intestinal Microflora in Health and Disease* (D. J. Hentges, ed.), Academic Press, New York, pp. 33–54.
60. Freter, R. (1990). Gastrointestinal flora homeostasis, in *Current Problems of Medical Gnotobiology and Orogastrointestinal Microflora*, Rostock, p. 4.
61. Freter, R., Stauffer, E., Cleven, D., Holdeman, L. V., and Moore, W. E. C. (1983). Continuous-flow cultures as in vitre models of the ecology of large intestine, *Infect. Immun., 39*: 666.
62. Fuller, R. (1975). Nature of the determinant responsible for the adhesion of lactobacilli to chicken crop epithelial cells, *J. Gen. Microbiol., 87*: 245.
63. Fuller, R., Houghton, S. B., and Brooker, B. E. (1981). Attachment of *Streptococcus faecium* to the duodenal epithelium of the chicken and its importance in the colonization of the small intestine, *Appl. Environ. Microbiol., 41*: 1433.
64. Galask, R. P., Warner, M. W., Petzold, R., and Wilbur, S. L. (1984). Bacterial attachment to the chorioamniotic membranes, *Am. J. Obstet. Gynecol., 148*: 915.
65. Geimberg, V. G. (1957). Importance of normal intestinal microflora for host, *Voprosy pitanija, 5*: 44 (in Russian).
66. Gibbs, R. S., Blanco, J. D., St. Clair, P. J., and Castaneda, I. S. (1982). Quantitative bacteriology of amniotic fluid from women with clinical intraamniotic infection at term, *J. Infect. Dis., 145*: 1.
67. Gibbs, R. S., Forman, J., St. Clair, P. J., and Baseman, J. B. (1987). Detection of enzyme-linked immunosorbent assay in women with intraamniotic infection, *Am. J. Obstet. Gynecol., 69*: 208.
68. Gilliland, S. E., Speck, M. L., and Morgan, C. G. (1975). Detection of L. acidophilus in faeces of humans, pigs and chickens, *Appl. Microbiol., 30*: 541.
69. Giorgi, A., Torriani, S., Dellaglio, F., Bo, G., Stola, E., and Bernuzzi, L. (1987). Identification of vaginal lactobacilli from asymptomatic women, *Microbiologica, 10*: 377.
70. Goldin, B. R., and Gorbach, S. L. (1984). The effect of milk and lactobacillus feeding on human intestinal bacterial enzyme activity, *Am. J. Clin. Nutr., 39*: 756.
71. Goldin, B. R., and Gorbach, S. L. (1987). *Lactobacillus* GG: a new strain with properties favorable for survival, adhesion and antimicrobial activity in the gastrointestinal tract, *FEMS Microbiol. Rev., 46*: 72.
72. Goldmann, D. A. (1988). The bacterial flora of neonates in intensive care-monitoring and manipulation, *J. Hosp. Inf. 11 (Suppl A)*: 340.
73. Golyanova, L. A. (1972). *Faecal lactoflora of healthy school-children*, Dissertation, Tartu (in Russian).
74. Goncharova, G. I., Semenova, L. P., Lyannaya, A. M., Kozlova, E. P., and Donskihh, E. E. (1987). Human bifidoflora, its normalizing and protective functions, *Antibiotiki, 3*: 179 (in Russian).
75. Gorbach, S. L. (1990). Lactic acid bacteria and human health. *Ann. Medicine, 22*: 37.
76. Gorbach, S. L., Plaut, A. G., Nahas, L., and Weinstein, L. (1967). Microorganisms of the small intestine and their relations to oral and faecal flora, *Gastroenterology, 53*: 856.

77. Gossling, J., and Slack, J. M. (1974). Predominant gram-positive bacteria in human feces: number, variety and persistence, *Infect. Immun., 9*: 719.
78. Gothefors, L. (1989). Effect of diet on intestinal flora, *Acta Paediatr. Scand. Suppl., 351*: 118.
79. Gracey, M. (1983). The contaminated bowel syndrome, in *Human Intestinal Microflora in Health and Disease* (D. J. Hentges, ed.), New York.
80. Gracey, M. (1986). Gastroenteritis in Australian children: studies on the aetiology of acute diarrhoea, *Microecol. Therapy, 16*: 47.
81. Graham, H., and Aman, P. (1987). The pigs as a model in dietary fibre digestion studies, *Scand. J. Gastroent., 22*: 55.
82. Graham, J. M. (1975). *An investigation into aerobic and anaerobic bacterial flora of normal and ill low-weight newborn babies, Dissertation*, University of London.
83. Gustafsson, B. E., and Norin, K. E. (1977). Development of germfree animal characteristics in conventional rats by antibiotics, *Acta Path. Microbiol. Scand. Sect. B., 85*: 1.
84. Haenel, H. (1980). Die gastointestinale Mikroflora in Abhangigkeit von Milieu und Ernahrung, *Mikrobielle Umwelt und Massnahmen, 4*: 17.
85. Haenel, H., and Bendig, J. (1975). Intestinal flora in health and disease, *Progr. Food Nutr. Sci., 1*: 21.
86. Haenel, H., Muller-Beuthow, W., and Scheunert, A. (1957a). Der Einfluss extremer Kostformen auf die faekale Flora des Menschen I Einleitung und Versuchsabsichten, *Zbl. Bakt. I Orig., 168*: 37.
87. Haenel, H., Muller-Beuthow, W., and Scheunert, A. (1957b). Der Einfluss extremer Kostformen auf die faekale Flor des Menschen I. *Zbl. Bakt. I Orig., Abt. A. 169*: 45.
88. Hall, M. A., Cole, C. B., Smith, S. L., Fuller, R., and Rolles, C. J. (1990). Factors influencing the presence of faecal lactobacilli in early infancy, *Arch. Dis. Child., 65*: 185.
89. Hanson, L. A., Adlerberth, I., Carlsson, B., Castrignano, S. B., Dahlgren, U., Jalil, F., Khan, S. R., Mellander, L., Eden, C. S., Svennerholm, A. M., et al. (1989). Host defense of the neonate and the intestinal flora, *Acta Paediatr. Scand. Suppl., 351*: 122.
90. Hanson, L. A., Ashraf, R., Cruz, J. R., Hahn-Zoric, M., Jalil, F., Nave, F., Reimer, M., Zaman, S., and Carlsson, B. (1990). Immunity related to exposition and bacterial colonization of the infant, *Acta Paediatr. Scand. Suppl., 365*: 38.
91. Heimdahl, A., and Nord, C. E. (1979). Effect of phenoxymethylpenicillin and clindamycin on the oral, throat and faecal microflora of man, *Scand. J. Infect. Dis., 11*: 233.
92. Hentges, D. J. (1983). Role of intestinal microflora in host defence against infections, in *Human Intestinal Microflora in Health and Disease* (D. J. Hentges, ed.), New York, p. 306.
93. Holdeman, L. V., Cato, E. P., and Moore, W. E. C. (1977). *Anaerobe Laboratory Manual*, Virginia Polytechnic Institute, Blacksburg, Virginia.
94. Holdeman, L. V., Good, I. J., and Moore, W. E. C. (1976). Human faecal flora: variation in bacterial composition within individuals and a possible effect of emotional stress, *Appl. Environ. Microbiol., 31*: 359.
95. Hoskins, L. C. (1981). Human enteric population ecology and degradation of gut mucins, *Digestive Dis. Sci., 26*: 769.

96. Hoverstad, T., Bjorneklett, A., Fausa, O., and Midtvedt, T. (1984). Short-chain fatty acids and bacterial overgrowth of the jejunum, *Scand. J. Gastroent., 19 (Suppl. 98)*: 43.
97. Hungate, R. E. (1984). Microbes of nutritional importance in the alimentary tract, *Proc. Nutr. Sci., 43*: 1.
98. Hwa, V., and Salyers, A. (1989). "Genetic analysis of the mucopolysaccharide utilization pathway in *Bacteriodes* thetaiotaomicron and assessment of its contribution to survival of *Bacteriodes* in the colon," Abstracts Int. Meeting SOMED, San Antonio, Texas.
99. Iams, J. D., Clapp, D. H., Contos, D. A., Whitehurst, R., Ayers, L. W., and O'Shaugnessy, R. W. (1987). Does extraamniotic infection cause preterm labor? Gas-liquid chromatography studies of amniotic fluid in amnionitis, preterm labor and normal controls, *Am. J. Obstet. Gynecol., 70*: 365.
100. Ivanov, N. A., Danilova, E. G., and Doroshenko, V. S. (1985). Characteristics of staphylococci isolated from newborns, *Z. Mikrobiol., 2*: 19 (in Russian).
101. Iwasaka, T., Wada, T., Kidera, Y., and Sugimori, H. (1986). Genital mycoplasma colonization in neonatal girls, *Acta Obstet. Gynecol. Scand., 65*: 269.
102. Johnson, J. L. (1980). Specific strains of *Bacteriodes* species in human fecal flora as measured by desoxyribonucleic acid homology, *Appl. Environ. Microbiol., 39*: 407.
103. Johnson, J. R. (1991). Virulence factors in *Escherichia coli* urinary tract infection, *Clin. Microbiol. Rev., 4*: 80.
104. Kaarma, A. I., Mikelsaar, M. E., and Lencner, A. A. (1974). Microflora of small and large intestine, Proc. Helmintologist's Soc. USSR, Vol. 26, pp. 26–31.
105. Kandler, O., and Weiss, N. (1986). Regular gram positive nonsporing rods, in *Bergey's Manual of Systematic Bacteriology, 2* (P. H. A. Sneath, N. S. Mair, M. S. Sharp, and J. G. Holt, eds.), pp. 1208–1234.
106. Kasesalu, R. H., Mikelsaar, M. E., Saag, H. V., and Peil, P. H. (1990). On anaerobic microflora of the vagina in women in their first and second pregnancy terms, in *Antibiotics and Colonization Resistance*, Moscow, pp. 47–54 (in Russian).
107. Kasesalu, R., Saag, H., Peil, P., Mikelsaar, M., and Lencner, A. (1991). The role of vaginal microflora in the first contamination of neonate, in *Medical Aspects of Microbial Ecology* (B. A. Shenderov, ed.), Moscow, pp. 102–108 (in Russian).
108. Kay, B., Fuller, R., Wilkinson, A. R., Hall, M., McMichaelo, J. E., and Cole, C. B. (1990). High levels of staphylococci in the faeces of breast-fed babies, *Microb. Ecol. Health Disease, 3*, 277.
109. Kearney, J. N., Harnby, D., Gowland, G., and Holland, K. T. (1984). The follicular distribution and abundance of resident bacteria on human skin, *J. Gen. Microbiol., 130*: 979.
110. Knoke, M., and Bernhardt, H. (1985). Mikrookologie des Menshen, in *Mikroflora bei Gesunden und Kranken*, Berlin, Akademie-Verlag.
111. Kolts, K. K., Mikelsaar, M. E., and Lencner, A. A. (1986). Adhesion of lactobacilli to mucosal of gastrointestinal tract of rhodents, in *Fundamental Aspects and Practical Usage in Medicine and Agriculture of Advance of Biology*, Tartu, pp. 60–63 (in Russian).

112. Korvjakova, E. P., and Gorohov, V. I. (1979). Influence of lysozyme on sensitivity of *Shigella* to furazolidone, *Chemotherapy of Bacterial Infections*, Moscow, pp. 26–27.
113. Kuhn, I., Tullus, K., and Mollby, R. (1986). Colonization and persistence of Escherichia coli phenotypes in the intestines of children aged 0 to 18 months, *Infection, 14*: 7.
114. Kuritza, A. P., Shaughnessy, P., and Salyers, A. A. (1986). Enumeration of polysaccharide-degrading *Bacteroides* species in human feces by using species-specific DNA probes, *Appl. Environ. Microbiol., 51*: 385.
115. Larsen, B., and Galask, R. P. (1980). Vaginal microbial flora: practical and theoretic relevance, *Obstet. Gynecol., 55*: 100(S).
116. Larsen, B., Goplerud, C. P., Petzold, C. R., Ohm-Smith, M. J., and Galask, R. P. (1982). Effect of estrogen treatment on the genital tract flora of postmenopausal women, *Obstet. Gynecol., 60*: 20.
117. Lee, A., and Gemmell, E. (1972). Changes in the mouse intestinal microflora during weaning: role of volatile fatty acids, *Infect. Immun., 5*: 1.
118. Leeming, J. P., Holland, K. T., and Cunliffe, W. P. (1984). Microbia ecology of pilosebaceous units isolated from human skin, *J. Gen. Microbiol., 130*: 803.
119. Lennox-King, S., O'Farrol, S., Bettelheim, K., and Shooter, R. (1976). *Escherichia coli* isolated from babies delivered by caesarean section and their environment, *Infection, 4*: 439–445.
120. Lenz, W. (1976). *Medizinische Genetik*, Thieme Verlag, Stuttgart.
121. Lencner, A. A. (1973). *Lactobacilli of human microflora*, Dissertation, Tartu (in Russian).
122. Lencner, H. P., Lencner, A. A. (1982). Lysozme activity and sensitivity to lysozyme of lactobacilli of human microflora, in *Mikrobielle Umwelt und antimikrobielle Massnahmen*, (H. Horn, W. W. Weuffen and H. Wiegert, eds.), J. A. Barth, Leipzig, pp. 73–81 (in Russian).
123. Lencner, A., Lencner, H., Brilis, V., Brilene, T., Mikelsaar, M., and Türi, M. (1987). Zur Abwehrfunktion der Lactoflora des Verdauungstraktes, *Nahrung*, 31: 405.
124. Lencner, A. A., Lencner, H. P., Mikelsaar, M. E., Shilov, V. M., Lizko, N. N., Syrych, G. D., and Legenkov, V. I. (1981). Investigation of gastrointestinal lactoflora of teams "Soyuz-13" i "Salyut-4", *Kosm. Biol., 4*: 36 (in Russian).
125. Lencner, A., Lencner, H., Mikelsaar, M., Türi, M., Toom, M., Valjaots, M., Silov, V., Lizko, N., Legenkov, V., and Reznikov, I. (1984). Die quantitative Zusammensetzung der Lactoflora des Verdauungstraktes vor und nach Kosmischen Flugen unterschiedlicher Dauer, *Nahrung, 28*: 607 (in Russian).
126. Lencner, A. A., Mikelsaar, M. E., and Kirch, R. A. (1982). Lactoflora of monozygotic twins and experimental animals of the same offspring, *Actual Questions of Immunodiagnostics and Immunoregulation*, Tallinn, pp. 210–212 (in Russian).
127. Lencner, A. A., Mikelsaar, M. E., Türi, M. E., Lencner, H. P., Chahava, O. V., and Shustrova, N. M. (1980). Interrelations of lactobacilli with *B. subtilis* and *V. cholerae* in intestine of germfree rats, in *Medical Faculty to Medicine*, Tartu, pp. 31–33 (in Russian).
128. Lencner, A. A., Shilov, V. M., Lizko, N. N., and Mikelsaar, M. E. (1973). Investigation of species composition of intestinal lactobacilli in cases of long-term isolation of man, *Kosm. biol., 4*: pp. 76–80 (in Russian).

129. Lerche, M., and Reuter, G. (1962). Das Vorkommen aerob wachsender grampositiver Stabchen als Genus Lactobacillus Beijerinek im Darminhalt erwachsener Menschen, *Zbl. Bakt. I Orig., 185*: 446.
130. Levin, R. J. (1968). The effects of hormones and the oestrous cycle on the transvaginal potential difference, *J. Physiol., 194*: 94p.
131. Lizko, N. N. (1987). Dysbacteriosis in extreme conditions, *Antibiotiki, 3*: 184 (in Russian).
132. Lizko, N. N., Shilov, V. M., and Syrych, G. D. (1984). Die besonderheiten der Bildung einer Dysbakteriose des Darms beim Menschen unter Extrembedingungen, *Nahrung, 28*: 599.
133. Lodinova-Zadnikova, Tlaskalova, H., and Kolesonova, L. (1985). Contribution of maternal milk to the protection of the newborn infant. Antigenic stimulation of the breast, in *Bacteria and the Host*, Prague, p. 167.
134. Long, S. S., and Swenson, R. M. (1977). Development of anaerobic faecal flora in healthy newborn infants, *J. Pediatr., 91*: 298.
135. Lorenz, A., Grutte, F.-K., Haenel, H., and Saunders, U. (1982). A mechanism of association of lactobacilli with the rat stomach epithelium, *Zbl. Bact. I Abt. Orig. A 252*: 9.
136. Luckey, T. D. (1977). Bicentennial overview of intestinal microecology, *Amer. J. Clin. Nutr., 30*: 1753.
137. Lundequist, B., Nord, C. B., and Winberg, J. (1985). The composition of the fecal microflora in breastfed and bottlefed infants from birth to eight weeks, *Acta. Paediatr. Scand., 74*: 45.
138. Lynch, M., and Poole, N. J. (1979). *Microbial Ecology: A Conceptual Approach*, Blackwell, London.
139. Maciejewski, Z., Melnyczuk, J., and Stanislawski, P. (1987). Presence of *Chlamydia* in the nasopharynx of newborn infants in the postpartum period and in the vagina of parturients in the first stage of labor, *Ginkeol. Pol., 58*: 628.
140. Mackie, R. J., and Wilkins, C. A. (1988). Enumeration of anaerobic bacterial microflora of the equine gastrointestinal tract, *Appl. Environ. Microbiol., 54*: 2155.
141. Mackowiak, P. A. (1982). The normal microbial flora, *New Engl. J. Med., 307*: 83.
142. Manso, E., Strusi, P., Stacchiotti, M. A., Vincenzi, R., Tiriduzzi, M., and Del Prete, U. (1986). Incidence and origin of Clostridium difficile in neonatology, *Boll Ist Sieroter Milan, 65*: 118 (in Italian).
143. Mata, L. J., and Urrutia J. J. (1971). Intestinal colonization of breast-fed children in a rural area of low socioeconomic level, *Ann. N.Y. Acad. Sci., 176*: 93.
144. Maye, D. P., Filthuth, I., Pugin, P., Waldvogel, E., and Herrman, W. L. (1983). Bacteriological study of amniotic fluid during labour, *Acta Obstet. Gynecol. Scand., 62*: 603.
145. McCormick, E. L., and Savage, D. C. (1983). Characterization of Lactobacillus sp. strain 100-37 from the murine gastrointestinal tract: ecology, plasmid content, and antagonistic activity toward *Clostridium ramosum* H1, *Appl. Environ. Microbiol., 46*: 1103.
146. McCroatry, J. A., Angotti, R., and Cook, R. L. (1988). *Lactobacillus* inhibitor production against *Escherichia coli* and coaggregation ability with uropathogens, *Can. J. Microbiol., 34*: 344.

147. McGregor, J. A. (1988). Prevention of preterm birth: new initiatives based on microbial-host interactions, *Obstet. Gynecol. Surv., 43*: 1.
148. McGregor, J. A., French, J. I., Richter, R., Franco-Buff, A., Johnson, A., Hillier, S., Judson, F. N., and Todd, J. K. (1990). Antenatal microbiologic and maternal risk factors associated with prematurity, *Am. J. Obstet. Gynecol., 163*: 1465.
149. McKay, J. C., and Speekenbrink, A. (1984). Implications of Freter's model of bacterial colonization, *Infect. Immun., 44*: 149.
150. Meijer-Severs, G. J., and van Santen, E. (1986). Variations in the anaerobic flora of ten healthy human volunteers with special references to the Bacteroides, *Zbl. Bakt. I Orig., Abt. A 261*: 43.
151. Mellander, L. (1985). *The development of mucosal immunity in relation to natural exposure and vaccination*, Dissertation, University of Goteborg.
152. Metschnikoff, E. (1908). *Prolongation of Life*, G. P. Putnam, New York.
153. Midtvedt, T. (1989a). Monitoring the functional state of the microflora, in *Recent Advances in Microbial Ecology*, Proc. 5th Int. Symp. Microbial Ecology, pp. 515–519.
154. Midtvedt, T. (1989b). Influence of antibiotics on biochemical intestinal microflora-associated characteristics in man and animals, in *The Influence of Antibiotics on the Host-Parasite Relationship III* (G. Gillissen, W. Opferbach, G. Peters, and G. Pulverer, eds.), Springer-Verlag, Berlin, Heidelberg, pp. 209–215.
155. Midtvedt, T. (1989c). The normal microflora, intestinal motility and influence of antibiotics. An overview, in *The Regulatory and Protective Role of the Normal Microflora* (R. Grubb, T. Midtvedt, and E. Norin, eds.), M. Stockton Press, Stockholm, pp. 149–167.
156. Midtvedt, T. (1990). Ecosystems: development, functions and consequences of disturbances, with special reference to oral cavity, *J. Clin. Periodontol., 17*: 474.
157. Mikelsaar, M. E. (1969). *Lactobacilli of human faecal microflora in some non-infectious diseases of gastrointestinal tract*, Dissertation, Tartu (in Russian).
158. Mikelsaar, M. (1986). The possibilities of balanced coexistence in the system: organism and its microflora, *Endocytobiosis Cell Res., 3*: 150.
159. Mikelsaar, M. E. (1988). Possible criteria of dysbiosis of intestine by faecal microflora, in *Human Autoflora in Norm and Pathology and Its Correction* (I. N. Blohina and K. Y. Sokolova, eds.), Gorkii, pp. 15–23 (in Russian).
160. Mikelsaar, M., Kolts, K., and Maaroos, H.-I. (1990). Microbial ecology of *Helicobacter pylori* in antral gastritis and peptic ulcer disease, *Microb. Ecol. Health Dis., 3*: 245.
161. Mikelsaar, M. E., and Lencner, A. A. (1982). Mucosal lactoflora of gastrointestinal tract, in *Mikrobielle Umwelt und/in antimikrobielle Massnahmen. Bd. 6. Mikrookologie des Magen-Darmkanals des Menschen* (H. Bernhardt and M. Knoke, eds.), Leipzig, pp. 89–96.
162. Mikelssar, M., Sepp, E., Kasesalu, R., and Kolts, K. (1989a). Some considerations on the formation of normal human microflora during the first year of life, *Wiss. Z. Ernst-Moritz-Arndt-Univ. Greifswald Med. Reihe, 38*: 27.
163. Mikelsaar, M. E., Siigur, U. H., and Tamm, A. O. (1989b). Microecological aspects of investigation of influence of antibacterial preparations in intestinal microflora, *Antibiotiki, 6*: 420.

164. Mikelsaar, M., and Türi, E. (1990). Effect of antibacterial drugs and dental surgery on the translocation of digestive tract microflora, *Microecology and Therapy, 20*: 93–97.
165. Mikelsaar, M. E., Türi, M. E., and Lencner, A. A. (1977). Possibilities of intestinal colonization with lactobacilli of white mice during several generations, Abstracts of Estonian III Congress of Epidemiologists, MIcrobiolgists, Infectionists and Hygienists, Tallinn, pp. 62–63, (in Russian).
166. Mikelsaar, M. E., Türi, M. E., and Lencner, A. A. (1986a). Relative amount of bifidobacteria in content and on mucosa of intestine, in *Bifidobacteria and Their Usage in Clinics, Medical Industry and Agriculture*, Moscow, pp. 25–29 (in Russian).
167. Mikelsaar, M. E., Türi, M. E., Lencner, H., Kolts, K., Kirch, R., and Lencner, A. (1987). Interrelations between mucosal and luminal microflora of gastrointestine, *Nahrung, 31*: 449.
168. Mikelsaar, M., Türi, M., Valjaots, M., and Lencner, A. (1984). Anaerobe Inhalts- und Wandmikroflora des Magen-Darm-Kanals, *Nahrung, 28*: 727.
169. Mikelsaar, M. E., Türi, M. E., Vytrishchak, L. V., Lencner, A. A. (1986b). Translocation of microorganisms from gastrointestinal tract into blood circulation in radiated conventional and associated germfree mice, in *Theoretical and Practical Problems of Gnotobiology*, Moscow, pp. 200–205 (in Russian).
170. Mikelsaar, M. E., Uibu, J.A., and Lencner, A. A. (1975). Long-term survey of stability of human faecal microflora, Proc. VI Congress of Estonian Therapeutists, Tallinn, pp. 207–209 (in Russian).
171. Miller, R. S., and Hoskins, L. C. (1981). Mucin degradation in human colon ecosystems. Faecal population densities of mucin-degrading bacteria estimated by a "most probable number" method, *Gastroenterology, 81*: 759.
172. Mitsuoka, T., Hayakawa, K., and Kimura, N. (1975). Die Faekalflora bei Menschen. III Mitteilung: Die Zusammensetzung der Laktobazillenflora der verschiedener Altersgruppen, *Zpl. Bakt. I Orig. Abt. A, 232*: 499.
173. Mitsuoka, T., and Kaneuchi, C. (1977). Ecology of the bifidobacteria, *Am. J. Clin. Nutr., 30*: 1799.
174. Mitsuoka, T., and Ohno, K. (1977). Die Faekalflora bei Menchen V. Die Schwankungen in der Zusammensetzung der Faekalflora gesunder Erwachsener, *Zbl. Bakt. I Orig. Abt. A, 238*: 228.
175. Moore, W. E. C., Cato, E. P., and Holdeman, L. V. (1979). Some current concepts in intestinal bacteriology, *Amer. J. Clin. Nutr., 31*: 33.
176. Moshchich, P. S., Chernyshova, L. I., Bernasovskaia, E. P., Selnikova, O. P., Sargsian (1989). Prevention of dysbacteriosis in the early neonatal period using a pure culture of acidophilic bacteria, *Pediatria, 3*: 25 (in Russian).
177. Neumann, von G. (1988). Regulationsfaktoren des vaginalen mikrookologishen Systems, *Zbl. Gynakol., 110*: 405.
178. Neut, C., Bezirtzoglou, E., Romond, C., Beerens, H., Delcroix, M., and Noel, A. M. (1987). Bacterial colonization of the large intestine in newborns delivered by caesarean section, *Zbl. Bakt. Mikrobiol. Hyg. (A), 266*: 330.
179. Nord, C. E., Lindmark, A., and Persson, I. (1989). Susceptibility of anaerobic bacteria to ALP 201, *Antimicrob. Chemother., 33*: 2137.

180. Norhagen, G., Engstron, P.-E., Hammarstrom, L., Smith, C. I. E., and Nord, C. E. (1990). The microbial flora of saliva and faeces in individuals with selective IgA deficiency and common variable immunodeficiency, *Microb. Ecol. Health Dis.*, *3*: 269.
181. Norin, K. E., Carlstedt-Duke, B., Hoverstad, T., Lingaas, E., Saxerholt, H., Steinbakk, M., and Midtvedt, T. (1988). Faecal tryptic activity in humans influence of antibiotics on microbial intestinal degradation, *Microb. Ecol. Health Dis.*, *1*: 65.
182. Oksanen, P. J., Salminen, S., Saxelin, M., Hamalainen, P., Ihantola-Vormisto, A., Muurasniemi-Isoviita, L., Nikkari, S., Oksanen, T., Porsti, I., Salminen, E., Siitonen, S., Stuckey, H., Toppila, A., and Vapaatalo, H. (1990). Prevention of travellers' diarrhoea by *Lactobacillus* GG, *Ann. Medicine*, *22*: 53.
183. Perdigon, G., Nader de Macias, M. E., Alvarez, S., Medici, M., Oliver, G., and Pesce de Ruiz-Holgado, A. (1986). Immunopotentiating activity of lactic bacteria administered by oral route. Favourable effect in infantile diarrheas, *Medicine (B. Aires)*, *46*: 751 (in Spanish).
184. Petrovskaya, V. G., and Marko, O. P. (1976). *Human Microflora in Norm and Pathology*, Moscow (in Russian).
185. Pollmann, D. S., Danielson, D. M., Wren, W. B., Peo, E. R., Shahani, K. M. (1980). Influence of *Lactobacillus acidophilus* inoculum on gnotobiotic and controversial pigs, *J. Animal Sci.*, *51*: 629.
186. Raibaud, P., Ducluzeau, R., Dubos, F., Hudault, S., Bewa, H., and Muller, M. C. (1980). Implantation of bacteria from the digestive tract of man and various animals into gnotobiotic mice, *Am. J. Clin. Nutr.*, *11*: 2440.
187. Rasmussen, H. S., Holtug, K., Ynggard, C., and Mortensen, P. B. (1988). Faecal concentrations and production rates of short chain fatty acids in normal neonates, *Acta Paediatr. Scand.*, *77*: 365.
188. Redondo-Lopez, V., Cook, R. L., and Sobel, J. D. (1990). Emerging role of lactobacilli in the control and maintenance of the vaginal bacterial microflora, *Rev. Infect. Dis.*, *12*: 856.
189. Reid, G., and Sobel, J. D. (1987). Bacterial adherence in the pathogenesis of urinary tract infection: a review, *Rev. Infect. Dis.*, *9*: 470.
190. Rettger, L. F. (1935). *Lactobacillus acidophilus. Its Therapeutic Application*, Yale University Press, New Haven.
191. Reuter, G. (1965). Untersuchungen uber die Zusammensetzung und die Beeinflussbarkeit der menschlichen Magen- und Darmflora unter besonderer Berucksichtigung der Laktobazillen, *Ernahrungsforschung*, *10*: 429.
192. Reuter, G. (1975). Position of the microflora of human small intestine and the behaviour of some microorganisms after oral intake, Proc. IAMS, Tokyo.
193. Robins-Browne, R. M., Path, F. F., and Levine, M. M. (1981). The fate of ingested lactobacilli in the proximal small intestine, *Am. J. Clin. Nutr.*, *34*: 514.
194. Robinson, I. M., Whipp, S. C., Bucklin, J. A., and Allison, M. J. (1984). Characterization of predominant bacteria from the colons of normal and dysenteric pigs, *Appl. Environ. Microbiol.*, *48*: 964.
195. Rogosa, M., and Sharpe, M. E. (1960). Species differentiation of human vaginal lactobacilli, *J. Gen. Microbiol.*, *23*: 197.

196. Ross, J. M., and Needham, J. R. (1980). Genital flora during pregnancy and colonization of the newborn, *J. Royal Soc. Medicine, 73*: 105.
197. Rotimi, V. O., and Duerden, B. I. (1981). The development of the bacterial flora in normal neonates, *J. Med. Microbiol., 14*: 51.
198. Rotimi, V. O., Olowe, S. A., and Ahmed, I. (1985). The development of bacterial flora of premature neonates, *J. Hyg. (London), 94*: 309.
199. Rudigoz, R. C., Bensoussan, G., Mallecourt, P., Delignette, M., Desmettre, O., Tigaud, S., and Salvat, J. (1988). Management of a pregnant woman with *Streptococcus* group B, *Rev. Fr. Gynecol. Obstet., 83*: 347, 351 (in France).
200. Salyers, A. A., and Leedle, J. A. Z. (1983). Carbohydrate metabolism in the human colon, in *Human Intestinal Microflora in Health and Disease* (D. J. Hentges, ed.), New York, pp. 142–143.
201. Salyers, A. A., West, S. E. H., Vercellotti, J. R., and Wilkins, T. D. (1977). Fermentation of mucins and plant polysaccharides by anaerobic bacteria from the human colon, *Appl. Environ. Microbiol., 34*: 529.
202. Saunders, J. M., and Miller, C. H. (1983). Neuraminidase-activated attachment of *Actinomyces naeslundii* ATCC 12104 to human buccal epithelial cells, *J. Dent. Res., 62*: 1038.
203. Savage, D. C. (1977). Microbial ecology of the gastrointestinal tract, *Ann. Rev. Microbiol., 31*: 107.
204. Savage, D. C. (1987). Microorganisms associated with epithelial surfaces and stability of the gastrointestinal microflora, *Nahrung, 31*: 383.
205. Savage, D. C. (1989). The normal human microflora-composition, in *The Regulatory and Protective Role of the Normal Microflora* (R. Grubb, T. Midtvedt, and E. Norin, eds.), M. Stockton Press, Stockholm.
206. Savage, D. C., Dubos, R., and Schaedler, R. W. (1968). The gastrointestinal epithelium and its autochthonous bacterial flora, *J. Exper. Medicine, 127*: 67.
207. Saxelin, M., Elo, S., Salminen, S., and Vapaatalo, H. (1991). Dose response colonisation of faeces after oral administration of *Lactobacillus casei* strain GG, *Microbial Ecology in Health and Disease, 4*: in press.
208. Shenderov, B. A. (1987). Normal human microflora and some questions of microecological toxicology, *Antibiotiki, 3*: 164.
209. Shenderov, B. A., Mitrohin, S. D., Gluhova, E. V., Vagina, I. M., Ivanova, L. V., Shchibanov, V. V., and Ivanov, A. A. (1989). Influence of antibiotics on excretion by faeces of some microbial metabolites, *Antibiotiki, 12*: 665 (in Russian).
210. Sherman, L. A., and Savage, D. C. (1986). Lipoteichoic acids in *Lactobacillus* strains that colonize the mouse gastric epithelium, *Appl. Environ. Microbiol., 52*: 302.
211. Shilov, V. M., Lizko, N. N., Bragina, M. P., Lencner, A. A., Mikelsaar, M. E. (1972). About the changes of intestinal microflora during long-term isolation of man in hermetically closed room, Abstracts of IV Scientific Conference of USSR, Moscow, pp. 235–238 (in Russian).
212. Shinar, E., Leitersdorf, E., and Yevin, R. (1984). *Lactobacillus plantarum* endocarditis, *Klin. Wochenschr., 62*: 1173.
213. Siigur, U., Tamm, A., and Tammur, R. (1991). The faecal SCFA and lactose tolerance in lactose malabsorbers, *Eur. J. Gastroenterol. and Hepatol., 3*: 321.

214. Silva, M., Jacobus, N. V., Deneke, C., and Gorbach, S. L. (1987). Antimicrobial substance from a human *Lactobacillus* strain, *Antimicrob. Chemother., 31*: 1231.
215. Simhon, A., Douglas, J. R., Drasar, B. S., and Soothill, J. F. (1982). Effect of feeding on infants' faecal flora, *Arch. Dis. Child., 57*: 54.
216. Simon, G. L., and Gorbach, S. L. (1984). Intestinal flora in health and disease, *Gastroenterology, 86*: 174.
217. Sims, W. (1964). The effect of mucin on the survival of lactobacilli and streptococci, *J. Gen. Microbiol., 37*: 335.
218. Skamene, E. (1983). Genetic regulation of host resistance to bacterial infection, *Rev. Infect. Dis., 6, Suppl. 4.*
219. Smolyanskaya, A. Z. (1987). Dysbacteriosis—infectious processis of mixed aetiology, *Antibiotiki, 3*: 186 (in Russian).
220. Solovyeva, I. V. (1986). About genital lacto- and bifidoflora of sick and healthy women, *Bifidobacteria and Their Usage in Clinics*, Moskva, pp. 29–32 (in Russian).
221. Solovyeva, I. V. (1987). *Characteristics of vaginal microflora in norm and pathology*, Dissertation, Moscow (in Russian).
222. Speck, M. L. (1976). Interactions among lactobacilli and man, *J. Dairy Sci., 59*: 338.
223. Stahl, C. E., and Hill, G. B. (1986). Microflora of the female genital tract, in *Infectious Diseases in the Female Patient. The Clinical Perspectives in Obstetrics and Gynecology* (R. P. Galask and B. Larsen, eds.), Springer-Verlag, New York, pp. 16–42.
224. Stark, L., Lee, A., and Parsonage, B. D. (1982). Colonization of the large bowel by *Clostridium difficile* in healthy infants: Quantitative study, *Infect. Immun., 35*: 895.
225. Starr, M. P., Stolp, H., Truper, H. G., Balows, A., Schegel, and H. G. (1981). *The Prokaryotes. A Handbook on Habitats Isolation and Identification of Bacteria*, Vol. I, Springer-Verlag, Berlin, Heidelberg, New York.
226. Stille, W. (1971). Die Mundhohle bei Infektionen und Infektionskrankheiten, *Therapiewoche, 21*: 3157.
227. Struve, J., Weiland, O., and Nord, C. E. (1988). *Lactobacillus plantarum* endocarditis in patient with benign monoclonal gammopathy, *J. Infect., 17*: 127.
228. Sycheva, T. B., Kashkovskaya, N. V., and Volpin, E. A. (1986). Search of effective aids and methods for prophylaxis of purulent conjunctivitis of newborns, in *Actual Problems of Nosocomial Infections*, Minsk, pp. 249–250 (in Russian).
229. Tannock, G. W. (1983). Effect of diet and environmental stress on the gastrointestinal microbiota, in *Human Intestinal Microflora in Health and Disease* (D. J. Hentges, ed.), New York, pp. 501–511.
230. Tannock, G. W., and Archibald, R. D. (1984). The derivation and use of mice which do not harbour lactobacilli in the gastrointestinal tract, *Canad. J. Microbiol., 30*: 849.
231. Tannock, G. W., Fuller, R., and Pedersen, K. (1990a). *Lactobacillus* succession in the piglet digestive tract demonstrated by plasmid profiling, *Appl. Environ. Microbiol., 56*: 1310.
232. Tannock, G. W., Fuller, R., Smith, S. L., and Hall, M. A. (1990b). Plasmid profiling of members of the family Enterobacteriaceae, lactobacilli, and bifidobacteria to study the transmission of bacteria from mother to infant, *J. Clin. Microbiol., 28*: 1225.

233. Tannock, G. W., Szylit, O., Duval, Y., and Raibaud, P. (1982). Colonization of tissue surfaces in the gastrointestinal tract of gnotobiotic animals by *lactobacillus* strains, *Can. J. Microbiol., 28*: 1196.
234. Tashjian, J. H., Coulam, C. B., and Washington, J. A. II (1976). Vaginal flora in asymptomatic women, *Mayo Clin. Proc., 51*: 557.
235. Thangkhiew, I., and Gunstone, R. F. (1987). Association of *Lactobacillus* plantarum with endocarditis [letter], *J. Infect., 16*: 304.
236. Tissier, H. (1905). Repartition des microbes dans l'intestin du nourrisson, *Ann. Inst. Pasteur (Paris), 19*: 109.
237. Torres-Alipi, B. I., Fragoso-Ramirez, J. A., Martinez-Limon, A. J., and Baptista-Gonzales, H. A. (1990). Bacterial colonization of the oral caviti in the newborn, *Bol. Med. Hosp. Infant Mex., 47*: 78 (in Spanish).
238. Tullus, K., Aronsson, B., Marcus, S., and Mollby, R. (1989). Intestinal colonization with *Clostridium difficile* in infants up to 18 months of age, *Eur. J. Clin. Microbiol. Infect. Dis., 8*: 390.
239. Türi, E., Lencner, H., Mikelsaar, M. E., Antalainen, K., Randloo, P. (1980). Investigation of odontogenic bacteremia, Proc. Tartu University, Tartu, pp. 93–98.
240. Türi, M. E., Türi, E. I., Lencner, A. A. (1982). Zur potentiellen Pathogenitat einiger Bakterien der Mikroflora des Gastrointestinaltrakts, in *Mikrobielle Umwelt und antimikrobielle Massnahmen*, (H. Horn, W. Weuffen, and H. Wiegert, eds.), Johann Ambrosius Barth, Leipzig, pp. 125–132, 279–280 (in Russian).
241. Tuschy, D. (1986). Verwendung von "Probiotika" als leistungsforderer in der Tierernahrung, *Obers. Tierernahrg, 14*: 157.
242. Ushima, T., and Ozaki, Y. (1988). Factors influencing potent antagonistic effects of *Escherichia coli* and *Bacteroides ovatus* on *Staphylococcus auerus* in anaerobic continuous flow cultures, *Can. J. Microbiol., 34*: 645.
243. Van der Merwe, J. P., Stegeman, J. H., and Hazenberg, M. P. (1983). The resident faecal flora is determined by genetic characteristics of host. Implications for Crohn's disease? *Antonie van Leeuwenhoek, 49*: 119.
244. Van der Waaij, D. (1985). The immunoregulation of the intestinal flora. Consequences of decreased thymus activity and broad spectrum antibiotic treatment, *Zbl. Bakt. Suppl. 13*: 73.
245. Van der Waaij, D. (1988). Evidence of immunoregulation of the composition of intestinal microflora and its practical consequences, *Eur. J. Microbiol. Infect. Dis., 7*: 103.
246. Van der Waaij, D., Berhuis de Vries, J. M., and Althes, C. K. (1973). Oral dose and faecal concentration of antibiotics during antibiotic decontamination in mice and in a patient, *J. Hyg. (Camb.)*, 197.
247. Varel, V. H., Pond, W. G., Pekas, J. C., and Yen, J. T. (1982). Influence of high-fiber diet on bacterial populations in gastrointestinal tracts of obese- and lean-genotype pigs, *Appl. Environ. Microbiol., 44*: 107.
248. Variyan Easwaran, P., and Hoskins, L. C. (1981). Mucin degradation in human colon ecosystems. Degradation of hog gastric mucin by faecal extracts and faecal cultures, *Gastroenterology, 81*: 751.
249. Voronina, M. N. (1968). *Lactobacilli of human gastric microflora*, Dissertation, Tartu (in Russian).

250. Wadström, T. (1988). Adherence traits and mechanisms of microbial adhesion in the gut, *Bailliere's Clin. Tropic. Med. Communicable Dis., 3*: 417.
251. Wadström, T., and Aleljung, P. (1989). Molecular aspects of bowel colonization, in *The Regulatory and Protective Role of the Normal Microflora* (R. Grubb, T. Midtvedt, and E. Norin, eds.), M. Stockton Press, Stockholm, pp. 35–46.
252. Wadström, T., Andersson, K., Sydow, M., Axelsson, L., Lindgren, S., and Gullmar, B. (1987). Surface properties of lactobacilli isolated from the small intestine of pigs, *J. Appl. Bacteriol., 62*: 513.
253. Wahbeh, C. J., Hill, G. B., Eden, R. D., and Gall, S. A. (1984). Intra-amniotic bacterial colonization in premature labor, *Am. J. Obstet. Gynecol., 148*: 739.
253a. Warner, C. M., Brownell, M. S., Ewoldsen, M. A. (1988). Why aren't embryos immunologically rejected by their mothers?, *Biol. of Repro., 38*: 117–129.
254. Weaver, G. A., Krause, J. A., Miller, T. L., and Wolin, M. J. (1989). Constancy of glucose and starch fermentation by two different human fecal microbial communities, *Gut, 30*: 19.
255. Wieken, A. J., Broady, K. W., Ayres, A., and Knox, K. W. (1982a). Production of lipoteichoic acid by lactobacilli and streptococci grown in different environments, *Infect. Immun., 36*: 864.
256. Wieken, A. J., Evans, J. D., Campbell, L. K., and Knox, K. W. (1982b). Teichoic acids from chemostat-grown cultures of *Streptococcus mutans* and *Lactobacillus plantarum, Infect. Immun., 38*: 1.
257. Zubrzycki, L., and Spaulding, E. H. (1962). Studies on the stability of the normal human fecal flora, *J. Bacteriol., 83*: 968.

10

Substrates and Lactic Acid Bacteria

Seppo Salminen
Valio Ltd., and University of Helsinki, Helsinki, Finland

Patricia Ramos
Sodima Centre de Recherche International André Gaillard, Ivry-sur-Seine, France

Rangne Fonden
Panova Ltd., Stockholm, Sweden

I. INTRODUCTION

In recent years, it has become evident that bifidobacteria and lactobacilli exist not only in the intestine of infants but also in healthy adults and various animals. They are among the predominant microorganisms in humans that contribute to the physiological well-being of the individual. It has been found that they are reduced or disappear with various diseases and with aging. For this reason, various attempts have been made to increase the numbers of lactic acid bacteria and bifidobacteria in the intestine.

There is an increasing consumer demand for natural products, particularly for food or food additives. This has created some possibilities for the use of slowly absorbable and nonabsorbable sugars as substrates for common dairy organisms.

The human colon contains a high concentration of bacteria (10^{12} per gram dry weight of colonic contents). It is assumed that the carbon and energy required to maintain such a large bacterial mass are derived from carbohydrates in host secretions or from dietary carbohydrates which are not digested in the small intestine. Recent studies show that bifidobacteria and lactobacilli are among the common anaerobic bacteria in the human colonic microflora, and may exert beneficial effects on the host. For example, in pediatric intractable diarrhea, the administration of a *Bifidobacterium* preparation normalizes the abnormal microflora and improves diarrhea. Similarly, during rotavirus diarrhea the use of a *Lactobacillus* preparation can reduce the duration of diarrhea by 50%.

For the multiplication of bifidobacteria in the human intestine dietary sugar sources are the main factors that we can influence. For example, the administration of nondigestible oligosaccharides such as raffinose, fructooligosaccharides, galactosyllactose, isomaltooligosaccharides, or transgalactosyl oligosaccharide (TOS) causes an increase in the number of endogenous bifidobacteria and some changes in lactic acid bacteria. However, the relationship between the changes and the dose of the oligosaccharides is not yet clear.

In order to increase the number of lactic acid bacteria and especially bifidobacteria in the intestinal tract, suitable slowly absorbable substrates are needed in the diet. The production of lactic acid and other organic acids by lactic acid bacteria as well as bifidobacteria is dependent on metabolism of carbohydrate substrates which have not been absorbed or metabolized in the upper digestive tract before reaching the large intestine or the colon. Such substrates have earlier been isolated from natural sources. Recent industrial processes have been developed to produce slowly absorbable substrates such as lactulose, lactitol and xylitol, and sorbitol and mannitol. Table 1 summarizes some studies on carbohydrate sources and their effect on the fecal microflora.

Table 1 Different Oligosaccharides, Disaccharides, and Polyols Influencing the Intestinal Microflora and Especially Bifidobacteria

Carbohydrate substrate	Promoted bacteria	References
Fructooligosaccharides	Bifidobacteria	Hidaka et al. (1986)
Transgalactosyl oligosaccharides (TOS)	Bifidobacteria	Kohmoto et al. (1988)
4′-Galactosyl-lactose	Bifidobacteria	Ohtsuka et al. (1989)
Isomaltooligosaccharides	Bifidobacteria	Homma (1988)
Galactooligosaccharides (Oligomate 50)	Bifidobacteria, lactobacilli	Ito et al. (1990)
Galactosyl-oligosaccharides	Bifidobacteria	Minami et al. (1983)
Soybean oligosaccharides	Bifidobacteria, some lactobacilli	Hayakawa et al. (1990), Kobayashi et al. (1984), Saito et al. (1992)
Xylooligosaccharides	Bifidobacteria	Okazaki et al. (1990)
Palatinose	Bifidobacteria	Kashimura et al. (1989)
Lactitol	?	Felix et al. (1990), Salminen and Salminen (1986)
Xylitol	Lactic acid bacteria	Salminen et al. (1985)
Lactulose	Bifidobacteria and lactic acid	Harju (1991), Terada et al. (1992)
Inulofructosaccharides	Bifidobacteria	Yamazaki and Dilawri (1990)

II. OLIGOSACCHARIDES

A. Raffinose

Raffinose is an oligosaccharide purified from several plants, white beet in particular. Since raffinose is considered to be indigestible in human intestines, it may, under circumstances of high dietary intake, be present in considerable concentrations in the lower intestine. It is well established that some intestinal bacteria, which consists of *Bifidobacterium* spp. and *Bacteroides* spp. as major genera of the fecal microbial population, are able to utilize it for their growth. Benno and co-workers have established that a four-week intake of 15 g/day of raffinose significantly increased numbers of bifidobacteria in human feces. Simultaneously, the numbers of *Clostridium* spp. and Bacteroifdaceae were decreased. Also, fecal pH values were slightly decreased during raffinose intake. This may have beneficial effects on constipation and other intestinal disturbances.

B. Soybean Oligosaccharides

Soybean oligosaccharides consist of stachyose, raffinose, and sucrose which are formed by galactose linked with sucrose. Stachyose incorporates one galactose to sucrose and raffinose on galactose to sucrose. The oligosaccharides themselves are manufactured from soybeans or soybean whey by extraction and purification. Unlike other oligosaccharides, soybean oligosaccharide production does not involve synthetic steps. Soybean oligosaccharides are heat-stable and acid-stable, and their stability is comparable with that of sucrose.

A short study on the effects of soybean oligosaccharides on human fecal microflora has been reported (Hayakawa et al., 1990). It reported that a purified stachyose and raffinose fraction of soybean oligosaccharides was fermented by *Bifidobacterium* spp. (in vitro). Also, ingestion of soybean oligosaccharides at the level of 10 g/day caused a significant increase in the number of bifidobacteria in the faeces of human volunteers. At the same time, a decrease of harmful intestinal bacteria was reported. These effects were at least partly related to the earlier observations of Yazawa et al. (1978) and Benno et al. (1987) that raffinose and stachyose as components of soybean oligosaccharides increase human fecal bifidobacteria.

C. Fructooligosaccharides

Fructooligosaccharides, containing 1-kestose (GF2), nystose (GF3), and fructofuranosylnystose (GF4), are found indigestible and to be selectively utilized by beneficial intestinal bacteria, particularly by Bifidobacteria. An increased population of beneficial intestinal bacteria would help to relieve constipation and to reduce the production of putrefactive substances.

A number of clinical studies have been done, and the results have shown that indigestible fructooligosaccharides are good for human health to relieve constipation or loose stool, to decrease the production of putrefactive substances in the large intestine, and to improve serum lipids in hyperlipidemia and reduce serum total cholesterol, triglycerides, blood glucose, and blood pressure (Hidaka et al., 1990).

Fructooligosaccharides (FOS) occur naturally in a variety of plant sources. The most abundant source appears to be Jerusalem artichoke tubers (*Helianthus tuberosus*) (Mitsuoka et al., 1987; Parekt and Margaritis, 1986). At present the industrial production of purified fructooligosaccharides is still in the developmental stage. There have been problems in the stability of fructooligosaccharides in industrial processes, and the stability has been poor compared to lactulose and soybean oligosaccharides.

D. Galactooligosaccharides

Galactooligosaccharides, which are oligosaccharides that stimulate the proliferation of bifidobacteria, are present in human milk, cow's milk, and commercial yogurt. A mixture of galactooligosaccharides from lactose by the enzymatic action of β-D-galactosidase from *Aspergillus oryzae* and *Streptococcus thermophilus* has been reported (Ito et al., 1990). The same study indicates that administration of galactooligosaccharides produced from lactose by the action of the above-mentioned enzymes had significant effects on the fecal flora. A linear dose-response between the galactooligosaccharides and fecal bifidobacteria was reported. Also, lactobacilli increased in numbers in the feces of volunteers. However, a daily dose of 10 g/day was required to achieve a significant change (Ito et al., 1990). Stool weights and frequencies were not altered during galactooligosaccharide administration.

E. Galactosyl Lactose

Galactosyl lactose (GL) is a trisaccharide found in breast milk. GL is also manufactured in Japan and added to commercial infant formulas to promote the growth of bifidobacteria in the gut of infants. It has been indicated that GL changes the fecal properties toward those of breast-fed infants. This is shown for instance in the proportion of organic acids in the feces of GL-fed infants when compared with regular formula or breast milk (Table 2).

GL appears to be suitable for infant formulas and other dairy products increasing the number of bifidobacteria as measured in fecal samples. No safety studies per se are available, but the experience from infant formulas support the safety of GL.

Table 2 The Percentage of Organic Acids and the pH in the Feces of Breast-Fed, Formula-Fed, and Galactosyl Lactose-Fed Infants

Organic acid	Percentage of breast-fed infants	Organic acid in feces formula-fed infants	GL formula-fed infants
Acetic	53.3	24.8	39.2
Lactic	20.0	11.1	19.4
Propionic	4.3	12.8	11.6
Succinic	6.2	19.7	14.4
Citric	4.8	7.9	5.5
Pyruvic	10.7	7.9	5.5
pH	5.59	6.66	5.74

Source: Adapted from Mitsuhashi et al. (1982).

F. Palatinose

Palatinose (6-O-α-D-glucopyranosyl-D-fructoruranose) is used in various food products as a noncariogenic nutritive sweetener. A mixture of palatinose and its condensates is obtained by heating palatinose-melt under suitable conditions. The condensates are classified as heterooligosaccharides composed of glucose and fructose residues. Palatinose is digestible, but digestibility of the condensates is not clear. In a human study the intake of palatinose has been associated with higher numbers of bifidobacteria in the feces (Kashimura et al., 1989).

III. DISACCHARIDES AND OTHER ALTERNATIVES

Lactulose, lactitol, xylitol, sorbitol, and mannitol have been used as alternatives for oligosaccharides. All these are slowly absorbable and offer fermentable substrates for lactic acid bacteria in the colon. Lactitol, lactulose, and xylitol have the best-documented good effects on intestinal microflora. Sorbitol and mannitol appear to promote Gram-negative flora. All the previously mentioned substances may cause transient diarrhea when given in larger amounts. However, gradual adaptation to larger doses increases the tolerance quite quickly.

The polyols have been evaluated by the WHO food additive group, and they have all been given an ADI value of "not limited," indicating that they are safe. Lactulose has been used as a pharmaceutical for several decades. No indication exists of any harmful effects. The acute toxicity of lactulose (23–26 g/kg/body weight) is of the same order as that for sucrose.

A. Neosugars

Neosugars include mainly fructooligosaccharides and transgalactoside oligosaccharides manufactured by chemical synthesis. They appear to be inert in the mouth and small intestine. However, they are digested in the colon and may therefore change the metabolic activity of the colon. No data is available on their safety, and their health effects are not sufficiently well studied yet.

IV. SAFETY OF SUBSTRATES

The safety aspects of oligosaccharides and substrates for lactic acid bacteria and bifidobacteria are not well studied. Apart from the polyols (xylitol, sorbitol, mannitol, and lactitol), no intensive toxicology testing has been completed. However, as indicated in Table 3, the polyols have been extensively tested and the data has been evaluated by the JECFA (Joint FAO WHO Expert Committee on Food Additives). Also, ADI values have been established for these polyols.

The other substrates including oligosaccharides have not been evaluated in extensive toxicological studies. It has been assumed that they pose no greater danger since they are either naturally occurring or made of natural components. Since no international evaluation is available, the practice of approving these compounds may greatly vary from country to country.

Most slowly absorbable sugars cause at least transient intestinal and abdominal problems. For polyols a gradual adaptation of larger doses occurs. However, even then abdominal distension and pain as well as flatulence and meteorisms are often experienced. At times also transient diarrhea may occur. Even though there are very few reports on the adverse effects of sugars and oligosaccharides, Table 3 can be used as a summary.

Table 3 Intestinal Problems Caused by the Consumption of Large Amounts of Slowly Absorbable Substrates

Sugar substitute	Transient diarrhea	Fullness	Abdominal pain	Meteorism
Xylitol[a]	+	+	+	+
Sorbitol[a]	+	+	+	+
Lactitol[a]	+	+	+	+
Galactooligosaccharides	+	+	+	+
Fructooligosaccharides	?	+	+	+
Soybean oligosaccharides	?	?	?	+

[a]Gradual adaptation to larger doses (in case of xylitol up to 150 g/day) have been observed.

V. POLYOLS

A. Sorbitol

Sorbitol is used mainly as a sweetening agent for dietic foods, where it combines moderate sweetening power, specific flavor characteristics, and pleasant viscosity in liquids. It is used in sugar-free candies and chewing gum and in diabetic foods. In mixtures with other sugars, sorbitol modifies the crystallization properties of foods. When added to syrups containing sucrose it reduces crystal deposition during storage. Sorbitol also has some uses as a humectant and stabilizer, and it can be used as a substitute for glycerol. Small amounts of sorbitol have been added to low-calorie drinks to mask the aftertaste of saccharin and to provide the normal mouth feeling (Grenby, 1983).Typical products in Europe include confectionery, pastilles, diabetic jam and cookies, ice cream, chocolates, and pastries.

The toxicology of sorbitol has been reviewed by WHO. The JECFA has given sorbitol an ADI of "not specified," which means that no health hazards are foreseen (WHO, 1982). Since sorbitol is absorbed slowly, foods sweetened with sorbitol are thought to be suitable for diabetics provided that the calories are taken into account. However, large amounts of sorbitol can cause flatulence, diarrhea, and abdominal distension. Gradual addition of sorbitol into the diet may increase the tolerance of the individual (Salminen et al., 1985).

In most countries sorbitol is approved as an ingredient when used for sweetening purposes. In the United States sorbitol is permitted for use as a food additive by the U.S. Food and Drug Administration. Sorbitol is also considered GRAS for use as a nutrient and dietary supplement (FASEB/SCGOS, 1973a).

B. Mannitol

Mannitol is a hexitol that is stereoisomeric to sorbitol. It is commonly found naturally in some plant foods, including beets, celery, olives, and seaweed. Mannitol has about 0.4–0.5 the sweetness of sucrose, and its properties are fairly similar to those of sorbitol. Only the solubility of mannitol is poor compared to sorbitol. Mannitol is produced from sucrose or dextrose and can also be obtained as a by-product of some fermentations.

Mannitol is used in sugar-free dietary foods, sugar-free chewing gum, sweets, and ice cream. In addition to its use as a sweetener, mannitol can be used as a texturizing agent, anticaking agent, or humectant. Sometimes it is used in breakfast cereals and frostings. At present mannitol is used mainly in sugar-free chewing gums as a sweetening agent and for dusting the chewing gum sticks.

Mannitol is slowly absorbed from the intestinal tract and may cause diarrhea and flatulence. In experimental animals an adaptation to mannitol can be seen (Salminen et al., 1985). In humans a laxative effect is observed after intakes of

20–30 g of mannitol. Toxicity studies have not indicated any adverse effects other than diarrhea. Therefore mannitol is considered safe for use in foods. Mannitol is also in the U.S. FDA GRAS list. An evaluation of its health effects has been conducted (FASEB/SCOGS, 1973b). An acceptable daily intake of "not specified" has been allocated for mannitol (WHO, 1982).

C. Lactitol

Lactitol is a disaccharide alcohol (1,4-galactosylglucitol) produced by the hydrogenation of lactose or lactulose. Lactitol can also be used as a raw material for oligosaccharides (Harju, 1988). Lactitol has been known since 1912, and its manufacture was mentioned by Aminoff in 1974, but attention has been directed to its properties only lately (Harju et al., 1990). Lactitol has a low sweetness (Table 3), and therefore its major use is not as a sweetener. More recent biological research cited by WHO (1983) indicates that lactitol may have a lower energy value than other carbohydrate sweeteners.

Commercially available lactitol has a molecular weight of 344 and crystallizes as a colorless odorless monohydrate with a pleasant mild sweetness (Saijonmaa et al., 1978; Linko et al., 1980). Since lactitol does not have a carbonyl group it cannot undergo Maillard reactions. In general use, lactitol is more stable than lactose. Its relative sweetness is about 50% that of glucose (Linko et al., 1980).

1. *Technological Uses*

Lactitol as well as most polyols can be applied to special dietary foods that can be consumed by diabetics provided that the calories are taken into account. Lactitol can be used as a sweetener in most foods, but due to its low sweetness it is not very attractive. However, it may have other uses as a sweet bulking or texturizing agent in the future because of its low energy value (WHO, 1983). The products that are manufactured with lactitol have excellent palatability, and no unpleasant aftertaste is associated with lactitol. Lactitol also has uses as a ingredient for the pharmaceutical industry. Apparently lactitol may also be utilized instead of lactulose for some therapeutic applications and special dietary foods (Patil et al., 1987). Lactitol has been indicated to change the intestinal microflora and to produce effects comparable to those of lactulose (Bird et al., 1990; Patil et al., 1987).

2. *Toxicology and Safety*

Lactitol has gone through extensive toxicological studies in Holland. All required toxicology studies have been completed, including long-term and carcinogenicity studies. Also, short studies on the gut microflora changes have been completed (Salminen and Salminen, 1986). These studies indicate that apart from diarrhea after consumption of large lactitol doses no toxicologically significant adverse effects have been noted (WHO, 1983). The EEC Scientific Committee on Food has accepted lactitol and most other polyols for use in food. However, it was pointed out that laxation may occur at high intakes.

Since lactitol has been made commercially available only recently, most countries have not classified it as an ingredient or a food additive. JECFA has given lactitol and ADI of "not specified," indicating that it is considered safe for food use (WHO, 1983). Food additive and ingredient petitions have been filed in most countries; lactitol has been approved as a food additive in most European countries, and a lactitol preparation is used as a pharmaceutical in Switzerland.

D. Xylitol

Xylitol is a pentitol that is found in most fruits and berries as well as vegetables. Commercially, xylitol is produced either from xylan containing plant material or by microbiological methods. Xylitol in larger doses causes significant changes in the intestinal flora (Salminen et al., 1985). It appears to promote Gram-positive acid producing bacteria and to lower the colonic pH. Xylitol is also known as an anticariogenic sweetener. Its properties were earlier reviewed by Salminen and Hallikainen (1991), and safety aspects were studied by the WHO (1982). Xylitol may also have future uses as a substrate for lactic acid bacteria.

VI. LACTULOSE

Lactulose (4-O-α-D-galactopyranosyl-D-fructofuranose) is a keto analog of lactose. Unlike lactose, it resists the hydrolytic action of intestinal β-galactosidases and therefore is not absorbed from the small intestine. Lactulose is mostly available in syrup containing 67% lactulose, but recently a crystalline form has also been produced. Lactulose has a relatively low sweetness, but it is stable in most foods.

Lactulose can be utilized in most liquid foods, but due to its low sweetness it does not have many food applications. In addition it has laxative properties that prevent its use in most common foods. Only special dietary foods that are intended for people suffering from constipation have utilized lactulose as a sweetener. It has been claimed that lactulose in infant foods and formulas enhances the development of intestinal flora containing *Bifidobacterium bifidum* that mimics the flora of breast-fed infants (Mendez and Olano, 1979). Lactulose may also act as a nonabsorbable substrate for colonic bacteria, thereby causing other favorable changes in the microflora and making lactulose suitable for special dietary foods in this area. Lactulose also has a number of applications in pharmaceutical preparations. The effect of dietary supplementation with lactulose (3 g/day for 2 weeks) in human volunteers has significantly increased the number of bifidobacteria in feces. It also decrease the mean fecal pH and increased fecal water content. Also fecal β-glucuronidase, nitroreductase, and azoreductase values decreased significantly (Terada et al., 1992).

The toxicity of lactulose has not been studied according to present guidelines. However, lactulose has been utilized for years in the treatment of chronic constipation and portosystemic encephalopathy, and patients have received large lactulose doses for long treatment periods. It is apparent that microflora changes in the intestinal tract occur during lactulose ingestion (Salminen and Salminen, 1986; Salminen et al., 1988). However, apart from meteorism and occasional diarrhea no other harmful effects have been observed. Both lactulose and lactitol appear to cause similar changes in the intestinal microflora. They have similar effects on protein digestion and metabolism in animal studies, and both have been proposed and used as therapeutic agents for portosystemic encephalopathy (Bird et al., 1990; Patil et al., 1987).

In most countries lactulose is utilized as a pharmaceutical preparation, and occasionally it is used in special dietary foods.

VII. SELECTION CRITERIA

The task for identifying suitable sugars which will increase the number of bifidobacteria and lactobacilli resident in the human intestinal tract is difficult. Based on earlier research it can be stated that the following points should be considered:

1. Slowly absorbable or nonabsorbable compounds are needed.
2. Nonreducing saccharides tend to be unavailable for many lactic acid bacteria.
3. Bifidobacteria can utilize di- or trisaccharides containing galactose without prior adaptation.
4. Oligosaccharides are commonly used by bifidobacteria and lactic acid bacteria.
5. Sugars consisting of galactose, glucose, and/or fructose can be utilized by bifidobacteria.
6. Inulin derivatives are utilized by bifidobacteria; however, an adaptation period may be needed.
7. Polyols need to be studied due to their slow absorption.
8. Decrease in the fecal pH is generally a desirable property for any substrate.
9. A combination of substrates may be desirable for promoting both lactobacilli and bifidobacteria.

REFERENCES

Benno, Y., Endo, K., Shiragmai, N., Sayanama, K., and Mitsuoka, T. (1987). Effect of raffinose on human fecal microflora, *Bifidobacteria Microflora, 6*: 59–63.

Bird, S. P., Hewitt, D., Ratcliffe, B., and Gurr, M. I. (1990). Effects of lactulose on protein digestion and metabolism in conventional and germ free animal models: relevance of their results to their use in the treatment of portosystemic encephalopathy, *Gut, 31*: 1403–1406.

FASEB/SCOGS (1973). Evaluatioon or sorbitol and mannitol as food ingrednets. Reports 9-10, NTIS PB 221-951 and 953.

Felix, Y., Hudson, M., Owen, R., Ratcliffe, B., van Es, A., van Velthuijsen, J., and Hill, M. (1990). Effect of dietary lactitol on the consumption and metabolic activity of the intestinal microflora in the pig and humans, *Microbial. Ecol. Health Dis., 3*: 259–267.

Harju, M. Lactose, its derivatives and their hydrolysis, *Finn. J. Dairy Sci., 49*: 1–47 (1991).

Harju, M. (1990). Lactobionic acid as a substrate for β-galactosidases, *Milchwissenschaft, 45*: 411–415.

Harju, M. (1988). Lactitol as a substrate for β-galactosidases. II: Results and discussion, *Milchwissenschaft, 43*: 148–152.

Hayakawa, K., Mizutani, J., Wada, K., Masai, T., Yoshihara, I., and Mitsuoka, T. (1990). Effects of soybean oligosaccharides on human faecal flora, *Microbioal. Ecol. Health Dis., 3*: 293–303.

Hidaka, H., Eida, T., Tazikawa, T., Tokunaga, T., and Tashiro, Y. (1986). Effect of fructooligosaccharides on intestinal flora and human health, *Bifidobacteria Microflora, 5*: 37–50.

Hidaka, H., Hirayama, M., Tokunaga, T., and Eida, T. (1990). The effects of undigestible fructooligosaccharides on intestinal microflora and various physiological functions on human health, in *New Developments in Dietary Fiber* (I. Furda and C. J. Brine, eds.), Plenum Press, New York, pp. 105–117.

Ito, M., Deguchi, Y., Miyamori, A., Matsumoto, K., Kikuchi, H., Matsumoto, K., Kobayashi, Y., Yajima, T., and Kan, T. (1990). Effects of administration of galactooligosaccharides on the human faecal microflora, stool weight and abdominal sensation, *Micr. Ecol. Health Dis., 3*: 285–292.

Kashimura, J., Nakajima, Y., Benno, Y., Endo, K., and Mitsuoka, T. (1989). Effects of palatinose and its condensate intake on human fecal microflora, *Bifidobacteria Microflora, 8*: 45–50.

Kobayashi, Y., Echizen, R., Mada, M., and Mutai, M. (1984). Effects of hydrol,ysates of konjac mannan and soybean oligosaccharides on intestinal flora in man and rats, in *Intestinal Flora and Dietary Factors* (T. Mitsuoka, ed.), Japan Scientific Societies Press, Tokyo, pp. 70–90.

Kohmoto, T., Fukui, F., Takaku, H., Machita, Y., Arai, M., and Mitsuoka, T. (1988). Effect of isomaltito-oligosaccharides on human fecal flora, *Bifidobacteria Microflora, 7*: 61–69.

Minami, Y., Kouhei, Y., Tamura, Z., Tanaka, T., and Yamamoto, T. (1983). Selectivity of Utilization of Galactosyl-oligosaccharides by bifidobacteria, *Chem. Pharmac. Bull. (Tokyo), 31*: 1688–1691.

Mitsuhashi, S., Yoshihama, M., Yahiro, M., Nishikawa, I., Deya, E., Ahiko, K., and Mitsuoka, T. (1982). Effects of oligosaccharides on intestinal flora and fecal characteristics of neonates, *Proc. III IPCR Symp. Intestinal Flora: Intestinal Flora and Nutrition* (T. Mitsuoka, ed.), Japan Scientific Societies Press, Tokyo.

Mitsuoka, T., Hidaka, H., and Eida, T. (1987). Effect of fructooligosaccharides on intestinal microflora, *Die Nahrung, 31*: 427.

Ohtsuka, K., Benno, Y., Endo, K., Ueda, H., Ogawa, O., Uchida, T., and Mitsuoka, T. (1989). Effect of 4′-galactosyllactose intake on human fecal microflora, *Bifidus, 2*: 143–149.

Okazaki, M., Fujikawa, S., and Matsumoto, N. (1990). Effect of xylooligosaccharide on the growth of bifidobacteria, *Bifidobacteria Microflora, 9*: 77–86.

Parekh, S. R., and Margaritis, A. (1986). Continuous hydrolysis of fructans in Jerusalem artichoke extracts using immobilized non-viable cells of *Kluveromyces marcianus, J. Food Sci., 51*: 854.

Patil, D. H., Westaby, D., Mahida, Y. R., Palmer, K. R., Rees, R., Clark, M. L., Dawson, A. M., and Silk, D. B. A. (1987). Comparative modes of action of lactitol and lactulose in the treatment of hepatic encephalopathy, *Gut, 28*: 255–259.

Saito, Y., Iwanami, T., and Rowland, I. (1992). The effects of soybean oligosaccharides on the human gut microflora cultured in *in vitro* culture, Abstracts XIV Annual Meeting of Intestinal Microecology, Helsinki, p. 9.

Salminen S., and Hallikainen, A. (1990). Sweeteners, in *Food Additives,* (Braven, A., Davidson, P., and Salminen, S., eds.), Marcel Dekker, New York, pp. 297–326.

Salminen, E., and Salminen, S. (1986). Lactulose and lactitol induced caecal enlargement and microflora changes in mice, *Proc. Euro Food Tox. II*, Institute of Toxicology, ETH, Zurich, pp. 313–317.

Salminen, S., Salminen, E., Bridges, J., and Marks, V. (1985). The effects of sorbitol on the gastrointestinal flora in rats, *Z. Ernahrungswiss. 25*: 91–95.

Salminen, S., Salminen, E., Marks, V., Bridges, J., and Koivistoinen, P. (1985). Gut microflora interactions with xylitol in the rat, mouse and man, *Food Chem. Toxicol., 23*: 985–990.

Tanaka, R., Takayama, H., Morotomi, M., Kuroshima, T., Takayama, H., Matsumoto, K., Kuroda, A., and Mutai, M. (1983). Effect of administration of TOS and Bifidobacterium breve 4006 on the human fecal flora, *Bifidobacteria Microflora, 2*: 17–24.

Terada, A., Hara, H., Kataoka, M., and Mitsuoka, T. (1992). Effect of lactulose on the composition and metabolic activity of the human fecal flora, *Microbial Ecol. Health Dis., 5*: 43–50.

WHO. (1982). *Toxicological Evaluation of Certain Food Additives.* WHO Food Additives Series.

Yamazaki, H., and Dilawri, N. (1990). Measurement of growth of bifidobacteria on inulo-fructosaccharides, *Lett. Appl. Microb., 10*: 229–232.

Yazawa, K., Imai, K., and Tamura, Z. (1978). Oligosaccharides and polysaccharides specifically utilizable by bifidobacteria, *Chem. Pharmac. Bull. (Tokyo), 26*: 3306–3311.

11

Toxicity of Lactic Acid Bacteria

D. C. Donohue, Margaret Deighton, and J. T. Ahokas
Royal Melbourne Institute of Technology, Melbourne, Victoria, Australia

Seppo Salminen
Valio Ltd., and University of Helsinki, Helsinki, Finland

I. INTRODUCTION

The oral safety of many lactic acid bacterial strains has been accepted after a long history of use and exposure in traditional foods. They include *Streptococcus thermophilus, Lactobacillus bulgaricus*, and many other strains that have been used in a number of countries around the world in cheese and fermented dairy products. These indigenous lactic acid bacteria are thought to be both safe and beneficial for human health and, therefore, toxicity studies have not been required. New strains of lactic acid bacteria are continually being introduced into fermented dairy products and other foods as well as into probiotic preparations. As there have been occasional reports of human infections from lactic acid bacteria, the potential of newly introduced strains to cause infections should be studied.

The toxicity of lactic acid bacteria is of increasing interest for regulatory purposes. In European countries the regulation on novel foods including ingredients and processes require both a safety assessment and nutritional evaluation of new and improved microorganisms (1). Interest focuses on two areas: the ability of defined strains to cause infection and the toxicity of orally administered bacteria. In Europe new guidelines for the use of lactic acid bacteria as novel foods, novel ingredients, or therapeutic products also require basic traditional testing for acute toxicity. For newer strains, especially those that are derived by biotechnological procedures, testing in an animal model has been proposed.

Very little data are available on the traditional toxicity testing of lactic acid bacteria. Japanese investigators have studied the toxicological properties of *Bifidobacterium longum*, which although closely related to lactic acid bacteria, is not classified as such (7). In their studies no acute toxicity was observed. However, the amounts ingested were extremely high indicating perhaps a low number of viable organisms. In another study, Ishihara and coworkers studied the acute toxicity of heat-treated cells of lactic acid bacteria in mice (5). The species used in this study included *Streptococcus faecium, Streptococcus equinus, Lactobacillus fermentum*, and *Lactobacillus salivarius*. No acute toxicity of the heat-killed cells was observed.

To further study the acute oral toxicity of viable lactic acid bacteria, we have selected three different strains of bacteria. *Streptococcus thermophilus* is a strain commonly used in commercial and domestic yogurt manufacture. *Lactobacillus helveticus* is commonly used in different types of cheese and cheese products.

Lactobacillus casei strain GG (*Lactobacillus* GG) is an improved natural strain closely resembling *L. casei* which has been used in dairy foods for some years and has been studied intensively for its probiotic properties. The effects of *Lactobacillus* GG have been studied in lethally irradiated mice, where no infections or deaths caused by the organism were observed (2). There is evidence that *Lactobacillus* GG preparations reduce the duration of diarrhea in premature babies and infants with rotavirus diarrhea (6). *Lactobacillus* GG has been shown to be useful in treating recurring diarrhea caused by *Clostridium difficile* (3).

The use of these three strains in dairy products and their ingestion by infants and immunocompromised hosts has not caused any adverse effects. The purpose of this study is to carry out an acute toxicity test for *Lactobacillus* GG using mice as the experimental animal. The oral toxicity of two organisms traditionally used in foods was assessed for comparison.

II. EXPERIMENTAL RESULTS

The present study was conducted in accordance with the 1987 OECD guidelines for acute toxicity testing and with the approval of the RMIT Animal Ethics Committee. The test bacteria and the number of colony-forming units/g of freeze-dried bacteria of each were *L. casei* strain GG (1.5×10^{11}), *S. thermophilus* (4×10^{10}), and *L. helveticus* 230 (2×10^{10}).

Male adult Swiss mice (Monash University Animal Service, Melbourne, Australia) were housed 10 per cage at 20° ± 3°C, relative humidity, 40–60% with a 12 hour light–dark cycle. Standard mouse ration and unlimited water were provided. Mice were acclimatised for 5 days before random assignation to test groups. A freeze-dried suspension of test organisms in 1 ml distilled water was given by gavage to fasted mice. Concentrations were adjusted to ensure a constant volume at all dose levels. Graduated doses of 1, 2, 4, or 6 g of test bacteria/kg body weight were administered to test groups of five animals. Controls received distilled water or remained untreated and were reassigned randomly for each treatment group. Animals were observed twice daily for 7 days, noting any changes in activity, behavior, condition, or food intake. Mice were weighed on days 1, 4, and 7 posttreatment, then humanely killed for necropsy. Organs were examined for gross pathological changes. Specific growth rates (SGR) for periods throughout the experiment were calculated using the formula SGR = Ln (weight Time x) – Ln (weight Time y) divided by (Time x – Time y) × 100. These values were used for regression analysis of SGR against either dose or treatment.

There were no treatment-related deaths. All mice except untreated controls showed lassitude and decreased appetite in the 24-hour period postdosing. Subsequently there were no observed differences in behavior or condition between groups.

Whole body weights were reduced slightly after dosing but weight gain resumed after 24 h. At necropsy no caecal enlargement was observed, and both stomach and liver appeared unremarkable.

No treatment-related signs of toxicity were observed between groups. Regression analysis showed no relationship between the dose or test bacteria and SGR over the period of the experiment (correlation coefficient = 0.044 test bacterium; 0.347 dose). There was no significant test bacterium or dose relationship with SGR either with individual growth periods or over the full experimental period (Figs. 1, 2, and 3).

III. DISCUSSION

The large number of recent publications is evidence of the interest in the possible beneficial effects of cultured dairy products to human health. The results from some studies have raised questions as to whether new bacterial strains used in cultured dairy products could have undesired side effects. The many types and strains of microorganisms involved in such products may differ from each other in their properties and effects, including toxicity. Therefore only limited data are available from toxicity studies as summarized in Table 1.

The initial loss of body weight which was observed in this study could be attributed to an initial loss of appetite due to the volume of dose. No toxic effects of the three strains of bacteria were observed at the highest dose of 6 g/kg body weight. However, a trend towards lower SGR was observed at the highest dose of *Lactobacillus* GG. Based on our study, the acute oral LD_{50} of each organism after seven days for male Swiss mice is greater than 6 g/kg body weight, and the tested strains can be considered nontoxic in an acute test system. This is in agreement

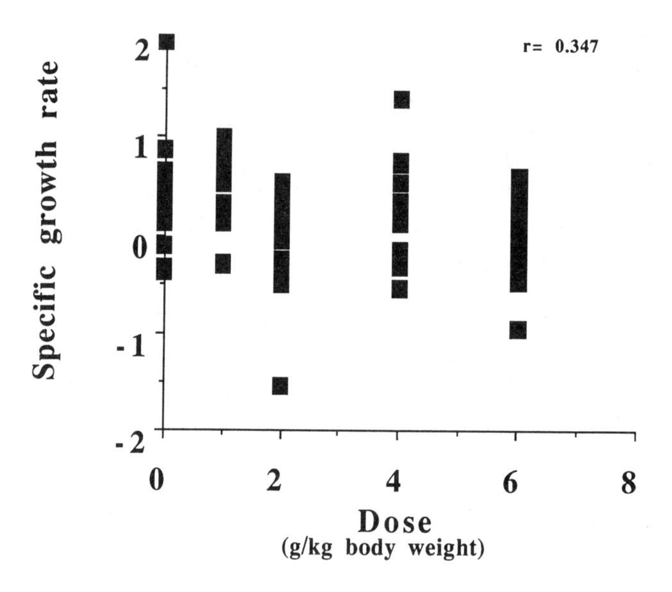

Figure 1 Specific growth rate of mice for the experimental period for all treatments.

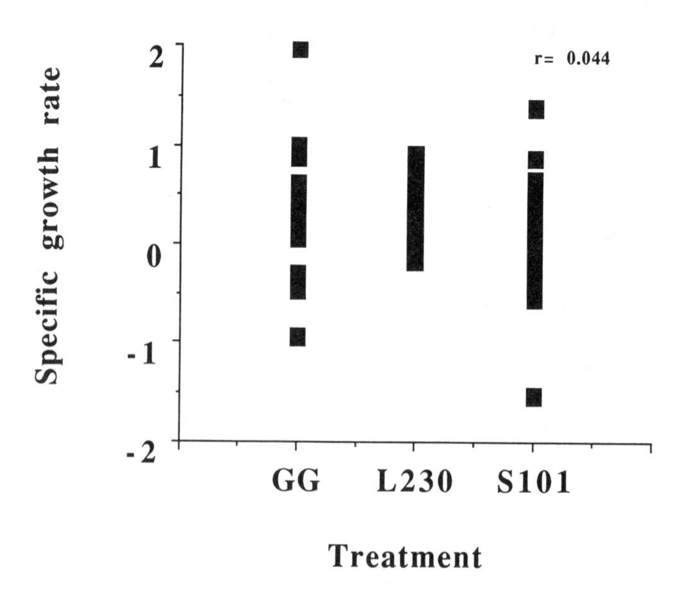

Figure 2 Specific growth rate of mice for the experimental period for all doses.

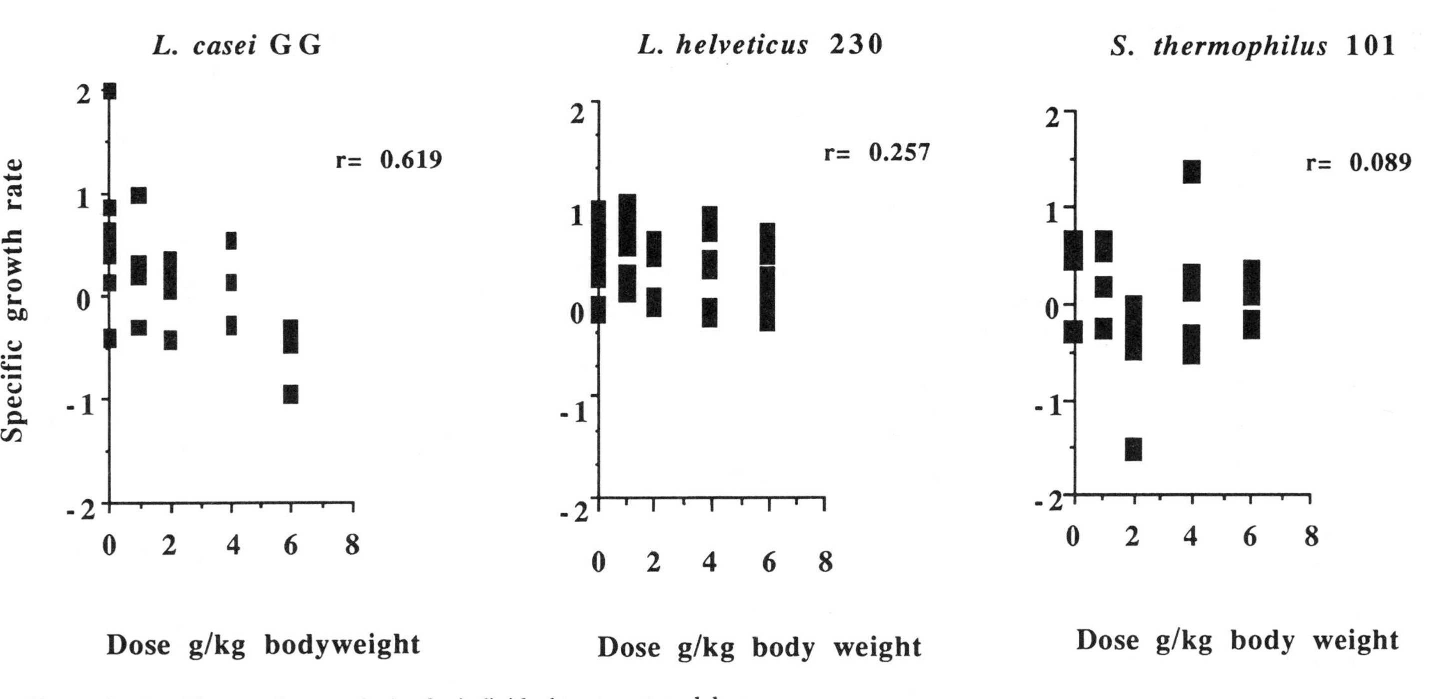

Figure 3 Specific growth rates of mice for individual treatments and doses.

Table 1 The LD_{50} Values for Heat-Treated and Viable Lactic Acid Bacteria and Bifidobacteria Observed in Different Studies

Lactic acid bacteria	LD_{50} (g/kg of body weight)
Streptococcus faecium AD1050[a]	>6.6
Streptococcus equinus[a]	>6.39
Lactobaccilus fermentum AD0002[a]	>6.62
Lactobacillus salivarius AD0001[a]	>6.47
Lactofacillus casei GG	>6.00
Lactobacillus helveticus	>6.00
Lactobacillus bulgaricus	>6.00
Bifidobacterium longum	25

[a]Heat-treated preparations
Source: Combined from Refs. 4 and 5 and the present study.

with the previously reported values for a number of nonviable (heat-treated) lactic acid bacteria strains and bifidobacteria (5,7).

Although extrapolation of oral LD_{50} values from animals to humans has limited validity, the values observed in this study would correspond to a dose of more than 420 g of washed freeze-dried bacteria for a 70 kg human. The normal daily intake from fermented milks vary from 10^9 to 10^{10} cfu (usually 10^6 to 10^9 cfu/ml in fermented milks) corresponding to 150 to 500 g of yogurt with live cultures while the lactic acid bacteria and propionic acid bacteria content of Swiss cheese and Edam cheese vary in the range of 10^6 to 10^7 cfu/g cheese. Thus with an average daily consumption of 20 g of cheese and 400 g of yogurt products the intake of lactic acid bacteria would be of the order of 10^{10} viable bacteria or about 1–2 g of freeze-dried bacteria.

Momose et al. (1979) in an acute toxicity study of *Bifidobacterium longum* BB-536 in mice found the oral LD_{50} to be more than 50 g/kg body weight and hence concluded that BB-536 was not toxic orally. It is of interest that Momose and co-workers could administer such a high dose, as in our study the maximum dose was restricted by the mass of culture which could be suspended in a volume of 1 ml. Ishihara and co-workers were able to administer 6.3–6.6 g of heat-treated dead cells of lactic acid bacteria.

Data from animal studies including our own indicate that most of the lactic acid bacteria do not possess acute toxicity. It may therefore be more useful to concentrate future efforts on other aspects of toxicity. It is most important, however, to test new strains and genera for their ability to cause infections including bacteremia. The method of Dong et al. (1987) in which lethal radiation is used could be recommended for animal studies. It is of importance to evaluate both the safety aspects and the nutritional effects of new strains of lactic acid bacteria as indicated in the novel food regulations (1,4).

REFERENCES

1. ACNFP. The Advisory Committee on Novel Foods and Processes. Her Majesty's Stationery Office, London, 1990.
2. Dong, M., Chang, T., and Gorbach, S. (1987). Effects of feeding *Lactobacillus* GG on lethal irradiation in mice, *Diag. Microbiol. Infect. Disease. 7*: 1–7.
3. Gorbach, S., Chang, T., and Goldin, B. (1987). Successful treatment of relapsing *Clostridium difficile* colitis with *Lactobacillus* G, *Lancet, 2*: 1519.
4. ILSI Europe. Working group on the evaluation of novel foods and processes, Brussels, 1991.
5. Ishihara, K., Miyakawa, H. Hasegawa, A., Takazoe, I., and Kawai, Y. (1985). Growth inhibition of *Streptococcus mutans* by cellular extracts of human intestinal lactic acid bacteria. *Infect. Immun. 49*(3): 692–694.
6. Isolauri, E., Juntunen, M., Rautanen, T., Sillanaukee, P., and Koivula, T. (1991). A human *Lactobacillus* strain (*Lactobacillus casei* sp. strain GG) promotes recovery from acute diarrhoea in children, *Paediatrics, 88*(1).
7. Momose, H., Igarashi, M., Era, T., Fukuda, Y., Yamada, M., and Ogasa, K. (1979). Toxicological studies on *Bifidobacteriurn longum* BB-536. *Pharmacometrics, 17*(5): 881–887.

12

Lactic Acid Bacteria as Animal Probiotics

Juha Nousiainen and Jouko Setälä
Valio Ltd., Helsinki, Finland

I. INTRODUCTION

The composition and metabolism of the gastrointestinal microflora affects the performance of farm animals in many ways, especially young ones subjected to environmental stress. The indigenous flora which is established after birth interacts with the digestive and immune systems of the body and its activities can be both beneficial and harmful to the host. The colonization of the different compartments of the gut by specific commensal bacteria, partly by means of association with the mucus layer or adhesion to the surface or epithelial cells, serves as a first defense barrier against invading microorganisms or toxic substances in the diet. In some species, especially in adult ruminants, the digestion of fibrous diet is mainly based on the fermentative action of the bacteria in the rumen. In addition to the digestive aid, the gut flora may produce substances or reprocess the refluxed host metabolites which are absorbed and utilized or excreted (Savage, 1977, 1986).

In the healthy animals each part of the intestines is colonized by a typical microflora, which is adapted to grow in a beneficial symbiosis with the host. Due to the intensive management methods of today, the farm animals are very susceptible to enteric bacterial imbalance, leading to inefficient digestion and absorption of nutrients and retarded growth. To overcome these difficulties diets have usually been supplemented with antibiotics, which have indeed proved to be very effective in decreasing diarrhea and promoting growth (Armstrong, 1984, 1986; Parker and Armstrong, 1987). However, the development of resistant strains of harmful bacteria may interfere with the use of veterinary antibiotics (Linton et al., 1988; Hedges and Linton, 1988) and decrease the efficiency of antibiotics per se. Possible residues in the animal products and cross-resistance with human pathogens might also result in health risks which are so far not completely understood (Mee, 1984; Hanson, 1985).

For the above reasons there is wide interest in replacing feed antibiotics with more natural feed additives—probiotics. This term was used by Lilly and Stillwell (1965) to mean a substance secreted by one microorganism which stimulated the growth of another. Parker (1974) stated that probiotics are organisms or substances contributing to optimal intestinal microbial balance. Fuller (1989) refocused the point "organisms or substances" so that they should be live microbial feed supplements, to exclude the possibility that the "substances" could mean antibiotics. As a synthesis of the former expressions, the authors consider animal probiotics to live indigenous microorganisms or nonantibiotic substances, which decrease the number of intestinal infections and/or increase production and/or improve food hygiene by contributing to a better gastrointestinal environment.

Since the early studies of Metchnikoff (1903, 1908) of the favorable effects of soured milk products in man, the most beneficial part of the intestinal flora is suggested to be lactic acid bacteria. These organisms are also most often found in commercial probiotic preparations (Tuschy, 1986; Anonymous, 1990), but spore-forming bacilli or bifidobacteria may be used as well. The purpose of this overview is to introduce the composition and the activities of the gut microflora in farm animals and to condense the latest knowledge of lactic acid bacteria as potential performance enhancers. Later in this context the term "lactic acid bacteria" (LAB) means members of the genera *Lactobacillus* and *Enterococcus* (formerly *Streptococcus*). The following discussion will be concentrated on pigs and young cattle, which are developing ruminant behavior. Studies of other species, such as poultry, will be referenced only when necessary to help understand the probiotic concept. For non-LAB probiotics (bacilli, yeasts, etc.) and other animal species, the reader is referred to reviews of Tournut (1989), Kozasa (1989), and Vanbelle et al. (1990).

II. GENERAL ASPECTS OF THE GUT MICROFLORA IN PIGS AND CALVES

A. Composition

The composition of the gut flora is known to vary due to many host-specific and environmental factors. Age and the gut site or the diet of the animal may be the most important examples of the former or the latter, respectively. Microbial communities in a certain part of the gut can be found in the lumen (attached to the feed particles or exist freely in the fluid), in association with the mucous epithelium, or in the bottom of the crypts (Savage, 1986). A detailed discussion of the colonization factors believed to affect the gut microbiota is given by Savage (1987a) and Tannock (1988), among others.

1. Pigs

The GI tract of pigs is first inoculated with bacteria occurring in the reproductive tract of the dam and then by those existing in the immediate environment (Ratcliffe, 1985). The stomach of the neonatal pig has been shown to be colonized by at least lactobacilli, streptococci, and coliforms within 48 h after birth, and strictly anaerobic organisms, such as bacteroides, can also be detected in the faeces when the pig is a few days old (Dulcuzeau, 1985). In the suckling period, bacteria which can utilize the components of milk predominate in the upper tract (Fuller et al., 1978; Barrow et al., 1980), and the milk constituents evidently largely determine which microbes can be implanted in the intestines. After the piglets start to consume creep feed and are finally weaned, an adult type of flora begins to develop in the upper (stomach and anterior small intestine) and lower (ileum, caecum, and colon) tract. At the same time the main site of bacterial fermentation changes from the stomach to the large intestine. In fact, the colonic flora of adult pigs resembles that of the rumen except for the lack of protozoa (Russel, 1979; Allison, 1989).

The adhering LAB on the nonsecretory squamous epithelium of the pigs stomach is believed to serve as a source of inoculum for the lumen. *Lactobacillus fermentum* and *Streptococcus salivarius* are the predominant strains on this area (Fuller et al., 1978; Barrow et al., 1980), reaching the level of 10^8 cfu cm^{-2}. A similar layer of bacteria can be detected in the crop of chickens and in the stomach of humans and rats, although the adherent strains differ between the species. Recent data reveal that the LAB population in the pig's stomach may undergo succession during its life (Pedersen and Tannock, 1989; Tannock et al., 1990). This means that the strains isolated from the young pig before weaning possibly cannot colonize the stomach of adult pigs, or vice versa.

Microbes other than LAB (*E. coli*, yeasts) often found in the stomach of pigs might be considered as transient, nonindigenous organisms, since they evidently cannot colonize the squamous area (Savage, 1977). According to Blomberg and Conway (1989) the increased *E. coli* (K 88) growth in the anterior porcine gut is connected to changes in the LAB population of the squamous area.

The microbiota of the small intestine (SI) is affected by the bile salts and fast passage rate, but the same microbial groups as in the stomach can be cultured. The number of bacteria increases posteriorly due to slower flow rate and possibly to lowered concentration of deconjugated bile acids. Jonsson (1985) noted that the pig SI flora may be transient because evidence for adhesion is lacking. Indeed, Muralidhara et al. (1977) found coliforms and lactobacilli up to 10^7 and 10^8 cfu g^{-1} luminal contents, the figures being lower for the mucosal homogenates (10^5 cfu g^{-1}). However, Wadström et al. (1987) demonstrated the in vitro adhesion of several strains of lactobacilli and streptococci, isolated from the SI wall homogenates to the SI epithelial cells. Fuller et al. (1981) reported the adhesion of *Streptococcus faecium* on the duodenal epithelium of chickens; the amount of attached bacteria was markedly higher in the macerated tissue homogenate than in the luminal contents.

Due to slow passage, the densest microbial population in pigs is found in the large intestine, the total being 10^{10}–10^{11} cfu g^{-1} wet contents. Bacteroides, lactobacilli, and bifidobacteria are the most numerous representatives, but enterococci and coliforms can also be found in high numbers (Jonsson, 1985). The colonic flora has been mainly studied by sampling feces, but many workers have argued against this method (Savage, 1977).

2. *Calves*

At an early age, the rumen of a calf has good physiological resources for the development of microflora. The inoculation of the contents in the forestomachs is a natural event, and is obtained from air, mother, etc. (Bryant and Small, 1960).

Cellulolytic and methanogenic bacteria can be found at the age of three days in the reticulorumen of the calf (Anderson et al., 1987). At the age of one to three weeks cellulolytic and lactate-fermenting bacteria and coliforms are present in the microflora (Bryant et al., 1958; Ziolecki and Briggs, 1961). Lactate-fermenting bacteria are decreased after this period, and at the age of 9 to 13 weeks the ruminal flora of the calf is similar to that of an adult ruminant.

Lengemann and Allen (1959) found that in milk-fed calves the development of cellulolytic bacteria or microflora generally was slower than in calves fed also with dry feeds. Nieto et al. (1985) suggested that rearing of the calf from milk to dry feeds caused a more rapid appearance of protozoa in the rumen. Protozoa could be found at the age of 8 days. Moreover, the artificial inoculation did not affect the establishment of the culturable bacteria, but after inoculation with rumen contents protozoa (Entodinia, Diplodinia, and Holotrichs) appeared at the age of six weeks (Bryant and Small, 1958). In a normal situation, ciliate protozoa were not found in the rumen earlier than 13 weeks old (Bryant et al., 1958).

After birth, milk or liquid milk replacers are the main feeds for a nonruminant. Liquid feed also passes the reticulorumen via esophageal groove to omasum, abomasum, and further into the small intestine. Therefore it is natural that one of the first groups of microorganisms in the rumen is lactic acid bacteria, and that rumen microflora has no great effect on feed digestion at an early age of the nonruminant. This means that in the nonruminant calf disorders in the digestive tract can be treated as with piglets. However, Marounek et al. (1988) suggested that metabolic products of some rumen microorganisms might have probiotic-type effects even at the early age of the calf.

The small and large intestinal flora of calves resembles that of the rumen (reviewed by Jonsson, 1985) and is affected also by diet and age of animal. Marshall et al. (1982) has isolated adherent lactobacilli from the epithelium of the esophageal groove, omasum, abomasum, and duodenum at levels of 10^4–10^7 cm^{-2}. According to Gilliland et al. (1980), the numbers of lactobacilli and coliforms in the small and large intestines vary between 10^6–10^7 and 10^8–10^9 and 10^5–10^7 and 10^8–10^9 per gram dry weight of the gut contents, respectively.

B. Digestion

1. *Carbohydrates*

Cranwell et al. (1976) has shown, for example, that in the stomach of piglets, large amounts of lactic acid are produced by LAB, mainly from lactose and glucose (Table 1). This may be essential for pH regulation and formation of an acid barrier, because HCl production is still limited in the suckling period. In a later study (Cranwell, 1985), heavy lactate production was observed to inhibit HCl secretion, indicating that the regulation is based on the hydrogen ion concentration and not on the acid per se. Lactate constitutes 80–90% of the total organic acids in the stomach (50–80 mmol L^{-1}) of a suckling pig (Friend et al., 1963), the proportion being much lower (50%) in older animals on a creep diet (Argensio and Southworth, 1974; Clemens et al., 1975). Sugars are also the most likely substrates for small intestinal bacteria, but the real quantity of the fermentation may be limited. However, organic acids up to 50–100 mmol L^{-1} can be found, lactate predominating and acetate accounting for most of the VFA. Part of the organic acids in the small gut may be contributed by the digesta emptying from the stomach (Ratcliffe, 1985).

The large intestine allows efficient bacterial fermentation to take place, especially in adult pigs. Degradation of plant cell-wall carbohydrates, mucin, and other endogenous secretions results in organic acids in a series of reactions, just like in the rumen (Argensio and Stevens, 1984). The amount of organic acids in the chyme varies between 150 and 200 mmol L^{-1} (Argensio and Southworth, 1974; Clemens et al., 1975), acetate, propionate, and butyrate accounting for 60, 30, and 15 mol%, respectively, and lactate occurring only occasionally in trace amounts. VFA absorbed from the large gut can represent 15% of the total net energy requirement of growing swine (Dierick et al., 1989).

Table 1 Contribution of Microbes to the Digestion of the Host in Different Compartments of the Gut

	End products of the bacterial digestion		
Substrate	Stomach	Small intestine	Cecum + colon
Carbohydrates	Lactic acid	Lactic acid; acetate	Volatile fatty acids
Protein and N compounds	Ammonia Amines (amino acids)	Ammonia Amines (amino acids)	Ammonia Amines
Lipids	Fatty acids	Fatty acids; deconjugated bile; modified cholesterol	De novo synthesis of fat; hydrogenated fatty acids; modified cholesterol

2. *Protein and N Compounds*

Bacteria unlikely possess any significant proteolytic activity, especially in the upper tract, but amino acids, peptides, and urea are used as N sources (Table 1). Almost all amino acids can be deaminated or decarboxylated to yield ammonia and amines (Dierick et al., 1986). Hill et al. (1970a,b) reported that *E. coli* is the main amine producer, and that feeding LAB significantly reduced amine formation in young pigs. In general, ammonia and amine formation is seen as a harmful process of the gut microbes, and the prophylactic effect of antibiotics is believed to be based on lowered production of these N compounds (Armstrong, 1986).

Urea is formed in the N turnover of tissues and secreted into all parts of the GI tract in digestive juices or straight across the luminal wall (Bergner et al., 1987), and probably the microbes in close association with the epithelium degrade urea and liberate ammonia in the lumen. Ammonia may be incorporated into bacterial N or absorbed and converted back to urea by the liver for urinal excretion or recycling into the gut. This enterohepatic circulation of urea (Deguchi and Namioka, 1989) demands energy from the host and may irritate the gut mucosa. However, in the cases of low N intake this process might conserve nitrogen for the host (Ratcliffe, 1985).

3. *Lipids*

The intestinal flora contributes to lipid metabolism of the host in two different ways; first, bacteria can digest dietary and endogenous lipids by lipases and hydrogenate the free fatty acids, and second they can deconjugate bile acids and modify cholesterol metabolism (Table 1). The apparent digestibility of fats may be decreased by the activity of bacteria, since hydrogenated fatty acids are less absorbable than unsaturated ones, but evidently also due to the fat synthesis de novo in the large gut.

Free bile acids are conjugated with taurine or glycine or maybe with sulfate or glucuronide. The primary bile acids are deconjugated by the gut flora, especially by lactobacilli (Gilliland and Speck, 1977), to less soluble and absorbable secondary products (Eyssen and van Eldere, 1984). The secondary bile acids are partly converted to tertiary bile products by the gut flora and hepatic enzymes and they may be toxic to the host.

The gut flora is also involved in the cholesterol metabolism, because bile acids are synthesized from cholesterol, although the mechanism has not thoroughly been studied in pigs or calves. Evidence from the other species suggests that the gut flora decreases the body pool of cholesterol by catabolizing and making it less absorbable (van Eldere and Eyssen, 1984). Many attempts have been made to reduce the serum cholesterol, considered as an additional risk of heart disease, by intake of LAB products. Some trials have been successful, but the response has been lacking in others.

C. Gut-Wall Function

The gut wall in all parts of the intestine is organized in a special way. The folded mucosa is clothed by fingerlike projections, villi, which are in turn clothed by the absorbing enterocytes or mucus-secreting goblet cells (Magee and Dalley, 1986). Between the villi exist the crypts of Lieberkühn, where the epithelial cells are proliferated, extending down to the lamina propria (Fig. 1). During migration from the bottom of the crypts to the villus tips, the mucosal cells differentiate and mature, and the amount of digestive enzymes increase during this process (Smith, 1985). The results obtained from studies comparing germ-free and conventional animals show a marked interaction between flora and structure of the intestinal mucosa (Kenworthy, 1967; Coates, 1980). In general, germ-free animals possess longer villi, shallower crypts, and, as a result of these morphological differences, higher enzyme activity than their conventional counterparts. According to the suggested mechanism, the gut bacteria per se or their metabolites increase the rate of mucosal cell renewal. According to Visek (1978, 1984), the products of bacterial N metabolism, ammonia and amines, have a harmful influence on the mucosal structure. Hampson (1986a,b), Miller et al. (1986), and Cera et al. (1988) have demonstrated in piglets a sharp decrease in villus length and an increase in crypt depth after weaning, and evidently the change of the gut flora contributes to this phenomenon.

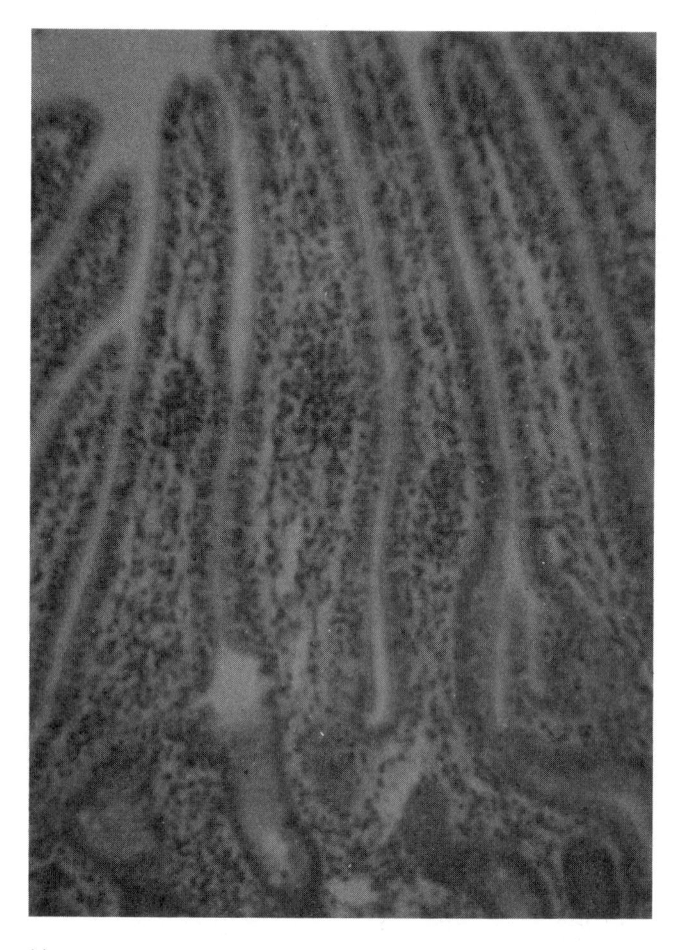

(a)

Figure 1 Photographed jejunal and ileal mucosa of a weaned pig describing mucosal architecture. Tissue slices stained for light microscopy (×16) with hematoxylin and eosin. (a) jejunum, (b) ileum.

Sakata (1987) demonstrated in rats that intraluminally infused VFA accelerate the crypt cell production rate and increase gut-wall mass. The stimulation was most efficient with butyrate. The effect may be systemic rather than local, since caecally administered VFA enhanced crypt cell production rate in jejunal samples. Roediger (1980) reported that in rats and humans butyrate is even preferred to

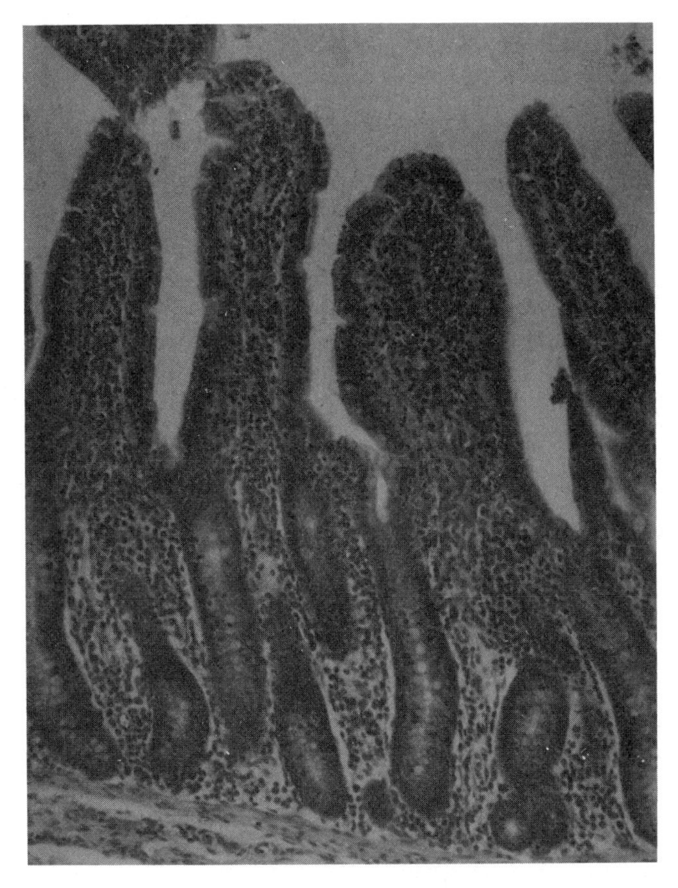

(b)

glucose as an energy source for colonocytes. Hill and Cowley (1990) demonstrated longer crypts and lower number of mature goblet cells in the colons of mice, equipped with a normal microflora, in comparison to their germ-free counterparts. On the other hand, dietary antibacterials, which obviously decrease the activity of the gut flora, have been shown to decrease the gut-wall mass and stimulate nutrient absorption (Visek, 1978; Parker and Armstrong, 1987). Yen et al. (1985), among others, noted that in young pigs the gut-wall mass is reduced by dietary antibacterials. They later speculated (Yen et al., 1987) that this may lead to lower fasting energy consumption because the intestinal epithelium is one of the most active tissues in the body.

D. Gut Immune System

Gut-associated lymphatic tissue (GALT), existing immediately under the outer cell layers of the mucosa, forms the first host-specific defense barrier against the antigen exposure of harmful bacteria and other antigens of the diet (for reviews see, for example, Berg and Savage, 1972; Porter and Barrat, 1987; Nagura, 1990). The GALT secretory immunoglobulins (mainly sIgA, sIgG also present) are complexed with the goblet cell mucine and are the main specific protecting mechanism. The lymphocytes which secrete IgA arise in Peyer's patch lymphoid regions. Secretory IgA is in the dimeric form, possessing a specialized peptide "j-chain," which binds the heavy chains of the immunoglobulin. It is believed that the IgA system is activated by the local antigens near the mucosal surfaces. The epithelial cells synthesize a receptor for this peptide known as a secretory component, facilitating the binding of IgA to the mucin and distribution over the exteral mucosa as a protective layer. Macrophages and cytotoxic T cells are responsible for the cell-mediated immune reactions of the gut. It is obvious that the gut flora and GALT interact together in an important manner; antibodies regulate the colonization of pathogens on the epithelium. On the other hand, certain indigenous gut bacteria can exist in close association with this mucin-antibody painting, evidently contributing to the defense effect. Moreover, the normal flora of the gut is even believed to stimulate the immune defense of the GALT. It is not known how the GALT system can distinguish indigenous bacteria and harmful pathogenic bacteria, which do not belong to the normal habitat of the gut. Savage (1987a) speculated that somehow indigenous flora has common antigens with the host.

III. THE POTENTIAL AND POSSIBLE MODES OF ACTION OF LACTIC ACID BACTERIA AS BIOLOGICAL PERFORMANCE ENHANCERS—PROBIOTICS

A. Competitive Exclusion

The pioneering evidence of the competitive exclusion concept was obtained from poultry chickens by Nurmi and Rantala (1973). The newly hatched birds do not obtain the normal gut flora of the adult, because of modern management methods. Since normal flora is lacking, the intestines of the birds are easily colonized by pathogens, most often by salmonella or coliforms. It is rare that the infected broilers get sick due to salmonella, or even show decreased growth, but as opportunistic organisms salmonella might contaminate poultry food products. When the chickens were inoculated just after birth by the caecal contents of an adult bird, the frequency of salmonella infections was radically reduced and the number of salmonella needed to colonize the caeca of the birds increased.

After the basic establishment of the concept, much work has been directed to describe the exact components of the competitive exclusion (for reviews see, for example, Pivnick and Nurmi, 1982 or Bailey, 1987). After Impey and Mead (1989) the main factors are competition for the receptor sites on the gut wall, production of VFA and/or other antibacterial substances by the anaerobic flora, and competition among different bacteria for limiting nutrients. The specific role of LAB as a probiotic for live poultry has been extensively discussed by Juven et al. (1991). The method of competitive exclusion has not been studied in pigs or calves as such, partly due to the management differences between these animal species, but treatment with selected gut bacteria, mainly LAB, has been examined in detail during the past decades. Fuller (1989) listed the possible modes of action of such selected probiotics as follows: (1) suppression of the viable counts of pathogens and harmful bacteria, (2) alteration of microbial metabolism (enzyme activity), and (3) stimulation of the immune response (see also Table 2).

Table 2 Proposed Mechanisms of Beneficial and Detrimental Effects of LAB Probiotics

Response	Proposed mechanisms	Main site of action
Beneficial		
Suppression of harmful bacteria	(1) production of antibacterial compounds	S, SI
	(2) competition for nutrients	S, SI, LI
	(3) competition for colonization sites	S, SI
Microbial/host metabolism	(1) production of enzymes which support digestion (e.g., lactase)	S, SI
	(2) decreased production of ammonia, amines or toxic enzymes	SI, LI
	(3) improved gut-wall function	SI, LI
Improved immune response of host	(1) increased antibody levels	SI, (LI)
	(2) increased macrophage activity	SI, (LI)
Detrimental		
Competition for nutrients with host	(1) consumption of glucose	S, SI
	(2) consumption of amino acids	S, SI

S = stomach; SI = small intestine; LI = large intestine.
Source: Data adapted according to Fuller (1989) and Impey and Mead (1989).

1. Antagonism: Production of Organic Acids or Specific Antibacterials

Lactic acid bacteria produce many kinds of metabolites which might affect the other microbes in the gut. Lactic acid produced both by homolactic and heterolactic strains reduce pH in the luminal contents, which is most obvious in the stomach of neonatal piglets (White et al., 1969; Cranwell et al., 1976). Moreover, acetic acid excreted by heterolactic strains and H_2O_2 may be toxic to some other bacteria (Spillman et al., 1978). It is well documented that organic acids and H_2O_2 produced by LAB are inhibitory against coliforms, salmonella, and clostridia in vitro, but convincing in vivo evidence is still lacking.

Several high molecular antibacterials, such as acidophilin, acidolin or reuterin, and nicin, have been described as being produced by lactobacilli and streptococci in vitro, respectively (Tagg et al., 1976; Juven et al., 1991). However, there is limited evidence that such substances can really be active in the intestines, and many researchers believe that the inhibitory effects are accounted for the lower pH, organic acids, or hydrogen peroxide. Klaenhammer (1982) suggested that the significance of *Lactobacillus* bacteriocins against undesirable intestinal organisms is questionable because of the narrow range of activity of these compounds.

Although the mechanisms of antagonistic properties of LAB are somewhat uncertain, there is some evidence that such a phenomenon really takes place in the gut (Table 3). Muralidhara et al. (1977) treated piglets immediately after birth with a human isolate of *Lactobacillus lactis*. A clear coliform suppressing response in fecal samples was noted, but the number of lactobacilli was not affected. Moreover, after treatment was stopped, a continued reduction of coliform numbers was observed. Ratcliffe et al. (1986) fed piglets from two days of age with *L bulgaricus* or *L. reuteri* fermented milk of nonfermented control milk. Both types of fermented milks decreased the numbers of coliforms and pH throughout the intestines. Since lactic acid added to the control milk gave similar results, the authors concluded that the favorable effects of fermented diets were due to the lower pH produced by lactic acid. Underdahl (1983) inoculated gnotobiotic pigs with three virulent strains of *E. coli*, which all developed severe diarrhea. Treatment with *Streptococcus faecium* reduced the severity of diarrhea, and pigs recovered earlier and gained weight normally compared to their untreated littermates. The better performance of treated pigs was associated with a lower number of both organisms in the small intestine and caecum. Ozawa et al. (1983) treated piglets and calves, reared on an antibiotic-containing diet, with *Streptococcus faecalis* and noted increased numbers of lactobacilli, streptococci, and bifidobacteria in the faeces of experimental animals. In addition, yeast flora and salmonella were suppressed due to the treatment. Barrow et al. (1980) showed decreased counts of *E. coli* in the stomach of piglets fed a combination of *S. salivarius* and *L. fermentum*. Treatment with *L. bulgaricus* in pigs showed that

Table 3 Data Supporting the Antagonistic Properties of Lactic Acid Bacteria Against Harmful Gut Bacteria

Target host	Organism	Response	Authors
Small piglets	*L. lactis*	Decreased *E. coli* in feces	Muralidhara et al. (1977)
Small piglets	*L. reuteri* *L. bulgaricus*	Decreased pH and *E. coli* in the gut	Ratcliffe et al. (1986)
Gnotobiotic pigs, *E. coli* challenged	*S. faecium*	Decreased *E. coli* Less scours	Underdahl (1983)
Small piglets	*S. faecalis*	Increased LAB and bifidobacteria Suppresed salmonella and yeasts	Ozava et al. (1983)
Pigs	*L. fermentum* *S. salivarius*	Decreased *E. coli* in stomach	Barrow et al. (1980)
Pigs, calves	*L. bulgaricus*	Neutralized *E. coli*-toxin	Mitchell and Kenworthy (1976) Schwab et al. (1980)
Calves	*L. acidophilus*	Suppressed *E. coli*	Gilliland et al. (1980)

the organism produced a substance which seemed to neutralize the effect of enterotoxin released from coliforms (Mitchell and Kenworthy, 1976). Additional evidence of the antienterotoxic property was obtained from trials with calves by Schwab et al. (1980). Gilliland et al. (1980) observed declined numbers of coliforms in the ileum of calves fed with host-specific *L. acidophilus*.

In spite of many supporting observations of the antagonistic properties of LAB, negative results also occur. For example, Pollmann et al. (1980a,b) did not observe any effect in the faecal flora of *Lactobacillus acidophilus* treated piglets. It is noteworthy, however, that the fecal sample is probably not a valid indicator of the intestinal ecosystem.

2. *Adhesion*

Adhesion or close association of the LAB probiotics to the epithelial cells may further contribute to competitive exclusion. Firstly, LAB that grow relatively slowly but attach to the gut wall can colonize and inoculate the luminal contents. This seems to be obvious, for example, in the stomach of pigs (Barrow et al., 1980) and in the crop and caecum of chickens (Fuller, 1977; Stavric et al., 1987). Secondly, if LAB occupy the adhesion receptors on the surface, the harmful

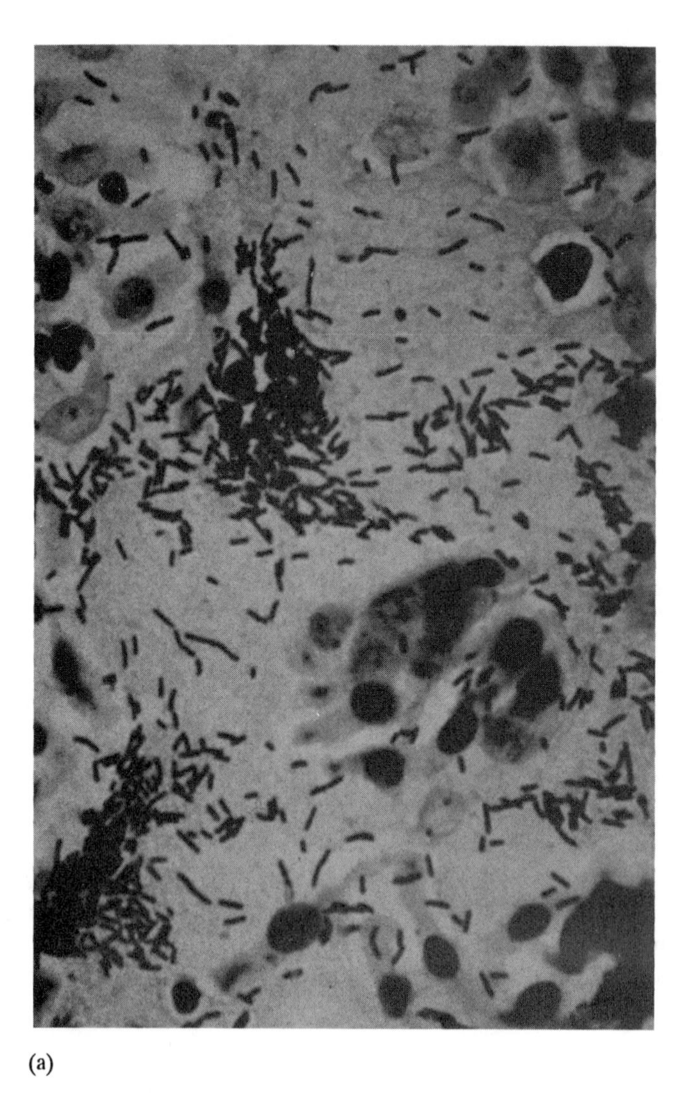

(a)

Figure 2 Two host-specific *Lactobacillus* strains differing in their ability to adhere to the epithelial cells of the small intestine of a pig. (a) adherent, (b) nonadherent.

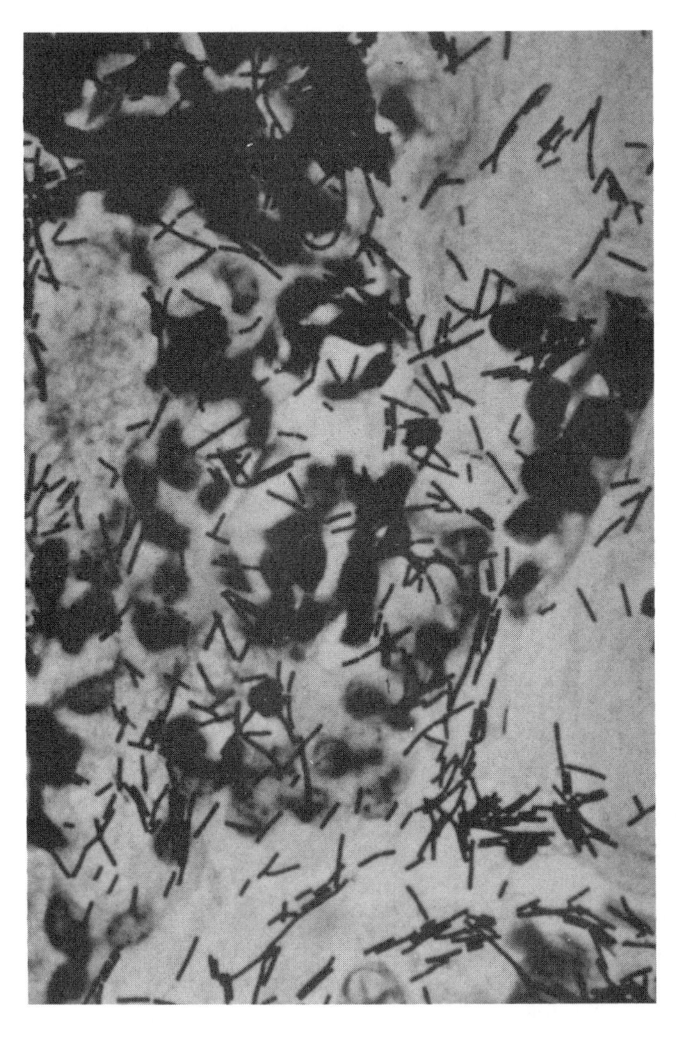

(b)

bacteria relying on them will be eliminated from the gut (Fig. 2). This is of course a valid principle only if pathogens and LAB have parallel attachment mechanisms. Davidson and Hirsch (1976) could block the colonization of pathogenic *E. coli* K88 with a nonpathogenic *E. coli* strain. Similarly, lactobacillus cells or cell-wall fragments were reported to prevent adhesion of *E. coli* on human uroepithelial cells (Chen et al., 1985).

The association mechanisms of intestinal bacteria in general and those of LAB on the gut surfaces have been discussed, for example, by Savage (1987b). Gram-negative bacteria, for example, pathogenic *E. coli*, attach to the target cells via proteinaceous projections (pili), but lactobacilli seem to adhere to the gut wall with extracellular substances containing polysaccharides, proteins, lipids, and lipoteichoic acids. The role of the latter has been discussed in detail by Tannock (1990). Lipoteichoic acids are glycerolphosphate polymers of the cell wall of lactobacilli, covalently linked with glycolipids, containing both hydrophilic and hydrophobic regions. Sherman and Savage (1986) detected macromolecular protein complexes rich in lipoteichoic acids in *Lactobacillus* strains, some of them known to associate with epithelial surfaces. Appearance of acidic carbohydrate-rich material between attached bacteria and epithelium was also supported earlier by Brooker and Fuller (1975) with electron microscopy. Lipoteichoid acids may also participate in attachment of streptococci to mammalian cells (Tannock, 1990).

Wadström et al. (1987) observed a number of *Lactobacillus* strains from the small intestine of pigs containing carbohydrate capsule polymers and possessing high hydrophobicity. Heat and protease treatment impaired these surface functions. They considered capsule formation to be the most important determinant of the intestinal colonization of lactobacilli in pigs. Henriksson et al. (1991) reported that the adhesive determinants of *Lactobacillus fermentum* on the porcine gastric squamous epithelium are proteinaceous, although also carbohydrates seemed to be involved. Attachment ability of the rough and smooth variants of *L. fermentum* differed; the former was lacking the adhesion protein. However, it was hypothesized that during colonization the rough variant is needed to reach the epithelium and the environmental conditions induce the growth of smooth variant which in turn binds efficiently to the mucosa.

Knowledge of the adhesion properties of LAB has been markedly increased during the latest years, but many microbe-epithelium interactions other than those mentioned above certainly exist in the gut. For example, the role of the components of goblet cell mucin, for example, sialic acid, has been speculated to be the key factor in the mucosal association of nonpathogenic intestinal bacteria. Also the general importance of adhesion for competitive exclusion and probiotic concept has to be evaluated further.

B. Alteration of Microbial and Host Metabolism*

Lactobacilli are claimed to affect the cholesterol metabolism of the host. Gilliland et al. (1985) treated pigs with a *Lactobacillus acidophilus* strain, selected for its

*See also Table 4.

Table 4 Data Supporting the Beneficial Shifts in Microbial or Host Metabolism by Feeding LAB Probiotics

Target host	Probiotic	Response	Authors
Pigs	*L. acidophilus*	Decreased serum cholesterol	Gilliland et al. (1985)
Calves	*L. acidophilus*	Formation of inhibitory bile acids	Gilliland and Speck (1977)
Pigs Humans	*L. acidophilus*	Decreased amine production	Hill et al. (1970a) Goldin and Gorbach (1984)
Humans	*L. acidophilus*	Decreased production of carcinogenic N compounds	Goldin et al. (1980) Goldin and Gorbach (1984a)
Rats Chicks Pigs	*L. bulgaricus* *L. acidophilus*	Hydrolytic enzymes which improve digestion	Garvie et al. (1984) Champ et al. (1983) Jonsson and Hemmingsson (1991)
Pigs	*Lactobacillus* sp.	Increased activity of brush-border enzymes	Collington et al. (1990)

ability to grow well in the presence of bile and to assimilate cholesterol in vitro. The treatment inhibited the increase in serum cholesterol on a high-cholesterol diet. Similar results was obtained by Danielson et al. (1989), who treated mature boars with *L. acidophilus*, screened in vitro for anticholesteremic and antimicrobial activities.

Deconjugation of bile acids by lactobacilli, as reported by Gilliland and Speck (1977), might be inhibitory to some other intestinal bacteria inhabitating the lower small intestine and colon. Fernandes et al. (1988) reported that addition of physiological concentrations of free bile acids to the growth medium decreased the growth and antimicrobial activity of *L. acidophilus*. Observations of Tannock et al. (1989) revealed that bile salt hydrolase activity in the ileum of mice was reduced by 86% or 98% in the absence of lactobacilli or both lactobacilli and enterococci, respectively, compared to conventional animals.

Besides affecting cholesterol and bile acid metabolism, LAB are claimed to reduce the intestinal production of harmful N-compounds. Pigs fed *L. acidophilus*–fermented milk showed less intestinal amine production than the pigs fed untreated control milk (Hill et al., 1970a). Also the major site of amine production changed from the small intestine to the caecum in the treated pigs. When various gut bacteria were tested in vitro, *E. coli* was noted to be the most efficient amine producer in pigs (Hill et al., 1970b). The carcinogenic faecal enzymes, β-glucuronidase, nitroreductase, and azoreductase, were reported to decrease in humans on a *L. acidophilus* containing diet (Goldin et al., 1980; Goldin and Gorbach, 1984a). Intestinal production of free amines was noted to decrease in rats administered aromatic nitro and azo compounds as well as an amine-glucuronide compound (Goldin and Gorbach, 1984b).

Ingested LAB produce and release hydrolytic enzymes, which might aid digestion in farm animals, particularly during the early life of calves and piglets. For example, humans suffering from β-galactosidase deficiency may digest lactose in yogurt better than the same amount of lactose in milk. Indeed, rats fed yogurt had increased β-galactosidase activity in their small intestine and the enzyme seemed to be of bacterial origin (Garvie et al., 1984).

Some observations suggest that lactobacilli could contribute also to the digestion of more complex carbohydrates than lactose. Champ et al. (1983) isolated three *Lactobacillus* strains from chicken crop, which showed amylolytic activity. The best amylolytic strain resembled *L. acidophilus*, producing maltose, maltotriose, and traces of glucose from amylopectin. Optimum pH and temperature of the amylase were 5.5 and 55°C, respectively. Jonsson and Hemmingsson (1991) found β-glucan degrading lactobacilli up to 10^8 cfu g^{-1} from the feces of 3- and 35-day-old piglets fed a creep diet containing 2% β-D-glucan. Glucanolytic probiotics might be very useful in the diets of poultry and pigs containing barley and oats, because host enzymes evidently cannot degrade β-D-glucan and because it interferes with starch digestion.

Probiotics containing LAB might also affect the levels of the host brush border enzymes, as speculated by Parker (1990) in his review. Collington et al. (1990) fed piglets with antibiotics or LAB probiotics, and reported increased lactase and sucrase activities in the small intestine mucosa with both treatments. It may be speculated that the effect of both treatments was not direct, but was due to lower production of harmful bacteria metabolites, which irritate the mucosa and affect the life span of the enterocytes. In contrast, Whitt and Savage (1987) found no direct influence with several indigenous lactobacilli on the enzyme activities of the duodenal enterocytes in germ-free and ex-germ-free mice. This led the authors to conclude that any beneficial effects of probiotic bacteria on the function of gut mucosa may result from the interaction with the whole microflora inhabiting the lumen (e.g., pathogens).

C. Stimulation of Immunity

An example of the interaction between microbes and the immune system is furnished by conventional animals equipped with a complete indigenous flora, which have higher immunoglobulin levels and phagocytic activity than their germ-free counterparts (Bealmer et al., 1984). Therefore it has been suggested by many authors on the basis of encouraging research results that probiotic bacteria could enhance immunity both locally on the mucosal surfaces and at the systemic level. *E. faecium* used as a monoassociate to germ-free mice reduced the counts of salmonella (intravenous challenge) in the spleen, implying a systemic response (Roach and Tannock, 1980). On the contrary, Kluber et al. (1985) did not observe any responses in in vivo cell-mediated immunity in artificially reared piglets treated with *E. faecium*. Perorally administration of *L. acidophilus* and *L. casei* increased phagocytic function of macrophages in mice (Perdigon et al., 1986). Moreover, *L. acidophilus*- and *S. thermophilus*-activated macrophages and lymphocytes whether given perorally or intraperitoneally (Perdigon et al., 1987). Macrophage activation was also noted by Saito (1988) with *L. casei* in mice. Lessard and Brisson (1987) fed piglets with rehydrated skim milk powder fermented with a mixture of lactobacilli and reported slightly increased serum IgG levels. A local immune enhancement with LAB was lately reported by Perdigon et al. (1990) in mice. Oral intake of *L. casei* increased the IgA production secreted to the intestinal lumen, providing mucosal defense against *Salmonella typhimurium*. These examples suggest that LAB can really modulate immunity. However, there is little straight evidence if this kind of response can be obtained in commercial circumstances where animals harbor a complex flora in their intestines.

IV. SELECTION CRITERIA FOR LAB TO BE USED AS PROBIOTICS

Many potential improvements in the animal performance may be achieved with LAB probiotics, as demonstrated above by research examples, but the responses obtained in the field trials may be variable. Much of this controversy between basic concepts and real life is obviously accounted for by the characteristics of the strains used, and therefore one aiming to develop a good probiotic has to carefully evaluate the selection criteria.

LAB probiotics marketed until today are freeze-dried bacteria, often belonging to the genera *Lactobacillus* spp. or *Enterococcus* spp. Usually the strains are suitable intestinal bacteria or dairy starters, but products based on host specificity also exist. There are a number of criteria which a successful probiotic must fulfill. The main criteria which are used in the authors' laboratory are listed in Table 5

Table 5 Criteria Used in the Authors' Laboratory for Screening Probiotics

Method	Basis
1. Acid tolerance	Survival during the passage through the stomach and duodenum
2. Bile tolerance	Survival during the passage through the upper small intestine
3. Acid production (glucose and lactose)	Production of efficient "acid barrier" in the upper gut
4. Production of antimicrobial substances	Competition with pathogens
5. Adhesion to gut epithelial cells	Efficient colonization, exclusion of other microbes from adhesion sites
6. Heat tolerance	Survival during pelleting of creep feed
7. Tolerance of feed antimicrobials	Use possible with medicated feed

(see also Gilliland, 1979; Fuller, 1989). Firstly, it must be a nonpathogenic representative of the normal intestinal flora, most preferable host specific, and it must maintain its activity in the presence of high acidity in the stomach and high concentration of bile salts in the small intestine. Secondly, a good probiotic must be able to grow and metabolize rapidly and exist in high numbers in the gut. Thirdly, an ideal probiotic strain may colonize some part of the tract, due to which adhesion to the epithelial surface is desirable. Fourthly, it must produce efficiently organic acids and might have specific antimicrobial properties against harmful bacteria. Finally, it must be easy to produce, survive growth in a large scale and retain its viability under storage and field conditions and be cost-effective to use for farm animals.

There hardly exists a strain, an ideal probiotic, which completely fulfills all these criteria. Nevertheless, much variation can be observed among isolated gut lactobacilli in these selection variables, as judged by our own experience (Table. 6). In practice, the choice of economically feasible probiotic is always a compromise between microbiological, production, and performance promoting properties of the strains tested. However, some points of the selection criteria need to be discussed briefly.

A. Acid and Bile Tolerance

High acidity in the stomach and high concentration of bile components in the proximal small intestine are the first host attributes, which affect the strain selection. Conway et al. (1987), among others, incubated LAB strains in the

Table 6 Selection Characteristics of Some LAB Strains Isolated from the Intestines or from Feces of Pigs in the Authors' Laboratory

Strain type	*N*	Lactic acid production[a]	% L/D lactate	Heat tolerance[b]	% Bile[c] tolerant	% Acid[d] tolerant
Lactobacillus acidophilus	15	0.39–1.33	50–100	60.4–69.5	40	30
Lactobacillus fermentum	25	0.38–1.68	47–93	62.5–69.5	44	88
Lactobacillus delbrueckii	4	0.60–1.20	47–63	64.0–65.4	—	—
Lactobacillus sp.[e]	13	1.03–2.07	39–70	65.5–69.0	39	92

[a]MRS broth, 1% glucose.
[b]Maximum temperature which is fully tolerated for 6 min.
[c]MRS broth, 0.3% Oxgall, no inhibition.
[d]MRS broth, pH 4.00, good growth.
[e]Adhesion to small intestinal cells of pig observed (Mäyrä-Mäkinen et al., 1983).

phosphate-buffered saline at pH 1, 3, and 5 for 0 to 4 h at 37°C to screen the human strains for their ability to survive in the stomach. Aspirated stomach juice obtained through a nasogastric tube after a 4-hour fast was used for the same purpose. The strains detected showed a variable survival with this method, and thus it was reported to be a valid tool to find potential microbes. Jonsson et al. (1985) used fistulated pigs to test the in vivo survival of orally fed lactobacilli during transit through the upper tract. The cannulas were inserted 1 m distal to the pylorus and in the terminal ileum.

Gilliland et al. (1984) observed a great variability among *Lactobacillus acidophilus* strains isolated from calf intestinal contents in their ability to grow in vitro in the presence of bile salts. When a strain exhibiting low tolerance to bile and another strain exhibiting high tolerance to bile were administered orally to calves, the more resistant strain caused greater increase in numbers of facultative lactobacilli than the one possessing low tolerance.

B. Production of Antimicrobial Substances

The in vivo evidence for the production of specific bacteriocins by LAB is limited, and subsequently the spectra of these substances seem to be quite narrow. Moreover, due to methodological difficulties (e.g., pH elimination), the preferred method might be to screen the candidate strains for their ability to produce organic acids (see Chapter 2, Sect. III; Chapter 3, Sect. II). Most commonly, simple sugars, such as glucose or lactose, are used as carbon sources. Also production of H_2O_2 can be used as was done, for example, by Jonsson and Olsson (1985).

C. Adhesion and Growth in the Gut

The problem of adhesion stems to the basic controversy whether the strains are fed continuously or just one time. In the latter case, the adhesion ability is a crucial property for the colonization of the probiotic. Additionally, attachment of probiotics to the gut wall may block the colonization of harmful bacteria on the mucosa. The adhesion test is usually made by incubating the strain and intestinal cell suspension together, and then verifying the binding by visual judgment with a microscope (see also Fig. 2). Since visual judgment is not an objective method, radiolabeled cultures have been used by some authors. The test methods have been described by Fuller et al. (1978) and Conway et al. (1987), among others.

Although the existence of mucosally attached microbes has been experimentally proved, and thus adhesion provides a sound basis for the development of probiotics, much arguments has been directed against its use. Firstly, during harvesting of the epithelial cells variable amounts of mucin are bound to the cells and thus interfere with the assay. In contrast, mucin may be essential for the gut-wall association of some bacteria. Secondly, some results strongly suggest that adhesion is a host-specific feature and, which is more important for the concept, the adhesive strains on the mucosa might vary according to the age and diet of the animal. For example, Jonsson (1986) did not observe permanent establishment of a host-specific *Lactobacillus* strain in pigs, although it adhered in vitro to the squamous epithelial cells. Finally, it may be theorized that if the inoculation of the animal is not done immediately postpartum, the indigenous flora developed near the mucosa resist the attachment of the probiotic bacteria.

Because of the above options, it would seem that the best method of using probiotics is continuous inoculation. However, even with continuous feeding it is still important to screen the probiotic candidates according to their ability to survive and grow in the gut, and attachment ability is also a recommended feature. There exists limited knowledge of the minimum effective dosage of probiotics when they are administered continuously. The fact that viable counts of probiotic organisms are found in the faeces is not sufficient proof to conclude on proliferation or metabolism during passage through the tract. Although it appears that according to the literature and our experimental data 10^6–10^7 cfu g^{-1} feed is necessary for a consistent effect, it may be reasonable to conclude that the effective dosage appears to be a strain-bound feature depending on the survival properties and on the specific growth rate of the organism.

D. Feed Antimicrobial Resistance

Probiotics are often mentioned as natural substitutes to feed antibiotics, but in some cases it may be feasible to combine the probiotic and antibiotic treatments to obtain an extra advantage. As stated above, natural flora resist the invasion of both harmful and probiotic bacteria. If the natural flora are weakened by the use of a feed antimicrobial, the probiotic bacteria may be more easily established in the guts of target animals. There exist some preliminary results in the authors' laboratory which support this idea. On the other hand, by combined treatment the level of antibiotics needed could decrease.

According to Pollmann et al. (1980b) a *Lactobacillus* culture in combination with lincomycin may have an additive effect. Harper et al. (1983) treated growing swine with a *Lactobacillus* probiotic or virginiamycin or both, but no interaction between treatments occurred. However, the viability of the cultures in the medicated feed was not monitored. Dutta and Devriese (1981) investigated the minimal inhibitory concentrations of some commonly used feed antimicrobial agents against lactobacilli isolated from pigs, cattle, and poultry. The percentage of resistant strains of all isolates varied in pigs, cattle, and poultry between 2 and 70, 10 and 95, and 8 and 83, respectively, depending on the drug and suggesting a potential for the combined treatment of antibiotics and LAB probiotics.

E. Technological Properties

The production process of probiotics involves mass growth in fermentors, concentration, and subsequently, in most cases, freeze-drying steps. Probiotics may be used as high-activity (10^8–10^{10} cfu g^{-1}) dry preparations with a dosage of a few grams per animal per day, or they may be mixed at the rate of 10^6–10^7 cfu g^{-1} in the meal diet, often followed by pelletizing. In the latter case the probiotics must tolerate heat (60–80°C, 5–10 min) and extremely high physical pressure. As far as the authors' know, there hardly exists a *Lactobacillus* strain which tolerates pelletizing in an economically feasible way. However, enterococci (e.g., *E. faecium*) having smaller cell size and being easier to produce are much more resistant than lactobacilli to pelletizing. This may be one reason why most commercial probiotics to date include enterococci (Tuschy, 1986).

Pollmann and Bandyk (1984) determined the stability of three commercial *Lactobacillus* products in nonmedicated and medicated (lincomycin) piglet starter feed stored for three months in different environments. The samples stored in a refrigerator maintained their stability relatively well during the trial. However, the activity of the samples stored in a pig nursery dropped substantially within the first week, and at the end of the trial there was no viability left in some samples. It was also noted that activity loss was slightly greater in the medicated feed.

Alaeddinoglu et al. (1989) studied the activity-loss kinetics of freeze-dried *Lactobacillus* cultures and pointed out the importance of optimizing the type and concentration of cryoprotectants during drying. Kearney et al. (1990) improved viability of *Lactobacillus plantarum* inoculum after lyophilization and rehydration by immobilizing the cells in calcium-alginate beads containing cryoprotectants. Many other attempts, for example, microencapsulation (Lyons, 1987), have been developed to improve survival of probiotic preparates, but none of them has been published in detail.

F. Mixed Probiotics

Although the probiotic concept is theoretically a sound method for supporting animal performance without antibiotics, it is not simple to introduce the right bacteria strains at the right time to the right animal in a biologically and economically efficient way. Another way of thinking is to find substrate(s) which could create selective pressure on the normal gut flora, especially in the large intestine. This means that the animals select their "probiotics" in situ.

Lactulose and lactitol are synthetic disaccharides made industrially from lactose by several isomerization and hydrogenation steps, respectively. The molecules of lactulose and lactitol are composed of galactose with fructose and sorbitol, respectively, connected with a specific β-galactosidic linkage. Many studies suggest (reviewed by Harju, 1988a) that small intestinal β-galactosidases of mammalian origin split the linkage inside the lactulose or lactitol molecules very little. As a consequence, lactulose and lactitol escape small intestinal digestion and absorption and are fermented by colonic bacteria into organic acids and gases, which lowers the colonic pH and ammonia content. This fermentation may also cause a positive shift in the colonic flora, because enzymes of certain strains of lactic acid and bifidobacteria have been shown to degrade lactulose and lactitol better than those of coliforms and clostridia (Harju, 1988b). We have extensively studied the possibility of using these sugars as probiotics in piglets and calves as such as a mixture or in combination with selected strains of lactobacilli and enterococci (Nousiainen, 1988, 1990). Such mixed probiotics may provide several advantages compared to simple usage of LAB strains: (1) The effect of the treatment extends to the whole tract; (2) the preparates are not too sensible to the feed manufacturing processes; and (3) the competition force of the probiotic strains can be improved by screening them for their ability to use nonabsorbable sugars as carbon sources. According to the data available until now, these sugar-LAB combinations appear to provide extra advantage compared to simple LAB probiotics in the treatment of young farm animals.

V. ZOOTECHNICAL TRIALS WITH LAB AS PROBIOTICS

A. Pigs

The three main probiotic types tested in pigs are nonviable and viable *Lactobacillus* (mainly *L. acidophilus*) or viable *Enterococcus faecium*. The probiotics were mixed in the diet (starter piglets, growing-finishing pigs) at the rate of 10^4–10^7 cfu g^{-1}, or the liquid diet (milk) was fermented with the probiotic strain. In very few cases more than one strain was used, and most of the strains were lacking host-specificity. Of the 26 piglet trials, summarized in Table 7, in 16 cases positive responses over the control were obtained with probiotics, but a significant difference were recorded only twice ($p < .05$). In contrast, negative results were monitored in nine trials, two being statistically significant at the 5% level. Feed conversion was improved in eight trials, and in nine trials probiotic treated piglets showed worsened feed efficiency. Decrease in feed efficiency has been often associated with fermentation of milk.

Far fewer zootechnical trials with LAB as probiotics have been made with growing-finishing pigs, which is not surprising because adult pigs digest their feed better, have improved immunity, and are more resistant to intestinal disorders than young piglets. It appears, however, that slight improvements in performance can be obtained also in adult pigs, but the responses (negative or positive) are of lower magnitude than in piglet trials (Table 8). From Tables 7 and 8, it is not possible to conclude whether *Lactobacillus* products are better than *Enterococcus*, or vice versa.

In the trials conducted in our laboratory the use of host-specific strains of *L. fermentum* and *E. faecium* alone or in combination with lactulose and lactitol for piglets have been tested on several commercial farms. In five of the seven trials summarized in Table 9, the probiotic treatments gave positive responses over controls in terms of improved daily gain, although in trials 3–7 a chemical growth promoter was included in the diet. Overall improvement was about 5.5% if all trials are interpreted together. Consequently, a slight decrease in mortality was recorded, the mean values being 7.7 and 9.9% for the probiotic and control groups, respectively. According to this data, it can be concluded that mixed probiotics are superior to simple ones. In trial 3/1991 all the piglets were treated with sulfa just after weaning, which evidently weakened the barrier effect of indigenous flora and greatly improved the competitiveness of probiotic bacteria. Markedly better gain and lower mortality in the probiotic group could in this particular trial be explained by the competitive exclusion concept.

Table 7 The Effect of *Lactobacillus* and *Enterococcus* Probiotics on the Performance of Unweaned Sucking and Weaned Starter Piglets

Type and dosage of probiotic	Animals	Performance (% of control) Gain	Feed/gain	Reference
L. acidophilus in feed	Starter fed	+10.8	–7.2	Baird (1977)
L. acidophilus in feed	Starter fed	No resp.	No resp.	Noland et al. (1978)
L. acicdophilus, nonviable	Starter fed	+4.7	–6.4[a]	Hale and Newton (1979)
L. acidophilus 4×10^6 cfu g^{-1} feed	Starter fed	+7.2	±0.0	Pollman et al. (1980b)
L. acidophilus 750 mg/kg (trial 1)	Starter fed	+4.5	–6.7	Pollmann et al. (1980c)
L. acidophilus 750 mg/kg (trial 2)	Starter fed	+9.7	–21.4[a]	
S. faecium 1250 mg/kg (trial 2)	Starter fed	–7.6	–8.7*	
L. acidophilus 4×10^6 kg^{-1}	Starter fed	–1.6	±0.0	Harper et al. (1983)
L. acidophilus, nonviable (trial 1)	Milk replacer fed	–8.2	+4.0	Pollmann et al. (1984)
(trial 2)		–6.8[a]	9.9	
S. faecium, 10^6 (*per os*)				
single	0–3 wk	+0.8	+39.8[a]	Kluber et al. (1985)
3 days	Milk replacer	–8.3 to +1.3	–9.5 to +11.2[a]	
continuous	fed	–5.7 to –1.6	+0.0 to +3.0	
S. faecium, 10^6 g^{-1} feed	Sucking	+9.7	No data	Gualtieri and Betti (1985)
L. fermentum 10^9 d^{-1} (host-specific)	0–9 wk	–7.1	No data	Jonsson (1986)
S. faecium 2×10^8 cfu d^{-1}	Starter fed	11.1[a]	No data	Mordenti (1986)
L. bulgaricus + *S. thermophilus*	Fermented milk	–8.8	18.6[a]	Ratcliffe et al. (1986)
L. bulgaricus + *S. thermophilus*	Fermented milk	–21.0[a]	+19.5	
L. reuteri Host-specific	Fermented milk	–22.5[b]	+48.3	
S. faecium				
10^6 cfu g^{-1}	Starter fed	+2.3	–1.8	Roth and Kirchgessner (1986)
Antibiotic		+6.3	±0.0	
10^6 cfu g^{-1}+antibiotic		+4.0	±0.0	

Table 7 (Continued)

Type and dosage of probiotic	Animals	Performance (% of control)		Reference
		Gain	Feed/gain	
S. faecium 10^6 cfu g^{-1}				
trial 1	Starter fed	−15.4	No data	Danek (1987)
trial 2	Starter fed	+23.3	−10.4	
L. acidophilus nonviable (0.1%)	Starter fed	+10.4[a]	±0.0	Lessard and Brisson (1987)
S. faecium, 10^6 g^{-1} feed	Starter fed	Improved	Improved	Wu et al. (1988a)
S. faecium, 10^6 g^{-1} feed	Starter fed	+13	+2	
Antibiotic		+24	+11	
Antibiotic + *S. faecium*		+33	+11	
L. acidophilus 10^9 cfu + *S. faecium* 10^9 cfu 3 days orally postpartum	Sucking	+14.3[a]	No data	Tournut (1989)

[a]Statistically different compared to control ($p < .05$).
[b]Statistically different compared to control ($p < .01$).

B. Calves

In Table 10 the results of 17 calf trials conducted between 1978 and 1990 testing the effectiveness of LAB probiotics are summarized. Ten experiments showed positive results if judged by the daily gain figures, and in three cases the improvement in gain was significant. Interestingly, the overall 4.6% improvement is of the same magnitude as in the reviewed piglet trials (see Table 7). It is also impossible to compare the efficiencies of *Lactobacillus* and *Enterococcus* products, because in many cases the exact nature of the treatment was not reported.

Table 8 The Effect of *Lactobacillus* and *Enterococcus* Probiotics on the Performance of Growing-Finishing Pigs

Type and dosage of probiotic	Animals	Performance (% of control)		Reference
		Gain	Feed/gain	
L. acidophilus in feed	Growing-finishing	+8.4	–5.8	Baird (1977)
L. acicdophilus, nonviable	Growing-finishing	+4.3	–0.7	Hale and Newton (1979)
L. acidophilus 750 mg/kg	Growing-finishing	–1.2	–0.9	Pollmann et al. (1980c)
S. faecium 1250 mg/kg	(35–95 kg)	–1.2	–0.9	
L. acidophilus 4×10^6 kg^{-1} (trial 1)	Growing-finishing	–5.8[a]	+3.1	Harper et al. (1983)
L. acidophilus 4×10^6 kg^{-1} (trial 2)	(17–100 kg)	+1.3	±0.0	
L. acidophilus	Growing-	Improved[a]	No response	Maxwell et al. (1983)
S. faecium	finishing	Improved[a]	No response	
S. faecium, 10^6 g^{-1} feed	Growing-finishing	Improved	Improved	Wu et al. (1988b)

[a]Statistically different compared to control ($p < .05$).

C. Ruminants

The addition of lactic acid bacteria in the diet of an adult ruminant has been investigated only in a few trials (Table 11). The complex ruminal microflora form a barrier, which can overcome the probiotic strain, and therefore the effects in the rumen fermentation are variable and difficult to interpret.

McCormick (1984) observed changes in rumen fermentation in one trial, but not in the other trial, when steers were treated with *Lactobacillus acidophilus*. Variable data were also presented by Hoyos et al. (1987) and Rust et al. (1989) of the performances of dairy and beef cattle, respectively.

Nevertheless, the most promising part in the digestive tract of ruminants for the function of probiotics might be the reticulorumen. As indicated by McCormick

Table 9 The Effect of Host-Specific *Lactobacillus* and *Enterococcus* Probiotics on the Performance of Small Piglets (Trials Conducted on Commercial Farms Between 1987 and 1991 in the Authors' Laboratory)

Trial/ year	No. of animals	Probiotic type and dosage	Growth response (% of control)	Mortality (treated vs. control)
1/1987	108 0–6 wk	*E. faecium* + *L. fementum* (10^9 cfu d^{-1}) Lactulose + lactitol (top-dressed)	+14.2[a]	0.0 vs. 1.1
2/1987	102 0–5 wk	*Lactobacillus* sp. (10^9 cfu (d^{-1}) Top-dressing to starter	+1.8	3.9 vs. 6.8
3/1991	98 0–11 wk	*E. faecium* + *L. fermentum* 10^6–10^7 cfu g^{-1} starter	+15.0[a]	6.8 vs. 15.9[a]
4/1991	102 0–11 wk	*E. faecium* + *L. fermentum* 10^6–10^7 cfu g^{-1} starter	+5.5	6.6 vs. 10.7
5/1991	189 0–10 wk	*E. faecium* + *L. fermentum* 10^6–10^7 cfu g^{-1} starter	–5.0	11.6 vs. 12.6
6/1991	106 0–11 wk	*E. faecium* + *L. fementum* (10^9 cfu d^{-1}) Lactulose + lactitol (top-dressed)	–3.5	15.6 vs. 19.5
7/1991	130 0–10 wk	*E. faecium* + *L. fermentum* 10^6–10^7 cfu g^{-1} starter	+4.7	9.7 vs. 2.7
8/1991	50[b] 0–7 wk	*E. faecium* + *L. fementum* (10^9 cfu d^{-1}) Lactulose + lactitol	+11.2[a]	
		L. fermentum + (10^9 cfu d^{-1}) *L. acidophilus* (10^9 cfu d^{-1})	+6.0	

[a]Statistically different compared to control ($p < .05$).
[b]Agricultural Research Centre, Swine Research Station (Nousiainen and Suomi, 1991).
Note: In trials 3–7 both the control and probiotic diets were pelletized and medicated (50 ppm carbadox) commercial creep starters.

(1984), changes in the rumen fermentation pattern and digestion can be obtained followed by the feeding of LAB. Moreover, the massive microbial population in the rumen greatly affects the energy and protein utilization and hence the performance of the ruminant.

Table 10 The Effect of *Lactobacillus* and *Enterococcus* Probiotics on the Performance of Calves

Type and dosage of probiotic	Animals	Performance (% of control)		Reference
		Gain	Feed/gain	
L. acidophilus + *L. lactis* (ferm. milk)	Small calves (1–5 wk)	−11.6	No response	Morril et al. (1977)
L. acidophilus 10^6 cfu l^{-1} milk	Small calves (0–3 wk)	No response	No response	Ellinger et al. (1978)
L. acidophilus	Young bulls	No response	No response	Hutcheson et al. (1980)
Lactobacillus fermentation product	Small calves (0–8 wk)	+5.3	No response	Schwab et al. (1980)
	Small calves (0–10 wk)	+28.3[a]	−1.5	
Lactobacillus fermentation product	Weaned calves (28 days)	−8.5	No response	Dew and Thomas (1981)
Killed *Lactobacillus*	Weaned calves (35 days)	No response	Impaired	Kiesling and Lofgreen (1981)
	Transported calves (28 days)	Improved	Improved	
Viable *Lactobacillus*	Transported calves (28 days)	No response	No response	Kiesling et al. (1982)
E. faecium, 0.5×10^7 g^{-1} diet	Small calves	Improved[a]	Improved[a]	Burgstaller et al. (1983)
Lactobacillus sp. 0–21 days 0–9 wk	Small calves	−0.3	No data	Owen and Larson (1984)
E. faecium, 1.0×10^7 g^{-1} diet	Small calves	No response	No response	Burgstaller et al. (1985)
L. .acidophilus 10^9–10^{10} cfu/d/calf	Small calves 0–7 wk	−7.6	+1.8	Jonsson and Olsson (1985)

Table 10 (Continued)

Type and dosage of probiotic	Animals	Performance (% of control) Gain	Feed/gain	Reference
E. faecium 10^6 cfu g^{-1} in MR	Small calves 0–8 wk	+4.1	–7.6	Havrevoll et al. (1988)
L. acidophilus 10^6 cfu g^{-1} in MR	Small calves 0–10 wk	+0.5	–4.0	
E. faecium, 10^{10} cfu g^{-1} in MR (0–5 days)	Small calves 30 days	+20[a]	No data	Tournut (1989)
Lactobacillus sp. 0.8–8.0 × 10^6 cfu g^{-1} in MR	Small calves	+6–7	–4–5	Vanbelle et al. (1989)
Enterococcus sp. 0.8–8.0 × 10^6 cfu g^{-1} in MR	Small calves	+3–4	–23	
L. fermentum[b] + *L. delbrüeckii*[b] + lactitol	Transported calves 4–10 wk	+15	No data	Nousiainen (1991) (unpublished)

[a]Statistically different compared to control ($p < .05$).
[b]10^9 cfu d^{-1}.
MR = milk replacer.

The role of LAB in the lower digestive tract of the ruminants is difficult to estimate, because practically no data are available. It is possible that a ruminant reacts for the fermentation in the intestine in a different way than nonruminant animals. For example, the preliminary data in our laboratory indicate that fermentation products (lactic acid, VFA) in the hind gut could decrease the feed intake of the ruminant.

VI. CONCLUDING REMARKS

The complex gut microflora-host interaction was discussed in terms of recent knowledge. Several hypotheses of the beneficial effects of lactic acid bacteria probiotics on the symbiosis between the host and microbes were summarized. It may be concluded that the whole probiotic concept as well as the components of the competitive exclusion have been demonstrated in many studies using, for

Table 11 The Use of Lactic Acid Bacteria as Probiotics for Ruminants

Treatment	Advantage	Author(s)
1. *L. acidophilus*	Variable changes in acetic/ propionic-ratio in rumen VFA	McCormick (1984)
2. *L. acidophilus* +	Lowered acetic/propionic-ratio, higher VFA and fiber digestion	McCormick (1984)
3. *L. acidophilus* + *S. faecium* + *S. cerevisiae*	Increased milk yield and milkfat content	Hoyos et al. (1987)
4. *L. acidophilus* *Bifidobacterium* sp. *L. faecalis*	No advantages in feed intake, daily gain or health (beef cattle)	Rust et al. (1989)
5. *L. lactis* *L. acidophilus* *B. subtilis*	No advantages in feed intake, daily gain or health (beef cattle)	Rust et al. (1989)

1, 2. 2×10^7 cfu animal^{-1} day^{-1}
3. Dosage not reported
4. 6×10^9 cfu animal^{-1} day^{-1}
5. 3.3×10^9 cfu animal^{-1} day^{-1}

example, gnotobiotic models or specific in vitro techniques in a scientifically sound way. However, the results obtained from animal trials testing the effect of LAB probiotics on the growth and health of farm animals have been quite variable. The overall efficiency of probiotics up to date does not perhaps reach the level which can be obtained with feed antibiotics, although in some trials positive effects can be noted even when LAB are added to medicated feed.

One reason for the variable responses of probiotics in field trials is certainly accounted for by the complicity of the phenomenon itself. No improvements in performance can be expected when the animals are equipped with a well-functioning gut microflora adapted to grow in a beneficial symbiosis with the host. In contrast, in the presence of any kind of environmental stress (management methods, diet) causing imbalance in the intestinal ecosystem, high-quality probiotics certainly have the potential to boost animal performance. On the other hand, the properties and dosing methods of the preparates used in trials may not have been in line with the basic probiotic concepts. For example, probiotic dosages too low to overcome the barrier effect of the indigenous flora or the wrong types of bacteria might have been used. Additionally, the sensitivity of the isolated gut bacteria to the industrial cultivation and the processes of compound feed technology can be a real problem. It has been suggested that LAB may lose, for example, their adhesion ability due to disappearance of plasmids during long-term technological usage. Therefore, much attention has to be given to improving the tolerance and viability of the probiotic bacteria, especially the most promising *Lactobacillus* strains.

To overcome the complicity of the probiotic concept based on LAB strains alone, we propose a new idea of "mixed probiotics" based on a combination of strains and nonabsorbable sugars being effective in the whole tract. Future experience will reveal the validity of this method as a tool for solving the problems stated above and for obtaining more consistent responses with probiotics in commercial circumstances.

REFERENCES

Alaeddinoglu, G., Güven, A., and Özilgen, M. (1989). Activity-loss kinetics of freeze-dried lactic acid bacteria, *Enzyme Microb. Technol., 11*:765.

Allison, M. J. (1989). Characterization of the flora of the large bowel of pigs: A status report, *Anim. Feed Sci. Technol. 23*:79.

Anderson, K. L., Nagaraja, T. G., Morril, J. L., Avery, T. B., Galitzer, S. J., and Boyer, J. E. (1987). Ruminal microbial development in conventionally or early-weaned calves, *J. Anim. Sci., 64*:1215.

Anonymous, (1990). Probiotics: A review of some of the products currently available to compound feed manufacturers, *The Feed Compounder, 5*:58.

Argensio, R. A., and Southworth, M. (1974). Sites of organic acid production and absorption in gastrointestinal tract of the pig, *Am. J. Physiol., 228*:454.

Argensio, R. A., and Stevens, C. E. (1984). The large bowel—a supplementary rumen? *Proc. Nutr. Soc., 43*:13.

Armstrong, D. G. (1984). Antibiotics as feed additives for ruminant livestock, in *Antimicrobials and Agriculture* (M. Woodbine, ed.), Butterworths, London, pp. 331–347.

Armstrong, D. G. (1986). Gut-active growth promoters, in *Control and Manipulation of Animal Growth* (P. J. Buttery, D. Lindsay, and N. B. Haynes, eds.), Butterworths, London, pp. 21–37.

Bailey, J. S. (1987). Factors affecting microbial competitive exclusion in poultry, *Food Technol., July*: pp. 88–92.

Baird, D. M. (1977). Probiotics help boost feed efficiency, *Feedstuffs, 49*:11.

Barrow, P. A., Brooker, B. E., Fuller, R., and Newport, M. J. (1980). The attachment of bacteria to the epithelium of the pig and its importance in the microecology of the intestine, *J. Appl. Bact., 48*:147.

Bealmer, P. M., Holtermann, O. A., and Mirand, E. A. (1984). Influence of the microflora on the immune responce. 1. General characteristics of the germ free animal, *The Germ Free Animal in Biomedical Research* (Coates, B. E. and Gustafsson, B. E., eds.), Laboratory Animals Ltd., London, pp. 335–346.

Berg, R. D., and Savage, D. C. (1972). Immunological responces and microorganisms indigenous to the gastrointestinal tract, *Am. J. Clin. Nutr., 25*:1364.

Bergner, H., Simon, O., Zebrowska, T., and Münchmeyer, R. (1987). Studies on the secretion of amino acids and of urea into the gastrointestinal tract of pigs. 3. Secretion of urea determined by continuous intravenous infusion of ^{15}N-urea, *Arch. Anim. Nutr., 36*:479.

Blomberg, L., and Conway, P. (1989). An in vitro study of ileal colonization of resistance to *Escherichia coli* strain Bd 1107/75 08 (K88) in relation to indigenous squamous gastric colonization in piglets of varying ages, *Microb. Ecol. Health Disease, 2*:285.

Brooker, B. E., and Fuller, R. (1975). Adhesion of lactobacilli to the chicken crop epithelium, *J. Ultrastruct. Res., 52*:21.

Bryant, M. P., and Small, N. (1960). Observations on the ruminal micro-organisms of isolated and inoculated calves, *J. Dairy Sci., 43*:654.

Bryant, M. P., Small, N., Bouma, C., and Robinson, I. (1958). Studies on the composition of the ruminal flora and fauna of young calves, *J. Dairy Sci., 41*:1747.

Burgstaller, G., Ferstl, R., and Alps, H. (1985). Milchsaurebakterien (*Streptococcus faecium* M 74) in Kombination mit Avoparcin in Milchaustauschfuttermittel für Mastkalber, *Züchtungskunde, 57*:278.

Burgstaller, G., Ferstl, R., and Peschke, W. (1983). Zum Einsatz von Lactiferm in der Kalbermast, *Züchtungskunde, 55*:48.

Cera, K. R., Mahan, D. C., Cross, R. F., Reinhart, G. A., and Whitmoyer, R. E. (1988). Effect of age, weaning and postweaning diet on small intestinal growth and jejunal morphology in young swine, *J. Anim. Sci., 66*:574.

Chen, R. C. Y., Reid, G., Irvin, R. T., Bruce, A. W., and Costerton, J. W. (1985). Competitive exclusion of uropathogens from human uroepithelial cells by *Lactobacillus* whole cells and cell wall fragments, *Infection and Immunity, 47*:84.

Champ, M., Szylit, O., Raibaud, P., and Ait-Abdelkader, N. (1983). Amylase production by three *Lactobacillus* strains isolated from chicken crop, *J. Appl. Bacteriol., 55*:487.

Clemens, E. T., Stevens, C. E., and Southworth, M. (1975). Sites of organic acid production and pattern of digesta movement in the gastrointestinal tract of swine, *J. Nutr., 105*:759.

Coates, M. E. (1980). The gut microflora and growth, in *Growth in Animals* (T. L. J. Lawrence, ed.), Butterworths, London, pp. 175–227.

Collington, G. K., Parker, D. S., and Armstrong, D. G. (1990). The influence of inclusion of either an antibiotic or a probiotic in the diet on the development of digestive enzyme activity in the pig, *Br. J. Nutr., 64*:59.

Conway, P. L., Gorbach, S. L., and Goldin, B. R. (1987). Survival of lactic acid bacteria in the human stomach and adhesion to intestinal cells, *J. Dairy Sci., 70*:1.

Cranwell, P. D. (1985). The development of acid and pepsin (EC 3.4.23.1) secretory capasity in the pig; the effects of age and weaning. 1. Studies in anaesthetized pigs, *Br. J. Nutr., 54*:305.

Cranwell, P. D., Noakes, D. E., and Hill, K. J. (1976). Gastric secretion and fermentation in the suckling pig, *Br. J. Nutr., 36*:71.

Danek, P. (1987). Effectiveness of the lactic acid bacteria *Streptococcus faecium* M-74 in feed mixtures for early weaned piglets, *Nutr. Abstr. Rev. (Series B), 57*(6):364.

Danielson, A. D., Peo, E. R., Shahani, K. M., Lewis, A. J., Whalen, P. J., and Amer, M. A. (1989). Anticholesteremic property of *Lactobacillus acidophilus* yoghurt fed to mature boars, *J. Anim. Sci., 67*:966.

Davidson, J. N., and Hirsch, D. C. (1976). Bacterial competition as a means of preventing neonatal diarrhoea in pigs, *Infection and immunity, 13*:1773.

Deguchi, E., and Namioka, S. (1989). Synthesis ability of amino acids and protein from non-protein nitrogen and role of intestinal flora on this utilization in pigs, *Bifidobacteria Microflora, 8*(1):1.

Dew, R. K., and Thomas, O. O. (1981). *Lactobacillus* fermentation product for post-weaned calves, *J. Anim. Sci., 53*(Suppl. 1):483.

Dierick, N. A., Vervaeke, I. J., Decuypere, J. A., and Henderickx, H. K. (1986). Influence of gut flora and of some growth-promotive feed additives on nitrogen metabolism in pigs. I. Studies in vitro, *Livest. Prod. Sci., 14*:161.

Dierick, N. A., Vervaeke, I. J., Demeyer, D. I., and Decuypere, J. A. (1989). Approach to the energetic importance of fibre digestion in pigs. I. Importance of fermentation in the overall energy supply, *Anim. Feed Sci. Technol., 23*:141.

Dulcuzeau, R. (1985). Implantation and development of the gut flora in the newborn piglet, *Pig News Inform., 6*:415.

Dutta, G. N., and Devriese, L. A. (1981). Sensitivity and resistance to growth promoting agents in animal lactobacilli, *J. Appl. Bact., 51*:283.

Ellinger, D. K., Muller, L. D., and Glantz, P. J. (1978). Influence of fermented colostrum and *Lactobacillus acidophilus* on faecal flora and selected blood parameters of young dairy calves, *J. Dairy Sci., 61*(Suppl. 1):126.

Eyssen, H., and van Eldere, J. (1984). Metabolism of bile acids, *The Germ-Free Animal in Biomedical Research* (M. E. Coates and B. E. Gustafsson, eds.), Laboratory Animals, London, pp. 291–316.

Fernandes, C. F., Shahani, K. M., and Amer, M. A. (1988). Effect of nutrient media and bile salts on growth and antimicrobial activity of *Lactobacillus acidophilus, J. Dairy Sci., 71*:3222.

Friend, D. W., Cunningham, H. M., and Nicholson, J. W. G. (1963). Volatile fatty acids and lactic acid in the alimentary tract of the young pig, *Can. J. Anim. Sci., 43*:174.

Fuller, R. (1977). The importance of lactobacilli in maintaining normal microbial balance in the crop, *Br. Poultry Sci., 18*:85–94.

Fuller, R. (1989). Probiotics in man and animals, *J. Appl. Bact., 66*:365.

Fuller, R., Barrow, P. A., and Brooker, B. E. (1978). Bacteria associated with the gastric epithelium of neonatal pigs, *Appl. Environ. Microbiol., 35*:582.

Fuller, R., Houghton, S. B., and Brooker, B. E. (1981). Attachment of *Streptococcus faecium* to the duodenal epithelium of the chicken and its importance in colonization of the small intestine, *Appl. Environ. Microbiol., 41*:1433.

Garvie, E. I., Cole, C. B., Fuller, R., and Hewitt, D. (1984). The effect of yoghurt on some components of the gut microflora and the metabolism of lactose in the rat, *J. Appl. Bacteriol., 56*:237.

Gilliland, S. E. (1979). Beneficial interrelationships between certain micro-organisms and humans: Candidate micro-organisms for use as dietary adjuncts, *J. Food Protect., 42*:164.

Gilliland, S. E., Bruce, B. B., Bush, L. J., and Stanley, T. E. (1980). Comparison of two strains of *Lactobacillus acidophilus* as dietary adjuncts for young calves, *J. Dairy Sci., 63,D:964.*

Gilliland, S. E., Nelson, C. R., and Maxwell, C. (1985). Assimilation of cholesterol by *Lactobacillus acidophilus, Appl. Environ. Microbiol., 49*:377.

Gilliland, S. E., and Speck, M. L. (1977). Deconjucation of bile acids by intestinal lactobacilli, *Appl. Environ. Microbiol., 33*:15.

Gilliland, S. E., Staley, T. E., and Bush, L. J. (1984). Importance of bile tolerance of *Lactobacillus acidophilus* used as dietary adjunct, *J. Dairy Sci., 67*:3045.

Goldin, B. R., and Gorbach, S. L. (1984a). The effect of milk and *Lactobacillus* feeding on human intestinal bacterial enzyme activity, *Am. J. Clin. Nutr., 39*:756.

Goldin, B. R., and Gorbach, S. L. (1984b). Alterations of the intestinal microflora by diet, oral antibiotics, and *Lactobacillus*: Decreased production of free amines from aromatic nitro compounds, azo dyes, and glucuronides, *J. Natl. Cancer Inst., 73*:689.

Goldin, B. R., Swenson, L., Dwyer, J., Sexton, M., and Gorbach, S. L. (1980). Effect of diet and *Lactobacillus acidophilus* supplements on human fecal bacterial enzymes, *J. Natl. Cancer Inst., 64*:255.

Gualtieri, M., and Betti, S. (1985). Effects of *Streptococcus faecium* on suckling pigs, *Nutr. Abstr. Rev. (Series B), 55*:(6):344.

Hale, O. M., and Newton, G. L. (1979). Effects of nonviable *Lactobacillus* species fermentation product on performance of pigs, *J. Anim.Sci., 48*:770–775.

Hampson, D. J. (1986a). Alterations in piglet small intestinal structure at weaning, *Res. Vet. Sci., 40*:32.

Hampson, D. J. (1986b). Attempts to modify changes in the piglet small intestine after weaning, *Res. Vet. Sci., 40*:313.

Hanson, D. J. (1985). Human health effects of animal feed drugs unclear, *Chem. Eng. News, 63*(7):7.

Harju, M. (1988a). Lactitol as a substrate for β-galactosidase. II. Literature review and methods, *Milchwissenschaft, 43*:76.

Harju, M. (1988b). Lactitol as a substrate for β-galactosidases. II. Results and discussion, *Milchwissenschaft, 43*:148.

Harper, A. F., Kornegay, E. T., Bryant, K. L., and Thomas, H. R. (1983). Efficacy of virginiamycin and a commercially-available *Lactobacillus* probiotic in swine diets, *Anim. Feed Sci. Technol., 8*:69.

Havrevoll, O., Matre, T., Pestalozzi, M., Storro, K., and Holland, S. (1988). Probiotics in feeds for calves, Proc. VI World Conf. Animal Production, Helsinki, p. 383.

Hedges, A. J., and Linton, A. H. (1988). Olaguindox resistance in the coliform flora of pigs and their environment: an ecological study, *J. Appl. Bact., 64*:329.

Henriksson, A., Szewzyc, R., and Conway, P. L. (1991). Characteristics of the adhesive determinants of *Lactobacillus fermentum* 104, *Appl. Environ. Microbiol. 57*(2):499.

Hill, R. H., and Cowley, H. M. (1990). The influence of colonizing micro-organisms on development of crypt architecture in the neonatal mouse colon, *Acta Anat., 137*:137.

Hill, J. R., Kenworthy, R., and Porter, P. (1970a). Studies of the effect of dietary lactobacilli on intestinal and urinary amines in pigs in relation to weaning and post-weaning diarrhoea, *Res. Vet. Sci., 11*:320–326.

Hill, J. R., Kenworty, R., and Porter, P. (1970b). The effect of dietary lactobacilli on in-vitro catabolic activities of the small-intestinal microflora of newly weaned pigs, *J. Med. Microbiol., 3*:593–605.

Hoyos, G., Garcia, L., and Medina, F. (1987). Effects of feeding viable microbial feed additives on performance of lactating cows, *J. Dairy Sci., 70*(Suppl. 1):P341.

Hutcheson, D. P., Cole, N. A., Kreaton, W., Graham, G., Dunlap, R., and Pittman, K. (1980). The use of a living, nonfreeze-dried *Lactobacillus acidophilus* culture for receiving feedlot calves, *Proc. Ann. Am. Soc. Anim. Sci. West. Sect., 31*:213.

Impey, C. S., and Mead, G. C. (1989). Fate of salmonellas in the alimentary tract of chicks pre-treated with a mature caecal microflora to increase colonization resistance, *J. Appl. Bacteriol., 66*:469.

Jonsson, E. (1985). Lactobacilli as probiotics to pigs and calves. A microbiological approach, Swedish University of Agricultural Sciences, Department of Animal Nutrition and Management, Uppsala, Report 148, 65 pp.

Jonsson, E. (1986). Persistence of *Lactobacillus* strain in the gut of suckling piglets and its influence on performance and health, *Swedish J. Agric. Res., 16*:43.

Jonsson, E., Björck, L., and Claesson, C. O. (1985). Survival of orally administered *Lactobacillus* strains in the gut of cannulated pigs, *Livest. Prod. Sci., 12*:279.

Jonsson, E., and Hemmingsson, S. (1991). Establishment in the piglet gut of lactobacilli capable of degrading mixed-linked B-glucans, *J. Appl. Bacteriol.*, 1991, in press.

Jonsson, E., and Olsson, I. (1985). The effect on performance, health and faecal microflora of feeding *Lactobacillus* strains to neonatal calves, *Swed. J. Agric. Res., 15*:71.

Juven, B. J., Meinersmann, R. J., and Stern, N. J. (1991). Antagonistic effects of lactobacilli and pediococci to control intestinal colonization of human enteropathogens in live poultry, *J. Appl. Bact., 70*:95.

Kearney, L., Upton, M., and McLoughlin, A. (1990). Enhancing the viability of *Lactobacillus plantarum* inoculum by immobilizing the cells in calcium-alginate beads incorporating cryoprotectants, *Appl. Environ. Microbiol., 56*(10):3112.

Kenworthy, R. (1967). Influence of bacteria on absorption from the small intestine, *Proc. Nutr. Soc., 26*:18.

Kiesling, H. E., and Lofgreen, G. P. (1981). Selected fermentation products for receiving cattle, *J. Anim. Sci., 53*(Suppl. 1):483.

Kiesling, H. E., Lofgreen, G. P., and Thomas, J. D. (1982). A viable *Lactobacillus* culture for feedlot cattle, *J. Anim. Sci., 55*(Suppl. 1):490.

Klaenhammer, T. R. (1982). Microbial considerations in selection and preparations of *Lactobacillus* strains for use as dietary adjuncts, *J. Dairy Sci., 65*:1339.

Kluber, E. F., Pollman, D. S., and Blecha, F. (1985). Effect of feeding *Streptococcus faecium* to artificially reared pigs on growth, hematology and cell mediated immunity, *Nutr. Rep. Int., 32*:57.

Kozasa, M. (1989). Probiotics in animal use in Japan, *Rev. Sci. Tech. Off. Int. Epiz., 8*(2):517.

Lengemann, F. W., and Allen, N. N. (1959). Development of rumen function in the dairy cattle. II. Effect of diet upon characteristics of the rumen flora and fauna of young calves, *J. Dairy Sci., 42*:1171.

Lessard, M., and Brisson, G. J. (1987). Effect of a *Lactobacillus* fermentation product on growth, immune responce and fecal enzyme activity in weaned pigs, *Can. J. Anim. Sci., 67*:509.

Lilly, D. M., and Stillwell, R. J. (1965). Probiotics: growth promoting factors produced by micro-organisms, *Science, 147*:747.

Linton, A. H., Hedges, A. J., and Bennet, B. M. (1988). Monitoring of resistance during the use of olaquindox as a feed additive on commercial pig farms, *J. Appl. Bact., 64*:311.

Lyons, T. P. (1987). Probiotics: an alternative to antibiotics, *Pig News Inform., 8*(2):157.

Magee, D. F., and Dalley, A. F. (1986). *Digestion and the Structure and the Function of the Gut*, S. Karger AG, Basel, 359 pp.

Marounek, M., Jehlickova, K., and Kmet, V. (1988). Metabolism and some characteristics of lactobacilli isolated from the rumen of the young calf, *J. Appl. Bacteriol., 65*:43.

Marshall, V. M., Philips, S. M., and Turvey, A. (1982). Isolation of hydrogen peroxide-producing strain of *Lactobacillus* from calf gut, *Res. Vet. Sci., 32*:259.

Maxwell, C. V., Buchanan, D. S., Owens, F. N., Gilliland, S. E., Luce, W. G., and Vencl, R. (1983). Effect of probiotic supplementation on performance, fecal parameters, and digestibility in growing-finishing swine, *Oklahoma Agric. Exp. Stat., Anim. Sci. Res. Rep., 114*:157.

Mäyrä-Mäkinen, A., Manninen, M., and Gyllenberg, H. (1983). The adherence of lactic acid bacteria to the columnar epithelial cells of pigs and calves, *J. Appl. Bact., 55*:241.

McCormick, M. E. (1984). Probiotics in ruminant nutrition and health, Proc. Georgia Nutrition Conf. Feed Industry, pp. 62–69.

Mee, B. J. (1984). The selective capacity of pig feed additives and growth promotants for coliform resistance, in *Antimicrobials in Agriculture* (M. Woodbine, ed.), Butterworths, London, pp. 349–358.

Metchnikoff, E. (1903). *The Nature of Man: Studies of Optimistic Philosophy*, Heineman, Lodres.

Metchnikoff, E. (1908). *Prolongation of Life*, G. Putnam's Sons, New York.

Miller, B. G., James, P. S., Smith, M. W., and Bourne, F. J. (1986). Effect of weaning on the capasity of pig intestinal villi to absorb nutrients, *J. Agric. Sci. Camb., 107*:579.

Mitchell, I. G., and Kenworthy, R. (1976). Investigations on a metabolite from *Lactobacillus bulgaricus* which neutralise the effect of enterotoxin from *Escherichia coli* pathogenic for pigs, *J. Appl. Bacteriol., 41*:163.

Mordenti, A. (1986). Probiotics and new aspects of growth promoters in pig production, *Inform. Zootechnol., 32*(5):69.

Morril, J. L., Dayton, A. D., and Mickelsen, R. (1977). Cultured milk and antibiotics for young calves, *J. Dairy Sci., 60*:1105.

Muralidhara, K. S., Sheggeby, G. G., Elliker, P. R., England, D. C., and Sandine, W. E. (1977). Effect of feeding lactobacilli on the coliform and *Lactobacillus* flora of intestinal tissue and feces from piglets, *J. Food Protection, 40*:288.

Nagura, H. (1990). Mucosal defense mechanism and secretory IgA system, *Bifidobacteria Microflora, 9*:17.

Nieto, N., Cabarello, A. G., and Martinez, A. (1985). Comparative study of the ruminal protozoal fauna in calves under different methods of rearing, *Nutr. Abstr. Rev. (Series B), 56*:5347.

Noland, P. R., Campbell, D. R., Johnson, Z. B., and Williams, R. (1978). Effect of weaning age and a cultured bacterial product on performance of young pigs, *Arkansas Farm Res., 27*:7.

Nousiainen, J. (1988). The use of lactose derivatives as feed additives to nonruminant farm animals, Proc. VI World Conf. Animal Production 27.6–1.7, Helsinki, p. 236.

Nousiainen, J. (1990). Effect of antimicrobials and probiotics on the gastrointestinal environment and performance of piglets, Licenciate dissertation, Department of Animal Husbandry, University of Helsinki, 145 pp.

Nousiainen, J., and Suomi, K. (1991). Comparative observations on selected probiotics and olaquindox used as feed additives for piglets around weaning. 1. Effect on the bacterial metabolites along the intestinal tract, blood values and growth, *J. Anim. Physiol. Anim. Nutr. 66*:212.

Nurmi, E., and Rantala, M. (1973). New aspects of salmonella infection in broiler production, *Nature, 241*:210.

Owen, F. G., and Larson, L. L. (1984). Effect or probiocin and starter preparations on calf performance, *J. Dairy Sci., 67*(Suppl. 1):139.

Ozawa, K., Yabu-uchi, K., Yamanaka, K., Yamashita, Y., Nomura, S., and Oku, I. (1983). Effect of *Streptococcus faecalis* BI0-4R on intestinal flora of weanling piglets and calves, *Appl. Environ. Microbiol., 45*:1513.

Parker, D. S. (1974). Probiotics, the other half of the antibiotic story, *Animal Nutr. Health, 29*:4.

Parker, D. S. (1990). Manipulation of the functional activity of the gut by dietary and other means (antibiotics/probiotics) in ruminants, *J. Nutr., 120*:639.

Parker, D. S., and Armstrong, D. G. (1987). Antibiotic feed additives and livestock production, *Proc. Nutr. Soc., 46*:415.

Pedersen, K., and Tannock, G. W. (1989). Colonization of the porcine gastrointestinal tract by lactobacilli, *Appl. Environ. Microbiol., 55*:279.

Perdigon, G., Alvarez, S., Nader de Macias, M. E., Roux, M. E., and Pesce de Ruiz Holdago, A. A. (1990). The oral administration of lactic acid bacteria increase the mucosal immunity in response to enteropathogens, *J. Food Prot., 53*:404.

Perdigon, G., Nader de Macias, M. E., Alvarez, S., Oliver, G., and Pesce de Ruiz Holdaga, A. A. (1986). Effect of perorally administered lactobacilli on macrophage activation in mice, *Infection and Immunity, 53*:404.

Perdigon, G., Nader De Macias, M. E., Alvarez, S., Oliver, G., and Pesce de Ruiz Holdaga, A. A. (1987). Enchangement of immune responce in mice fed with *Streptococcus thermophilus* and *Lactobacillus acidophilus, J. Dairy Sci., 70*:919.

Pivnick, H., and Nurmi, E. (1982). The Nurmi concept and its role in the control of Salmonellae in poultry, in *Developments in Food Microbiology* (R. Davis, ed.), Vol. 1, Applied Science, Barking, England, p. 41.

Pollmann, D. S., and Bandyk, C. A. (1984). Stability of viable lactobacillus products, *Anim. Feed Sci. Technol. 11*:261.

Pollmann, D. S., Danielson, D. M., and Peo, E. R., Jr. (1980a). Effects of microbial feed additives on performance of starter and growing-finishing pigs, *J. Anim. Sci., 51*:577.

Pollmann, D. S., Danielson, D. M., Wren, W. B., Peo, E. R., Jr. and Shahani, K. M. (1980b). Influence of *Lactobacillus acidophilus* inoculum on gnotobiotic and conventional pigs, *J. Anim. Sci., 51*:629.

Pollmann, D. S., Danielson, D. M., and Peo, E. R. Jr. (1980c). Effect of *Lactobacillus acidophilus* on starter pigs fed a diet supplemented with lactose, *J. Anim. Sci., 51*:638.

Pollmann, D. S., Kennedy, G. A., Koch, B. A., and Allee, G. L. (1984). Influence on nonviable *Lactobacillus* fermentation product on artificially reared pigs, *Nutr. Rep. Int., 29*:977.

Porter, P., and Barrat, M. E. J. (1987). Immunity, nutrition, and performance in animal production, in *Recent Advances in Animal Nutrition 1987* (W. Haresign and D. J. A. Cole, eds.), Butterworths, London, pp. 107–116.

Ratcliffe, B. (1985). The influence of the gut microflora on the digestive processes, Proc. 3rd Int. Sem. Digestive Physiology in the Pig (A. Just, H. Jorgensen, and J. A. Fernandez, eds.), *Beretning fra Statens Husdyrbrugsforsog*, No. 580, pp. 245–267.

Ratcliffe, B., Cole, C. B., Fuller, R., and Newport, M. J. (1986). The effect of yoghurt and milk fermented with a porcine intestinal strain of *Lactobacillus reuteri* on the performance and gastrointestinal flora of pigs weaned at two days of age, *Food Microbiol., 3*:203.

Roach, S., and Tannock, C. W. (1980). Indigenous bacteria that influence the number of *Salmonella typhimurium* in the spleen of intravenously challenged mice, *Can. J. Microbiol. 26*:408.

Roediger, W. E. W. (1980). Role of anaerobic bacteria in the metabolic welfare of the colonic mucosa in man, *Gut, 21*:793.

Roth, F. X., and Kirchgessner, M. (1986). Zur nutritiven Wirksamkeit von *Streptococcus faecium* (Stamm M 74) in der Ferkelaufzucht, *Landwirtsch. Forshung, 39*:198.

Russel, E. G. (1979). Types and distribution of anaerobic bacteria in the large intestine of pigs, *Appl. Environ. Microbiol., 37*:187.

Rust, S. R., Lutchka, L. J., Yokoyama, M. T., and Ritchie, H. D. (1989). Microbial cultures in receiving programs for stressed cattle, *J. Anim. Sci., 67*(Suppl. 1):28.

Saito, H. (1988). Enhancement of host resistance to bacterial and viral infections by *Lactobacillus casei, Bifidobacteria Microflora, 7*:17.

Sakata, T. (1987). Stimulatory effect of short-chain fatty acids on epithelial cell proliferation in the rat intestine: a possible explanation for trophic effects of fermentable fiber, gut microbes and luminal trophic factors, *Br. J. Nutr., 58*:95.

Savage, D. C. (1977). Microbiol ecology of the gastrointestinal tract, *Ann. Rev. Microbiol., 31*:107.

Savage, D. C. (1986). Gastrointestinal microflora in mammalian nutrition, *Ann. Rev. Nutr., 6*:155.

Savage, D. C. (1987b). Factors affecting the biocontrol of bacterial pathogens in the intestine, *Food Technology, July*:82–87.

Savage, D. C. (1987b). Microorganisms associated with epithelial surfaces and stability of the indigenous gastrointestinal microflora, *Die Nährung, 31*(5):383.

Schwab, C. G., Moore, J. J., Hoyt, P. M., and Pretience, J. L. (1980). Performance and fecal flora of calves fed a nonviable *Lactobacillus bulgaricus* fermentation product, *J. Dairy Sci., 63*:1412.

Sherman, L. A., and Savage, D. C. (1986). Lipoteichoic acids in *Lactobacillus* strains that colonize the mouse gastric epithelium, *Appl. Environ. Microbiol. 52*:302.

Smith, M. W. (1985). Expression of digestive and absorptive function in differentiating enterocytes, *Ann. Rev. Physiol., 47*:247.

Spillman, H., Puhan, Z., and Banhequi, M. (1978). Antimikrobielle Aktivität thermophiler Laktobasillen, *Milchwissenschaft, 33*:148.

Stavric, S., Cleeson, T. M., Blanchfield, B., and Pivnick, H. (1987). Role of adhering microflora in competitive exclusion of Salmonella from young chicks, *J. Food Protect., 50*:928.

Tagg, J. R., Dajant, A. S., and Wannamaker, L. W. (1976). Bacteriocins of Gram-positive bacteria, *Bacteriol. Rev., 40*:722.

Tannock, G. W. (1988). Mini review: Molecular genetics: A new tool for investigating the microbial ecology of the gastrointestinal tract, *Microb. Ecol., 15*:239.

Tannock, G. W. (1990). The microecology of lactobacilli inhabiting the gastrointestinal tract, *Adv. Microb. Ecol., 11*:147.

Tannock, G. W., Dashkevicz, M. P., and Feighner, S. D. (1989). Lactobacilli and bile salt hydrolase in the murine intestinal tract, *Appl. Environ. Microbiol., 55*:1848.

Tannock, G. W., Fuller, R., and Pedersen, K. (1990). *Lactobacillus* succession in the piglet digestive tract demonstrated by plasmid profiling, *Appl. Environ. Microbiol., 56*:1310.

Tournut, J. (1989). Applications of probiotics to animal husbandry, *Rev. Sci. Tech. Off. Int. Epiz., 8*(2):551.

Tuschy, D. (1986). Verwendung von "Probiotika" als Leistungsförderer in der Tierernährung, *Übers. Tierernährg., 14*:157.

Underdahl, N. R. (1983). The effect of *Streptococcus faecium* upon Escherichia coli induced diarrhea in gnotobiotic pigs, *Prog. Fd. Nutr. Sci., 7*:5.

Vanbelle, M., Teller, E., and Focant, M. (1990). Probiotics in animal nutrition: A review, *Arch. Anim. Nutr., 7*:543.

van Eldere, J., and Eyssen, H. (1984). Metabolism of cholesterol, in *The Germ-Free Animal in Biomedical Research* (M. E. Coates and B. E. Gustafsson, eds.), Laboratory Animals, London, pp. 317–332.

Visek, W. J. (1987). The mode of growth promotion by antibiotics, *J. Anim. Sci., 46*:1447.

Visek, W. J. (1984). Ammonia: Its effects on biological systems, metabolic hormones and reproduction, *J. Dairy Sci., 67*:481.

Wadström, T. W., Andersson, K., Sydow, M., Axelsson, L., Lindgren, S., and Gullmar, B. (1987). Surface properties of lactobacilli isolated from the small intestine of pigs, *J. Appl. Bact., 62*:513.

White, F., Wenham, G., Sharman, G. A. M., Jones, A. S., Rattray, E. A. S., and McDonald, I. (1969). Stomach function in relation to scour syndrome in the piglet, *Br. J. Nutr., 23*:847.

Whitt, D. D., and Savage, D. C. (198). Lactobacilli as effectors of host functions: No influence on activities of enzymes in enterocytes in mice, *Appl. Environ. Microbiol., 53*:325.

Wu, M. C., Wung, L. C., Cheng, S. Y., and Kuo, C. C. (1988a). Study on the feeding value of *Streptococcus faecium* M-74 for pigs. 1. Large scale feeding trial of *Streptococcus faecium* on the performance of weaning pigs, *Nutr. Abstr. Rev. (Series B), 58*:682.

Wu, M. C., Wung, L. C., Cheng, S. Y., and Kuo, C. C. (1988b). Study on the feeding value of *Streptococcus faecium* M-74 for pigs. 2. Effects of adding *Streptococcus faecium* on the performance of growing-finishing pigs, *Abstr. Rev. (Ser. B), 58*:682.

Yen, J. T., Nienaber, J. A., and Pond, W. G. (1987). Effect of neomycin carbadox and length of adaptation to calorimeter on performance, fasting metabolism and gastrointestinal tract of young pigs, *J. Anim. Sci., 65*:1243.

Yen, J. T., Nienaber, J. A., Pond, W. G., and Varel, V. H. (1985). Effect of carbadox on growth, fasting metabolism, thyroid function and gastrointestinal tract in young pigs, *J. Nutr., 115*:970.

Ziolecki, A., and Briggs, C. A. E. (1961). The microflora of the rumen of the young calf. II. Source, nature and development, *J. Appl. Bacteriol., 24*:148.

13

Bifidobacteria and Probiotic Action

Jean Ballongue

Université de Nancy 1, Vandoeuvre-les-Nancy, and Centre de Recherche International André Gaillard, Ivry-sur-Seine, France

The medical world has long been interested in the nutrient properties of yogurt. The theory of Metchnikoff (1908), which holds that milk fermented with *Lactobacillus* has a favorable influence on the endogenous intestinal flora, was challenged in 1915 by Rahe, who demonstrated that these microorganisms do not survive passage through the stomach and small intestine. Subsequently, numerous studies have been carried out on *Lactobacillus*. The frequently contradictory findings are due to the unreliability of the methods for isolating and identifying bacteria from stools.

Nutritionists subsequently turned their attention to other microorganisms. According to Gurr et al. (1984), "the microorganisms with the best chance of passing through the stomach and small intestine and colonizing the medium are those endogenous to the species consuming the fermented product." Research has been focused on the genus *Bifidobacterium* which, unlike the bacteria of yogurt which are not obtained from human ecosystems, are isolated from animals and humans.

The probiotic effects of *Bifidobacterium*, already alluded to when they were first discovered in 1899 (Tissier, 1990), were demonstrated by Manciaux in 1958. The therapeutic properties of this genus of bacterium led the Japanese to introduce it to their diet (Mitsuoka, 1982; Yamazaki et al., 1985; Ebissawa et al., 1987). Since 1986, the traditional microflora of yogurt: *Streptococcus salivarius* ssp.

thermophilus and *Lactobacillus delbrueckii* ssp. *bulgaricus* has been enhanced by a third bacterium belonging to the genus *Bifidobacterium* and sometimes associated with *Lactobacillus acidophilus.* This new product with pleasant organoleptic qualities has aroused considerable interest from consumers, who were soon followed by dairy industrialists and medical teams.

I. THE BIFIDOBACTERIA: DISCOVERY AND HISTORY

In 1899, at the Institut Pasteur, Tissier observed and isolated from stools of infants a bacterium with a very unusual and hitherto unknown Y-shape. The problem of the place of this bacterium within the classification system was then raised.

At the beginning of the century, taxonomy was based entirely on morphological criteria and Tissier (1990) named this bacterium *Bacillus bifidus communis.* At about the same time, but in Italy, Moro (1900a) discovered in similar conditions a bacterium which he recognized as being different from that of Tissier and which he identified as belonging to the genus *Lactobacillus.* Despite the differences between these two bacteria, Holland (1920) proposed a common name: *Lactobacillus bifidus,* which was to develop and gain precision as time passed in parallel with the progress in biology.

Orla-Jensen (1924), using new methods, was responsible for a decisive shift in the direction of the history of taxonomy. The classification and identification of microorganisms which had hitherto been based entirely on their morphology, henceforth took into account new criteria: the physiology, nutritive requirements of the energy metabolism and above all metabolic and enzymatic characteristics of the organism. Thus, in 1967, De Vries and Stouthamer (1967) demonstrated the presence in bifides of fructose-6-phosphate phosphoketolase (F6PPK) and the absence of aldolase and glucose-6-phosphatase dehydrogenase, two enzymes found in the lactobacilli. They therefore conclude that the classification of the bifidobacteria in the genus *Lactobacillus* is not justified.

Two trends were distinguished: the French school, which was for the separation of the genuses *Lactobacillus* and *Bifidobacterium* to combine all bifides bacteria under the single classification of *B. bifidum* (Holland, 1920; Orla-Jensen, 1919, 1924; Prevot, 1940, 1955), and the Anglo-Saxon school which preferred to integrate the bifidobacteria in the genus *Lactobacillus.* Table 1 summarizes the various names proposed for this bacterium since its discovery (Rasic and Kurmann, 1983) and Table 8 lists the species isolated to date.

The advent of chemotaxonomy in the 1960s marked the beginning of another period in bacterial taxonomy. Research into the biochemistry of the prokaryotes has shown that analysis of the cell constituents could become an essential tool in the classification and identification of bacteria. The development of instruments of analysis made it possible to obtain accurate and reproducible data, minimize

Table 1 Chronology of the Taxonomy of *Bifidobacterium*

Name	Author	Year
Bacillus bifidus	Tissier	1900
Bacteroides bifidus	Castellani and	1919
	Chalmers	1923–1934
Lactobacillus bifidus	Bergey's Manual	
	eds. 1–4	1920
Bifidobacterium bifidum	Holland	1924
Bacterium bifidum	Orla-Jensen	1927
Tisseria bifida	Lehmann and Neumann	1929
Nocardia bifida	Pribram	1931
Actynomices bifidus	Vuillemin	1934
Actinobacterium bifidum	Nannizzi	1937
Lactobacillus acidophilus	Puntoni	1938
var *bifidus*	Weiss and Rettger	
Lactobacillus parabifidus		1938
Bifidobacterium bifidum	Weiss and Rettger	1938
Lactobacillus bifidus	Prevot	1939–1957
Cohnistreptothrix bifidus	Bergey's Manual	
	eds. 5–7	1944
Corynebacterium bifidum	Negrovi and Fischer	1949
Lactobacillus bifidus	Olsen	1950
Lactobacillus bifidus	Norris et al.	1953
var *pennsylvanicus*		1957
Five groups of bifidus bacteria	György	
Description of human species	Dehnert	1963
New animal species	Reuter	1969
New animal species	Mitsuoka	1969
New animal species	Scardovi	1972
Creation of the genus *Bifidobacterium*	Holdeman and Moore	1974
constituted by 11 species	Bergey's Manual ed. 8	

Source: From Rasic and Kurmann (1983).

errors in individual research and eliminated subjective judgments (Dellaglio, 1989). It was necessary to wait till 1965 and the progress of molecular genetics for the teams of Sebald et al. (1965) and Werner et al. (1966) to show that the percentage of G + C in the DNA of *Bifidobacterium* differed from that of *Lactobacillus, Corynebacterium* and *Propionibacterium*. In 1974, the VIIIth edition of *Bergey's Manual of Determinative Bacteriology* recognized Bifidobacterium as a genus in its own right consisting of 11 species (Buchanan and Gibbons, 1974). Today, this genus which belongs to the *Actinomycetaceae* group (Scardovi,

1986) includes 24 species which are grouped according to their ecological origin: 15 are isolated only from animals and 9 colonize the natural cavities of man (Scardovi, 1981).

II. MORPHOLOGY

The bacteria of the genus *Bifidobacterium* present a globally bacillar form, show Gram-positive staining, and are immobile and nonsporulate.

These rods, with an irregular outer wall, are usually concave and their extremities generally swollen to form "lumps," which may have one or more ramifications. It is, however, not unusual to encounter more rounded shapes as well as very long or short bacilli of varying widths. Gram staining reveals a frequently irregular distribution of chromatin which often accumulates in the bifurcations or lumps (Mayer and Moser, 1950; Raynaud and Guintini, 1959).

However, this polymorphism cannot be assimilated to degeneration since these forms can generate the initial forms once more (Mayer and Moser, 1950). It would appear rather that the composition of the culture medium is responsible for these V-, Y-, or X-shaped forms encountered in the genus *Bifidobacterium*.

Several medium constituents may influence the shape of these bacteria:

The concentration of *N*-acetylglucosamine, which is involved in the synthesis of peptidoglycan (Fig. 1) affects the shape of *B. bifidum* var. *pennsylvanicus* (Glick et al., 1960).

Various amino acids (alanine, aspartic acid, glutamic acid and serine) (Husain et al., 1972).

Ca^{2+} ions (Kojima et al., 1968, 1970a,b).

The lower the levels of *N*-acetylglucosamine and amino acids, the more highly branched are the shapes. In contrast, in a favorable medium the bacilli are longer (Mayer and Moser, 1950).

III. PHYSIOLOGY

A. Respiratory Type

The bifidobacteria are strictly anaerobic microorganisms. However, the degree of tolerance of oxygen depends on the species and culture medium (De Vries and Stouthamer, 1969).

Three types of responses are observed during the switch from anaerobiosis to aerobic conditions:

Aerobic growth without the accumulation of H_2O_2: a strain of *B. bifidum* which is relatively aerotolerant, forms small quantities of H_2O_2 by NADH oxidation. The

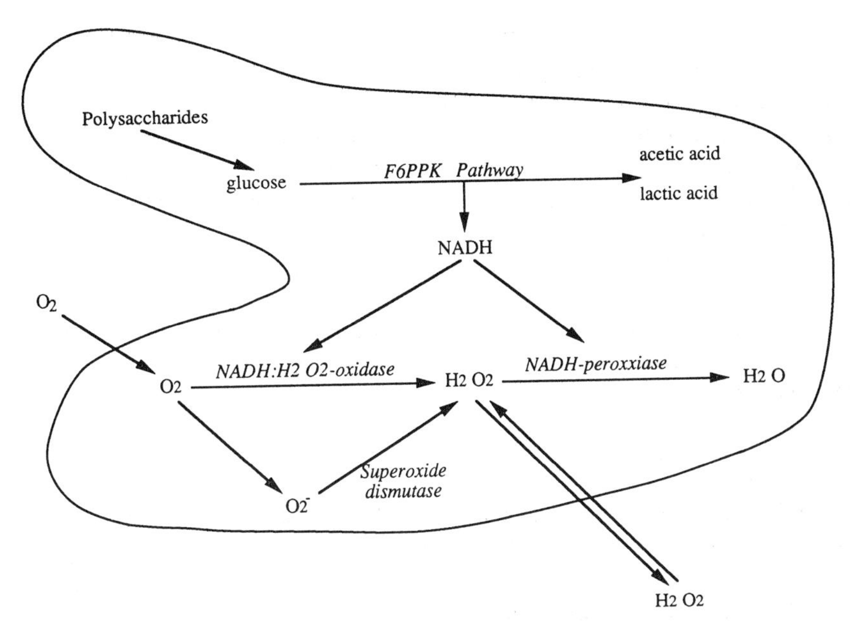

Figure 1 Oxygen dissimilation in *Bifidobacterium*.

absence of H_2O_2 seen in liquid aerobic culture devoid of catalase of NADH peroxidase activity can be explained by an unknown peroxidase system which could destroy the H_2O_2.

Limited growth with the accumulation of H_2O_2, the accumulation of hydrogen peroxide is considered to be toxic for the key enzyme in the sugar metabolism of *Bifidobacterium*: fructose-6-phosphate phosphoketolase (F6PPK) (Rasic and Kurmann, 1983).

No growth without the accumulation of H_2O_2: the strains tested require a low redox potential for growth and fermentation.

In the presence of CO_2, the sensitivity to oxygen varies considerably depending on the strain. Amongst the strains able to develop in the presence of oxygen, some remain catalase negative, others become catalase positive and for others still the presence of catalase is linked to the presence of hemin in the medium (Scardovi, 1986).

A study of the absorption of oxygen by five strains of *Bifidobacterium* of human origin has shown that the partial pressure of oxygen falls in the medium during the multiplication of these strains. The endogenous absorption of oxygen is linked to the presence of NADH oxidase. It takes place even in the absence of glucose and appears to depend directly on the quantity of polysaccharides accumulated in the cells. Furthermore, all strains accumulated hydrogen peroxide which is subsequently reduced by NADH peroxidase, but the activity of this enzyme varied depending on the strain investigated. The strains most sensitive to oxygen had low NADH peroxidase activity, resulting in an accumulation of toxic hydrogen peroxide. Another possibility would be the prevention of multiplication by the presence of active oxygen such as superoxide. These conclusions are summarized in Fig. 1 (Ishibashi, 1989).

The mutants of some strains identified at the time as *B. bifidum* characterized by the loss of the strictly aerobic character, have been isolated (Mayer and Moser, 1950; Norris et al., 1950) but these early studies should be repeated in view of the difficulty in identifying species of *Bifidobacterium* at the time these studies were performed.

B. Temperature and pH

The optimum temperature for the development of the human species is between 36 and 38°C. In contrast, that for the animal species is slightly higher, about 41 to 43°C and may even reach 46.5°C. There is no growth below 20°C and the bacteria of this type have no thermoresistance above 46°C: *B. bifidum* dies at 60°C (Rasic and Kurmann, 1983).

The initial optimum growth pH is between 6.5 and 7.0. No growth can occur below 5.0 or above 8.0 (Scardovi, 1986).

IV. METABOLISM

A. Sugar Metabolism

In the genus *Bifidobacterium*, hexoses are degraded exclusively and specifically by the fructose-6-phosphate pathway described by Scardovi and Trovatelli (1965). Aldolase and glucose-6-phosphate dehydrogenase are absent, whereas fructose-6-phosphate phosphoketolase (F6PPK) is found (De Vries and Stouthamer, 1967) (Fig. 2).

B. Metabolites

The fermentation of two moles of glucose leads globally to three moles of acetate and two moles of lactate. In reality, pyruvic acid can be broken down along two pathways: The first is the reduction of the pyruvate to form L(+) lactate by L(+) dehydrogenase (E.C. 1.1.1.27), an enzyme the activity of which is controlled by fructose-1,6-diphosphate.

The second pathway involves the splitting of the pyruvate by phosphoroclastic enzyme to form formic acid and acetyl phosphate, a portion of which is

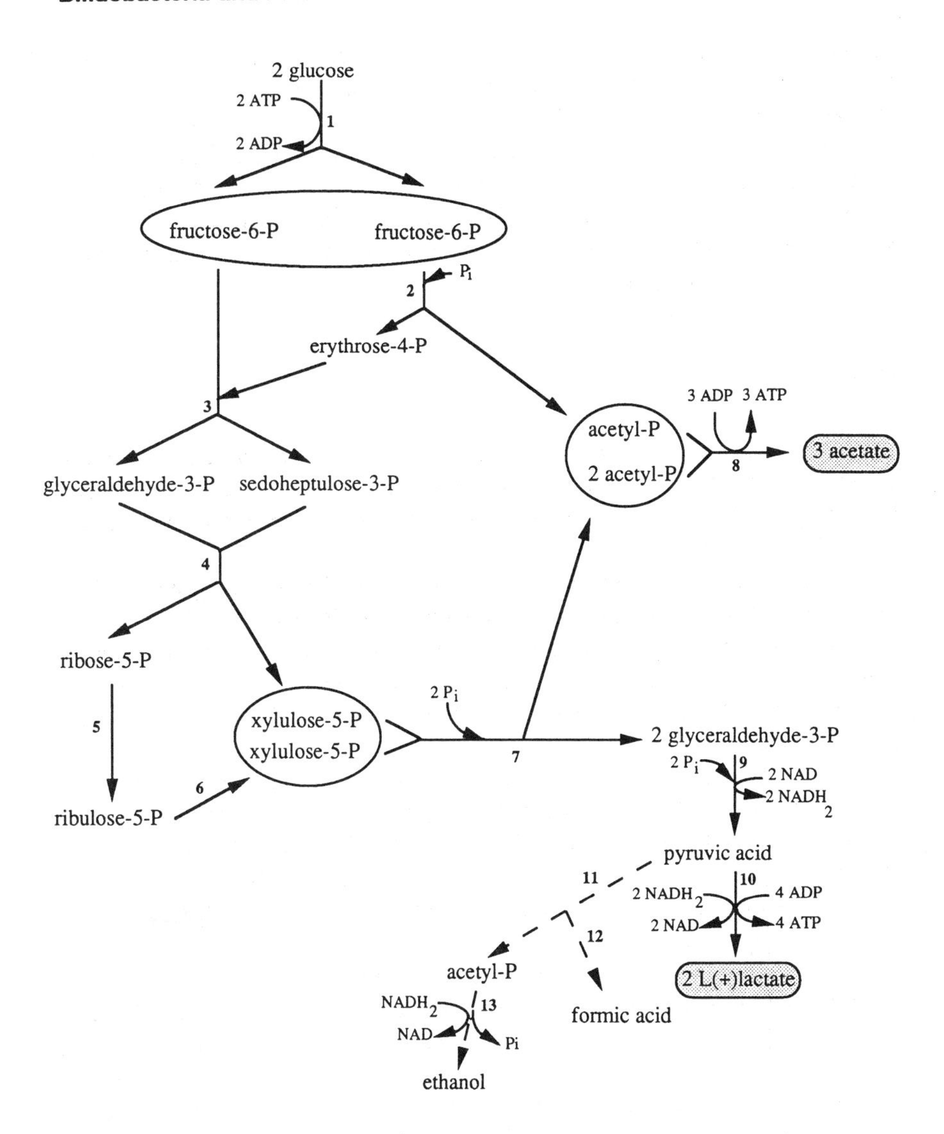

Figure 2 Metabolic pathway of *Bifidobacterium*. 1 = hexokinase and glucose-6-phosphate isomerase; 2 = fructose-6-phosphate phosphocetolase; 3 = transaldolase; 4 = transketolase; 5 = ribose-5-phosphate isomerase; 6 = ribulose-5-phosphate epimerase; 7 = xylulose-5-phosphate phosphocetolase; 8 = acetate kinase; 9 = homofermentative pathway enzymes; 10 = L(+) lactate dehydrogenase; 11 = phosphoroclastic enzyme; 12 = formate dehydrogenase (EC 1.2.1.2); 13 = alcohol dehydrogenase (EC 1.1.1.1).

subsequently reduced to form ethyl alcohol and so regenerate NAD. However, tests carried out to detect phosphoclastic enzyme have been unsuccessful (De Vries and Stouthamer, 1968).

The proportions of the final fermentation products vary considerably from one strain to another and even within the same species (De Vries and Stouthamer, 1968; Lauer and Kandler, 1976). Small quantities of succinic acid are produced by some strains and a small amount of CO_2 may be produced during the degradation of gluconate (Scardovi, 1986).

C. Enzymes

The final fermentation products are formed by the sequential action of transaldolase, transketolase, xylulose-5-phosphate phosphoketolase and enzymes belonging to the Embden-Meyerhoff-Parnas pathway, which act on glyceraldehyde-3-phosphate.

The characteristic enzyme of the sugar metabolism by the genus *Bifidobacterium* is *fructose-6-phosphate phosphoketolase* (F6PPK, EC 4.1.2.22). This enzyme, which is specific to the genus, is absent in the anaerobic bacteria which could be morphologically confused with the bifidobacteria, that is, *Lactobacillus, Arthrobacter, Propionibacterium, Corynebacterium,* and *Actinomycetaceae* (Scardovi and Trovatelli, 1965).

Biavati et al. (1986) have demonstrated using electrophoresis in starch gel (zymogram) followed by comparison of electrophoretic metabolism that are three different types of this enzyme depending on the ecological source of the strain: mammalian, bee or man (Scardovi et al., 1971a). The F6PPKs of *B. globosum* (animal type) and *B. dentium* (human type) have been purified (Sgorbati et al., 1976).

Using the same method, 14 isoenzymes of transaldolase (EC 2.2.1.2) and 29 isoenzymes of 6-phosphogluconate dehydrogenase (6 PGD) (EC 1.1.1.44) have been identified (Scardovi et al., 1976a). Transaldolase is an apparently essential enzyme and characteristic of the fructose-6-phosphate shunt, but 6PGD is apparently nonfunctional in the bifidobacteria, at least in cells cultured on glucose which are generally deficient in detectable glucose-6-phosphate dehydrogenase (Scardovi and Trovatelli, 1965).

Some research has been carried out on other less characteristic enzymes:

1. Miura et al. (1979) determined the following activities by HPLC:
   arylsulfatase (R-sulfate + $H_2O \rightarrow$ R–OH + sulfate)
   β-glucuronidase (R-glucuronide + $H_2O \rightarrow$ R–OH + glucuronic acid) and
   β-glucosidase (hydrolysis of the aryl- or alkyl-β-glycosides).

Galactokinase (EC 2.7.1.6) (galactose + ATP → galactose-1-phosphate + ADP) of *B. bifidum* purified and characterized after growth on galactose (Lee et al., 1980).

2. Tochikura et al. (1986) purified β-D-galactosidase from *B. longum* (lactose + H_2O → galactose + glucose)
3. Desjardins and Roy (1990) used API ZYM systems to determine the 22 strains of human origin which they tested which possess α- and β-galactosidases and α-glucosidase activities. In contrast, β-glucosidase has not been detected in either *B. bifidum* or *B. longum*. This method can however be used in a preliminary study. These studies were confirmed in the same year by Chevalier et al. (1990). β-glucosidase, β-glucuronidase, and *N*-acetyl-glucosaminidase activities have also been demonstrated.

D. The Vitamins Produced

Deguchi et al. (1985) were interested in the synthesis of six vitamins by *Bifidobacterium* of human origin: thiamine (B1), riboflavine (B2), pyridoxine (B6), folic acid (B9), cyanocobalamine (B12), and nicotinic acid (PP). Five of these vitamins (with the exception of riboflavin) are synthesized by most of the strains examined and a large proportion of each (B6, B9, and B12) is excreted. These authors also note that with regard to thiamine, folic acid, and nicotinic acid, *B. bifidum* and *B. infantis* are good producers whereas *B. breve* and *B. longum* release small quantities and *B. adolescentis* do not synthesize any of these vitamins.

The production of vitamins B2 and B6 by *B. longum* is exceptional. *B. breve* and *B. infantis* are characterized by a high level of production of vitamins PP and H respectively. The results are shown in Table 2.

Table 2 Vitamin Production by *Bifidobacterium*

	B. breve	*B. infantis*	*B. longum*	*B. bifidum*	*B. adolescentis*
Thiamin (B1)	+	+++	+	+++	+
Riboflavin (B2)	+	+	+++	++	+
Pyridoxine (B6)	++	++	+++	+	++
Folic acid (B9)	+	+++	+	++	+
Cobalamin (B12)	+	++	+++	+	+
Ascorbic acid (C)	++	++	+++	++	+
Nicotinic acid (PP)	+++	+++	+	+++	+
Biotin (H)	++	+++	++	++	++

E. Nutrient Requirements

1. Nitrogenous Matter

Most strains of *Bifidobacterium* are able to use ammonium salts as their only source of nitrogen (Hassinen et al., 1951). However, *B. suis, B. magnum, B. choerinum*, and *B. cuniculi* develop only in the presence of organic nitrogen. In vitro and in the absence of any organic source of nitrogen, the bifidobacteria may synthesize large quantities of amino acids. *B. bifidum*, for example, produces alanine, valine, and aspartic acid and up to 150 mg/L of threonine (Matteuzi et al., 1978). According to Hatanaka et al. (1987a,b), the glutamine synthetase and glutamate dehydrogenase of the *Bifidobacterium* may be involved in the assimilation of nitrogenous compounds by these microorganisms.

2. Trace Elements

B. bifidum grows only in the presence of magnesium, manganese and above all iron. Iron may be assimilated by *B. bifidum* in both oxidation forms depending on the acidity of the medium (Bezkorovainy et al., 1986; Bezkorovainy and Topouzian, 1981a, 1983; Ueda et al., 1983).

Fe^{2+} ferrous iron is used at pH 5. Transport depends on a membrane ATPase (and its incorporation may be competitively inhibited by zinc (Bezkorovainy et al., 1986).

Fe^{3+} ferric iron is used only at neutral pH. Through the intermediary of ferro-enzymes, iron is involved in the production of acetic acid by *B. bifidum*.

3. Vitamins

It is impossible to draw up any rule for the genus *Bifidobacterium* with regard to vitamin requirements. Strains of human origin seem to need thiamine (B1), pyridoxine (B6), folic acid (B9), cyanocobalamine (B12), and nicotinic acid (PP) for their growth (Teraguchi et al., 1984; Deguchi et al., 1985).

4. Growth Factors

Poch and Bezkorovainy (1988) supplemented an entirely synthetic minimum base medium with growth factors in order to identify those essential to the development of the various species of *Bifidobacterium*. Only *B. adolescentis* and *B. longum* were able to develop in the unsupplemented medium. All the other species required the presence of growth factors of various types.

a. Bifidigenic Factors. In 1953, Györgi (1953) discovered a strain of *B. bifidum* (then known as *L. bifidus*) which was to develop only in the presence of human milk and more specifically in the presence of derivatives of *N*-acetyl-glucosamine (Gauhe et al., 1954; Montreuil, 1957; Raynaud, 1959) and showed soon afterwards that the strain *B. bifidum* Tissier required protein factors and not N-acetylated sugars for its development.

In fact, the species *B. bifidum* can be divided into two variants: the "A" variant or *B. bifidum,* which Tissier found in adult human beings and the "B" variant or *B. bifidum* var. *pennsylvanicus* which György isolated from infants. These observations suggest that the various strains of the same species *B. bifidum* have differing nutritive requirements. *B. bifidum* var. appears to be insensitive to *N*-acetylglucosamine derivatives and to require protein factors in the same way as *B. longum* and *B. infantis,* whereas the "B" variant of *B. bifidum* requires the sugar factors from human milk in varying quantities depending on the strains (Neut et al., 1981; Romond et al., 1980; Beerens et al., 1980).

Most species of the genus *Bifidobacterium* are unable to develop in a totally synthetic medium and require complex biological substances such as bovine casein digestate, lactoserum of bovine milk, porcine gastric mucin or yeast extract (Poch and Bezkorovainy, 1988; Petschow and Talbott, 1990).

These growth factors required for the development of the *Bifidobacterium* are known as bifidigenic factors. We can now distinguish three main groups of bifidigenic factors which differ depending on the species with which we are concerned (Modler et al., 1990): the BB factors (BF1, BF2, and glycoproteins) and the BI and BL factors (Table 3).

BB Factors. The factors BBa and BBb are characterized as the elements in human milk which do not lose their stimulant activity for *B. bifidum* var. a and *B. bifidum* var. b respectively after heating or irradiation. The BBa factors are found mainly in yeast extracts, liver extracts, lyophilized milks, bovine casein hydrolysate and porcine mucin (Raynaud, 1959; Nichols et al., 1974; Bezkorovainy et al., 1979) whereas colostrums, human milk and rat milk (György et al., 1954a,b), human casein hydrolysates (Nichols et al., 1974) and porcine mucin (Raynaud, 1959) contain BBb factors (Table 3).

Three groups of natural BB factors can be distinguished.

1. *György's bifidus factor I or BF1.* This is factor BF1 found in milk and colostrum and in the form of gynolactose which is active particularly on variant B. It would seem that the presence of an *N*-acetylglucosamine structure in the oligosaccharide structure is essential but not sufficient to the expression of bifidigenic activity (Seka Assy, 1982). In addition, *B. bifidum* var. *pennsylvanicus* has *N*-acetyl-D-glucosaminidase activity which is considerably greater than that found for other bifidobacteria (Desjardins and Roy, 1990).

Native human casein (Seka Assy, 1982) or its trypsin hydrolysate (Bezkorovainy et al., 1979) consisting of glycoproteins may be effective versus *B. bifidum* var. b. The trypsin or chymotrypsin hydrolysis of native human K-casein gives rise to fractions containing 60–70% of sugars such as galactose, glucosamine and galactosamine, which are themselves active.

The mucins (glycoproteins of mucus) are produced and secreted by the mucus cells of the salivary glands, the esophagus, the stomach, the small intestine and

Table 3 Characteristics of the Main Bifidigen Factors

Bifidigen factor		Species concerned	Source	Resistance			Active structure
				Heat	Ray.	Lyoph.	
BB	BF1	*B. bifidum* var. b	Milk and colostrum Human casein hydrolysate Mucins	+	+		*N*-acetylglucosamine glycoproteins
	BF2	*B. bifidum* var. a	Casein hydrolysate	+	+		Nonglycosyled peptides
	Glycoproteins	*B. bifidum* var. a *B. bifidum* var. b	Human milk and colostrum				Glucidic part
BI		*B. infantis*	Plant extracts Liver extracts Milk	±	±	–	Proteic part
BL		*B. longum*	Plant extracts Liver extracts Milk	–	–		Proteic different from BB factors

colon. Their molecular weight exceeds one million daltons (Allen, 1984). The mucins, which are the major constituents of mucus (Forstner et al., 1984) consist of 70–80% sugar (Allen, 1981). The oligosaccharide chains contain between 2 and 20 monosaccharide residues which may be the following: galactose, fucose, *N*-acetylgalactosamine, *N*-acetylglucosamine, and sialic acid.

These oligosaccharide chains are linked to peptide segments accounting for 20% of the weight of the molecule and consisting of more than 70% of proline, serine and threonine.

Porcine gastrointestinal mucins and the meconium are an abundant source of BB factors. The activity of the meconium in vitro is 1.2 to 2 times greater than that of human milk (György, 1953). Mild hydrolysis of the mucins give rise to oligosaccharides similar to those in human milk (Kuhn et al., 1953; Tomarelli et al., 1954).

2. *BF2 factors.* Their nature has been described essentially by Raynaud (1959) from a strain of *B. bifidum* var. a. They appear to consist of nonglycosylated peptides obtained by the action of a protease on casein.

3. *Glycoproteins.* The glycoproteins isolated form human colostrum and milk lactoserum appear to be effective versus both variants and this type of activity appears to be related to their sugar fraction (Seka Assy, 1982).

BI and BL Factors. The BI factor which stimulates the growth of *B. infantis* is destroyed by lyophilization whereas the BL factor, which activates the growth of *B. longum*, is sensitive to heating and irradiation. These BI and BL factors are abundant in many plant extracts as well as liver and milk extracts. The BI factors from human milk are of two types: thermo- and radiolabile BI and thermo- and radiostable BI. These factors are proteins, as are the BL factors of human milk (Beerens et al., 1980).

Active constituents of the bifidigenic factors. The factors with general activity are hydrolysates of bovine casein and yeast extracts rather than human milk lactoserum. The other growth factors: human or bovine milk lactosera, porcine gastric mucin and bovine albumin serum digestate are active with regard to certain species only (Poch and Bezkorovainy, 1988).

In fact, the disulfide/sulfhydryl residues of K casein are important biologically active compounds responsible for this phenomenon in *B. bifidum* and *B. longum*. The growth-promoting activity resides in the p-K. casein portion and not in the carbohydrate portion after trypsin digestion. It appears that the combination of disulfide/sulfhydryl residues with something else is the basis of the microbial growth-promoting activity in hydrolysates of casein, porcine gastric mucin, and yeast extract (Poch and Bezkorovainy, 1991).

Role of the bifidigenic factors and conclusion. In vivo, the administration of a dairy-based food supplemented with BF1 or BF2 factor to infants restores the bifidum flora partially (Levesque et al., 1959). These early studies should, however, be repeated on the basis of recent understanding of the taxonomy of *Bifidobacterium* and the biochemistry of the bifidigenic factors, which appear to be highly complex. The correlation between a given factor and a given species appears to be an important aspect of the study of the bifidigenic factors. It would be important to explore the specificity of these factors with regard to the species of *Bifidobacterium* which colonize the intestine, since the degree of difference between the collection strains and strains encountered in nature is doubtless considerable. Furthermore, the studies conducted were carried out in vitro or in vivo on axenic or monoxenic animals and these require extrapolation to man. However, what is the influence of the "bifidigenic" factors on the other bacterial genera of the intestinal flora and particularly of the human flora?

b. Lactoferrin. Lactoferrin and its three metal complexes (Fe, Cu, Zn) have a promoting effect on the growth of eight species of *Bifidobacterium*, five of human origin and three of animal origin, at the beginning of the logarithmic growth phase. Furthermore, these lactoferrin-metal complexes demonstrate an antibacterial activity versus *E. coli* and *Staphylococcus aureus* (Shimamura, 1989).

c. Lactulose and lactitol. Lactulose (4-O-β-D-galactopyranosyl-D-fructose) is not metabolized by human or animal species. Lactulose, which is not detected in raw milk, is present in dairy products subjected to heat treatment. Manciaux (1958) reports that Petuely, in the 1930s, isolated lactulose from human milk. In vivo, lactulose can increase the development of *B. bifidum*. However, this factor is not active in vitro and is not present in the free state in mother's milk. According to Petuely (1956), its action is due to the fact that it resists better than lactose to degradation by the lactases in the digestive tract and can therefore be used massively by the bifidobacteria. Lactulose is not used specifically by the bifidobacteria and may be metabolized by other intestinal bacteria, as must be the case for the bifidigenic factors (Yazawa et al., 1978). Lactitol is considered to be a bifidigenic factor with a less marked effect (Mitsuoka et al., 1987).

d. Oligoholosides and polyholosides

Raffinose, stachyose, and inulin (polyfructose) with molecular weights of less than 4500 are used only by *B. infantis* and not by other intestinal bacteria such as *E. coli*, *L. acidophilus*, and *S. faecalis*.

Oligosaccharides higher than the trisaccharide of inulin and the tri- to pentasaccharides of dextran are also metabolized specifically by *B. infantis*.

In contrast, the oligosaccharides of amylose and cellulose are not specific to *B. infantis* and *B. breve* (Yazawa et al., 1978).

e. Fructooligosaccharides (FOS). These polymers of fructose with a degree of polymerization of between 2 and 35 have a stimulant effect on the growth of bifidobacteria (Hidaka et al., 1986). They are metabolized by bifidobacteria and also by other types of bacteria and are not degraded by human digestive enzymes nor generally by undesirable microorganisms within the digestive tract. The most important source of FOS is the Jerusalem artichoke tuber (Mitsuoka et al., 1987). Today it is easier to prepare a similar substance by enzymatic route from sugar (Hidaka et al., 1986, 1988) than to purify the FOSs from natural sources (Yamasaki and Dilawri, 1990).

V. RESISTANCE TO ANTIBIOTICS

Knowledge of the antibiotics to which the bifidobacteria are resistant is of dual interest:

1. It offers the possibility of maintaining the bifidobacteria in the digestive tract without aggression, particularly during antibiotic treatment.
2. It makes it possible to incorporate antibiotics as selective agents in culture media for the isolation of bifidobacteria from complex flora derived from medical or dietary samples.

The sensitivity of the bifidobacteria has been the subject of little research and the work done before the publication of an international standard are difficult to compare because experimental conditions vary. However, we can accept the following points: Most bifidobacteria are resistant to numerous antibiotics and notably to nalidixic acid, gentamicin, kanamycin, metronidazole, neomycin, polymyxin B, and streptomycin, but the sensitivity of the species varies from 10 to 500 or more μg antibiotic/mL (Lavergne et al., 1959) in *B. bifidum* (Miller and Finegold, 1967; Matteuzi et al., 1983). In contrast, ampicillin, bacitracin, chloramphenicol, clindamycin, erythromycin, lincomycin, nitrofurantoin, oleandomycin, penicillin G, and vancomycin strongly inhibit most species (Scardovi, 1986). Sensitivity to tetracycline varies from one species to another and even from one strain to another (Mateuzzi, unpublished; Lavergne et al., 1959).

VI. CULTURE MEDIA AND CULTURE PARAMETERS

Three types of medium can be distinguished which have been designed for the isolation, culture and characterization of bifidobacteria: complex media, semi-synthetic media, and synthetic media.

A. Complex Media

These richly varied media are prepared from liver or meat extracts, a wide range of peptones, yeast extract, tomato juice, horse blood, or human milk and permit the growth of as many strains of *Bifidobacterium* as possible. In addition, they are supplemented with substances with a low redox potential: cysteine, cystine, ascorbic acid, or sodium sulfite (Rasic and Kurmann, 1983).

A wide range of complex culture media have been proposed in recent years. We will consider the following:

BL agar medium described by Ochi et al. (1964) and then by Mitsuoka et al. (1965, 1982) and finally slightly modified by Teraguchi et al. (1978) is considered to be the optimum culture medium for the detection of bifidobacteria.

Scardovi's tryptone phytone yeast medium (TPY) (1981, 1986) can be used for the culture and isolation of bifidobacteria but also of other lactic bacteria from all habitats.

Mention should also be made of the YN-6 medium (Resnick et al., 1981a) and YN-17 medium (Mara et al., 1983). These media are not very efficient since YN-6 medium inhibits some species of *Bifidobacterium* but allows other genera to develop (Carillo et al., 1985) whereas YN-17 medium inhibits some of the bifidum population (Munoa et al., 1985).

B. Semisynthetic Culture Media

Complex constituents of known composition are included in these media. We will note particularly the following:

Tomarelli's medium (1949) for the culture of *B. bifidum*

Norris's medium (Norris et al., 1950), which is a modification of the Tomarelli medium

György's medium (1954b), which is also a modified Tomarellia medium

C. Entirely Synthetic Culture Media

All the constituents of these media are chemically defined.

Petuely (1956) was the first to propose a synthetic culture medium.

Hassinen medium (Hassinen et al., 1951).

Gyllenberg modified the Petuely medium (Gyllenberg and Niemelä, 1959).

Tanaka and Mutai (1980).

Ueda et al. (1983) developed a synthetic medium for the culture of the ES5 strain of *B. bifidum*.

Poch and Bezkorovainy (1988).

D. Selective Culture Media

The media listed above are efficient for the maintenance of pure strains but are less effective for isolating them from complex flora since they often permit the growth of other genera.

Since the physiological requirements of bifidobacteria are extremely varied, it is difficult to define a selective medium appropriate for all species (Scardovi,

1986). This is why, since the recent enthusiasm for incorporating bifidobacteria in fermented dairy products, several selective media have been proposed in order to differentiate between *Bifidobacterium* and other lactic bacteria and to isolate bifidobacteria from the intestinal flora.

Initially, ascorbic acid and sodium azide were used as selective substances.

Beck (1967) isolated *Bifidobacterium* on a medium containing added bifidigenic growth factors (*N*-acetyl-D-glucosamine). The relatively low pH of this medium (5.8) and anaerobiotic culture makes it possible to eliminate most enterobacteria.

Chang et al. (1983) modified the MRS agar medium, the new medium containing cysteine, azide and China ink in order to isolate numerous species of *Bifidobacterium*.

Matteuzzi et al. (1983) suggested the addition of 80 µg of kanamycin/mL to the medium. However, intraspecific variations of resistance are so great that the isolation of unknown strains with a medium of this type would be unreliable.

Ushiima et al. (1985) were able to selectively isolate *B. adolescentis* from a complex gastric flora by using selective agents: polymyxin, propionate, and linoleate.

Sonoike et al. (1986) based themselves on the fact that bifidobacteria are able to metabolize carbohydrates such as fructo- and galactosyl-oligosaccharides. Twenty-two species of *Bifidobacterium* are able to develop on a medium containing *trans*-galactosylated oligosaccharides as a carbon source.

Munoa and Pares (1988) attempted to quantify *Bifidobacterium* from water on a new selective medium: *Bifidobacterium* Iodoacetate Medium 25 (BIM-25). This medium is a reinforced clostridial medium (RCM) containing added antibiotics (nalidixic acid, polymyxin B, kanamycin), iodoacetic acid and 2,3,5-triphenyltetrazolium chloride (TTC). Iodoacetate, which inhibits glyceraldehyde-3-phosphate dehydrogenase, considerably reduces the growth of nonbifidum colonies. TTC makes it possible to differentiate between *Bifidobacterium* and other species since the bifidobacteria develop in large white colonies.

Poch et al. (1988) supplemented a synthetic base medium with various substrates in order to identify the growth factors necessary for each species of *Bifidobacterium*. The base medium was similar to Norris's medium (Poupard et al., 1973).

Mitsuoka used propionate as one of the selective agents added to BL agar medium for the selective counting of intestinal bifidobacteria. The BS agar medium thus obtained is not entirely satisfactorily for the detection of bifidum in the stools.

Beerens (1990) proposed a selective and elective medium by adding 5 g/L of propionic acid to a base medium and adjusting the pH to 5.0. This medium is not universally accepted.

E. Culture Parameters

The appearance of the colonies of *Bifidobacterium* cultured on agar medium under anaerobic conditions may vary in function of the medium and the species used, but also within a given species. In general, the colonies formed are round, dull or glossy and of variable diameter, but Scardovi (1986) and Boventer (1938) distinguished two differing types of colony for *B. bifidum*. Some colonies were smooth, convex, white, and shiny, whereas other colonies were rough, with uneven edges and map.

VII. COMPOSITION OF THE WALL

The main constituent of the wall of the bacteria of the genus *Bifidobacterium* (Gram +) is mucopeptide, peptidoglycan or murein. This is a macromolecule consisting of linear polysaccharide chains which are linked with each other by tetrapeptide bridges associated with peptides. The polysaccharide chain consists of alternating *N*-acetylglucosamine (NAG) and *N*-acetylmuramic acid (NAM). The tetrapeptides are linked to the NAM residues and to each other through the intermediary of peptides (Fig. 3). The amino acids constituting the various

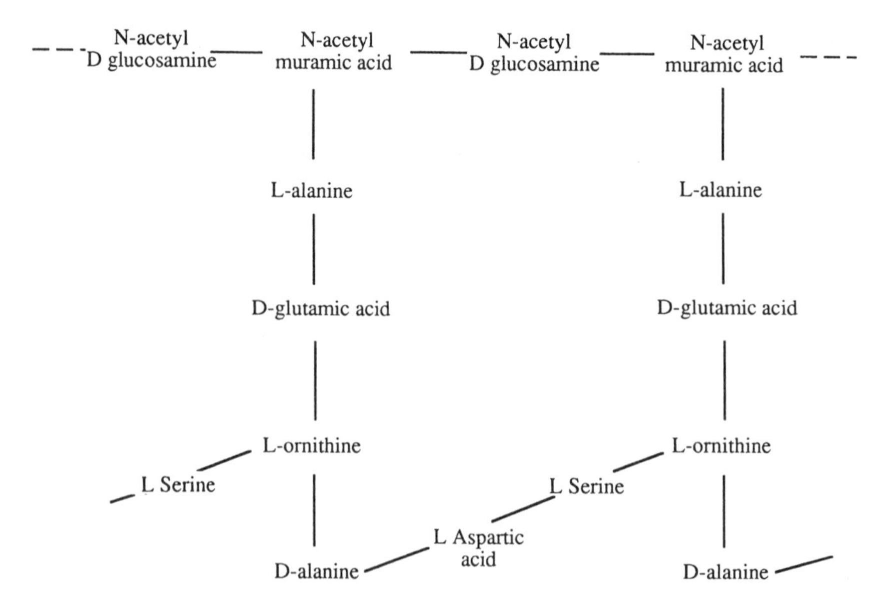

Figure 3 Peptidoglycane structure of *Bifidobacterium bifidum*. (From Ballongue, 1989.)

peptides are alanine, glutamic acid and ornithine or lysine associated with one or two of the following amino acids: glycine, serine, aspartic acid, or threonine (Cummins et al., 1957; Veerkamp et al., 1965).

The amino acids may be the same or may differ from one strain to another, but even if they are the same their sequence within the tetrapeptide and their types of cross linkage may vary (Kandler and Lauer, 1974), that is, this may consist of a simple amino acid, a dipeptide or even a tripeptide. *B. longum*, for example, possesses an ornithine-type tetrapeptide and the link peptide is L Ser–L Ala–L Thr–L Ala (Rasic and Kurmann, 1983). This macropeptide is covalently linked to other macromolecules such as:

Polyosides: glucose, galactose, and rhamnose composing the polysaccharide portion of the wall
Teichoic acids, which are polymers of glycerol phosphate

These teichoic acids are attached to the NAG-NAM skeleton of the peptidoglycan.

VIII. CHARACTERISTICS OF THE GENOME

A. DNA Base Composition

The G + C percentage (G + C %) of the bacteria of the genus *Bifidobacterium* is between 57.2 and 64.5% (Sebald et al., 1965; Scardovi, 1986).

B. Plasmids

Of the 24 species of *Bifidobacterium*, only 5 species have plasmids (Sgorbati et al., 1982, 1986):

1. *B. longum* contains 13 model plasmids (1.25–9.5 MDa) (Sgorbati et al., 1983, 1986).
2. *B. globosum* contains a plasmid belonging to each of the three molecular weight categories (13.5, 24.5, and 46 MDa).
3. *B. asteroides* has 14 types of plasmids (1.2–22 MDa) which are structurally very varied (Sgorbati, unpublished).
4. *B. indicum*: 60% of the strains isolated contain a single 22 MDa plasmid.
5. *B. breve*: Iwata et al. (1989) have demonstrated plasmids in 40% of the strains of this species, even though Sgorbati et al. (1982) had not found any plasmid in *B. breve*.

In some strains, these plasmids may be temperate phages, but this appears to be unusual (Sgorbati et al., 1983). Very curiously, *B. infantis*, which constitutes a continuum with *B. longum*, as we shall see below (Neut et al., 1981; Sgorbati and London, 1982) has no plasmid. No phenotypic characteristic has so far correlated with the presence of plasmids (Sgorbati et al., 1982).

IX. *BIFIDOBACTERIUM* ECOLOGY

Of the 24 species of *Bifidobacterium* so far recognized, 9 are isolated essentially from humans: *B. bifidum, B. longum, B. infantis, B. breve, B. adolescentis, B. angulatum, B. catenulatum, B. psuedocatenulatum*, and *B. dentium*.

A. Implantation in the Neonate

It is generally admitted that until the time of birth the fetus is surrounded by a completely sterile environment. After birth, the digestive tract is rapidly colonized by bacteria (Moro, 1900b; Mitsuoka et al., 1974; Bezirtzoglou, 1985). Forty-eight hours after birth, the colon contains 10^9 to 10^{10} bacteria per gram of stools consisting mainly of enterobacteria, staphylococci, and streptococci (Mitsuoka et al., 1974; Bezirtzoglou, 1985; Moreau et al., 1986).

The bifidobacteria appear only between day 2 and day 5 (Mitsuoka et al., 1974) and become dominant (10^{10}–10^{11} per gram of stools) barely one week after birth. They reach a level of 99% of the fecal flora, whereas the levels of other bacterial (*E. coli*, lactobacillus, enterococcus) decline sharply by about 1000-fold (Frisel, 1951; Mayer, 1956; Hoffmann, 1966; Neuissen-Verhage et al., 1987). Anaerobes such as *Bacterioides* and *Clostridium* and other putrefying bacteria are enormously reduced and may disappear.

1. *Origin of Colonization*

This rapid invasion of the sterile digestive tract at birth raises the question of the origin of the bacteria, and more particularly of the anaerobes such as *Bifidobacterium*, which survive only precariously in the atmosphere. Do the bacteria invade from the digestive tract from the mouth or from the rectum? Several studies tend to disprove the hypothesis that the digestive tract is colonized by the rectal route (Naujoks, 1921; Lauter, 1921) and tend to demonstrate that, on the contrary, bifidobacteria enter the body of the neonate by oral route:

Kleinschmidt (1925, 1949) detected *B. bifidum* in the upper segment of the digestive tract of a child operated due to the absence of an anal perforation.

Boventer (1938, 1949) clearly showed that the implantation of *Bifidobacterium* in the digestive tract occurs by descending route, since the rectum remains sterile until colonization is complete.

Mutai and Tanaka (1987) isolated and observed in the mouth of 23 neonates *Bifidobacterium, Propionibacterium, Bacteroides, Peptostreptococcus, Fusobacterium, Enterococcus, Lactobacillus*, and *Enterobacteriaceae* 10 min after the birth via the genital tract, whereas following cesarian, only *Propionibacterium* and *Enterococcus* were isolated in 8 out of 9 neonates investigated.

Another weighty argument in favor of the hypothesis of colonization by the vaginal or fecal flora of the mother is the observation by many authors that invasion of the digestive tract of the neonate occurs much more rapidly after birth by the genital tract than after birth by cesarian (Walch, 1956). Mitsuoka et al. (1974) isolated bifidobacteria in 41% of births by genital route, whereas Bezirtzoglou (1985) found them in 21% of infants aged 4 days and born by cesarian (41% after 15 days).

2. *Factors Influencing Colonization*

In addition to the method of delivery, which we have just seen directly affects the speed of invasion by bifidobacteria, several other factors also influence colonization.

a. Prematurity. This is a cause of difficult implantation of *Bifidobacterium* due to the lack of receptors and/or endogenous substrates, whereas enterobacteria and *Bacteroides* readily to colonize the colon (Frisel, 1951; Mayer, 1956; Hoffmann, 1966; Stark and Lee, 1982; Stevenson et al., 1985).

b. Method of Feeding. Tests for bifidobacteria in mother's milk have always been negative (Bezirtzoglou, 1985) with the exception of the findings of Mayer and Moser (1950) who isolated *B. bifidum* from the colostrum and milk of a woman before breast-feeding commenced.

Influence of the type of feeding on the composition of the intestinal flora. The effects of breast-feeding and bottle-feeding have been compared in very numerous studies. At the beginning of the century, Tissier wrote that the flora of a neonate raised on the breast consisted entirely of *Bifidobacterium*, whereas lactobacilli were predominant in the flora of bottle-fed infants. From subsequent research, it has emerged that there is no difference in the qualitative distribution of species between these two types of feeding. The difference lies in the quantitative level in the proportion of *Bifidobacterium* and other species, with clearly higher levels of *Bifidobacterium* for children breast-fed who show high levels of *Bifidobacterium* which are generally higher than those of enterobacteria and *Bacteroides* within the first week of life (see Table 4 and Figure 4) (Bourrillon et al., 1980; Neut et al., 1980a,b, 1984; Braun, 1981; Mitsuoka, 1982; Yuhara et al., 1983; Benno et al., 1984; Moreau et al., 1986).

The stools of a breast-fed child are also characterized by a granular appearance, slightly vinegary smell and marked acidity (pH 4.8 to 5.2) (Neimann et al., 1965) which is probably due to the abundance of *Bifidobacterium* and therefore to considerable acetate production (Neimann et al., 1965; Mata and Urrutia, 1971; Mitsuoka et al., 1974; Bezkorovainy and Topousian, 1981b). In contrast, the stools of children fed artificially are more similar to those of adults (pH 6.4 to 7.0) (Romond et al., 1980) which indicates the present of putrefying organisms.

Thus, in artificial feeding the fundamental difference lies in the maintenance of high levels of optional aerobic species (*E. coli* and streptococci) which initially colonized the digestive tract and the development of anaerobes (*Bacteroides,*

Table 4 Comparison Between Fecal Flora of Breast-Fed and Bottle-Fed Infants

Bacteria	Breast-fed infants[a]	Bottle-fed infants[a]
Enterobacteriaceae	8.6	9.5
Streptococcus	7.9	9.8
Staphylococcus	5.8	5.5
Lactobacillus	7.0	5.9
Bifidobacterium	10.7	10.0
Eubacteria	3.1	7.3
Bacteroidaceae	6.1	9.9
Peptococcaceae	2.4	7.9
Cl. perfringens	1.0	6.4
Clostridium	1.3	0.9
Veillonellae	5.8	5.9

[a]log cfu.

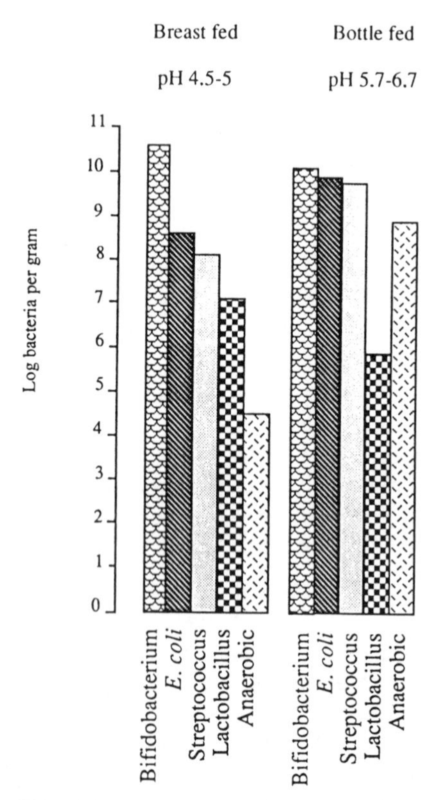

Figure 4 Comparison of fecal flora between breast-fed and bottle-fed infants.

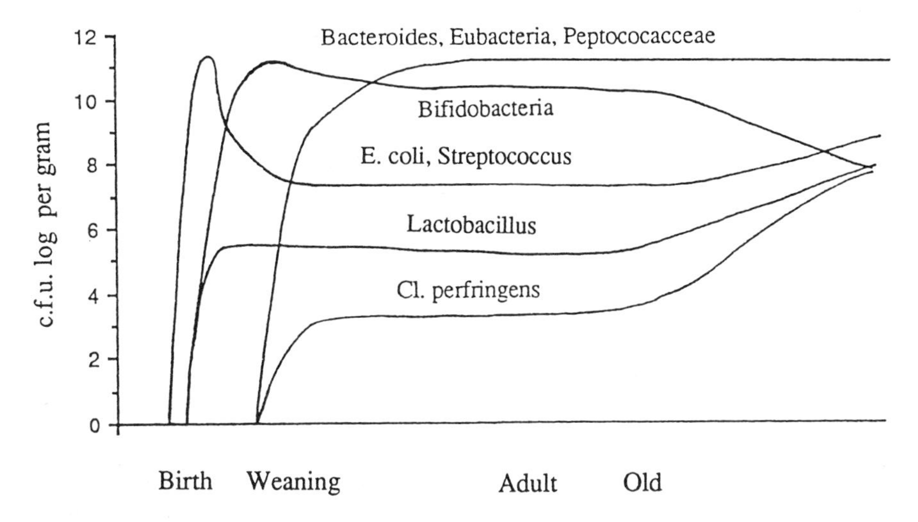

Figure 5 Changes of intestinal flora from birth to old age (from Mitsuoka, 1984).

Clostridium, Eubacteria, Peptostreptococcaceae, *Clostridia*, Enterobacteriaceae, *Streptococcus*) (Mitsuoka and Kaneuchi, 1977; Mitsuoka, 1989). *Bifidobacterium* appears fairly late and is found in lower proportions in the stools (Neut et al., 1981).

Some authors (Rose, 1984; Lundequist et al., 1985), on the contrary, note that breast-feeding does not increase the level of *Bifidobacterium* in the first few days after breast-feeding, the predominant population consisting of enterobacteria and not of strictly anaerobic species. Benno et al. (1984) note that in both types of feeding, *Bifidobacterium* constitutes the predominant genus (Fig. 5).

At weaning, a sudden change occurs in the fecal flora following the first bottle-fed (Neut et al., 1985) or solid food. In some children, the bifidium flora falls sharply; in others, however, it remains stable. There is a remarkable proliferation of *Bacteroides, Eubacterium, Peptostreptococcus*, and *Clostridium* (Mitsuoka, 1989). In all cases, the ratio (*Bifidobacterium*/"putrefying flora") falls and is reversed. To date, no industrial dairy formula has made it possible to maintain the same equilibrium as that found during breast-feeding.

Influence of the type of feeding on the species of *Bifidobacterium*. The proportions of the various species of *Bifidobacterium* also vary with the type of feeding (Braun, 1981). *B. bifidum* appears to be the dominant species during breastfeeding (Neut et al., 1980b; Beerens et al., 1980; Yuhara et al., 1983).

In contrast, in Italy, Biavati et al. (1984) did not observe any change in the distribution of the species in breast- or bottle-fed infants:

B. bifidum	3%
B. longum	8%
B. infantis	12%
B. breve	11%

The discrepancies between these findings are probably to be explained by the differing techniques used to identify the strains; the French and Japanese workers use carbohydrate fermentation, whereas the Italian authors identify species using DNA-DNA hybridization.

Constituents of human milk affecting the equilibrium of the infant's flora. Humanized milks have an organoleptic composition which is similar to that of human milk, that is, a high concentration of lactose, they contain lactoferrin and lactulose, low concentrations of the proteins used by the putrefaction organisms and a low buffering potential (Bullen et al., 1977), but are still incapable of providing the conditions favorable to the bifidobacteria in breast-fed infants. Human milk provides factors which are essential to the intestinal development of *Bifidobacterium.*

In 1930, Polonowski and Lespagnol (1930) isolated oligosaccharides other than lactose from human milk which they named gynolactose (it consists of low molecular weight molecules) and allolactose [B-D-galactopyranosyl-(1,6)-D-glucopyranose].

Levesque et al. (1960) administered relatively large quantities of *N*-acetyl-glucosamine to infants with low levels of *B. bifidum* in the flora. They then observed the appearance of this species in the stools and its disappearance if the administration was stopped. In children fed on a milk containing added porcine mucin, the pH of the stools fell and the bifidum population rose (Inoue and Nagayama, 1970).

Numerous substrates therefore appear to be involved in maintaining the equilibrium of the infant's intestinal flora. To date, no definitively active molecule has been identified.

c. Endogenous Substrates. Endogenous substrates are substances which exist within the digestive tract without a dietary source. They are produced by the host and may be used by bacteria. Some strictly anaerobic bacteria produce enzymes able to degrade the blood group antigens and mucin oligosaccharides. These bacteria include species of the genus *Bifidobacterium: B. bifidum* and *B. infantis* (Hoskins and Boulding, 1976; Hoskins et al., 1985). They are able to remove the *N*-acetyl-D-galactosamine residues from the blood group A factors and also secrete a-L-fucosidases, sialidases and β-glycosidases (Neutra et al., 1987).

d. Environment. The country, hospital and even the unit within which the delivery takes place influence the rapidity of colonization by *Bifidobacterium.* The species present and the biotypes found also vary with time. Mitsuoka et al.

(1974) isolated mainly *B. infantis* from the stools of Japanese infants but 10 years later *B. breve* was recognized as the dominant species.

Numerous observations suggest that the environment and, in particular, obstetrical and therapeutic customs (increasingly frequent use of antibiotics) play a far from negligible role in the colonization of neonates by bifidobacteria. It would even appear that very strict conditions of hygiene delay the implantation of *Bifidobacterium* (Simhon et al., 1982; Yoshioka et al., 1984; Lundequist et al., 1985).

B. Change From Weaning to Adulthood

1. *Influence of Age*

Many authors observe that the number of *Bifidobacterium* falls significantly in adult stools and particularly in those of the elderly, whereas the numbers of *Bacteroides, Eubacterium, Peptococcus, Clostridium*, Enterobacteriaceae, *Streptococcus*, and *Lactobacillus* increase (Orla-Jensen et al., 1945; Haenel and Muller-Beuthow, 1963; Hoffmann, 1966; Drasar and Hill, 1974; Mitsuoka, 1984). This change is usually due to a reduction in the gastric secretion in this age group. The fall-off in *Bifidobacterium* is accelerated in the elderly where there is an increase in *Clostridium* and optional aero-anaero species. Figure 5 shows the change in the intestinal flora throughout life.

Table 5 Distribution of *Bifidobacterium* Species in Human Colon

Population	Predominating species	Minor species
Breast-fed infants	*B. longum* *B. infantis* *B. breve*	
Bottle-fed infants	*B. adolescentis*	*B. bifidum* biovar b
Children	*B. infantis,* *B. breve,* *B. bifidum* biovar b, *B. longum*	
Adults	*B. adolescentis* biovars a and b, *B. longum*	*B. bifidum* biovar a
Old aged	*B. adolescentis* biovar b, *B. longum*	

The proportions of the various species of the genus *Bifidobacterium* also vary with age and each age group has its characteristic species (see Table 5).

Thus, in children under 7 months of age, Mitsuoka et al. (1974) and Biavati (1980) isolated *B. infantis, B. breve, B. longum* biovar. b, and *B. bifidum* var. b. These species, with the exception of *B. longum* biovar. b, are not present in young children and disappear in the flora of older children and adults. *B. infantis* appears to be specific to neonatal infants (Biavati et al., 1986) and to show a higher incidence in the stools of breastfed infants.

In children and adults, *B. longum* biovar. a, *B. adolescentis* and species with similar fermentation characteristics (*B. catenulatum, B. pseudocatenulatum*) are well represented. In contrast, *B. infantis* and *B. breve* are not present in this age group (Mitsuoka, 1974; Biavati et al., 1984, 1986), *B. adolescentis* is the species characteristic of the adult flora and the proportion of *B. adolescentis* type b increases considerably in the elderly (Mitsuoka et al., 1974, 1984).

2. Influence of Diet

The high degree of variability of the intestinal flora in infants depending on diet contrasts with the apparent stability of the flora in adults, despite differences in diet. It would seem that foodstuffs have little impact on the constitution of the dominant intestinal flora (Haenel, 1970; Stark et al., 1982). However, Mitsuoka (1984) notes a reduction in the level of *Bifidobacterium* after the administration of a Western diet to individuals used to Japanese food.

3. Pathogenicity

B. dentium (formerly *Actinomyces eriksonii*) is isolated from dental caries or abscesses and this species may be confused with four other species, notably *B. adolescentis* (see Section 11 and Table 13).

X. IDENTIFICATION BY PHENOTYPE INVESTIGATION

When the bifidobacteria were discovered by Tissier at the beginning of this century, taxonomy was based entirely on morphological observations. This lack of differentiation criteria explains the numerous debates which preceded the creation of the genus *Bifidobacterium*. Taxonomy subsequently based itself on increasingly numerous phenotype characteristics and today can make use of progress in genotyping.

A. Identification of the Genus *Bifidobacterium*

Until the 1960s, the only identification criteria were phenotype characteristics. We could mention the following:

1. Morphology

Since the branched appearance can also be seen in other bacterial genera (*Arthrobacter, Propionibacterium, Corynebacterium,* and *Actinomyces*), it cannot be considered to be a specific characteristic but only an indicative criterion (Scardovi and Trovatelli, 1965).

2. Culture Conditions

Bifidobacterium develop under anaerobic conditions at 37°C in species of human origin or 42°C and higher for species of animal origin and require an incubation time of 48 h (Scardovi, 1986).

3. Metabolites

The determination by gas chromatography of the organic acids produced at the end of fermentation and notably of the acetic acid/lactic acid ratio of about 3/2 provides an excellent identification criterion for the genus *Bifidobacterium.* In addition, it is important to note that the bifidobacteria produce the L+ isomer of lactic acid.

4. Enzyme Tests

The association of a branched shape with the presence of fructose-6-phosphate phosphoketolase (F6PPK) in a strain indicates that it belongs to the genus *Bifidobacterium.* The detection of F6PPK can be completed by a test for α-galactosidase. The API ZYM system indicates a-galactosidase activity in bifidobacteria (Roy et al., 1989) but not in lactobacilli (Lee et al., 1986). This test can therefore be used as an identification indicator (Chevalier et al., 1990).

5. Study of the Electrophoretic Patterns

All soluble cell protein electrophoretic patterns show a band which migrates over the same distance, with the exception of *B. boum* for which this band is located at a slightly greater distance from the anode (Biavati et al., 1982). The presence of this band therefore appears to provide an appreciable criterion for the identification of the genus.

6. Lipids and Constituents of the Cell Wall Membrane

Bifidobacteria have the following fatty acids: $C_{14:0}$ (myristic acid), $C_{16:0}$ (palmitic acid), $C_{16:1}$ (palmitoleic acid), $C_{18:0}$ (stearic acid), and $C_{18.1}$ (oleic acid). In addition, though the genera *Bifidobacterium* and *Lactobacillus* both contain diphosphatidylglycerol and phosphatidylglycerol, only *Bifidobacterium* possesses polyglycerolphospholipids and their lyso-derivatives, alanylphosphatidylglycerol, and the lyso-derivatives of diphosphatidylglycerol (Exterkate et al., 1971). Analysis of the cell composition in terms of lipids and phospholipids therefore provides a good criterion for distinguishing between the genus *Bifidobacterium* and the *Lactobacillaceae*. It should be noted that the growth temperature and the composition of the culture medium have a marked influence on the distribution of lipids and phospholipids, although the peptoglycan structure of *Bifidobacterium* is closer to that of the *Lactobacillaceae* than the *Actinomycetaceae* (Kandler, 1970; Kandler and Lauer, 1974).

7. Other Identification Criteria

Other tests can be used to identify the genus *Bifidobacterium*, notably (Beerens et al., 1957; Prevot, 1961; Rasic and Kurmann, 1983):

The rapid and complete coagulation of milk without the formation of gas
The fermentation of glucose, lactose, levulose, fructose, and galactose, accompanied by marked acidification
No acid production from rhamnose, sorbose, adonitol, dulcitol, erythritol, or glycerol
The development of these microorganisms in peptone water
Negative catalase
No reduction of nitrates
No indole formation
No liquefaction of gelatin
No fermentation of glycerol
No attack of coagulated proteins

B. Identification of the Species

It is obvious from the multiple taxonomic revisions which have taken place within a few decades to understand how difficult species identification is within the genus *Bifidobacterium*.

1. Sugar Fermentation

This criterion is certainly the one which has been used most frequently to identify and define new species. Until 1957, most researchers classed all bifidobacteria together as a single species: *Bifidobacterium bifidum*. In 1957, Dehnert (1957) was the first to demonstrate the presence of several *Bifidobacterium* biotypes and used 24 sugar fermentation processes to classify the various species into five groups. A few years later, Reuter (1963, 1971) associated serological properties to sugar fermentation to identify new human derived species isolated from the stools of adults and children and their various biotypes.

It was also using fermentation profiles and the ability to grow at 46.5°C that enabled Mitsuoka (1969b) to separate human strains from animal strains (pig, chicken, calf, sheep, rat, mouse, guinea pig, and bee). He proposed two new species: *B. thermophilum* var. a, b, c, and d, *B. pseusolongum* var. a, b, c, and d, and a new variant: *B. longum* subsp. *animalis* a and b. *B. ruminale* (synonym of *B. thermophilum*) and *B. globosum* and then *B. asteroides, B. indicum*, and *B. corneforme* were isolated in the same year (Scardovi et al., 1969; Scardovi and Trovatelli, 1969).

The ability of a strain to ferment certain sugars is certainly the test used first to identify species. Numerous sugars have been tested and the results obtained have been compared with the identification tables produced by Mitsuoka (1982, 1984) and Scardovi (1986) (Table 6). This method presents no major operating problems but does not have several drawbacks: it is lengthy and tedious because to be interpretable a panel of 30 sugars must be studied for 10 days. In addition, the interpretation of the results using identification tables remains controversial, and can at best give an indication of an identification based on the fundamental characteristics which are not open to doubt, for example:

B. longum ferments melezitose, whereas *B. animalis* is unable to ferment this sugar.
B. pseudolongum ferments pentoses and starch, whereas *B. thermophilum* does not ferment pentoses but does ferment starch.
B. breve ferments ribose, mannitol, esculine, and amygdaline but does not ferment arabinose or xylose.
B. infantis does not ferment arabinose, whereas *B. longum* ferments arabinose and melezitose.

Roy et al. (1989) claim to have developed a rapid method for identifying bifidobacteria species based on the fermentation of seven sugars: arabinose, cellobiose, lactose, mannose, melezitose, ribose, and salicin. The utilization of a mixture of these seven sugars is monitored by gas chromatography and should make it possible to identify six to eight typical strains of *Bifidobacterium* in less than 24 h.

2. *Study of the Isoenzymes of F6PPK*

The test using a colorimetric reaction following starch gel electrophoresis for three isoenzymes of F6PPK can give an indication of a species identity (Biavati et al., 1986). These isoenzymes catalyze the same reaction but are distinguished by differing electrophoretic patterns. The migration distances are linked to the ecological origin of the species: human (15 cm), mammalian (10 cm), or bee (Scardovi, 1986) (Table 7). In addition, purified preparations of F6PPK from *B. globusum* (mammalian origin) and *B. dentium* (human origin) demonstrate activities which vary with regard to optimum pH, the identity of the metal inducing maximum activity, heat inactivation, molecular weight, and affinity toward the substrate (Sgorbati et al., 1986).

3. *Study of the Protein Profiles*

A bacterial strain cultured under standard conditions always gives the same protein profiles. The sequence of the amino acids, the molecular weight and the

Table 6 Sugar Fermentation by *Bifidobacterium* sp.

	D-ribose	L-arabinose	Lactate	Cellobiose	Melezitose	Raffinose	Sorbitol	Starch	Gluconate	Xylose	Mannose	Fructose	Galactose	Sucrose	Maltose	Trehalose	Melibiose	Mannitol	Inulin	Salicin
B. bifidum	–	–	+	–	–	–	–	–	–	–	–	+	+	v	–	–	v	–	–	–
B. longum	+	+	+	–	+	+	–	–	–	v	v	+	+	+	+	–	+	–	–	–
B. infantis	+	–	+	–	–	+	–	–	–	v	v	+	+	+	+	–	+	–	v	–
B. breve	+	–	+	v	v	+	v	–	–	–	+	+	+	+	+	v	+	v	v	+
B. adolescentis	+	+	+	+	+	+	v	+	+	+	v	+	+	+	+	v	+	v	v	+
B. angulatum	+	+	+	–	–	+	v	+	–	+	+	+	+	+	+	–	+	–	+	+
B. catenulatum	+	+	+	+	–	+	+	–	v	+	–	+	+	+	+	v	+	v	v	+
B. pseudocatenulatum	+	+	+	v	–	+	v	+	v	+	+	+	+	+	+	v	+	–	–	+
B. dentium	+	+	+	+	+	+	–	+	+	+	+	+	+	+	+	+	+	+	–	+
B. globosum	+	v	+	–	–	+	–	+	–	v	–	+	+	+	+	–	+	–	–	–
B. pseudolongum	+	+	v	v	v	+	–	+	–	+	+	+	+	+	+	–	+	–	–	–
B. cuniculi	–	+	–	–	–	–	–	+	–		–	–	+	+	+	–	+	–	–	–
B. choerinum	–	–	+	–	–	+	–	+	–	–	–	–	+	+	+	–	+	–	–	–
B. animalis	+	+	+	v	v	+	–	+	–	+	v	+	+	+	+	v	+	–	–	+
B. thermophilum	–	–	v	v	v	+	–	+	–	–	–	+	+	+	+	v	+	–	v	v
B. boum	–	–	v	–	–	+	–	+	–	+	–	–	–	+	+	–	+	–	+	–
B. magnum	+	+	+	–	–	+	–	–	–	+	–	+	+	+	+		+	–	–	–
B. pullorum	+	+	–	–	–	+	–	–	–	+	+	+	+	+	+	+	+	–	+	+
B. suis	–	+	+	–	–	+	–	–	–	+	v	v	+	+	+	–	+	–	–	–
B. minimum	–	–	–	–	–	–	–	+	–	–	–	+	–	+	+	–	–	–	–	–
B. subtile	+	–	–	–	+	+	+	+	+	–	–	+	+	+	+	v	+	–	v	v
B. coryneforme	+	+	–	+	–	+	–	–	+	+	–	+		+	+	–	+	–		+
B. asteroides	+	+	–	+	–	+	–	–	v	+	–	+	v	+	v	–	+	–	–	+
B. indicum	+	–	–	+	–	+	–	–	+	–	v	+	v	+	v	–	+	–	–	+

+ = positive, – = negative, v = variable.

Table 7 Migration of F6PPK in *Bifidobacterium* sp.

Species	Migration (cm)
B. bifidum	15
B. longum	15
B. infantis	15
B. breve	15
B. adolescentis	15
B. angulatum	15
B. catenulatum	15
B. pseudocatenulatum	n.d.
B. dentium	15
B. globosum	10
B. pseudolongum	10
B. cuniculi	n.d.
B. choerinum	n.d.
B. animalis	10
B. thermophilum	10
B. boum	n.d.
B. magnum	10
B. pullorum	10
B. suis	10
B. minimum	10
B. subtile	10–15
B. coryneforme	16
B. asteroides	16
B. indicum	16

Note: n.d., not determined.

Source: From Scardovi (1986).

net electrical charge of each protein are determined by the sequence of nucleotides in the DNA. The protein profile of each strain is therefore a fingerprint of the genome. The cell proteins are dissolved using detergents such as SDS, but many studies have been carried out using only the soluble fraction of the disintegrated cells (Dellaglio, 1989). Two types of study can be envisaged for the comparison of *Bifidobacterium* species with each other:

1. Electrophoresis in starch gel (zymogram) of the 14 enzymes of transaldolase and the 19 isoenzymes of 6-phosphogluconate dehydrogenase (6PGD) can be used to compare the electrophoretic mobility of these enzymes in function of the original strain (Table 10). A colorimetric method applied to 3-phosphoglyceraldehyde dehydrogenase is able to identify other strains (Scardovi et al., 1979a). The electrophoretic migration distances for F6PPK appear to

be linked to the ecological origin of the species but the same is not true for the other glucose metabolism enzymes of *Bifidobacterium*, that is, transaldolase, transketolase, 6-phosphogluconate dehydrogenase and aldolase (Scardovi and Sgorbati, 1974).

2. Electrophoresis in a polyacrylamide gel of the lysate of a strain provides electrophoretic profiles of the soluble cell proteins. The distribution of the protein bands is then compared with those for a reference strain (Moore et al., 1980; Biavati et al., 1982, 1986). This method is doubtless the most discriminating and is both reliable and sensitive since it is able to distinguish between strains with DNA-DNA homology levels of up to 80% (Biavati et al., 1982) but is an onerous method, requiring reference strains and is difficult to interpret.

Use of these two types of electrophoresis has given the following results:

1. The homology between *B. dentium* (the only species thought to be pathogenic) and *B. eriksonii* (formerly *Actinomyces eriksonii*) was established by comparing their electrophoretic patterns (zymograms) (Scardovi et al., 1979a). This identity complies with the high percentage DNA-DNA homology (80–100%).
2. Electrophoresis in polyacrylamide gel enabled Biavati et al. (1982) to recognize four new species: *B. minimum, B. subtile, B. cornyeforme*, and *B. globosum*, which is now distinguished from *B. pseudolongum*.
3. *B. adolescentis* and *B. dentium*, which have identical phenotype profiles, can also be differentiated by their zymograms (starch gel electrophoresis) which differ (Scardovi et al., 1979a). Polyacrylamide gel electrophoresis also confirms these findings (Biavati et al., 1982).
4. The comparison of the zymograms and protein electrophoretic patterns in polyacrylamide gel of *B. infantis* and *B. longum* is interesting.

These two species have the same isoenzymes of transaldolases, that is, three different isoenzymes which migrate to a distance of 5, 6, or 8 units. The only difference is the incidence within the strain: the isoenzyme migrating to a distance of 5 occurs more frequently in *B. infantis* and that which goes furthest, to 8, is found most frequently in *B. longum* (Scardovi et al., 1979a). The bands obtained on the electrophoretic diagrams of these two species show an identical distribution. Only the concentrations of the proteins differ (Biavati et al., 1982, 1986).

We are therefore faced by a very unusual phenomenon: these strains, although they belong to different species, present similar profiles, thus defining a "continuum." Table 8 shows that the transaldolases and 6PGD of *B. adolescentis* with electrophoretic mobilities of 8 and 5 respectively are found in 50% of *B. longum* and in many strains of *B. infantis*, thus highlighting what is doubtless a very close degree of relatedness between these species. In contrast, the electrophoresis patterns of the cell proteins for these three strains differ, confirming that they are indeed three separate species (Scardovi et al., 1971; Biavati et al., 1982).

Table 8 Migration of Transaldolase and 6PGD in *Bifidobacterium* sp.

Species	Electrophoretic pattern	
	Transaldolase	6PGD
B. bifidum	7	7-(8)
B. longum	(5)-6-8[a]	5-(6[a])
B. infantis	5-(6)-(8[a])	(3)-4[a]-(5)
B. breve	6	(5)-6-7
B. adolescentis	8	5
B. angulatum	5	5
B. catenulatum	5	6[a]-8
B. pseudocatenulatum	4[a]-(5)	1[a]-3
B. dentium	4	(2)
B. globosum	2	(3)-(4)-(5)-6-(7)
B. pseudolongum	2	7
B. cuniculi	1	4
B. choerinum	3	4
B. animalis	5	8-9[a]
B. thermophilum	(7)-8[a]	7-8-9[a]
B. boum	6	8[a]-9
B. magnum	5	7
B. pullorum	2	Absent
B. suis	6	5-8
B. minimum	10	6
B. subtile	3	2
B. coryneforme	6	6
B. asteroides	(6)-(7)-8-(9)	(9)-10[a]-(11)-(12)-(13)
B. indicum	(6)-7-8-9[a]	6-(7)-8-(9[a])

[a]Number of isoenzymes for type strains.
( ) Number of isoenzymes in less than 10% of strains.
Source: From Scardovi (1986).

The zymogram of *B. bifidum* is in contrast highly specific. This species is the only one to show a transaldolase and F6PPK migration distance of 7 units. Only a few strains of *B. thermophilum* of bovine origin are similar. In this case, the differentiation is based on the mobility of the 3-phosphoglyceraldehyde dehydrogenase.

Study of the protein patterns provides valuable information about a given strain and the numerical analysis of the patterns of a large number of strains makes it possible to achieve:

Rapid grouping of the strains.
Archiving of a large number of models in a reference data bank.

The attribution of unknown bacteria to their group and their possible identification.
A quick method of determining whether two colony types in a culture are due to variation or contamination.
Determining the homogeneity or heterogeneity of the taxa.

When the preparation conditions for the extracts and their electrophoresis are standardized, a high degree of reproducibility (in excess of 96%) can be obtained (Kersters and De Ley, 1975).

4. Transaldolase Serology

An immunological approach by investigating the serology of the transladolases can also be used to differentiate between species within the genus *Bifidobacterium*. Figure 6 shows the conclusion of the studies performed (Sgorbati and Scardovi, 1979; Sgorbati, 1979; Sgorbati and London, 1982).

This method consists of preparing immunosera against the highly purified transaldolases of eight species of *Bifidobacterium: B. infantis, B. angulatum, B. globosum, B thermophilum, B. suis, B. cuniculi, B. minimum*, and *B. asteroides* and to test them against 21 bacterial species of the genus in order to determine

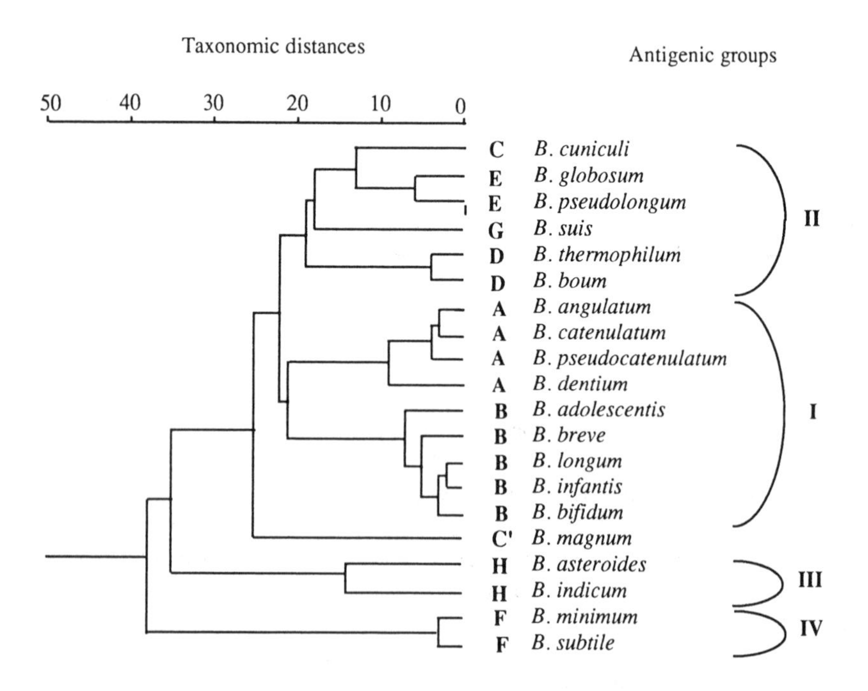

Figure 6 Transaldolases, phylogenic relations in *Bifidobacterium*.

their immunological distances. These results, expressed as taxonomic distance, are shown in the dendogram (Fig. 6). This diagram illustrates the interrelationships existing within the genus and shows that the seven groups defined by Sgorbati and London (1982) (A, B, C, D, E, F, and G) which are detected by this model can be split into four distinct groups closely linked at the ecological origin of the species. These four antigenic groups (I, II, III, and IV) coincide with the arrangement of the species of *Bifidobacterium* based on electrophoretic mobility (Scardovi et al., 1971; Sgorbati, 1979) and are also confirmed by DNA-DNA hybridization studies.

In addition, this dendogram is able to distinguish two subgroups, A and B, among the strains of human origin (I): (1) *B. angulatum, B. catenulatum, B. pseudocatenulatum*, and *B. dentium*; and (2) *B. adolescentis, B. breve, B. longum, B. infantis* and *B. bifidum* (Sgorbati, 1979). The species associated with mammalian animal habitats (II) belong to groups C, D, E, and G. The two species found only in the bee (group H) are antigenetically very distant from the other members of the genus (III). The most widely separated species (IV) (group F) were isolated from waste water.

5. *Enzymes*

B. breve is one of the few species to produce β-glucuronidase activity (Desjardins and Roy, 1990). It is suspected that this enzyme may convert procarcinogens into carcinogens (Kurmann, 1988). In addition, *B. longum* is the only species to have neither β-glucosidase nor *N*-acetylglucosaminidase activity.

6. *Composition of the Wall*

The sugar composition of the wall varies with strain, particularly with regard to the percentage of rhamnose and glucose (Rasic and Kurmann, 1983) (Table 9). The sequence of amino acids in the peptidoglycan may vary from one species to another, thus making it possible to separate species from which are relatively close to each other, such as *B. boum* from *B. thermophilum* or *B. minimum* from *B. subtile* (Kandler and Lauer, 1974).

In addition, Bezirtzoglou (1985) notes that the only species *B. bifidum* has a poly-(1,2)-glycerophosphate skeleton in the lipoteichoic acids which is substituted in the end position by a polysaccharide.

Op Den Camp et al. (1985) prepared antibodies to the lipoteichoic acids of *B. bifidum* by coupling with an immunogenic protein. They were specified towards the polyglycerol phosphate core (essentially poly 1, 2) and to a small extent to the polysaccharide portion.

Crossed reaction tests with phenolic extracts of lipoteichoic acids of *Bifidobacterium* and *Lactobacillus* have shown that only the former react, thus making it possible to envisage a serogroup with lipoteichoic acids as group antigens.

Table 9 Cell Wall Composition of *Bifidobacterium* sp.

Species	Origin	Kind of peptidoglycan	Polysaccharide		
		Cross linkage	Glucose	Galactose	Rhamnose
B. *bifidum*	Adult stool	Orn-Ser-Asp-Ala	+	+	+
B. infantis	Infant stool	Lys-Gly	+	+	+
B. breve	Infant stool	Lys-Gly	+	+	+
B. liberorum	Infant stool	Lys-Gly	+	+	+
B. parvulorum	Infant stool	Lys-Gly	+	+	+
B. asteroides	Bee	Lys-Gly	+	+	–
B. suis	Pig	Orn or (Lys)-Ser-Ala-Thr-Ala	+	+	(+)
B. longum	Adult stool	Orn or (Lys)-Ser-Ala-Thr-Ala	+	+	+
B. thermophilum	Pig	Orn-(Lys)-Glu	+	+	+
B. adolescentis	Adult stool	Lys or (Orn)-Asp	+	+	–
B. indicum	Bee	Lys-Asp	–	+	+
B. pseudolongum	Pig	Orn or (Lys)-Ala	+	+	+

7. *Processing of the Results*

All these phenotype identification criteria give responses which must be classified and interpreted. Most of the problems involved in processing data tables resulting from the examination of the physiological and biochemical properties of the bacteria are solved by the use of computer-assisted numerical taxonomy. The usefulness of numeric taxonomy depends on several criteria (Dellaglio, 1989):

1. Strain selection
2. Number of characters examined (greater than 50)
3. Rigorous standardization of the methods of analysis
4. Weight attributed to each characteristic in the evaluation
5. Classification of the reactions as positive, negative, or noncomparable
6. Type of software used with the computer

XI. IDENTIFICATION BY STUDY OF THE GENOME

A. Identification of the Genus *Bifidobacterium*

The DNA base composition of the chromosome of *Bifidobacterium* differs from that of *Lactobacillus* (Sebald et al., 1965) and from that of other lactic bacteria.

Genus	GC%
Lactobacillus	34.7–50.8
Streptococcus	33–44
Leuconostoc	39–42
Bifidobacterium	57.2–64.5

However, two organisms with similar G + C (GC%) analysis are not necessarily closely related. The GC% cannot therefore be considered as an exclusion characteristic in bacterial taxonomy.

B. Species Identification

1. DNA Base Composition

B. longum is distinguished from *B. animalis* and *B. pseudolongum* by its GC%. The GC% of *B. longum* is 58 whereas that of *B. animalis* and *B. pseudolongum* is 60.

2. DNA-DNA Hybridization

The DNA-DNA homology is a mean measurement of similarity in which the entire genome of one organism is compared with that of another. A fragment of denatured DNA from a reference strain is labeled and then used as a probe to hybridize with a single strand of DNA from the strain to be identified. The more DNA base pairing there is between two strains, the closer they are genetically.

The use of DNA-DNA hybridization methods has advanced the taxonomy of the bifidobacteria (Scardova et al., 1970, 1971b). On the basis of the DNA-DNA homology percentages, 11 different species can be described: *B. indicum, B. coryneform, B. asteroides, B. ruminale, B. globosum, B. suis, B. pullorum, B. magnum, B. catenulatum, B. dentium,* and *B. angulatum.* (Scardovi and Trovatelli, 1969; Matteuzi et al., 1971; Scardovi and Crociani, 1974; Scardovi and Zani, 1974; Trovatelli et al., 1974). The fact that *B. bifidum* and *B. longum* belonged to two different species has been confirmed, but the following species have been combined to form a single species:

B. lactentis, B. liberorum, and *B. infantis* under the name of *B. infantis*
B. ruminale and *B. thermophilum*
B. breve and *B. parvulorum* under the name of *B. breve*
B. psuedolongum and *B. globosum* (Scardovi et al., 1971b)

In the following year and using the same methods, Holdeman and Moore (1972) divided *B. infantis* into three subspecies and *B. adolescentis* into four groups. In addition, they described two new species: *B. cornutum* and *B. eriksonii.*

Table 10 DNA-DNA Homology in the Genus *Bifidobacterium*

	B. bifidum	*B. longum*	*B. infantis*	*B. breve*	*B. adolescentis*	*B. angulatum*	*B. catenulatum*	*B. psedocatenulatum*	*B. dentium*
B. bifidum	100	*	*	*	0		*	*	
B. longum	/	88	63	*	0		*	*	
B. infantis	/	65	88	*			*	*	
B. breve	*	/	/	100			*	*	
B. adolescentis	*		*	*	86	*	*	*	*
B. angulatum	*	*	*	*	*	88	*	*	*
B. catenulatum	*	*	*	*	/	*	90	65	*
B. pseudocatenulatum							67	97	
B. dentium	*	*	*	*	*	*	*	*	90
B. globosum	*		*	*				*	
B. pseudolongum	*		*	*				*	
B. cuniculi	*		*	*				*	
B. choerinum	*		*	*				*	
B. animalis	*	*	*	*	*	*	*	*	*
B. thermophilum	*		*	*				*	
B. boum	*		*	*				*	
B. magnum	*	*	*	*	*	*	*	*	*
B. pullorum	*		*	*				*	
B. suis	*		*	*				*	
B. minimum	*	*	*	*	*	*	*	*	0
B. subtile	*	*	*	*	*	*	*	*	0
B. coryneforme				*					
B. asteroides				*				*	
B. indicum				*					

*Value between 40% and 60%.
/Value less than 40%.

B. globosum	*B pseudolongum*	*B. cuniculi*	*B. choerinum*	*B. animalis*	*B. thermophilum*	*B. boum*	*B. magnum*	*B. pullorum*	*B. suis*	*B. minimum*	*B. subtile*	*B. asteroides*	*B. indicum*
0		*	*	*	*	*	*	*	0	*	*	0	0
0		*	*	*	0	*	*	*	0	*	*	0	
0		*	*	*	*	*	*	*	0	*	*	0	
0		*	*	*	*	*	*	*	0	*	*	0	
		*	*	*		*	*	*		*	*		
*		*	*	*	*	*	*	0	*	*	*		
*		*	*	*	*	*	*	*	*	*	*		
		*	*			*							
*		*	*	*	*	0	*	0	*	*	0		
100	75	/	/	/	*	*	*	*	*	*	*	*	*
67	100	*	/			*			*				
/	*	98	*	*	*	*		*					
/	/	*	98	*	*	*		*					
*	*	*	*	86	*	*	*	0	*	*	*	*	0
*	*	*	*	*	98	/	*	*	*	*	*	0	0
		*	*	*	/	82		0					
*		*	*	*	*	*	90	*	*	*	*		
		*	*	*		*		102					
*	0	*	*	*	*	*	*	*	100	*	*		
*		*	*	*	*	0	*	*	*	102	*	*	0
*		*	*	*	*	0	0	*	*	*	85	*	*
*		*	*	0				0		*	0	*	/
0		*	/		0	0		*	0	0	*	100	*
0		*	*		10			*		0	0	*	100

In 1974, Scardovi and Trovatelli (1974) detected genetic differences between *B. longum* ss *longum* and *B. longum* ss *animalis* and proposed raising these subspecies to rank of species: *B. animalis, B. longum*, and *B. animalis.*

A few years later, some new species were described by Scardovi (1981): *B. pseudocatenulatum, B. cuniculi, B. choerinum*, and *B. boum* and then, in 1982, by Biavati et al. (1982): *B. minimum* and *B. subtile.*

The results of the DNA-DNA hybridizations between the various species of *Bifidobacterium* are shown in Table 10. From this table it can be seen that two continua can be defined from the human-defined strains: *B. infantis* and *B. longum* (50–76% homology), which had already been detected by the zymogram, electrophoresis of soluble proteins, and growth factor requirements (Neut et al., 1981) and suspected by Scardovi et al. (1971b). The other continuum is that of *B. catenulatum* and *B. pseudocatenulatum* (60–80% homology) (Scardovi et al., 1979b).

These genotypic findings nonetheless raise some questions:

1. The electrophoretic mobility of the transaldolases and 6PGD had suggested proximity of *B. adolescentis, B. longum*, and *B. infantis*. However, although the DNA-DNA hybridizations between these strains confirm the relationship between *B. longum* and *B. infantis, B. adolescentis*, on the contrary, appears to be genetically very distant from *B. longum* (0% hybridization).
2. The serology of the transaldolases makes it possible to separate the strains of human origin into two subgroups and *B. adolescentis* appears to belong to the same subgroup as *B. breve, B. longum, B. infantis*, and *B. bifidum* (and not to the subgroup containing *B. angulatum, B. catenulatum, B. pseudocatenulatum*, and *B. dentium*). The hybridizations challenge this finding since, genetically, *B. adolescentis* appears to be distant from these four species (less than 20% homology) but close to the species in the other subgroup and in particular to *B. dentium* (some strains showing up to 49–57% homology). In contrast, the results obtained by electrophoresis of soluble proteins and the zymograms appear to distance *B. adolescentis* and *B. dentium.*

3. *Plasmid Tests*

The number of plasmids detected in a strain of *Bifidobacterium* may suggest the identification of certain species because only some of them possess plasmids (Sgorbati et al., 1983, 1986). This technique is particularly useful for the separation of *B. longum* and *B. infantis.*

4. *New Methods*

Today, hope is based on the use of new methods which should soon lead to further changes:

1. The whole bacterial DNA can be processed using restriction enzymes (Bove and Saillard, 1979). But even if the genome is relatively small, there are too many restriction fragments for the electrophoretic pattern to be easily read. This method can demonstrate that two strains are different, but can in no case be used to confirm the similarity of two bacteria (Colmin et al., 1991). The successive use of several restriction enzymes should make it possible to pinpoint identification.
2. Grimont and Grimont (1986) suggest that the parameters used to describe species and strains should include the size of the DNAr restriction fragments following agar medium electrophoresis. The DNA fragments carrying genes (DNAr) coding for ribosomal RNA are then localized on filters by hybridizing with either a DNAr 16 + 23S from *Escherichia coli* or with a DNA probe which codes for the well conserved portions of ribosomal RNA 5S or 16S.
3. Many more studies are required before a new level of identification can be defined which could be achieved by sequencing the RNAr 5S and 16S which are very well conserved molecules throughout the animal kingdom. Comparison of these sequences with those of reference strains should give information which is extremely useful for identification (Grimont and Grimont, 1986).
4. DNA-DNAr hybridization is also a method the value of which improving the classification of bifidobacteria remains to be demonstrated.
5. Pulsed field electrophoresis requires large fragments of DNA. Studies have been carried out in order to identify restriction enzymes able to cut DNA at infrequent restriction sites.

C. Conclusion

The identification of species within the genus *Bifidobacterium* is difficult. The kits and commercial tests which provide phenotype identification information are not appropriate for these bacteria and their use does not provide a sufficiently reliable answer. Even with 100 phenotype characters, the identification is not definite.

Genotyping today provides a more accurate and reliable approach to bacterial taxonomy. DNA-DNA hybridization is widely used and is a particularly appropriate method since it makes it possible to study the entire bacterial genome including those parts which do not code for proteins. The new genome methods now under investigation hold the promise of a more accurate and quicker identification method.

Table 11 shows the accuracy and precision of identification obtained by means of genotyping. Several species which cannot be differentiated by phenotyping can be distinguished from their genetic characteristics. DNA-DNA hybridization methods take into account the entire genome, whereas phenotype analysis involves only the segments of the genome of which the phenotype expression can be measured (Owen and Pitcher, 1985), which amounts to about 10%.

Table 11 Relation Between DNA-DNA Homology and Phenotypy

Genotype	Phenotype
B. minimum	*B. minimum*
B. bifidum	*B. bifidum a* and *b*
B. thermophilum	
B. boum	*B. thermophilum*
B. choerinum	
B. subtile	
B. pseudolongum	*B. pseudolongum*
B. globosum	
B. animalis	
B. magnum	*B. animalis*
B. suis	
B. pullorum	
B. cuniculi	*B. cuniculi*
B. longum	*B. longum a* and *b*
	B. infantis ss. *infantis a* and *b*
B. infantis	*B. infantis* ss. *liberorum*
	B. infantis ss. *lactentis*
B. breve	*B. breve* ss. *breve a* and *b*
	B. breve ss. *parvulorum a* and *b*
B. adolescentis	
B. dentium	
B. catenulatum	*B. adolescentis a, b, c,* and *d*
B. pseudocatenulatum	
B. angulatum	
B. asteroides	*B. asteroides*
B. coryneforme	
B. indicum	*B. indicum*
B. gallinarum	*B. gallinarum*

Note: DNA-DNA homology higher than 50%.

XII. RELATIONSHIP BETWEEN THE INTESTINAL FLORA AND INTESTINAL CELLS. ADHESION AND COLONIZATION METHODS

The study of the mechanisms of adhesion has been facilitated by the development of investigation methods and can be carried out today in vivo by taking a human or animal colon biopsy sample which is then examined under electron microscopy or in vivo by the culture of intestinal cells or tissues under survival conditions and exposed to bacteria before being examined under the electron microscope.

A. Colonization Conditions

To obtain a probiotic or, on the contrary, pathogenic effect depending on the invasive bacterial species concerned, the bacteria must adhere to the cell surfaces of the digestive tract. Two conditions must be present to allow the implantation of a bacterium:

1. It must be able to multiply, which is related to the presence of a substrate and a redox potential appropriate for its growth requirements.
2. It must be able to reside in situ, i.e., adhere to the cells or mucus and avoid expulsion (Savage, 1984).

B. Mechanism of Adhesion

There are numerous conflicting hypotheses concerning the possible implantation of ingested microorganisms. These bacteria may either simply pass through or they may colonize the intestine by adhering to the cell wall and in this case belong to the subdominant flora.

In most cases, there is a uniform interstice between the bacterium and the host cell measuring less than 40 nm which is filled with fibrillar material (Guerina and Neutra, 1984). The bacterium may be completely encircled by the membrane of the apical pole of the cell or only partially surrounded and in some cases the association takes place without penetration. This adhesiveness depends on three factors: adhesins, adhesin receptors, and mucus.

1. The Adhesins

These structures which are implied in the attachment process and are present in some bacteria may be of two types:

a. Protein Type. These are either external membrane proteins or individualized structures such as the pili or fimbriae of Gram-negative bacteria (Costerton et al., 1981). The pili may be bound to the glycoproteins and glycolipids of the cell membrane of many vertebrate species (Guerina and Neutra, 1984; Savage and Levitt, 1984).

These nonprotein structures may also be fibrillar structures other than fimbriae. They have been described in the superficial adhesion process of enteropathogenic *E. coli* (Knutton et al., 1987).

Recently, elements known as "fimbriosomes" have also been detected in hyperadhesive mutant strains. Fimbriosomes are rounded structures which are closely associated with the fimbriae (Abraham et al., 1988). It has not yet been demonstrated whether these structures potentiate the effects of the fimbriae or whether they act separately.

b. Nonprotein Type. These are the polysaccharides of the capsule or slime. The polysaccharide fraction of *B. infantis* is involved in the adhesion of this bacterium to the epithelial cells of the ileum or the lipotechoic acids (LTA) of Gram+ bacteria (Sato et al., 1982; Contrepois, 1988).

Many studies have demonstrated that the LTA of Gram+ bacteria have a high binding affinity to the membranes of the epithelial cells of mammals (Beachey et al., 1979, 1981; Simpson et al., 1980; Courtney et al., 1981). Binding occurs spontaneously through the intermediary of the lipid fraction of the LTAs (Courtney et al., 1981). The binding of the TLAs of the bifidobacteria to human epithelial cells in culture is dependent on cell concentration and time and appears to be reversible. LTA is bound through the fatty acids which are themselves bound to the esters (Op Den Camp et al., 1985).

An important phenomenon which should be highlighted is "phase variation," which allows bacterium to modify its surface and thus its adhesive potential, depending on its phase. This phenomenon is familiar in *E. coli* both in vitro and in vivo (Guerina and Neutra, 1984).

2. Adhesin Receptors

The most probable receptor on the membrane of the intestinal epithelial cells is a protein or glycoprotein with fatty acid binding sites (Op Den Camp et al., 1985). However, the adhesion of *E. coli* to uroepithelial cells and isolated human colon cells appears to involve the intermediary of glycolipids (glycolipids account for about 20% of the membrane lipids of the enterocyte) containing gal a 1–4 Gal β residues, as well as through the intermediary of receptors which are sensitive to mannose and which could consist of one or several glycoproteins (Wadolowski et al., 1988). It appears that the Gal a 1–4 Gal β receptors are indeed irregularly distributed amongst the intestinal cells (Krivan et al., 1986). The surface receptors trap the bacteria in mucus secretions.

The membrane receptivity of the host cells is strongly influenced by cell maturity (Walker, 1985), the age of the host (Abraham et al., 1985; Darfeuille-Michaud, 1987) and the portion of the digestive tract involved. This results from differences of structure and composition of the enterocytes of the host which are mature in adults and immature in young subjects (Cheney and Boedeker, 1984). The differences found in the composition of the intestinal flora in infants, children, and adults may be related to this change of receptors in function of age. The type and quantity of specific glycolipid receptors is also determined genetically (Itoh et al., 1988).

3. Mucus

Mucus covers the entire mucosa of the digestive tract from the stomach to the colon, and it consists of a sort of protective elastic and viscous gel consisting of glycoproteins. Its most obvious function is to provide specific protection against bacterial penetration but its presence is also essential in the mechanisms of bacterial adhesion. Using electron microscopy, Croucher et al. (1983) have confirmed the close association of bacteria with the mucus layer.

C. Analysis of the Adherent Flora

Compared to the diversity of the intraluminal flora, the flora adhering to the epithelium is generally limited to a few species of bacteria. The bacterial concentrations are generally lower by one or two log factors than the intraluminal populations. If biopsies are taken from various points of the duodenum and jejunum, the bacterial population appears to be equally distributed along a given segment of the gastrointestinal tract (Jones, 1977). Despite the marked predominance of anaerobic bacteria in the lower portion of the gastrointestinal tract, their relationship with the epithelium of the colon or rectum has not really been studied. Only the adhesive capacities of the optional anaerobes (such as enterobacteria) have been studied, and few studies have been devoted to this topic. The adhesion of *E. coli* and other enterobacteria has been demonstrated by electronmicroscopy of biopsies (Hartley et al., 1979; Bergogne-Berezin et al., 1986). The density of bacteria ranges from 10^6 to 10^7 cfu/g of fresh tissue.

Examination of colonic biopsies allowed Hartley et al. (1979) to demonstrate that the same strain of *E. coli* is closely associated with the intestinal wall throughout the length of the colon.

D. Adhesion of the Bifidobacteria

Sato et al. (1982) have demonstrated the role played by the polysaccharides in the adhesion of the bifidobacteria by inhibiting this phenomenon using antibodies targeted against polysaccharides. The Gram staining becomes negative in cell-bound *Bifidobacterium*. This phenomenon reflects an increase in the membrane permeability of these bacteria. The bonds which permit this adhesion are probably strong.

The characteristic of the "epithelial cell of the human colon/lipoteichoic acids (LTA of *Bifidobacterium*" complex have been investigated by determining the radioactivity of carbon-14 labeled LTA after interaction with suspended colonocytes. The most probable receptor on the colonocyte membrane is a protein or glycoprotein with fatty acid binding sites (Huub et al., 1985).

The adhesion of *Bifidobacterium* and more particularly the hydrophobic interactions are promoted by a high level of fatty acids in the LTA, resulting in a high level of hydrophobicity of the bacterium. The strong electrostatic charge of the polysaccharides of Gram-positive bacteria also favors adhesion (Savage, 1984).

E. Modifications of Adhesion

Several studies have shown that some bacterial species may modify the receptor sites of epithelial cells, resulting in an inhibition of the adhesion of the microorganisms which use them. Thus, extracellular enzymes of *Bifidobacterium* may degrade specific sites of pathogenic organisms or their toxins (Guerina and Neutra, 1984; Savage, 1984; Hofstad and Kalvenes, 1985). These enzymes, some of which are glycosidases, may degrade the receptors within the cell or mucus, and also eliminate any bacteria which were bound to them (Walker, 1985).

XIII. THE INTESTINAL FLORA

Knowledge of the intestinal flora has developed simultaneously with the methods of investigation both with regard to sampling technique and analysis of the flora.

A. Methods of Evaluation

Conventional stool collection allows only investigation of the terminal flora of the digestive tract. The dominant flora consists of highly anaerobic microorganisms which are therefore difficult to isolate and keep alive. More elaborate sampling methods, such as biopsy of the intestinal mucosa and collection of luminal aspirate fluid using a Camus probe or weighted tube, have made it possible to carry out detailed exploration of all portions of the digestive tract. These new sampling methods and the design of more appropriate culture media for the various species now make it possible to use a good direct approach to understanding the intestinal bacterial flora by identifying and counting the species samples.

The importance and role of the microbial flora within the host can also be assessed indirectly by determining the bacterial metabolites produced:

The breath test or measurement of the hydrogen expired by a subject after ingesting lactulose.

Measurement of the activities of various bacterial enzymes in intestinal samples.

Tests for fatty acids excreted in the stools by gas chromatography reflecting bacterial metabolism.

Counting of a given species of *Bacteroides* from a mixture of bacterial DNA isolated directly from the stools and exposed to a species-specific labeled DNA probe has been suggested (Kuritza et al., 1986; Salyers, 1989). However, this method would not make it possible to differentiate between viable and nonviable bacteria and a confirmation of the percentage of live bacteria would still have to be carried out using conventional dish culture methods.

B. Composition

The intestinal flora of a man weighs between 1 and 2 kg, that is, roughly the same weight as organs such as liver, brain, or lungs. The digestive tract houses about 10^{14} bacteria, which means there are more living entities in the flora than there are cells in a normal body and the bacteria consist of 500 species. This flora can be divided into two categories:

1. The dominant population, the effects of which on the host were the first to be understood;
2. The subdominant population, which accounts for less than 1% of the total bacterial population but which, according to recent studies (Hoskins et al., 1985) may play a non-negligible role in the equilibrium of the intestinal ecosystem.

Numerous studies have been carried out to define the composition of the intestinal flora. The main results, shown in Table 12, show some discrepancies which can easily be explained from the choice of culture methods, isolation

Table 12 Fecal Flora of Different Human Groups

Bacterian group	Infant from 1 to 4 days	Infant from 5 to 90 days	Infant from 4 to 6 years	Adult from 20 to 64 years	Adult from 65 to 86 years
Total bacteria[a]	10.1	10.5	10.8	10.8	10.5
Aerobic or facultative anaerobic					
Enterobacteria	9.3	8.8	8.0	8.2	7.8
Streptococcus	8.5	8.1	7.8	7.7	8.2
Lactobacillus	6.4	7.3	7.0	6.7	8.0
Staphylococcus	6.2	6.8	4.0	4.4	4.3
Yeast	3.5	4.0	4.2	3.7	4.6
Anaerobic					
Bacteroides	8.6	8.2	10.4	10.3	10.0
Eubacteria	0	9.7	9.9	9.9	9.5
Bifidobacterium	9.3	9.9	10.1	9.8	9.4
Peptococcus	0	9.0	8.1	8.9	7.7
Clostidium perfringens	5.9	6.9	5.7	4.8	6.6
Veillonella	5.6	6.3	5.2	4.8	6.1

[a]log cfu.
Source: From Mitsuoka (1984).

media, and counting methods. Generally, the most numerous population in adults is *Bacteroides* (about $10^{10.3}$ per gram of feces). Thereafter, in decreasing order come *Eubacterium, Bifidobacterium*, and then the *Peptococcacceae*. Of the aero-anaerobes, there are the enterobacteria ($10^{8.2}$) followed by *Streptococcus*, aerobic *Lactobacillus* and then finally *Staphylococcus* ($10^{4.4}$) (Mitsuoka, 1982).

C. Factors Affecting the Flora

1. Factors Ensuring the Equilibrium of the Intestinal Flora

The diversity of bacterial species and their quantity at various levels within the digestive tract can be preserved only by means of physical, chemical, and biological regulatory mechanisms.

Intestinal peristaltism results in the elimination of many microorganisms.

The acidity of the stomach maintains a low concentration of bacteria in the upper part of the digestive tract and destroys some pathogens.

The interactions which exist between various bacterial species are also important in maintaining the equilibrium of the intestinal microflora. It is possible to observe symbioses between species, as a result of the production of vitamins or amino acids or other metabolites which can be assimilated by other species, and also of antagonisms due to the release of antibiotics, bacteriocins, or factors such as the volatile fatty acids.

2. Location and Physiology

The composition of the intestinal flora varies depending on the rate of transit and luminal secretions but also of the intestinal segment (Abrams, 1980). Thus, a given well-defined resident flora corresponds to each portion of the tract. The various factors which are active along the digestive tract result in qualitative and quantitative differences in the digestive flora, as shown in Table 13. Thus, the flora present in the proximal small intestine (duodenum and jejunum) consists of aerobic Gram-positive microorganisms (streptococci and staphylococci) and a few yeasts.

This aero-anaerobic flora is subsequently replaced, within the ileum, by a flora consisting of *E. coli* and anaerobes such as *Clostridium, Fusobacterium*, and *Bacteroides* (10^6 total bacteria per ml) (Mitsuoka, 1982). This switch from a dominant aerobic population within the stomach to a strictly anaerobic population within the colon can be explained if we accept that the aero-anaerobic bacteria use any oxygen present, thus creating the redox conditions for the implantation of anaerobic species in more distal portions.

Finally, two parts should be distinguished within the colon: (1) the ascending colon, which contains mainly Gram-positive bacteria which have the primary role of sugar fermentation; and (2) the descending colon, in which the flora, known

Table 13 Human Gastrointestinal Flora

	Stomach	Jejunum	Ileum	Colon
Total microbial concentration	$0–10^{3}$[a]	$0–10^{5}$	$10^{3}–10^{7}$	$10^{11}–10^{12}$
Strict aerobic or facultative anaerobic bacteria				
Enterobacteria	$0–10^{2}$	$0–10^{3}$	$10^{2}–10^{5}$	$10^{4}–10^{10}$
Streptococcus	$0–10^{3}$	$0–10^{4}$	$10^{2}–10^{6}$	$10^{5}–10^{10}$
Staphylococcus	$0–10^{2}$	$0–10^{3}$	$10^{2}–10^{5}$	$10^{4}–10^{7}$
Lactobacillus	$0–10^{3}$	$0–10^{4}$	$10^{2}–10^{5}$	$10^{6}–10^{10}$
Fongy	$0–10^{2}$	$0–10^{2}$	$10^{2}–10^{3}$	$10^{2}–10^{6}$
Anaerobic bacteria				
Bacteroides	Rare	$0–10^{2}$	$10^{3}–10^{6}$	$10^{10}–10^{12}$
Bifidobacterium	Rare	$0–10^{3}$	$10^{3}–10^{7}$	$10^{8}–10^{12}$
Peptococcus	Rare	$0–10^{3}$	$10^{3}–10^{4}$	$10^{8}–10^{12}$
Clostridium	Rare	Rare	$10^{2}–10^{4}$	$10^{6}–10^{11}$
Fusobacterium	Rare	Rare	Rare	$10^{9}–10^{10}$
Eubacteria	Rare	Rare	$10^{3}–10^{5}$	$10^{9}–10^{12}$
Veillonellae	Rare	$0–10^{2}$	$10^{3}–10^{4}$	$10^{3}–10^{4}$

[a]Number per gram of intestinal contents.

as “putrefaction flora,” consists mainly of Gram-negative bacteria but also some Gram-positive bacteria (*Clostridium, Bacteroides*).

This conventional theory, associating a region of the colon with a bacterial function and consequently with particularly dominant species (Abrams, 1980) has been challenged by the work of Croucher et al. (1983). Their studies of human colon biopsies tend to demonstrate that there is no specific location of the various species within the colon.

D. Age and Diet

This change in the flora is closely linked with the maturation of the digestive system, once more highlighting the importance of the reciprocal host-bacteria relationship. Bacteriological examination of the feces shows that diet has little or no effect on the constitution of the dominant intestinal flora (Simon and Gorbach, 1986).

E. Role and Effect of the Intestinal Flora

“The gastrointestinal tract is a complex ecosystem with characteristics which depend at each moment on a dynamic equilibrium between the host and the native bacteria" (Abrams, 1980). Exogenous bacteria also influence all the bacteria within the intestinal flora. Some may be probiotic and others simply commensal, whereas others may be pathogens. The overall effect of the microbial flora on the host is generally evaluated by comparing an axenic animal with a holoxenic animal in which the flora has developed normally. This is also the tool which has been found to be most useful in demonstrating the effects of a given species or small group of species on the host.

1. Effect on the Physiology of the Intestinal Wall and the Immune Defense System

The intestinal flora modifies the morphology of the mucosa and the rate of turnover and differentiation of epithelial cells. It also follows enterocyte maturation and development of the velocities in the neonate (Luckey, 1965; Coates and Fuller, 1977; Abrams, 1980). In the axenic animal, Simon and Gorbach (1986) observed an increase in the activity of enterocytic enzymes and in particular in alkaline phosphatase, disaccharidase, and β-glucosidase. All these studies have demonstrated the importance of the role of the flora, since it determines the uptake of nutrients and permits the formation of the ecological site for other bacteria.

An important role played by the flora is its action in cell maturation observed in the normal development of Peyer's patches. These observations, which were first made as a result of histological investigation of the intestinal wall, have subsequently been confirmed by numerous studies which have demonstrated that resistance to various pathogens is conditioned by the presence of a flora. The intestinal bacteria ensure the maintenance of the immune status by providing repeated antigen stimulation throughout the human life span.

2. Bacteria as Nutrient Sources

The bacterial mass of the intestine is itself an important source of nutrients: thiamine, riboflavin, folic acid, vitamin B_{12}, pantothenic acid, short-chain fatty acids, amino acids, and proteins which are partially absorbed and used by the host (Rasic, 1983; Teraguchi, 1984; Deguchi et al., 1985).

3. Metabolic Effects

The bacterial flora produces a very large and varied quantity of enzymes which are used by the flora itself but also by the host. All the aspects of the intestinal metabolism of the host are influenced by the enzymatic activity of the bacteria which it shelters and more particularly the anaerobic bacteria. We will list below some samples of the effects of bacterial metabolism on the host.

a. Enzymatic Action. These bacteria are able to compensate for enzymatic deficiencies of the host if they are introduced in a sufficiently large number into the digestive tract. This is the case of lactobacilli ingested with yogurt, which can produce the lactase activity which is missing in lactose-intolerant subjects (Ducluzeau and Raibaud, 1979).

b. Detoxification. Another important action of the intestinal bacteria is their involvement in the enterohepatic cycle and the detoxification of numerous substances and drugs (Goldman, 1985). Thus, cholesterol is converted to form coprostanol, and the bile salts to form bile acids and then lipocholic acid and other derivatives conjugated with amino acids such as glycine and taurine, thus facilitating their detoxification and elimination (Rasic, 1983). Rowland and Grasso (1975) have investigated the degradation of the *N*-nitrosamines by the intestinal flora.

c. Production of Harmful Substances. In contrast, some microorganisms produce substances which are toxic for the host, notably histamine, tyramine, agmatine, cadaverine, ammonium, phenols, *N*-nitrosamines, and bacterial toxins.

4. Tumoral Action

Cancer of the colon is the second greatest cause of death in Great Britain and in the United States. It would appear that 90% of human cancers are due to the environment and could therefore be avoided. Major differences have been observed between cancer risks in one country and another, however, it would seem that neither the place where the population live nor their race is responsible, but that the etiology of the disease should be related to diet (Borriello, 1987). Considerable research has been carried out in an attempt to identify carcinogens.

A diet containing low fiber and high quantities of animal fats appear to promote the onset of cancer of the colon, but no directly active carcinogen has been isolated. Aries et al. (1969) therefore believe that carcinogens may be produced in situ, probably as a result of the enzymatic activity of the bacteria of the digestive flora on a harmless procarcinogen substrate derived from the diet. It is reasonable to think that the intestinal flora could produce or potentiate carcinogens or procarcinogens.

5. Effect of the Anti-Infectious Barrier Toward Pathogens

The microbial population of the gastrointestinal tract forms a barrier against proliferation of exogenous pathogens (Wilhelm et al., 1987). One explanation may be that the colonization of the endogenous flora maintains the pathogens at a subclinical level by preventing the colonization of the undesirable flora by competition for the substrate or epithelial receptors (Savage, 1977). This recently discovered role challenges the theory of the anti-infectious barrier effect suggested by Ducluzeau et al. (1979) which holds that the barrier effect can be observed only in bacteria belonging to the dominant flora, and that pathogenicity can be effective only above a certain colonization threshold of the invasive bacteria. In fact, the subdominant flora uses the endogenous substrates for its own metabolism, but also for the dominant population, thus preventing the proliferation of pathogenic bacteria and also preventing the adhesion of other organisms. The intestinal microorganisms inhibit the growth of the invasive pathogens by

Producing organic acids, particularly volatile fatty acids
Deconjugating bile acids, which inhibit pathogenic bacteria in their conjugated forms
Producing bacteriocin
Volatile acids which stimulate peristaltism (Faure et al., 1982; Okamura et al., 1986)

XIV. BIFIDUM-INTESTINAL RELATIONSHIPS. PROBIOTIC ROLE OF *BIFIDOBACTERIUM*

Tissier (1923) observed a close relationship between the immunity of the breast-fed child and his or her specific flora. Bifidobacteria had long been recognized as bacteria with probiotic, nutritive, and therapeutic properties. These properties have subsequently been clearly defined: the ingestion of fermented milk results in stimulation of the immune system (Perdigon et al., 1988; Lemonnier, 1988). The quantities of β-glucuronidase, azoreductase, and nitroreductase formed by the flora are reduced by the ingestion of *Lactobacillus adisophilus* (Goldin and Gorbach, 1977) which has also been shown to be active in the degradation of nitrosamines (Rowland and Grasso, 1975). A large portion of the world's adult population shows a deficit in galactosidase. Numerous studies have shown that deficient subjects do not show any intolerance toward yogurt (Lemonnier, 1988). Two explanations can be suggested for this: the activity of bacterial β-galactosidase may persist in the intestine where a fraction of the lactose present may be metabolized, or the ingestion of yogurt may stimulate any intestinal lactase which is still active.

In the case of the bifidobacteria, a probiotic effect can only occur if the bifidobacteria survive their transit through the stomach. Some strains of *Bifidobacterium* are able to resist gastric acidity, and this resistance is increased by the food bolus (Berrada et al., 1989). The bacterial production of organic acids, particularly lactic and acetic acids, of bacteriocins and even antibiotics as well as the secretion of enzymes, vitamins, and other growth factors are, together with the stimulation of the immune system and accumulation of specific metabolites, the determining factors in a probiotic action. We can now attribute the functions described below to the bifidobacteria.

A. Adhesion to the Intestinal Epithelium

The adhesion of the bifidobacteria to the epithelial cells permits the formation of ecological niches within which the growth of bacteria is maintained regardless of changes in the habitat, thus enabling them to produce a genuine effect. The bacterial biofilm, bound to the epithelial walls, maintains the mechanism of in situ production of various bacterial metabolites, which themselves have an effect on other bacterial genera and even on the host. In addition, it has a defense function against pathogenic bacteria.

B. Action on the Morphology and Physiology of the Digestive Tract Wall

Bifidobacteria influence the maturation and turnover cycle of the enterocyte, together with the development of the intestinal velocities (Abrams, 1980; Simon and Gorbach, 1986). They are also involved in the degradation and replacement of intestinal mucins (Salyers et al., 1977; Salyers and Leedle, 1983; Vercellotti et al., 1977; Miller and Hoskins, 1981; Hoskins et al., 1976, 1981, 1985). They may also have an action on the immune system appended to the digestive tract (Lemonnier, 1988).

C. Nutritional Effects

The production of vitamins (B_1, B_6, B_9, B_{12}, and PP), amino acids (alanine, valine, aspartic acid, and threonine), and the fact that they produce only L+ lactic acid, which is completely metabolizable by man, enhances the nutritional characteristics of the bifidobacteria (Rasic, 1983; Teraguchi et al., 1984; Deguchi et al., 1985).

D. Metabolic Effects

1. Suppression of Lactose Intolerance

Bifidobacteria, unlike *Lactobacillus delbruekii* ssp. *bulgaricus* and *Streptococcus salivarius* ssp. *thermophilus* are resistant to biosalts (Rao et al., 1989), and as a result can have an in situ effect on the metabolism of lactose.

2. Hypercholesterolemic Effect

Several studies have tended to demonstrate a relationship between the presence of a lactic microflora and a reduction in plasma cholesterol. The administration to hypercholesterolemic human subjects of fermented milks containing very large quantities of *Bifidobacterium* (10^9 bacteria per gram) results in a fall in the total cholesterol from 3 to 1.5 g/mL (Homma et al., 1967; Homma, 1988).

The consumption of fermented dairy products could lead to a reduction in serum levels of cholesterol. Bacteria-producing lactic acid also produces hydroxymethylglutaryl–Co A reductase which is involved in the synthesis of cholesterol (Mann and Spoerry, 1974). More recently, Rao et al. (1981) have shown that the metabolites produced from orotic acid during fermentation of fermented products could be responsible for this hypercholesterolemic effect. Jaspers et al. (1984) have shown that both orotic and hydroxymethylglutaric acids reduce serum cholesterol, whereas uric acid inhibits the synthesis of cholesterol. In vitro, bifidobacteria apparently affect the activity of HMG-CoA reductase (Homma, 1988). It is difficult to determine the role which should be attributed to the bifidobacteria and studies are in progress intended to demonstrate the involvement of bifidobacteria in the reduction of cholesterol levels.

3. Deconjugation of Bile Acids, Reduction of Nitrosamines, and Inhibition of the Reduction of Nitrates

Bile acids are secreted into the duodenum in their form of conjugates with glycine or taurine. Most strains of the genus *Bifidobacterium* are able to hydrolyze sodium taurocholate and glycocholate in the colon (Ferrari et al., 1980). The hydrolases involved are constitutive and extracellular (Rasic, 1983).

4. Other Metabolic Effects

The ingestion of milk fermented with *Lactobacillus acidiophilus* and *B. bifidum* for a period of three weeks has no effect on the production of hydrogen or methane or on fetal β-galactosidase activity, but does increase the activity of fecal β-glucosidase and these four parameters are good indicators of the fermentation capacity of the colonic flora (Marteau et al., 1990).

E. Effect of the Anti-Infectious Barrier to Pathogenic Bacteria

It has been possible for a long time to demonstrate close relationships between the probiotic and therapeutic effects of bifidobacteria, but few studies have been carried out in human subjects:

B. longum has a barrier effect versus *Escherichia coli* in the axenic rat (Faure et al., 1982).

Axenic mice monoassociated with *B. longum* live for longer than truly axenic mice after the intravenous or intragastric administration of high doses of viable *E. coli* (Yamazaki et al., 1982).

The intestinal flora of leukemia patients is modified by chemotherapy, and bacteria which are usually rare in healthy subjects multiply considerably. This disequilibrium of the intestinal microflora is countered by oral administration of *Bifidobacterium longum* (Kageyama et al., 1984).

In infants suffering from rotavirus-induced diarrhea, the concomitant administration of *B. longum* fermented milk with the antibiotic treatment results in a reduction in the number of stools, the number of *Bacteroides* and a more rapid regain of weight compared with treatment with the antibiotic alone (Romond, 1988).

Two hypotheses have been advanced to explain this probiotic effect:

1. Prevention of the colonization of the intestine by pathogens by competing for nutrients and for binding sites on the epithelial surfaces (Ducluzeau et al., 1979, 1980; Rasic, 1983; Okamura et al., 1986).
2. The production of lactic and acetic organic acids during the fermentation of carbohydrates by bifidobacteria which results in a reduction in the pH of the intestine and consequently inhibits the growth of the undesirable bacteria (Rasic, 1983).

The implantation of bifidobacteria is promoted in infants by breast-feeding. The permanent acidity of the intestinal contents which results from the development of bifidobacteria has a bacteriostatic effect versus *E. coli* and Gram-negative bacteria (Bullen et al., 1977).

Acidification has a bactericidal potential, especially versus Gram-negative bacteria. Acetic acid has a stronger antagonistic effect versus Gram-negative bacteria than lactic acid and it is produced in greater quantities by *Bifidobacterium* (Rasic, 1983). This difference appears to be due more to the quantity of undissociated acid than to the type of acid. The pK_a of acetic acid is 4.76, whereas that of lactic acid is 3.86. Acetic acid 8.4% and lactic acid 1.1% are present in an undissociated form at an intestinal pH of 5.8. Acetic and lactic acids are produced in the bifidobacteria in a ratio of 3/2 which results in about 11 times more undissociated acetic acid than undissociated lactic acid. The acidity stimulates the peristaltic movements of the intestine, which facilitates elimination of any pathogens present (Mayer, 1969; Savage, 1977).

Knocke et al. (1984) have investigated the in vitro interactions between *B. adolescentis* and *Bacteroides ovatus*. In a continuously renewed complex medium they observed that the inhibition of *Bacteroides* by the bifidobacteria appears to be due to the production of certain metabolites by the latter.

Inhibition of bacterial translocations which involve cell immunity. Yamazaki et al. (1985) have investigated the translocation capacity of *E. coli* in the axenic mouse. Intragastric inoculation of *E. coli* at sublethal doses results, two weeks later, in the appearance of this strain in the liver, spleen, kidneys and lungs of monoassociated animals. In contrast, previous implantation of *B. longum* in the axenic mouse allows the animal to survive and results in the disappearance of *E. coli* from all the organisms invaded within one week, although they remain present at high levels within the colon.

The most probable explanation for this phenomenon would be the reinforcement of the immune barriers. *B. longum* may be capable of affecting both humoral immunity and cellular immunity (Faure et al., 1982). The production of bacteriocins by bifidobacteria has been studied by Meghrous (1990) in 13 strains. The antimicrobial substance detected is of a protein type, heat-stable and active at pH values ranging from 2 to 10 versus other Gram-positive species including some strains of Clostridia.

F. Therapeutic Effects

1. Anti-tumor Effect

The research so far carried out has been essentially concerned with the direct or indirect anti-tumor action of streptococci, lactobacilli, and bifidobacteria have been mainly studied in animals and less frequently in man (Goldin et al., 1980, 1984). *B. longum* has a direct inhibitory effect on liver tumors in the mouse (Mitzutani and Mitsuoka, 1979). In the mouse, *B. infantis* has an undeniable antitumor effect (Kohwi et al., 1978). The number of tumors developed by mice with an intestinal flora including *E. coli, enterococcus faecalis*, and *Clostridium paraputrificum* is considerably reduced if *Bifidobacterium longum* is present (Mizutani and Mitsuoka, 1979). The anti-tumor action may be obtained as a result of (1) direct suppression of the procarcinogens (Hosono et al., 1990), (2) the reduction of indirect suppression of procarcinogens or bacterial enzymes which result in their formation, (3) activation of the host's immune system (Toida et al., 1990), and (4) reduction of the intestinal pH (Fernandes and Shahani, 1990).

2. Prevention and Treatment of Other Diseases

In 1966, Bamberg (1966) successfully treated digestive disorders induced by antibiotic treatment using a lyophilized culture of *B. longum*. Similarly, Haller and Kraüberg (1990) and Neumeister and Schmidt (1963) had obtained similar results in radiation-treated subjects. Bifidobacteria were successfully administered to premature infants with an intestinal flora which had been disturbed by the taking of antibiotics (Kozlova, 1976). Seki et al. (1978) have developed a treatment for constipation in the elderly which is based on *Bifidobacterium*.

The ingestion of milk fermented with *B. longum* is successful in regularizing digestive transit in pregnant women (reduced abdominal ballooning, diarrhea-type phenomena, or constipation) (Ebissawa, 1987). The intestinal flora and health status of children suffering from diarrhea were restored more rapidly after the ingestion of milk fermented with *B. breve* (Hotta et al., 1987). The ingestion of milk fermented with *B. longum* plays a role not only in prevention in the healthy subject by maintaining high levels of bifidobacteria in the flora, thus preventing diarrhea and constipation, but also plays a therapeutic role in patients suffering from diarrhea (Asselin, 1988). The intestinal disorders in 34 Soviet cosmonauts were successfully treated by the ingestion of bifidobacteria (Lizko, 1987).

This nonexhaustive review of the beneficial effects of bifidobacteria demonstrates that in all cases it was bifidobacteria of human origin and more particularly *B. longum* which had been used.

XV. CONCLUSION

The current use in fermented milks of probiotic bacteria and of bifidobacteria in particular presumes that the user will take all necessary precautions with regard to the following aspects.

A. Choice of Species

This is the most important factor if it is intended that the product should have genuine probiotic qualities. All the factors which we have just reviewed so that species of human origin have a better chance of producing a beneficial effect in man. Since *B. dentium* is recognized as being pathogenic, eight species could theoretically be used.

B. Identification of the Strain

B. bifidum, B. breve, B. infantis, and *B. longum* do not constitute any danger, but other bifidobacteria in this group could be confused with *B. dentium* if the identification of the strain used is not done using genetic methods (Table 11).

C. In Vitro and In Vivo Verification of the Probiotic Potential of the Strain

This potential can only be demonstrated through clinical and nutritional studies, and there is no strain which has all the qualities which have been demonstrated for the genus *Bifidobacterium* as a whole. In this field, the choice must still be based on the probiotic activities desired.

D. Technological Potential

The choice in this field must take into account several factors such as the acetate/lactate ratio, the relationships of the bifidobacteria with other species in the product, the tolerance of the strain of acidity and above all of oxygen. Furthermore, the distribution conditions for these products must be such as to ensure the survival of the bifidobacteria under good physiological conditions.

REFERENCES

Abraham, S. N., Goguen, J. D., and Beachey, E. H. 1985. Hyperadhesive mutant of type-1 fimbriated *E. coli* associated with formation of FIMH organelles (fimbriosomes), *Infect. Immun., 56*: 1023–1029.

Abrams, G. D. 1980. Morphological aspect of gastro-intestinal tract colonization, Les anaérobes, Symp., Paris, Masson, pp. 65–75.

Allen, A. 1981. The Structure and Function of Gastrointestinal Mucus, in *Basic Mechanisms of Gastrointestinal Mucosal Cell Injury and Protection*, (J. W. Harmon, ed.), Williams and Wilkins, Baltimore, pp. 351–67.

Allen, A. 1984. The structure and function of gastrointestinal mucus, in *Attachment of Organisms of the Gut Mucosa* (E. C. Boedeker, ed.), CRC Press, Boca Raton, FL, pp. 3–12.

Aries, V. C., Crowther, J. S., Drasar, B. S., Hill, M. J., and Williams, R. E. O. 1969. Bacteria and the aetiology of cancer in the large bowel, *Gut, 10*: 334.

Asselin, D. 1988. Effets de l'ingestion de laits fermentés au *B. longum* sur la flore intestinale humaine, *Thèse de l'Université de Caen en Sciences pharmaceutiques.*

Ballongue, J. 1989. Evolution de la taxonomie des bifidobactéries. *Bifidobacterium* et facteurs bifidigènes, Rôle en santé humaine, ARBBA, pp. 33–45.

Bamberg, H. 1966. Über die sanierung von typhus dauerausscheidern mit Ampicillin, *Med. Wellt., 39*: 2086.

Beachey, E. H. 1981. Bacterial adherence: adhesin-receptor interactions mediating the attachment of bacteria to mucosal surfaces, *J. Infect. Dis., 143*: 325–345.

Beachey, E. H., Dale, J. B., Grebe, S., Ahmed, A., Simpson, W. A., and Ofek, I. 1979. Lymphocyte binding and T cell mitogenic properties of group A streptococcal lipoteichoic acid, *J. Immunol., 122*: 189–195.

Beck, J. 1967. Un nouveau milieu de culture, d'isolement et d'entretien pour *B. bifidum. Ann. Biol. Clin., 25*(10–12): 1255–1260.

Beerens, H. 1990. An elective and selective isolation medium for *Bifidobacterium* spp. Beerens H, *Lett. Appl. Microbiol., 11*(3): 155–157.

Beerens, H., Gerard, A., and Guillaume, J. 1957. Etude de 30 souches de *Bifidobacterium bifidum* (*Lactobacillus bifidus*). Caractérisation d'une variété buccale. Comparison avec les souches d'origine fécale, Ann. Inst. Pasteur Lille, *9*: 77–85.

Beerens, H., Romond, C., and Neut, C. 1980. Influence of breast-feeding of the bifid flora of the newborn intestine, *Am. J. Clin. Nutr., 33*: 2434–2439.

Benno, Y., Sawada, and Mitsuoka, T. 1984. The intestinal microflora of infants: composition of faecal flora in breast-fed and bottle-fed infants, *Microbiol. Immun., 28*: 975.

Bergogne-Berezin, E., Cerf, M., Gaudin, B., Cazier, A., Bizet, J., and Feldman, G. 1986. Recherche des phénomènes d'adhésion bactérienne dans le tube digestif chez l'homme, *Rev. Inst. Pasteur Lyon, 19*: 93–103.

Berrada, N., Laroche, G., Lemeland, J. F., and Tonetti, H. 1989. Survie de bifidobactéries dans l'estomac de l'homme, in *Les laits Fermentés. Actualité de la Recherche*. John Libbey Eurotext, pp. 259–260.

Bezirtzoglou, E. 1985. Contribution à l'étude de l'implantation de la flore fécale anaérobie du nouveau-né mis au monde par césariene, Thèse, Paris-Sud.

Bezkorovainy, A., Grohlich, D., and Nichols, J. H. 1979. Isolation of a glycopolypeptide fraction with L. bifidus subspecies penn. growth promoting activity from whole human milk casein, *Am. J. Clin. Nutr., 32*: 1428–1432.

Bezkorovainy, A., and Topouzian, N. 1981a. The effect of metal chelators and other metabolic inhibitors on the growth of *B. bifidum* var. *penn, Clin. Biochem., 14*(3): 135–141.

Bezkorovainy, A., and Topouzian, N. 1981b. *B. bifidus* var. *penn.* growth promoting activity of human milk casein and its derivatives, *Int. J. Biochem., 13*: 585–590.

Bezkorovainy, A., and Topouzian, N. 1983. Aspects of iron metabolism in *B. bifidum* var. *penn, Int. J. Biochem., 15*(3): 361–366.

Bezkorovainy, A., Topouzian, N., and Miller Catchpole, R. 1986. Mechanisms of ferric and ferrous iron uptake by *Bifidobacterium bifidum* var. *pennsylvanicus, Clin. Physiol. Biochem., 4*: 150–158.

Biavati, B., Castagnoli, P., Crociani, F., and Trovatelli, L. D. 1984. Species of the genus *Bifidobacterium* in the feces of infants, *Microbiologica, 7*: 341–345.

Biavati, B., Castagnoli, P., and Trovatelli, L. D. 1986. Species of the genus *Bifidobacterium* in the feces of human adults, *Microbiologica, 9*: 39–45.

Biavati, B., Scardovi, V., and Moore, W. E. C. 1982. Electrophoretic patterns of proteins in the genus *Bifidobacterium* and proposal of four new species, *Int. J. System. Bacteriol., 32*(3): 358–373.

Borriello, S. P. 1987. *Clostridium difficile* and gut disease, in *Microbes and Infections of the Gut* (C. S. Goodwin, ed.),

Bourrillon, A., Boussougant, Y., Lejeune, C., and Paillerets, F. 1980. Etablissement et cinétique des anaérobies dans la flore fécale du nouveau-né, *Les Anaérobies* (Symp., Paris), Masson, pp. 96–107.

Bove, J. M., and Saillard, C. 1979. Cell biology of spiroplasma, in *The Mycoplasma* (R. F. Whitcomb and J. G. Tully, eds.), Academic Press, New York, pp. 83-153.

Boventer, K. 1938. Untersuchungen über das Bacterium bifidum, *Zentralbl. Bakteriol. I. Orig., 142*: 419–430.

Boventer, K. 1949. *Bacterium bifidum* und seine Bedentung, *Ergebn. Hyg., 26*: 193–234.

Braun, O. H. 1981. Effect of consumption of human milk and other formulas on intestinal bacterial flora in infants, in *Textbook of Gastroenterology and Nutrition in Infancy* (Lebenthal, ed.), Raven Press, New York, pp. 247–253.

Buchanan, R. E., and Gibbons, N. E. 1974. *Bergey's Manual of Determinative Bacteriology*, 8th ed., Williams and Wilkins, Baltimore.

Bullen, C. L., Tearle, P. V., and Stewart, M. G. 1977. The effect of "humanised" milks and supplemented breast feeding on the faecal flora of infants, *J. Med. Microbiol., 10*: 403–413.

Carrillo, M., Estrada, E., and Hazen, T. C. 1985. Survival and enumeration of fecal indicators *Bifidobacterium adolescentis* and *Escherichia coli* in a tropical rain forest watershed, *Appl. Environ. Microbiol., 50*: 468–476.

Chang, J. H., Kwon, I. K., and Kim, H. U. 1983. Studies on the bifidobacteria in breast-fed Korean infant gut, *Korean J. Dairy Sci., 5*: 111.

Cheney, C. P., and Boedeker, E. C. 1984. Appearance of the host intestinal receptors for pathogenic *E. coli* with age, in *Attachment of Organisms to the Gut Mucosa* (E. C. Boedeker, ed.), CRC Press, Boca Raton, FL, pp. 157–166.

Chevalier, P., Roy, D., and Ward, P. 1990. Detection of *Bifidobacterium* species by enzymatic methods, *J. Appl. Bacteriol., 68*: 619–624.

Colmin, C., Pebay, M., Simonet, J. M., and Descaris, B. 1991. A species-specific DNA probe *Streptococcus salivarius subsp. thermophilus* detects strain restriction polymorphism, *FEMS Microbiol. Lett., 81*: 123–128.

Contrepois, M. 1988. Les colibacilles pathogènes: adhérence et facteurs de colonisation des colibacilles entérotoxinogènes, in *L'Intestin Grêle* (J. C. Rambaud, and R. Modigliani, eds.), Excerpta Medica, Elsevier, pp. 138–150.

Costerton, J. W., Irvin, R. T., and Cheng, K. J. 1981. The bacterial glycocalyx in nature and disease, *Ann. Rev. Microbiol., 35*: 299–324.

Courtney, H. S., Ofek, I., Simpson, W. A., and Beachey, E. H. 1981. Characterization of lipoteichoic acid binding to polymorphonuclear leukocytes of human blood, *Infect. Immun., 32*: 625–631.

Croucher, S. C., Houston, A. P., Bayliss, C. E., and Turner, R. J. 1983. Bacterial populations associated with different regions of the human colon wall, *Appl. Environ. Microbiol., 45*(3): 1025–1033.

Cummins, C. S., Glendenning, O. M., and Harris, H. 1957. Composition of the cell wall of *Lactobacillus bifidus, Nature, 180*: 337–338.

Darfeuille-Michaud, A. 1987. Contribution à la recherche et à l'identification des facteurs d'adhésion d'*E. coli* responables de diarrhées; description de nouveaux facteurs d'adhésion, Thèse de doctorat en sciences de l'Université Clermont II.

Deguchi, Y., Morishita, T., and Mutai, M. 1985. Comparative studies on synthesis of water-soluble vitamins among human species of bifidobacteria, *Agric. Biol. Chem., 49*(1): 13–19.

Dehnert, J. 1957. Untersuchungen über die gram positive Stuhlflora des Brust-milchkindes, *Zentralbl. Bakteriol. Parasitenk. Infektionskr. Hyg. I. Abt. Orig., 169*: 66–83.

Dellaglio, F. 1989. Approches moléculaires et chemiotaxonomiques dans la classification des bactéries lactiques, in *Bif et Facteurs bifidigènes. Rôle en Santé Humaine*. ARRBA, pp. 55–72.

Desjardins, M. L., and Roy, D. 1990. Growth of bifidobacteria and their enzyme profiles, *J. Dairy Sci., 73*: 299–307.

Desjardins, M. L., Roy, D., Toupin, C., and Goulet, J. 1990. Uncoupling of growth acids production in *Bifidobacterium* ssp. *J. Dairy Sci., 73*(6): 1478–1484.

De Vries, W., and Stouthamer, A. H. 1967. Pathway of glucose fermentation in relation to the taxonomy of bifidobacteria, *J. Bacteriol., 93*(2): 574–576.

De Vries, W., and Stouthamer, A. H. 1968. Fermentation of glucose, lactose, galactose, mannitol and xylose by Bifidobacteria. *J. Bacteriol., 96*(2): 472–478.

De Vries, W., and Stouthamer, A. H. 1969. Factors determining the degree of anaerobiosis of *Bifidobacterium* strains, Arch. Mikrobiol., 65: 275–287.

Drasar, B. S., and Hill, M. J. 1974. *Human Intestinal Flora*, Academic Press, New York.

Ducluzeau, R., and Raibaud, P. 1979. *INRA. Actualités Scientifiques et Agronomiques. Ecologie Microbienne du Tube Digestif*, Masson.

Ducluzeau, R., Raibaud, P., Hudault, S., and Nicolas, J. L. 1980. Rôle des bactéries anaérobies stricyes dans les effets de barrière exercés par la flor du tube digestif, dans *Les anaérobies*, Symp. Paris, Masson, pp. 86–95.

Ebissawa, E., Assari, T., Takeda, S., Watanabe, A., Nihei, K., Tamashita, T., Wakiguchi, H., and Watanabe, S. 1987. Utilisation de lait fermenté additionné de Bifidus Actif chez la femme enceinte, premier colloque *Bifidobacterium longum* et santé, Monte Carlo, *Médecine et Chirurgie Digestive 16*(3): 9–11.

Exterkate, F. A., Otten, B. J., Wassenberg, H. W., and Veerkamp, J. H. 1971. Comparison of the phospholipid composition of *Bifidobacterium* and *Lactobacillus* strains, *J. Bacteriol., 106*: 824–829.

Faure, J. C., Schellenberg, D., Bexter, A., and Wurzner, H. P. 1982. Barrier effect of *Bifidobacterium longum* on *Escherichia coli* in the germ-free rat, *Int. J. Vit. Nutr. Res., 52*(2): 225–230.

Fernandes, C. F., and Shahani, K. M. 1990. Anticarcinogenic and immunological properties of dietary lactobacilli, *J. Food Protect., 53*: 704.

Ferrari, A., Pacini, N., and Canzi, E. 1980. A note on bile acids transformations by strains of *Bifidobacterium, J. Appl. Bacteriol., 49*: 193–197.

Forstner, G., Sherman, S. P., and Forstner, J. 1984. Mucus: function and structure, in *Attachment of Organisms of the Gut Mucosa* (E. C. Boedeker, ed.), CRC Press, Boca Raton, FL, pp. 13–21.

Frisel, E. 1951. Studies on *Bacterium bifidum* in healthy infants. A clinical bacteriological investigation, *Acta Paediat., Stockh., 40*(suppl. 80): 1–123.

Gauhe, A., György, P., Hoover, J. R. E., Kuhn, R., Rose, C. S., Ruelius, H. W., and Zilliken, F. 1954. Bifidus factor IV. Preparations obtained from human milk, *Arch. Biochem. Biophys., 48*: 214–224.

Glick, M. C., Sall, T., Zilliken, F., and Mudd, S. 1960. Morphological changes of *L. bifidus var. penn.* produced by a cell-wall precursor, *Biochim. Biophys. Acta, 37*(2): 363–365.

Goldin, B. R., and Gorbach, S. L. 1977. Alterations in fecal microflora enzymes related to diet age, *Lactobacillus* supplements and dimethylhydrazine, *Cancer, 40*: 2421–2426.

Goldin, B. R., and Gorbach, S. L. 1984. The effect of oral administration of *Lactobacillus* and antibiotics on intestinal bacteria activity and chemical induction of large bowel tumor, *Dev. Indust. Microbiol., 25*: 139–144.

Goldin, B. R., Swenson, L., Dwyer, J., Sexton, M., and Gorbach, S. L. 1980. Effect of diet and *Lactobacillus acidophilus* supplements on human faecal bacterial enzymes, *J. Natl. Cancer Inst., 64*(2): 255–261.

Goldman, P. 1985. Drug metabolism by the intestinal flora and its implications, *Microecol. Therapy, 15*: 257–260.

Grimont, F., and Grimont, P. A. D. 1986. Ribosomal ribonucleic acid gene restriction patterns as potential taxonomic tools, *Ann. Inst. Pasteur. Microbiol. 137B*: 165–175.

Guerina, N. G., and Neutra, M. R. 1984. Mechanism of bacterial adherence, *Microecol. Therapy, 14*: 183–199.

Gurr, M. I., Marshall, V. M. E., and Fuller, R. 1984. Laits fermentés, microflore intestinale et nutrition. *Bull. Fedération Int. Laiterie*. pp. 54–59.

Gyllenberg, N., and Niemelä, S. 1959. A selective method for the demonstration of Bifid-Bacteria (*Lactobacillus bifidus*) in materials tested for faecal contamination, *J. Sci. Agric. Soc. Finland, 31*: 94–97.

György, P. 1953. Hitherto unrecognized biochemical difference between human milk and cow's milk, *Pediatrics, 11*: 98–108.

György, P., Kuhn, R., Rose, C. S., and Zilliken, F. 1954a. Bifidus factor II: Its occurrence in milk from different Species and in other natural products, *Arch. Biochem. Biophys., 48*(1): 202–208.

György, P., Norris, R. F., and Rose, C. S. 1954b. Bifidus factor I. Variant of *Lactobacillus bifidus* requiring a special growth factor, *Arch. Biochem. Biophys., 48*(1): 193–201.

Haenel, H. 1970. Human normal and abnormal gastrointestinal flora, *Am. J. Clin. Nutr., 23*(1): 1433–1439.

Haenel, H., and Muller-Beuthow, W. 1963. Examination of german and bulgarian young men for intestinal eubiosis, *Zbl. Bakt., 188*: 70–80.

Haller, J., and Kräubig, H. 1960. The influence upon radiation of the intestines by the oral use of combination of living acidophilus-bifidus and colibacteria, *Strahlenther, 113*(2): 272.

Hartley, C. L., Neumann, C. S., and Richmond, M. H. 1979. Adhesion of commensal bacteria to the large intestine wall in humans. *Infect. Immun., 23*: 128–132.

Hassinen, J. B., Durbin, G. T., Tomarelli, R. M., and Bernhart, F. W. 1951. The minimal nutritional requirements of *Lactobacillus bifidus, J. Bacteriol., 62*: 771–777.

Hatanaka, M., Tachiki, T., Kumagai, H., and Tochikura, T. 1987a. Distribution and some properties of glutamine synthetase and glutamate dehydrogenase in bifidobacteria, *Agric. Biol. Chem., 51*(1): 251–252.

Hatanaka, M., Tachiki, T., Kumagai, H., and Tochikura, T. 1987b. Purification and some properties of glutamine synthetases from bifidobacteria, *Agric. Biol. Chem., 51*(2): 425–433.

Hidaka, H., Eida, T., Takizawa, T., Tokunaga, T., and Tashiro, Y. 1986. Effects of fructooligosaccharides on intestinal flora and human health, *Bifidobacteria Microflora, 5*: 37–41.

Hidaka, H., Hirayama, M., and Sumi, N. 1988. A fructooligosaccharide-producing enzyme from *Aspergillus niger* ATCC 20611, *Agric. Biol. Chem., 52*: 1181.

Hoffmann, K. 1966. Bakterielle besiedlung des menschlichen, Darmes Theoretische und klinische medizin in einzeldarstellungen, Dr. Alfred Hüthig, Heidelberg, German Federal Republic.

Hofstad, T., and Kalvenes, M. B. 1985. Adhesion of anaerobic gram-negative bacteria to mucosal surfaces, *Scand. J. Infect. Dis.* (suppl.), *46*: 33–36.

Holdeman, L. W., and Moore, W. E. C. 1972. Anaerobe Laboratory Manual, Virginia Polytechnic Institute and State University, Blacksburg, Virginia.

Holland, D. F. 1920. Generic index of the commoner forms of bacteria, *J. Bact., 5*: 215–229.

Homma, N. 1988. Bifidobacteria as a resistance factor in human beings, *Bifidobacteria microflora, 7*(1), 35–43.

Homma, N., Nishihara, K., and Isoda, K. 1967. Antifidus cocci, their biological properties and clinical significance, *Mschr. Kinderheilk, 115*(4): 296.

Hoskins, L. C., Agustines, M., Mc Kee, W. B., Boulding, E. T., Kriaris, M., and Niedermeyer, G. 1985. Mucin degradation in human colon ecosystems, *J. Clin. Invest., 75*: 944–953.

Hoskins, L. C., and Boulding, E. T. 1976. Degradation of blood group antigens in human colon ecosystems, *J. Clin. Invest., 57*: 63–82.

Hoskins, L. C., and Boulding, E. T. 1981. Mucin degradation in human colon ecosystems, *J. Clin. Invest., 67*: 163–172.

Hosono, A., Wardojo, R., and Otani, H. 1990. Inhibitory effects of lactic acid bacteria from fermented milk on the mutagenicities of volatile nitrosamines, *Agric. Biol. Chem., 54*(7): 1639.

Hotta, M., Sato, Y., Iwata, S., Yamashita, N., Sunakawa, K., Oikawa, T., Tanaka, R., Watanabe, K., Takayama, H., Yajima, M., Sekiguchi, S., Arai, S., Sakurai, T., and Mutai, M. 1987. Clinical effects of *Bifidobacterium* preparations on pediatric intractable diarrhea, *Keio J. Med.*, 298–314.

Husain, I., Poupard, J. A., and Norris, R. F. 1972. Influence of nutrition on the morphology of strain of *Bifidobacterium bifidum. J. Bacteriol., III*: 841–844.

Huub, J. M., Op Den Camp, H. J. M., Oosterhof, A., and Veerkamp, J. H. 1985. Interaction of bifidobacterial lipoteichoicacid with human intestinal epithelial cells, *Infect. Immun.*, *47*(1): 332–334.

Inoue, K., and Nagayama, T. 1970. Effect of mucin upon growth of low birth-weight infants and their intestinal microflora, *Acta. Paediat. Jap* (overseas ed.), *12*: 15–20.

Ishibashi, N. 1989. Absorption d'oxygène et métabolisme du *Bifidobacterium*, in *Bifidobacterium et Facteurs Bifidigènes. Rôle en santé humaine*, ARBBA, pp. 129–138.

Itoh, K., Matsui, T., Tsuji, K., Mitsuoka, T., and Veda, K. 1988. Genetic control in the susceptibility of germ free inbred mice to infection by *E. coli* 0115, a, c: K (B), *Infect. Immun., 56*: 930–935.

Iwata, M., and Morishita, T. 1989. The presence of plasmids in *Bifidobacterium breve. Letters in Appl. Microbiol. 9*: 165–168.

Jaspers, D. A., Massey, L. K., and Leudecke, L. O. 1984. Effect of consuming yogurts prepared with three culture strains on human serum lipoproteins, *J. Food Sci., 49*: 1178.

Jones, G. W. 1977. The attachment of bacteria to the surfaces of animal cells, in *Microbial interactions* (Reissing, ed.), Chapman and Hall, London, pp. 133–176.

Kageyama, T., Tanoda, T., and Nakano, Y. 1984. The effect of *Bifidobacterium* administration in patients with leukemia, *Bifidobacteria Microflora, 3*(1): 29–33.

Kalnitsky, G., and Guzman Barron, E. S. 1974. The effect of fluoroacetate on the metabolism of yeast and bacteria, *J. Biol. Chem., 170*: 83–95.

Kandler, O. 1970. Aminoacid sequence of the murein and taxonomy of the genera *Lactobacillus, Bifidobacterium, Leuconostoc* and *Pediococcus, Int. J. System. Bacteriol., 20*(4): 491–507.

Kandler, O., and Lauer, E. 1974. Modern concepts of the taxonomy of bifidobacteria, *Zbl. Bakt., 228*: 29–45.

Kersters, K., and De Ley, J. 1975. Identification and grouping of bacteria by numerical analysis of their electrophoretic protein patterns, *J. Gen. Microbiol., 87*: 33–342.

Kleinschmidt, H. 1925. Studien über die anaerobier des Säuglingsdarms, *Mschr. Kinderkeilk, 29*: 550–5654.

Kleinschmidt, H. 1949. Versuche zur anpassung der künstlichen an die natürliche nahrung des säuglings, *Osterr, Z. kinderheilk, 3*: 55–58.

Knocke, M., and Hannelore, B. 1986. Stimulation of intestinal biocenosis in continuous flow culture. *Microecology Therapy, 14*: 127–135.

Knutton, S., Lloyd, D. R., and McNeish, A. S. 1987. Identification of a new fimbrial structure in enterotoxigenic *Escherichia coli* (ETEC) serotype 0148: HL 28 which adheres to human intestinal mucosa: a potentially new human ETEC colonization factor, *Infect. Immun., 55*: 86–92.

Kohwi, Y., Imai, K., Tamura, Z., and Hashimoto, Y. 1978. Antitumor effect of *Bifidobacterium infantis* in mice, *Gann, 69*(5): 613–618.

Kojima, M., Suda, S., Hotta, S., and Hamada, K. 1968. Induction of pleomorphism in *Lactobacillus bifidus, J. Bact., 95*(2): 710–711.

Kojima, M., Suda, S., Hotta, S., and Hamada, K. 1970a. Induction of pleomorphology and calcium ion deficiency in *Lactobacillus bifidus, J. Bact., 102*: 217–220.

Kojima, M., Suda, S., Hotta, S., and Suganuma, A. 1970b. Necessity of calcium for cell division in *Lactobacillus bifidus, J. Bact., 104*(2): 1010–1013.

Kozlova, E. P. 1976. Intestinal microbial flora in premature infants with septicemia and its correction with *Bifidobacterium, Voprosy Okhr. Materin. Det., 21*: 50–53.

Krivan, H. C., Clark, G. F., Smith, D. F., and Wilkins, T. D. 1986. Cell surface binding site for Clostridium difficile enterotoxin: evidence for a glycoconjugate containing the sequence gal 1-3 Gal 1-4 glc NAc, *Infect. Immun., 53*(3): 573–581.

Kuhn, R., Gauche, A., and Baer, H. H. 1953. Über ein N-haltiges tetrasaccharid aus frauenmilch, *Chem. Berichte, 86*(6): 827–830.

Kuritza, A. P., Shaughnessy, P., and Salyers, A. A. 1986. Enumeration of polysaccharide degrading *Bacteroides* species in human feces using species-specific DNA probes, *Appl. Environ. Microbiol., 51*: 385–390.

Kurmann, J. A. 1988. Starter for fermented milks. Section 5. Starters with selected intestinal bacteria, *Int. Dairy Fed. Bull. 227*: 41.

Lauter, L. 1921. Über das vorkommen des *Bacillus bifidus* bein Neugeborenen, *Zentralbl. Bakteriol. L. Orig. A, 86*: 579–581.

La Vergne, E., Burdin, J. C., Schmitt, J., and Manciaux, M. 1959. Sensibilité de *B. bifidum* à onze antibiotiques, *Annales de l'Institut Pasteur, 97*(1): 104–107.

Lee, B. H., Hache, S., and Simard, R. E. 1986. A rapid method for differentiation of dairy lactic acid bacteria by enzyme systems, *J. Ind. Microbiol., 1*: 209–212.

Lee, L. J., Kinoshita, S., Kumagai, H., and Tochikura, T. 1980. Galactokinase of *B. bifidum, Agric. Biol. Chem., 44*(12): 2961–2966.

Lemonnier, D. 1988. Le yaourt, enjeu scientifique et industrial, *Recherche, 19*: 543–545.

Levesque, J., Aicardi, H., and Gautier, A. 1959. Rôle du facteur bifidus II dans l'établissement et le maintien de la flore bifide chez les nourrissons, Sem. Hop. Paris. Ann. Pédiat, Paris, pp. 30–36.

Levesque, J., Georges-Janet, L., and Raynaud, M. 1960. Recherches sur les régimes bifidogènes, *Arch. Péd., 17*: 553–540.

Lizko, N.N. 1987. Dysbacteriosis under extreme conditions, *Antibiot. Med. Biotekhnol. 32*(3): 184–190.

Luckey, T. D. 1965. Gnotobiologic evidence for functions of the microflora, *Ernährungsforschung, 10*(2–3): 192–250.

Lundequist, B., Nord, C. E., and Winberg, J. 1985. The composition of the faecal microflora in breastfed and bottlefed infants from birth to eight weeks, *Acta Paediatr. Scand., 74*: 45–51.

Manciaux, M. 1958. Bifidobacterium bifidum: Ses facteurs de crossance. Essais d'implantation chez le nourrisson, Thèse, Université de Nancy.

Mann, G. V., and Spoerry, A. 1974. Studies of a surfactant and cholesteremia in the Maasai, *Amer. J. Clin. Nutr., 27*: 464.

Mara, D. D., and Oragui, J. I. 1983. Sorbitol-fermenting bifidobacteria as specific indicators of human faecal pollution, *J. Appl. Bacteriol., 55*: 349–357.

Marteau, P., Pochart, P., Flourie, B., Pellier, P., Santos, L., Desjeux, J. F., and Rambaud, J. C. 1990. Effect of chronic ingestion of a fermented dairy product containing *Lactobacillus acidophilus* and *Bifidobacterium bifidum* on metabolic activities of the colonic flora in humans, *Am. J. Clin. Nutr., 52*(4): 658–688.

Mata, L. J., and Urrutia, J. J. 1971. Intestinal colonization of breast-feed infants in a rural area of low socio economic level, *Ann. N.Y. Acad. Sci., 176*: 93–98.

Matteuzi, D., Crociani, F., and Brigidi, P. 1983. Antimicrobial susceptibility of *Bifidobacterium*, *Ann. Microbiol., 134A*: 339–349.

Matteuzi, D., Crociani, F., and Enaldi, O. 1978. Aminoacids produced by bifidobacteria and some clostridia, *Ann. Microbiol. (Paris), 129*(2): 175–182.

Matteuzi, D., Crociani, F., Zani, G., and Trovatelli, L. D. 1971. *Bifidobacterium suis n. sp.*: a new species of the genus *Bifidobacterium* isolated from pig faeces, *Z. Allg. Mikrobiol., 11*: 387–395.

Mayer, J. B. 1956. Das bifidum problem, *Ergebn. Inn. Med. Kinderheilk, 7*: 429–452.

Mayer, J. B. 1969. Interrelationships between diet, intestinal flora and viruses, *Phys. Med. Rehabil., 10*(1): 16–23.

Mayer, J. B., and Moser, L. 1950. Die typendifferenzierung des thermobakterium Bifidum, *Z. Kinderheilk, 67*: 455–468.

Meghrous, J., Euloge, P., Junelles, A. M., Ballongue, J., and Petitdemange, H. 1990. Screening of *Bifidobacterium* strains for bacteriocin production, *Biotechnol. Lett., 12*(8): 575–580.

Metchnikoff, E. 1908. *The Prolongation of Life*. G. P. Putnam, 4.

Mevissen-Verhage, E. A. E., Marcelis, J. H., De Vos, M. N., Harmsen-Van-Amerongen, W. C. M., and Verhoef, J. 1987. *Bifidobacterium, Bacteroides* and *Clostridium* spp. in fecal samples from breast-fed and bottle-fed infants with and without iron supplement, *J. Clin. Microbiol., 25*(2): 285–289.

Miller, L. G., and Finegold, S. M. 1967. Antibacterial sensitivity of *Bifidobacterium (Lactobacillus bifidus)*, *J. Bact., 93*(1), 125–130.

Miller, R. S., and Hoskins, L. C. 1981. Mucin degradation in human colon systems, *Gastroenterology, 81*: 759–765.

Mitsuoka, T. 1969a. Vergleichende Untersuchungen über die Laktobazillen aus den Faeces von Menschen, Schweinen und Hühnern, *Zentralbl. Bakteriol. Parasitenk. I. Abt. Orig., 210*: 32–51.

Mitsuoka, T. 1969b. Comparative studies on bifidobacteria isolated from the alimentary tract of man and animals, including descriptions of *Bifidobacterium thermophilum nov. spec.* and *Bifidobacterium pseudolongum nov. spec.*, *Zentralbl. Bakteriol. Parasitenk. I. Abt. Orig., 210*: 52–64.

Mitsuoka, T. 1982. Recent trends in research on intestinal flora, *Bifidobacteria Microflora, 1*(1): 3–24.

Mitsuoka, T. 1984. Taxonomy and ecology of bifidobacteria, *Bifidobacteria Microflora, 3*(1): 11–28.

Mitsuoka, T. 1989. *A Profile of Intestinal Bacteria*. Yakult Honsha Co., Tokyo.

Mitsuoka, T., Hidaka, H., and Eida, T. 1987. Effect of fructooligosaccharides on intestinal microflora, *Die Nahrung, 31*: 427.

Mitsuoka, T., Hayakawa, K., and Kimura, N. 1974. Die faekal flora bei Menschen. II. Mitteilung: die Zuzammensetzung der Bifidobakterien-flora der verschiedenen Altersgruppen, *Zentralbl. Bakteriol. Parasitenk. I. Abt. Orig. A., 226*: 469–478.

Mitsuoka, T., and Kaneuchi, C. 1977. Ecology of the bifidobacteria, *Am. J. Clin. Nutr., 30*: 1799–1810.

Mitsuoka, T., Sega, T., and Yamamoto, S. 1965. Eine verbesserte Methodik der qualitativen und quantitativen Analyse des Darmflora von Menschen und Tieren, *Zbl. Bakt. (1. Abt. Orig.), 195*: 455–465.

Miura, K., Nakamura, H., Tanaka, H., and Tamura, Z. 1979. Simultaneous determination of multiple enzyme activities by high performance liquid chromatography, *Chem. Pharm. Bull., 27*(8): 1759–1763.

Mizutani, T., and Mitsuoka, T. 1979. Effect of intestinal bacteria on incidence of liver tumors in gnotobiotic C3H/HE male mice, *J. Natl. Cancer Inst., 63*: 1365–1370.

Mizutoni, T., and Mitsuoka, T. 1980. Inhibitory effect of some intestinal bacteria on liver tumorigenis in gnotobiotic C3H/He male mice. *Cancer Letters, II*: 89–95.

Modler, H. W., McKellar, R. C., and Yaguchi, M. 1990. Bifidobacteria and bifidogenic factors, *Can. Inst. Food Sci. Technol. J., 23*(1): 29–41.

Montreuil, J. 1957. Les glucides du lait de femme, *Bull. Soc. Chim. Biol., 39*: 395.

Moore, W. E. C., Hash, D. E., Holdeman, L. V., and Cato, E. P. 1980. Polyacrylamide slab gel electrophoresis of soluble proteins for studies of bacterial floras, *Appl. Environ. Microbiol., 39*: 900–907.

Moreau, M. C., Thomasson, M., Ducluzeau, R., and Raibaud, P. 1986. Cinétique d'établissement de la microflore digestive chez le nouveau-né humain en fonction de la nature du lait, *Reprod. Nutr. Develop., 26*(2B): 745–753.

Moro, E. 1990a. Über die nach Gram-färbbaren bacillen des säuglingsstuhles, *Wien. Lin. Wschr, 13*: 114.

Moro, E. 1990b. Über *B. acidophilus, Jb. Kinderheilk, 52*: 38.

Munoa, F. J., and Pares, R. 1988. Selective Medium for isolation and enumeration of *Bifidobacterium* spp., *Appl. Environ. Microbiol., 54*(7): 1715–1718.

Mutai, M. and Tanaka, R. 1987. Ecology of *Bifidobacterium* in the human intestinal flora, *Bifidobacteria Microflora, 6*(2): 33–41.

Naujoks, H. 1921. Das vorkommen des B. acidophilus bei schwangeren und Gebärenden und sein zeitlicher und vörtlicher Übergang auf den Neugeborenen, *Zentralbl. Bakteriol. Orig. A., 86*: 582–585.

Neimann, N., La Vergne, E., Manciaux, M., Sterlin, S., and Percebois, G. 1965. Clinical and bacteriological study of a milk bifidogenic because of its lactulose content, *Pediatrie, 20*: 139–145.

Neumeister, K. and Schmidt, W. 1963. Treatment of intestinal radiation reactions, *Med. Klin, 58*(20): 842–844.

Neut, C., Beerens, H., and Romond, C. 1984. La colonisation intestinale du nouveauné après césarienne: influence de l'environnement, *Microbiol. Alim. Nutr., 2*: 337–349.

Neut, C., Lesieur, V., Beerens, H., and Romond, C. 1985. Changes in the composition of infant fecal flora during weaning, *Microecol. Therapy, 15*: 303.

Neut, C., Romond, C., and Beerens, H. 1980a. Influence of breast-feeding on the bifidus flora in the newborn's intestine, *Microecol. Therapy, 10*: 127–137.

Neut, C., Romond, C., and Beerens, H. 1980b. Contribution à l'étude de la répartition des espèces de *Bifidobacterium* dans la flore fécale des nourrissons alimentés soit au sein, soit par des laits maternisés, *Reprod. Nutr. Develop., 20*: 1679–1684.

Neut, C., Romond, C., and Beerens, H. 1981. Identification des *Bifidobacterium* en fonction de leurs besoins nutritionnels, *Rev Institut Pasteur Lyon, 14*: 19–26.

Neutra, M. R., and Forstner, J. F. 1987. Physiology of the gastrointestinal tract, in *Gastrointestinal mucus: synthesis, secretion and function*, 2nd ed. (Leonard R. Johnson, ed.), Raven Press, New York, Chap. 34.

Nichols, J. H., Bezkorovainy, A., and Landau, W. 1974. Human colostral whey M-1 glycoprotein and their *L. Bifidus* var. penn. growth promoting activities, *Life Sci., 14*: 967–976.

Norris, R. F., Flanders, T., Tomarelli, R. M., and György, P. 1950. The isolation and cultivation of *Lactobacillus bifidus*. A comparison of branched and unbranched strains, *J. Bact., 60*: 681–696.

Ochi, Y., Mitsuoka, T., and Sega, T. 1964. Studies on the intestinal flora of chickens. III. The development of the flora from chicks till hens, *Zbl. Bakt. 1. Abt. Orig., 193*: 80–95.

Okamura, N., Nakaya, R., Yokota, H., Yanai, N., and Kawashima, T. 1986. Interaction of shigella with bifidobacteria, *Bifidobacteria Microflora, 5*(1): 51–55.

Op Den Camp, H. J. M., Peeter, P. A. M., Oosterhof, A., and Veerkamp, J. H. 1985. Immunochemical studies on the lipoteichoic acid of *B. bifidum* subsp. *penn, J. Gen. Microbiol., 131*: 661–668.

Orla-Jensen, S. 1919. *The Lactic Acid Bacteria*, Host, Copenhagen, Denmark, pp. 1–196.

Orla-Jensen, S. 1924. La classification des bactéries lactiques, *Le Lait, 4*: 468–480.

Orla-Jensen, S., Olsen, E. and Geill, T. 1945. Senility and intestinal flora. A reexamination of metchnikoff's hypothesis, *Biol. Skr. K. Danske Vidensk Selsk, 3*(4): 3–38.

Owen, R. J., and Pitcher, D. 1985. Currents methods for estimating DNA base composition and levels of DNA-DNA hybridization. Chemical methods in bacterial systematics.

Perdigon, G., De Macias, M. E. N., Alvarez, S., Olivier, G., and De Ruioz Holgado, A. P. 1988. Systematic augmentation of the immune response in mice by feeding fermented milks with *L. casei* and *L. acidophilus, Immunology, 63*(1): 17–23.

Petschow, B. W., and Talbott, R. D. 1990. Growth promotion of *Bifidobacterium* species by whey and casein fractions from human and bovine milk, *J. Clin. Microbiol., 28*: 287–292.

Petuely, F. 1956. Ein enfacher vollsynthetischer Selektivnährboden für den *Lactobacillus bifidus, Selektivnährboden für den Lactobacillus bifidus.*

Poch, M., and Bezkorovainy, A. 1988. Growth-enhancing supplements for various species of the genus *Bifidobacterium, J. Dairy Sci., 71*: 3214–3221.

Poch, M., and Bezkorovainy, A. 1991. Bovine milk K-casein trypsin digest is a growth enhancer for the genus *Bifidobacterium, J. Agric. Food Chem., 39*: 73–77.

Polonowski, M., and Lespagnol, A. 1930. Sur l'existence de plusieurs glucides dans le lactosérum de femme, *C.R. Soc. Biol. (Paris), 104*: 555–557.

Poupard, J. A., Husain, I., and Norris, R. F. 1973. Biololgy of the bifidobacteria. *Bacteriol. Rev., 37*: 136–165.

Prevot, A. R. 1940. Manuel de Classification et de Détermination des Bactéries Anaérobies, Masson, Paris, pp. 195–199.

Prevot, A. R. 1955. Concerning the distinction between *Bifidobacterium bifidum* and *Lactobacillus acidophilus, Ann. Inst. Pasteur, 88*: 229–231.

Prevot, A. R. 1961. *Traité de Systématique Bactérienne*, Tome 2, Dunod, Paris, p. 508.

Rao, D. R., Chawan, C. B., and Pulusani, S. R. 1981. Influence of milk and thermophilus milk on plasma cholesterol levels and hepatic cholesterogenesis in rats, *J. Food Sci., 46*: 1339.

Rao, A. V., Shiwnarain, N., and Maharaj, I. 1989. Survival of microencapsulated *Bifidobacterium pseudolongum* in simulated gastric and intestinal juices, *Can. Inst. Food Sci. Technol., 22*(4): 345–351.

Rasic, J. Lj. 1983. The role of dairy foods containing bifido- and acidophilus bacteria in nutritional and health. *North European Dairy J., 49*(4): 80–88.

Rasic, J. Lj. and Kurmann, J. A. 1983. *Bifidobacteria and Their Role*, Birkhaüser, Basel.

Raynaud, M. 1959. Le facteur bifidus 2, *Ann. Pédiatrie, 241*: 8–23.

Raynaud, M., and Guintini, J. 1959. Aspect microscopique du bifidus, *Ann. Pédiatrie, 241*: 24–26.

Resnick, I. G., and Levin, M. A. 1981a. Quantitative procedure for enumeration of bifidobacteria, *Appl. Environ. Microbiol., 42*: 427–432.

Reuter, G. 1963. Vergleichenden untersuchung über die Bifidus-flora im sauglings und Erwachsenenstuhl, *Zentralbl. Bakteriol. Parasitenk. Infectionskr. Hyg. I. Abt. Orig. 191*: 486–507.

Reuter, G. 1971. Designation of type strains for *Bifidobacterium* species, *Int. J. System. Bacteriol., 21*(4): 273–275.

Romond, A. F. 1988. Apport d'un lait fermenté à *B. longum* dans le traitement des diarrhées à rotavirus du nourrisson, Doctorat en médecine, Lille.

Romond, C., Beerens, H., Neut, C., and Montreuil, J. 1980. Contribution à l'étude de la maternisation des laits, *Ann. Microbiol. Inst. Pasteur, 131*: 309–314.

Rose, S. J. 1984. Bacterial flora of breast-fed infants, *Pediatrics, 74*(4): 563.

Rowland, J. R., and Grasso, P. 1975. Degradation of *N*-nitrosamines by intestinal bacteria, *Appl. Microbiol., 29*: 7–12.

Roy, D., Ward, P., and Chevalier, P. 1989. Rapid characterization of *Bifidobacterium* sp. using enzymatic and biochemical methods. Les laits fermentés. Actualité de la recherche, Syndifrais, p. 281.

Salyers, A. A. 1989. Molecular and biochemical approaches to determining what bacteria are doing in vivo, *Antonie van Leeuwenhoek, 55*: 33–38.

Salyers, A. A., and Leedle, J. A.Z. 1983. Carbohydrate utilization in the human colon, in *Human Intestinal Microflora in Health and Disesae* (D. J. Hentges, ed.), Academic Press, London, pp. 129–146.

Salyers, A. A., West, B., Vercelloti, S. E. H., and Wilkins, T. D. 1977. Fermentation of mucins and plant poly-saccharides by anaerobic bacteria from the human colon, *Appl. Environ. Microbiol., 34*: 529–533.

Sato, J., Mochizuki, K., and Homma, N. 1982. Affinity of the *Bifidobacterium* to intestinal mucosal epithelial cells, *Bifidobacteria Microflora, 1*(1): 51–54.

Savage, D. C. 1977. Interaction between the host and its microbes, in Clarke and Bauchop, pp. 277–310.

Savage, D. C. 1984. Overview of the association of microbes with epithelial surfaces, *Microecol. Therapy, 14*: 169–182.

Scardovi, V. 1981. The genus *Bifidobacterium*, in *The Prokaryotes* (M. P. Starr, H. Stolp, H. G. Trüper, A. Balows, and H. G. Schlegel, eds.), pp. 1951–1961.

Scardovi, V. 1986. *Bifidobacterium. Bergey's Manual of Determinative Bacteriology.* 9th ed., Williams and Wilkins, Baltimore.

Scardovi, V., Casalicchio, F., and Vincenzi, N. 1979a. Multiples electrophoretic forms of transaldolase and 6-phosphogluconate dehydrogenase and their relationships to the taxonomy and ecology of bifidobacteria, *Int. J. System. Bacteriol., 29*: 312–339.

Scardovi, V., and Crociani, F. 1974. *Bifidobacterium catenulatum, B. dentium et B. angulatum*. Three new species and their deoxyribonucleic acid homology relationships. *Int. J. System. Bacteriol., 24*: 6–20.

Scardovi, V., and Sgorbati, B. 1974. Electrophoretic types of transaldolase, transketolase and other enzymes in bifidobacteria, *Antonie van Leeuwenhoek., 40*: 427–440.

Scardovi, V., Sgorbati, B., and Zani, G. 1971a. Starch gel electrophoresis of fructose-6-P phosphoketolase in the genus *Bifidobacterium, J. Bacteriol., 106*: 1036–1039.

Scardovi, V., and Trovatelli, L. D. 1965. The fructose-6-phosphate shunt as peculiar pattern of hexose degradation in the genus *Bifidobacterium, Ann. Microbiol., 15*: 19–29.

Scardovi, V., and Trovatelli, L. D. 1969. New species of bifid bacteria from Apis mellifica L. and Apis indica F. A contribution to the taxonomy and biochemistry of the genus *Bifidobacterium, Zentralbl. Bakteriol. Parasitenk. Infektionskr. Hyg. I. Abt. Orig., 123*: 64–88.

Scardovi, V., and Trovatelli, L. D. 1974. *Bifidobacterium animalis* (Mitsuoka). Canb. nov. and the "minimum" and "subtile" group of new bifidobacteria found in sewage, *Int. J. System. Bacteriol., 24*: 21–28.

Scardovi, V., Trovatelli, L. D., Biavati, B., and Zani, G. 1979b. *B. cuniculi, B. choerinum, B. boum and B. pseudocatenulatum* four new species and their deoxyribonucleic acid homology relationships, *Int. J. System. Bacteriol., 29*: 291–311.

Scardovi, V., Trovatelli, L. D., Crociani, F., and Sgorbati, B. 1969. *Bifidobacterium: B. globosum* n. sp. and *B. ruminale* n. sp., *Arch. Mikrobiol., 68*: 278–274.

Scardovi, V., Trovatelli, L. D., Zani, G., Crociani, F., and Matteuzzi, D. 1971b. Deoxyribonucleic acid homology relationships among species of the genus *Bifidobacterium, Int. J. System. Bacteriol., 21*(4): 276–294.

Scardovi, V., and Zani, G. 1974. *Bifidobacterium magnum* sp. nov.: a large acidophilic *Bifidobacterium* isolated from rabbit feces, *Int. J. System. Bacteriol., 24*: 29.

Scardovi, V., Zani, G., and Trovatelli, L. D. 1970. Deoxyribonucleic acid homology among the species of the genus *Bifidobacterium* isolated from animals, *Arch. Mikrobiol., 72*: 318–325.

Sebald, M., Gasser, F., and Werner, H. 1965. DNA base composition and classification. Application to group of bifidobacteria and to related genera, *Ann. Inst. Pasteur, 109*: 251–269.

Seka Assy, N. 1982. Contribution â l'étude des facteurs bifidigènes présents dans le lait maternel, Thèse de 3ème cycle de l'Université des Sciences et Techniques de Lille I.

Seki, M., Igarashi, M., Fukuda, Y., Simamura, S., Kawashima, T., and Ogasa, K. 1978. The effect of *Bifidobacterium* cultured milk on the "regularity" among on aged group, *31*(4): 379–387.

Sgorbati, B. 1979. Preliminary quantification of immunological relationships among the transaldolase of the genus *Bifidobacterium, Antonie van Leeuwenhoek, 45*: 557–564.

Sgorbati, B., Lenaz, G., and Casalicchio, F. 1976. Purification and properties of two fructose-6-phosphate phosphoketolases in *Bifidobacterium, Antonie van Leeuwenhoek, J. Microbiol. Serol., 42*: 49–57.

Sgorbati, B., and London, J. 1982. Demonstration of phylogenetic relatedness among members of the genus *Bifidobacterium* using the enzyme transaldolase as an evolutionary marker, *Int. J. System. Bacteriol., 32*(1): 37–42.

Sgorbati, B., and Scardovi, V. 1979. Immunological relationships among transaldolases in the genus *Bifidobacterium, Antonie van Leeuwenhoek J. Microbiol. Serol., 45*: 129–140.

Sgorbati, B., Scardovi, V., and Leblanc, J. 1982. Plasmids in the genus *Bifidobacterium, J. Gen. Microbiol., 128*: 2121–2131.

Sgorbati, B., Scardovi, V., and Leblanc, J. 1986. Related structures in the plasmid profiles of *Bifidobacterium longum, Microbiologica, 9*: 415–422.

Sgorbati, B., Smiley, M. B., and Sozzi, T. 1983. Plasmids and phages in *Bifidobacterium longum, Microbiologica, 6*: 169–173.

Shimamura, S. 1989. Effet positif de la lactoferrine sur la croissance du *Bifidobacterium*, in *Bifidobacterium et Facteurs Bifidigènes. Rôle en Santé Humaine*. ARBBA, pp. 105–113.

Simhon, A., Douglas, J. R., Drasar, B. S., and Soothill, J. F. 1982. Effect of feeding on infant's faecal flora, *Arch. Dis. Child., 57*: 54–88.

Simon, G., and Gorbach, S. 1986. The human intestinal microflora, *Dig. Dis. Sci., 31*(9): 1475–1625.

Simpson, W. A., Ofek, I., Sarasohn, C., Morrison, J. M., and Beachey, E. M. 1980. Characterization of the binding of streptococcal lipoteichoic acid to human oral epithelial cells, *J. Infect. Dis., 141*: 457–462.

Sonoike, K., Mada, M., and Mutai, M. 1986. Selective agar medium for counting viable cells of bifidobacteria in fermented milk, *J. Food Hyg. Soc. Jap., 27*: 238–244.

Stark, P. L., and Lee, A. 1982. The bacterial colonization of the large bowel of preterm low birth weight neonates, *J. Hyg., 89*(1): 59–67.

Stevenson, D. K., Yang, C., Kerner, J. A., and Yeager, A. S. 1985. Intestinal flora in the second week of life in hospitalized preterm infants fed stored frozen breast milk or a proprietary formula, *Clin. Pediatr., 24*(6): 338–341.

Tanaka, R., and Mutai, M. 1980. Improved medium for selective isolation and enumeration of *Bifidobacterium, Applied Environ. Microbiol., 40*(5): 866–869.

Teraguchi, S., Kawashima, T., and Kuboyama, M. 1982. Test tube method for counting bifidobacteria in commercial dairy and pharmaceutical bacteria products, *J. Food Hyg. Soc. Japan, 23*: 29–35.

Teraguchi, S., Ono, J., Kiyosawa, I., Fukuwatari, Y., Araki, J., and Okonogi, S. 1986. Vitamin production by bifidobacteria originated from human intestine. *J. Jap. Soc. of Nutr. and Food Sci., 32*(2): 157–169.

Teraguchi, S., Uehara, M., Ogasa, K., and Mitsuoka, T. 1978. Enumeration of bifidobacteria in dairy products, *Jap. J. Bact., 33*(6): 753–761.

Tissier, H. 1990. Recherche sur la flore intestinale des nourrissons (etat normal et pathologique), Thèse de médecine de l'Université de Paris.

Tissier, H. 1923. La putréfaction intestinale, *Bull. Inst. Pasteur, 21*: 361, 409, 577, 625.

Tochikura, T., Sakai, K., Fujiyoshi, T., Taachiki, T., and Kumagai, H. 1986. *p*-Nitrophenyl glycoside-hydrolyzing activities in bifidobacteria and characterization of β-D-galactosidase of *B. longum* 401, *Agric. Biol. Chem., 50*(9): 2279–2286.

Toida, T., Sekine, K., Tatsuki, T., Saito, M., Kawashima, T., Hashimoto, Y., and Sakurai, Y. 1990. Biochemical characterization and antitumor activity of a new cell wall preparation, whole peptidoglycan (WPG) from *Bifidobacterium infantis, J. Can. Res. Clin. Oncol., 116*: 342–348.

Tomarelli, R. M., Hassinen, J. B., Eckardt, E. R., Clark, R. H., and Bernhart, F. M. 1954. The isolation of a crystalline growth factor for a strain of *Lactobacillus bifidus, Arch. Biochem. Biophys., 48*: 225–232.

Tomarelli, R. M., Norris, R. F., György, P., Hassinen, J. B., and Bernhart, F. W. 1949. The nutrition of variants of *Lactobacillus bifidus, J. Biol. Chem., 181*(2): 879–888.

Trovatelli, L. D., Crociani, F., Pedinotti, M. et Scardovi, V. 1974. *Bifidobacterium pullorum* sp. nov.: a new species isolated from chicken feces and a related group of bifidobacteria isolated from rabbit feces, *Arch. Microbiol., 98*: 187.

Ueda, M., Nakamoto, S., Naakai, R. and Takagi, A. 1983. Establishment of a defined minimal medium and isolation of auxotrophic mutants for *Bifidobacterium bifidum* es 5, *J. Gen. Appl. Microbiol., 29*: 103–114.

Ushijima, T., Takahashi, M., and Ozaki, Y. 1985. Fourteen selective media facilitate evaluation of populations of coexisting fixed bacterial strains of enteric pathogens and normal human faecal flora, *J. Microbiol. Meth., 4*: 189–195.

Vercellotti, J. R., Salyers, A. A., Bullard, W. D., and Wilkins, T. D. 1977. Breakdown of mucin and plant polysaccharides in the human colon, *Can. J. Biochem., 55*: 1190–1196.

Veerkamp, J. H., Lambert, R., and Saito, Y. 1965. The composition of the cell wall of *Lactobacillus bifidus var pennsylvanicus, Archs. Biochem. Biophys., 112*(1): 120–125.

Waldolowski, E. A., Laux, D. C., and Cohen, P. S. 1988. Colonization of the streptomycin treated mouse large intestine by a human fecal *E. coli* strain: role of adhesion to mucosal receptors, *Infect. Immun., 56*: 1036–1043.

Walch, E. 1956. Zur Frage deer beeinflub barkeit der darmflora des säuglings durch. *N*.acétyl-D-glucosamin-derivate, *Dtsch. med. Wschr., 81*: 661–664.

Walker, W. A. 1985. Role of the mucosal barrier in toxin/microbial attachment to the gastrointestinal tract, in *Microbial Toxins and Diarrhoeal Disease, CIBA Foundation* Symp. Pitman, London, pp. 34–56.

Werner, H., Gasser, F., and Sebald, M. 1966. DNA-basenbestimmungen an 28 Bifidus stämmen und anstämmen morphologish ähnlicher gattunger, *Zbl. Bak. I. Brig., 198*S: 504–516.

Wilhelm, M. P., Lee, D. T., and Rosenblatt, J. E. 1984. Bacterial, *Eur. J. Clin. Microbiol., 6*(3): 266–270.

Yamazaki, H., and Dilawri, N. 1990. Measurement of growth bifidobacteria on inulofructosaccharides, *Lett. Appl. Microbiol., 10*: 229–232.

Yamazaki, S., Kamimura, H., Momose, H., Kawashima, T., and Ueda, K. 1982. Protective effect of *Bifidobacterium*-monoassociation against lethal activity of *Escherichia coli, Bifidobacteria Microflora, 1*(1): 55–59.

Yamazaki, S., Machii, K., Tsuyuchi, S., Momose, H., Kawashima, T., and Ueda, K. 1985. Immunological responses to monoassociated *Bifidobacterium longum* and their relation to prevention of bacterial invasion, *Immunol., 56*: 43–50.

Yazawa, K., Imai, K., and Tamura, Z. 1978. Oligosaccharides and polysaccharides specifically utilizable by bifidobacteria, *Chem. Pharm. Bull., 26*(11): 3306–3311.

Yoshioka, H., Fujita, K., and Iseki, K. 1984. Bacterial flora of breast-fed infants: reply, *Pediatrics, 74*(4): 563.

Yuhara, T., Isojima, S., Tsuchiya, F., and Mitsuoka, T. 1983. On the intestinal flora of bottle-fed infant, *Bifidobacteria Microflora, 2*(1): 33–39.

14

Future Aspects in Research and Product Development on Lactic Acid Bacteria

Seppo Salminen
Valio Ltd., and University of Helsinki, Helsinki, Finland

Atte von Wright
University of Kuopio, Kuopio, Finland; University of Helsinki, Helsinki, Finland; and The Royal Veterinary and Agricultural University, Frederiksberg, Denmark

I. INTRODUCTION

Even a cursory look at the different chapters in this book should convince the reader that a very significant amount of research has been focused during the last few years on lactic acid bacteria and their traditional and modern applications. So many different groups in various institutions, laboratories, and commercial enterprises study these fascinating organisms from so many different points of view that several breakthroughs opening new possibilities for applications and products are sure to materialize in the near future. However, science is seldom quite predictable, and we, as editors, unfortunately do not own a crystal ball capable of showing us exactly in advance the birth of the next remarkable inventions in this field. The following attempts to deduce the foreseeable progress are based on our impressions from reading through the material and comparing the results and claims presented to our own research experience.

II. FUNDAMENTAL RESEARCH ASPECTS

The taxonomy of lactic acid bacteria has changed considerably during the last few years. One can only hope that the application of new molecular taxonomic criteria gradually calms the field down, and the degree of confusion, which many researchers now feel, changes to new confidence and better understanding of the organisms studied.

There have been significant developments both in the biochemistry and physiology of lactic acid bacteria, and the progress in the genetics of these organisms has been spectacular. There apparently now is a need to combine the genetic and molecular biological techniques with the physiological approach to fully understand the fundamental biology of these microbes. The studies on proteolysis, peptidase action, and peptide transport are first examples of the power of molecular genetic analysis of physiologically important phenomena in lactic acid bacteria.

The interaction of the lactic acid bacterial microflora with the physiology, nutrition, and metabolism of higher animals, as well as its role in health and disease, is an area of much speculation. However, new and rigorous studies have started to shed light even on this much contested field. Even though the results and studies reported in this book might have their weaknesses and may well need a revision in the future, we feel that a path is now open to get some undisputable facts resolved in the next few years. Needless to say, understanding these phenomena is the prerequisite of their efficient application.

III. FUTURE APPLICATIONS

The successful fundamental research is reflected in the manifold applications lactic acid bacteria already have, as well as in some totally new industrial or medical uses they might possess. In the following sections some future prospects are briefly reviewed. The list is by no means exhaustive, and contains only some of the most obvious ideas that the present state of knowledge allows. It is quite probable that the real breakthroughs are waiting their imaginative inventor outside the common and well-trod paths of research and product development.

A. Dairy Industry

In the dairy industry one of the main technological developments has been the gradual replacement of traditional starter systems by new starter concentrates. These starter preparations decrease the risk of contamination by potentially harmful bacteria or bacteriophages, eliminate the fluctuations in the strain composition of the starter culture, and thus ensure the stable quality of the product. Undoubtedly, advances in the knowledge of the physiology of the starter organisms will help to formulate the starter concentrates so that their functional properties are optimized.

Genetic research of lactic acid bacteria has many potential dairy applications. Development of phage-resistant starter strains is one of the most obvious. Control of the proteolysis, carbohydrate utilization, and flavor development should also be possible in the near future, provided that the genetic manipulation techniques reach the status where they are both ethically and legally acceptable and also approved by consumers.

B. Other Industrial Applications

The many kinds of antimicrobial substances produced by several species and strains of lactic acid bacteria could offer alternatives to chemical food additives used to control pathogenic or otherwise harmful contaminants in food. The success of nisin as a food additive is an encouraging example of the potential in this field. In fermented foods the strains producing antimicrobials could be used themselves, while in other food products isolated and purified antimicrobial substances could be added.

Knowing the genetic basis of the synthesis of bacteriocins opens new possibilities to increase their production to industrial scale. Although even these compounds, of course, need a safety evaluation before their use as food additives, the status of lactic acid bacteria as "generally regarded as safe" organisms should make this assessment less costly and laborious as with purely "man-made" food chemicals.

With the rapid development in the genetic modification of lactic acid bacteria, their potential as production hosts for heterologous proteins useful either in food and feed industry or for medical purposes should also be considered. It should be noted that lactic acid bacteria with new enzymatic activities could have advantages in many of their traditional applications, even though the production of enzymes themselves would not reach the level sufficient to make their production profitable in industrial scale. As with the dairy applications of genetically modified bacteria, it is of course necessary to ensure that the eventual processes and products are ethically and legally acceptable.

Advances in the genetical and physiological research of lactic acid bacteria should also improve the selection criteria for strains used in the different processes. As this approach does not involve any actual genetic engineering, but is a continuation of the traditional technological product development with modern methods, there should not be any regulatory delays to adapt these techniques.

C. Human and Animal Probiotics

It can be safely stated that the studies reported in this volume demonstrate some actual beneficial health effects associated with certain lactic acid bacterial strains. However, much work remains to be done before the mechanisms behind these phenomena are elucidated. Only then the selection of probiotic strains can reach the scientific status.

We have already mentioned the importance of finding out the interactions between the bacterial microflora and the physiology of higher animals. Work in this field is very laborious and time-consuming (as Mikkelsaar and Kasesalu demonstrate) and does not promise any immediate rewards either scientifically or economically. However, it should be done, and the studies in this field should get the attention of the research-funding bodies. This is all the more important, as the results probably have wider significance in terms of health and disease than only in the development of probiotic strains.

Although one then still needs a lot of caution in interpreting the final value of lactic acid bacteria as probiotics, there is no question about the relevance and potential studies in this field. Antibiotics in animal feeds present such risks that it might turn out that probiotics, although less effective, prove to be more profitable in the long run. The same can, to an extent, be applied to human therapeutic bacterial preparations aimed at special groups of patients such as patients with pseudomembraneous colitis, viral diarrhea, or recurrent gastrointestinal infections, to whom conventional medication might cause undue stress.

Many new probiotic or therapeutic bacterial preparations both for veterinary or medical purposes shall undoubtedly enter the market during the next few years. It is the responsibility of the regulatory bodies as well the scientific community and marketing companies to ensure that the claims associated with those preparations will be based on as rigorous scientific scrutiny as those of ordinary medicines. The effects are there. What remains is to find out the methods to select the beneficial strains or strain combinations as well as to define formulations preserving those properties.

IV. CONCLUSIONS

This book started with a statement about the ancient history of lactic acid bacteria in human cultural traditions. Up to our time these organisms have been performing basically similar tasks, differing only in scale and precision, as they did thousands of years ago. We hope that this volume convinces the reader that we might be on the verge of quite new applications based on a better understanding about the potential of lactic acid bacteria and especially of their role in health and disease. It is too early to say to what extent these expectations are going to be fulfilled. Anyway, they more than justify further multidisciplinary research in this field.

Index